Revolution in Science

REVOLUTION IN SCIENCE

I. Bernard Cohen

THE
BELKNAP PRESS OF
HARVARD UNIVERSITY PRESS
Cambridge, Massachusetts, and London, England

Library of Congress Cataloging in Publication Data

Cohen, I. Bernard, 1914–
Revolution in science.

Bibliography: p.
Includes index.
1. Science — History.   I.   Title.
Q125.C542 1985   509   84-12916
ISBN 0-674-76777-2 (cloth)
ISBN 0-674-76778-0 (paper)

Designed by Gwen Frankfeldt

To
Henry Guerlac,
friend and colleague
for almost half a century

To
Ernst Mayr,
mentor and friend

And to
friends and fellow students
of revolution in science,
Rupert and Marie Hall
Paolo Rossi

Contents

————◆————

IV *Changing Concepts of Revolution in the Eighteenth Century*

V *Scientific Progress in the Nineteenth Century*

VI *The Twentieth Century, Age of Revolutions*

Preface

----◆----

Revolution in Science is a histori-
cal and analytic study of a concept through the course of four centuries.
Such a complex topic, covering a broad range of events, individuals, and
ideas, has seemed to require a number of approaches from different
angles. The first of these is an analysis of the stages by which revolutions
in science progress from the inception of a revolutionary idea to the
acceptance and use of a new science by a sizable number of scientists.
Whether a particular set of events in science does or does not constitute
a revolution is necessarily a personal judgment, but I have developed a
set of criteria — based on historical evidence — for the occurrence of a
scientific revolution. These stages and criteria (outlined in chapters 2
and 3) constitute the analytic framework of the book.

I have used this framework to examine critically some of the major
revolutions in science which have occurred during the four centuries
modern science has existed. For each of those centuries, an introductory section on the political or social revolutions of that age and the
images of revolution then current is presented, because I have found

that occurrences of the word 'revolution' in the context of science have always reflected current theories concerning political and social revolution, as well as the awareness of actual revolutions that have taken place. Thus, the thinking about each revolution in science that I discuss is set against the background of social and political revolution.

A distinction must be made between the historical perception of revolution and the historian's perception. The former comprises the judgments made at the time of the revolution and during succeeding ages and are the objective facts or data of history; but the latter are present-day subjective judgments. Of course, in the case of each revolution discussed in this book, I have made a subjective historian's judgment. Yet in every example, I have stressed the historical evidence. In almost every case there is a confluence of the two; those revolutions that pass the test of historical evidence tend to be those that in the judgment of today's historians (and scientists) are also revolutions. But the comparison of historical evidence and the judgment of historians has also disclosed some fascinating anomalies.

In particular, the study of historical evidence shows that the concept of revolution in science, like the concept of revolution itself, is not and has not been static. For example, this book documents the changing views of scientists and historians on whether the progress of science is primarily gradual and incremental or is the result of a succession of revolutions. In addition to alterations in the general viewpoint toward revolutions in science, there have also been shifts of judgment concerning the revolutionary character of particular events. A case in point is the Copernican revolution. The notion that a revolution in astronomy attended the publication of Copernicus's *De Revolutionibus* in 1543 was a fanciful invention of eighteenth-century historians of astronomy; it was popularized to such an extent that the Copernican revolution became the paradigmatic revolution in science. But critical examination of the evidence by historians has shown that the revolution was not at all Copernican, but was at best Galilean and Keplerian.

The changing perspective of time has also produced radical alterations of the sense and significance of even great political revolutions. In *The Rights of Man* (1791), Thomas Paine explained how the American and French revolutions had introduced a new kind of revolutionary thinking into political science. Known primarily for his pamphleteering during the American Revolution — *Common Sense* and *The Crisis* are his most notable productions — Paine wrote *The Rights of Man* in reply to Edmund Burke's *Reflection on the Revolution in France* (1790). Paine's exposition of the new view of revolution resulting from the events in America and in France is a classic example of the way in which political concepts arise in relation to events and not mere theory:

> What we formerly called Revolutions were little more than a change of persons, or an alteration of the local circumstances. They rose and fell like things of course, and had nothing in their existence or their fate that could influence beyond the spot that produced them. But what we now see in the world, from the Revolutions of America and France, are a renovation of the natural order of things, a system of principles as universal as truth and the existence of man, and combining moral with political happiness and national prosperity.

Less than a half-century later, however, in 1835, Giuseppe Mazzini no longer considered the French Revolution to be a sound model for progressive political action. "The progress of France," he wrote (1907, 251) "depends upon its power to emancipate itself from the eighteenth century and the old Revolution." He argued that the French Revolution should be "considered, not as a programme, but as a summary: not as the initiative of a new age, but as the last formula of an expiring age." In the nineteenth century, and even in the twentieth, revolutionaries aimed to achieve what the French Revolution had failed to do, as may be seen clearly in the writings of Marx and Engels and many twentieth-century theorists of revolution.

English political history provides two clear-cut examples of the way in which the passage of time changes the way events may be perceived as revolutions. In other words, revolutions in science are not the only revolutions that undergo successive changes in their image as revolutions. The Glorious Revolution of 1688 was the paradigmatic political revolution for eighteenth-century historians and political theorists, but today it does not appear to have been very revolutionary. And the same is true for the American Revolution, often now called the Revolutionary War or the War of Independence. Contrariwise, the English Revolution of the mid-seventeenth century was not generally conceived to have been a revolution at all until some two hundred years later. This English Revolution was, according to some nineteenth- and twentieth-century commentators, an abortive social revolution rather than a political revolution like the Glorious Revolution. The concept of what constitutes a revolution also differs greatly from one age to the next, as may be seen by reading the literature of revolution from the late seventeenth and eighteenth centuries, from the half-century or so between the French Revolution and Marx, from Marx's era to Lenin's, from the decades following the Russian Revolution of 1917, and from the 1950s, 60s, 70s, and 80s. Not surprisingly, discussions of revolution in science reflect these changes.

A historical discussion of the origins and successive meanings of the term 'revolution' (whether scientific or political) may seem abstract and innocent of partisanship, but a single example will show that this is not necessarily the case. In his essay "Islamic Concepts of Revolution," Bernard Lewis (1972, 37–38) discussed the occurrence in classical Arabic of "a number of words to denote rebellion or insurrection," among them the word 'thawra'. "The root *th-w-r* in classical Arabic," he wrote, "meant to rise up (e.g., of a camel), to be stirred or excited, and hence . . . to rebel." Lewis then explained that the word "is often used in the context of establishing a petty, independent sovereignty" and that the noun form "at first means excitement, as in the phrase . . . wait till this excitement dies down" — which Lewis said was "a very apt recommendation." Edward Said replied to Lewis (1978, 315) by asking why "introduce the idea of a camel rising in an etymological root for modern Arab revolution except as a clever way of discrediting the modern?" Said alleged that "Lewis's reasoning" had the obvious aim of "bringing down revolution from its contemporary valuation to nothing more noble (or beautiful) than a camel about to raise itself from the ground." We may see the force of Said's critique by imagining a reverse situation, in which an oriental scholar might disparagingly criticize the Western European or American concept of revolution because the term itself had developed from a cyclical idea of return or ebb and flow. Said actually interpreted Lewis's version of etymology to be tinged with a style of thought he has named "Orientalism," a "Western style for dominating, restructuring, and having authority over the Orient." Said considered Lewis's etymological discussion to be an expression of his political and social position, leading him to associate "*thawra* with a camel rising and generally with excitement" rather than "with a struggle on behalf of values." This was also notably the case for the editor's introduction to the volume in which Lewis's essay appears. For here it is said that "struggles for independence and radical movements in the Middle East, *coups d'état,* insurrections and rebellions" are not proper revolutions, as this term is understood in the West (Valikiotis 1972, 11). The alleged reason is that "Western notions of the right to resist bad government are alien to Islamic thought."

I began my book originally as an inquiry into the origins and successive uses of two concepts: the Scientific Revolution (of the sixteenth and seventeenth centuries) and revolution in science as a mode of scientific progress. I found that many historians, and even historians of science, believed that both of these concepts arose in our own days and that historians of science who used them were anachronistically attempting to force events of the past into a twentieth-century mold. The reader may well imagine my astonishment as my research began to produce examples of discussions of revolution in science from each of the past four centuries, and of references to the Scientific Revolution at least as long ago as the early nineteenth century. Because this material is not at all familiar to historians—and to scientists, philosophers, and sociologists—a large part of the book serves as a chronological record of these usages.

My first findings were presented in an article in the *Journal of the History of Ideas* (1976, 37: 257–288), which I had intended to expand into a small monograph. But as Thomas Mann (in the preface to his Joseph series) and many authors have said, "Fata sua habent libelli" ("Books have their own fate"). The overwhelming accumulation of evidence has led to this more ambitious book. Even it by no means exhausts my findings; I could easily have written a volume three or four times as large. References to revolution in science since World War I and the Russian Revolution could, alone, have been the subject of a monograph. Of necessity, I have given only some carefully selected samples that seem to me either to be typical of current expressions of opinion or to have special interest.

This work is part of a broad research program with a double aim. In part I am concerned with exploring and elucidating the creative process by which a practitioner in one discipline uses the ideas (concepts, methods, theories, tools) of another discipline. I gave an earnest of this investigation in my book *The Newtonian Revolution* (1980). There I stressed the doctrine of 'transformation of ideas' as a key ingredient in the revolutionary process. Here, however, I have restrained my use of this concept of transformation, so as not to put off readers primarily interested in the broad historical chronicle and the analysis of revolutions in science. I reserve for later study the further analysis of conceptual transformations in revolutions in science. The second goal of my research is to define and analyze the interactions between the natural and exact sciences and the social and behavioral sciences. This work combines historical and analytical studies. Its purpose is not only to identify and study in particular cases the general process of transformation that occurs whenever an idea from one discipline is used in another; additionally, I am concerned with analyzing the 'scientific' basis of the social sciences and examining how they have used the sciences to validate applications of their findings in matters of public policy. Although it is generally believed that ideas tend to flow from the natural and exact sciences to the social and behavioral sciences, there are many significant cases in which the influence has been in the other direction. The present book on revolutions is related to this theme because the concept and name 'revolution' arose in the sciences (astronomy and geometry) and then entered the discourse of political and social change, undergoing a significant initial transformation. As the book documents, this changing concept of revolution was then transferred back from the social sciences and the literature of political theory and action to discussions of scientific change. Thus the book explores an area of the relations between these two worlds of discourse.

The interactions between the concepts of political or social revolution and revolution in science are mentioned throughout this book, but I am fully aware that this topic merits a much more complete exploration. As early as the seventeenth century, even before the modern noncyclical concept of revolution had become universal, various authors sought to explain scientific advance by political analogies. There is also a countertheme, which I have mentioned but not explored, of the possible influence of science and of scientific revolutions on political revolutions. It is well known that Marx and more particularly Engels saw their revolutionary movement as 'scientific'. The terms 'scientific socialism' and 'scientific communism' occur frequently in the Marxian (especially the Soviet) literature, but I know of no critical assessment of the degree to which this use of 'scientific' depends on the use of science as commonly understood in national scientific communities.

Although the theme of change in the concept of scientific revolution is woven throughout this book and is indeed its major thread, many readers will find the case histories of particular revolutions to be of greatest interest. These case histories, which make up a large part of the book, describe some of the great revolutions that mark the development of modern science, and display in specific examples the stages of revolution which I have developed and the evidence for considering a particular series of events to be a revolution in science. These case histories also indicate how the recognition of revolution in the sciences has been (and is) conditioned by the image of political revolutions and by current revolutionary theory. A striking example occurs in relation to ideas of revolutions in the earth before and after the French Revolution. Another is the effect of T. S. Kuhn's writings on scientists who have identified and chronicled a revolution in the earth sciences growing out of new ideas of plate tectonics and continental drift.

In most of the case histories I have tended to quote expressions of revolution by creating or participating scientists, or even by nonparticipating observers, without in each instance attempting to define precisely what that person may have had in mind. But I have usually given the context or the current state of ideas concerning revolution in general. The problem here is twofold. First of all, we do not know exactly what a particular scientist may have had in mind; second, many scientists (a number of examples appear throughout the book) make very explicit statements about a particular revolution in science or about revolutions in science in general, but without having necessarily developed a carefully worked-out theory of revolutions or even of the modes of scientific change at large. It is tempting, for example, to link Albert Einstein's 1905 and 1906 remarks about revolution in a scientific context to the events of 1905, notably the failed Russian revolution and the idealistic hopes for a radical reform of Russian society; similarly, his statements against revolution in the context of relativity theory could be interpreted as a reaction against the excesses of the 1917 Russian revolution and the immediately post–World War II abortive revolutions in Germany, including the fighting in the streets of Berlin. But Einstein's reaction against the newspapers' extravagant attribution to him of a revolution must also be factored in; it certainly helped incline him to a view that his work had been evolutionary rather than revolutionary. In assessing Einstein's views on revolution in science, one must bear in mind that all of Einstein's statements about revolution and evolution occur in single isolated sentences or phrases, very often responses to a statement by somebody else; I do not know of a single instance of a complete essay or letter or even a completely developed full paragraph on the processes by which science advances, much less on revolutions in science. And the case is much the same for other scientists of the past three centuries who have expressed themselves on particular revolutions in science or even on revolutions in science in general. Hence I have in each case given the reader the actual expressions used in relation to revolution. But the reader will easily be aware that there is no warrant for assuming that the implications of the word 'revolution' are necessarily identical in every statement about revolution by a single person or by different persons in relation to a particular scientific theory.

Finally, I have often referred to my findings in what may be too positive a manner. I am aware that in many cases the phrases 'so far as I know' or 'so far as my research has shown' should have been inserted. Are there earlier examples than I have found? I would be the last person to assume that my research has been exhaustive, a conclusion precluded by the nature of the topic. And so I hope that readers who have access to further information will inform me so that I can make corrections in any later editions.

Readers will naturally wish to know how this book is related to T. S. Kuhn's *Structure of Scientific Revolutions* and other writings. As many readers will be aware, Kuhn's work has been of fundamental importance in reorienting the thinking of scientists and historians of science, converting them to (or making them mindful of) the notion that revolutions are a regular feature of scientific change. Hence, Kuhn's writings constitute a major event in my history of the concept of revolutions in science. A major theme in Kuhn's analysis is that scientific changes of all kinds, including revolutions, are not the result of a contest of ideas, as Ernst Mach and others have supposed, but rather of scientists who accept or believe in ideas. I address this theme by analyzing four stages of development which I find to be characteristic of all revolutions in science. Finally, I accept Kuhn's general notion of revolutions as a shift in a set of scientific beliefs — in 'paradigms', to use that original term introduced in this context by Kuhn but later (unfortunately, in my opinion) abandoned by him when it was shown that he had used this term ambiguously and even in a number of quite different senses.

In my book I do not, however, discuss some particular features that Kuhn has assigned to the "structure of scientific revolutions." For example, I do not explore the theme (which I find to have too many exceptions to be useful) that revolutions in science are necessarily precipitated by crises. And it is the same for other details of his schema. Nor do I go into the question of Kuhn's changing distinctions among 'paradigm', 'exemplar', and 'disciplinary matrix'. It is an interesting fact of record that whereas Kuhn's schema has been subject to considerable discussion, criticism, and approval by historians of science, the latter (including Kuhn himself) have tended not to make use of a Kuhnian framework in their actual writings. Hence Kuhn's influence appears to be stronger among philosophers and sociologists of science (and scholars in wholly different areas such as political theory) than among scientists and practicing historians of science. An exception, however, must be made for historians of the recent revolution in the earth sciences. (For a first-rate — and good-naturedly irreverent — analytical presentation of Kuhn's system and the history of its reception by the community of historians of science, see Reingold 1980.)

Kuhn refers again and again to smaller revolutions and great revolutions. The latter are those generally accepted as revolutions in scientific discourse — those associated with Copernicus, Newton, Lavoisier, Darwin, and Einstein. But Kuhn's smaller revolutions may involve no more than a couple of dozen scientists replacing an accepted exemplar by a new one. In public discussions and writings Kuhn stresses the ubiquitous nature of these smaller revolutions. In my book, however, I have tended to concentrate on the larger or more visible revolutions. One of the reasons is that my formulation of an objective means of determining when a revolution occurs applies directly to revolutions in science which are more analogous with political revolutions.

Readers will discern, furthermore, that I am neither philosopher nor sociologist of science. As a historian, I have aimed more to produce a critical and analytical historical study than to debate the merits of Kuhn's system or the systems of other philosophers or sociologists of science. In short, my purpose and Kuhn's are not parallel but necessarily intersect. This book is not another discussion of Kuhn's "structure"; it is instead an attempt to examine the subject of revolution in science from a new and strictly historical viewpoint.

I have earlier quoted a Latin phrase used by Thomas Mann and others to indicate the well-known phenomenon that books tend to have a life of their own, that they develop by an internal logic of research and writing. Just as this book was going to press, however, I encountered the complete quotation of which this is an extract. Composed by Terentianus Maurus *(De litteris syllabis et metris Horatii,* line 1286), it reads in full: "Pro captu lectoris habent sua fata libelli." Who could possibly disagree that the fortunes of books depend on their reception by the reader? I hope that this book will find both critical and sympathetic readers, so that it may stimulate further research and thought. If this fascinating subject of revolutions can attract the attention of scholars, it will achieve the potential it so richly deserves.

I. Bernard Cohen

Acknowledgments

———◆———

I am grateful to many scholars — colleagues, friends, students — who, over the years, have either brought to my attention some examples of discussion of revolution in science during the past four centuries or have answered queries. This company includes, among others, James Adler, Peter Buck, Lorraine J. Daston, Joy Harvey, Michael Heidelberger, Joseph Dauben, Stillman Drake, Henry Guerlac, Pierre Jacob, Gerard Jorland, Robert Proctor, Barbara Reeves, Joan Richards, Shirley Roe, and Frank Sulloway. A number of kind friends and scholarly colleagues helped me by giving one or more chapters a critical reading: Jed Z. Buchwald, Peter Galison, Owen Gingerich, John Heilbron, Gerald Holton, Ursula Marvin, Arthur Miller, and Noel Swerdlow. Additionally I have profited from the comments of three scholars who gave the penultimate version a complete reading: Joseph Dauben, Richard Kremer, and Roy Porter.

In the preparation of this book, I have depended heavily on the constant support of Julia Budenz, who has worked with me on every aspect of the research and the final text. It is difficult for me to imagine how I could have brought this long and complex book to completion without her help. Anne Miller Whitman, as ever, gave me the great benefit of her wisdom and insight. The contribution of these two old and dear friends, co-workers in many projects, was all the more significant in that this is the first work I have had to complete without the loving support and creative criticism of my wife Frances Davis.

During the early stages of research Kristie I. Macrakis served as research assistant. I am greatly indebted to three students who have helped me check and complete the text and references: Diane Q. Webb, Deborah Coon, and Kristin Peterson. The final assembling and checking of the bibliography was done by Bertha Adamson and Diana L. Barkan. Sarah Tracy prepared the index.

I especially record my debt to Arthur Rosenthal, director of Harvard University Press, who held out an encouraging 'carrot' at just the moment when my flagging spirit needed revival. It has been a constant joy to work with Susan Wallace, an editor of rare taste and discernment, whose critical judgment bettered the book at every stage.

Finally, I gratefully acknowledge the support of the Spencer Foundation in the preliminary stages of the research on which the book is based. The major support of my research during the years of exploring the topic of revolutions in science and of the larger subject which it forms in part came from the Alfred E. Sloan Foundation; it is difficult to imagine any foundation that could have been more considerate to its grantees.

Science

and

Revolution

———◆◆———

I

1

Introduction

———◆———

Today we are apt to take it for granted that science and its associated technology progress through a series of revolutionary leaps—giant steps forward that give us an altogether new perspective on the natural world. But has revolution always been so familiar and acceptable a way to describe the advance of science? Could such innovative scientific thinkers as Kepler, Galileo, and Harvey conceive of their own work as being revolutionary in the sense that we use the word today? Did the contemporaries of Darwin, Freud, and Einstein see the theories of these scientists as creating a revolution, or did they prefer to view scientific progress in a less dramatic light? What effect have such social and political upheavals as the French Revolution and the rise of Marxism had on the way scientists, philosophers, and historians think about revolutions in science? For all of their emphasis on the great scientific revolutions of the past, surprisingly few scholars have addressed these sorts of questions—questions having to do with the historical development of the idea of revolution as a feature of scientific change. It was my own curiosity about these problems that led to the writing of this book.

The main body of the presentation deals with the chronological history and successive transformations of the concept of revolution in science in the seventeenth, eighteenth, nineteenth, and twentieth centuries — with illustrations taken from some major revolutions in each of these periods. These revolutions have been chosen either because of their intrinsic historical importance (as in the case of the Copernican, Newtonian, Darwinian, and Einsteinian revolutions) or because of their relevance in clarifying or exemplifying what I see as the central characteristics of all revolutions in science.

In declaring that certain historical episodes constitute revolutions in science, I have not relied solely on my own personal evaluations, or even on the consensus of qualified historians, but rather on historical evidence, both the judgments of contemporaneous observers and participants and the continuing tradition. For example, it is a historical fact that in the early 1700s Fontenelle said *expressis verbis* that the invention of the calculus was a revolution in mathematics, that in 1773 Lavoisier declared that his research program would lead to a revolution, that in 1859 Charles Darwin hailed Lyell's revolution in geology and predicted that the acceptance of his own ideas would produce a "considerable revolution in natural history." Contemporaneous documents show that the radical innovations of Lavoisier and Darwin and both relativity and quantum theory were very quickly acknowledged to be revolutions. Furthermore, almost all scientists and historians of science today agree with the opinion of the past that certain momentous restructurings of scientific thought were revolutions. Such a consensus, of course, does not make these events revolutions; we shall see in chapter 3 that additional tests can be applied to help us decide what is and what is not to be considered a revolution in science, and (in chapter 2) that distinct stages in the development of a revolutionary idea can indicate whether or not a true scientific revolution has occurred. These questions aside, there can be no debate concerning the overall historical record: it shows that for some three hundred years, ever since the first coming-of-age of modern science, great events in the development of science have been seen as revolutions in thought and practice. The main burden of this book is to delineate and to analyze those events and the interpretation of them as revolutions.

Defining 'Revolution in Science'

The problem of defining 'revolution', which plagues almost every discussion of political and social revolutions, has penetrated the literature on revolutions in science. I have not attempted in this book to display a strict definition of 'revolution' or of 'revolution in science', although I have discussed certain features of all revolutions in science: their stages of development, evidential tests for their occurrence, and transformations of ideas in the production of revolutionary innovations. While there would be little disagreement that the examples I have discussed in this book are to be considered revolutions — at least among all scholars who believe that scientific revolutions do occur — there would be no consensus concerning the exact defining characteristics that all of these revolutions have in common.[1] Discussions of what constitutes a revolution and how to define revolution are part of philosophy, however related to history. As a historian, and aware that I am not a philosopher, I have always very carefully held in check any temptation to ultracrepidate. In the Medawars' *Aristotle to Zoos: A Philosophical Dictionary of Biology* there is a discussion of definitions (1983, 66) which is very illuminating:

> In certain formal contexts — mathematical logic, for example, in which a definition is a rule for substituting one symbol for one or more others — definitions are crucially important, but in everyday life and in sciences such as biology their importance is highly exaggerated. It is simply not true that no discourse is possible unless all technical terms are precisely defined; if that were so, there would be no biology.

The life sciences differ in this regard from the exact sciences — mathematics, rational mechanics, theoretical physics and astronomy, and parts of chemistry — fields that have developed a long tradition in which definition has come to be of basic importance. But if exact definitions are not required of all sciences, then surely there is no ground for insisting that history of science must be like one portion of the sciences rather than the other.

It is a matter of record that the word 'revolution' first gained general currency as a technical term in the exact sciences, where it long had (and still has) a meaning very different from that of a sudden dramatic shift. Revolution means to return again, to go through a cyclical succession, as in the seasons of the year, or to ebb and flow, as in the motion of the tides. In the sciences, revolution thus implies a constancy within all change, an endless repetition, an end that is a beginning all over again. This is the meaning we have in mind in such a phrase as 'the revolutions of the planets in their orbits'. The expression 'scientific revolution' or 'revolution in science', however, conveys no such sense of continuity and permanence; rather, it implies a break in continuity, the establishment of a new order that has severed its links with the past, a sharply defined plane of cleavage between what is old and familiar and what is new and different. It is the historian's task to find out how and when an innocent scientific term that implies permanence and recurrence became transformed into an expression for radical change in political and socioeconomic affairs, and then to discover the way in which this altered concept was applied to science itself. This set of transformations embodies more than a mere shift in terminology. It suggests that there has been a profound conceptual change in our analysis of human and social action and in our image of the scientist and of scientific activity.

From the eighteenth century to our own, many scientists have written of their own scientific creations as revolutionary, but neither Copernicus nor Newton did so. Newton and his predecessors did not conceive of themselves as revolutionaries in part because their work was produced before the term 'revolution' had become generally applied to the sciences. But there is a deeper reason; we shall see that during the first century or so of modern science, many of the great creative scientists tended to think of themselves (and even to be viewed by their contemporaries) as revivers or rediscoverers of ancient knowledge, even as innovators who improved or extended knowledge, but not as revolutionaries in the sense in which we would commonly use this expression today.

Early in the eighteenth century, soon after Fontenelle recognized that there had been a revolution in mathematics, Newton's *Principia* was seen to constitute a revolution in physics, and before long Robert Symmer was proclaiming that he had made a revolution in the science of electricity. These events occurred when revolution in the political sense had a benign connotation, before the French Revolution had reached such extremes in the Terror that 'revolution' became a frightening word rather than an expression of rapid progress. Joseph Priestley, who suffered political persecution for his espousal of the French Revolution and who emigrated to the United States in 1794, shows us how the attitude toward revolution changed by the end of the eighteenth century. In a letter to Robert R. Livingston, statesman and inventor, who worked with Robert Fulton on the development of the steamboat, Priestley congratulated his correspondent on his "most valuable discovery relating to the fabrication of *paper*" (Schofield 1966, 300). "If you can succeed in bleaching it," Priestley wrote, "you will produce a complete revolution in the whole manufacture." The date was 1799, and Priestley — mindful of the generally negative feeling about revolutions — hastened to add a note of regret that Livingston's innovation must not "be called a *revolution* in these times. That alone would discredit it, tho ever so useful. It is not, however, the less acceptable to me."

The publication of *The Communist Manifesto* in the middle of the nineteenth century, the revolutions of 1848, and the formation of the first 'International', with its plans for world-wide revolution, conveyed anew a sense of violence associated with rapid change. Because these negative aspects of revolution were in the forefront of the thoughts of most people living in the 1850s, it is a little surprising to find English and Irish scientists like Darwin and Hamilton categorizing their respective reconstructions of science as revolutionary in the old benign sense, as if the new political urgencies had no significance for the image of scientific change. On the Continent, the reactions of scientists were markedly different.

In the twentieth century, the stark drama of the Russian Revolution and the specter of a possibly imminent world communism caused some scientists and nonscientists to be aghast at the alleged 'bolshevism' of such radical physics as Einsteinian relativity. Closer to our own time, the doctrines of Mao and the Chinese Revolution with its attendant 'cultural revolution' have in their turn affected the concept and image of revolutionary activity.

Comparison of Political and Scientific Revolutions

Political theories and events that involve rapid change in the social structure have had a pervasive influence on concepts of scientific revolution since the seventeenth century. Therefore we might profitably ask which specific features of political revolutions (and theories about them) have been incorporated into the concept of scientific revolution that most of us recognize today, and which other ones have proved to be inapplicable. A comparison of the two types of revolution reveals a closer degree of concordance than might at first be imagined. (A historical perspective on the comparisons of political and scientific revolutions is provided in §1.1., below.)

One of the features that all political revolutions have in common is the element of 'newness,' as Hannah Arendt (1965) has insisted. "The modern concept of revolution," she writes, is "inextricably bound up with the notion that the course of history suddenly begins anew." Revolution thus implies "that an entirely new story, a story never known or told before, is about to unfold." We shall see, however, that in revolutions in science there are links of transformation between the old and the new.[2] This is also true of political revolutions, although perhaps to a lesser degree. Paradoxically, this feature does not diminish the net intensity or effect of either scientific or political revolutions.

Clearly, one must make a judgment about the extent or profundity of the novelty in deciding whether or not a sequence of events 'really' constitutes a revolution. Or perhaps, as Pettee (1938, ii) suggests, there is a continuous gradation from "the great revolutions," such as the French and Russian revolutions, to "such palace revolutions as Macbeth's murder of Duncan." But others would see _coups d'état_ or palace revolutions as 'revolts', not involving any fundamental political (that is, constitutional) or social change. To some extent, then, the designation of a particular event as a revolution depends not only on objective criteria concerning the kind of change (whether or not there is a constitutional change) but on personal judgment concerning the degree of change.[3] This last factor weakens any attempt at a universal definition.

Anyone who studies revolutions in science will quickly find that these events — like social, political, and economic revolutions — can have different orders of magnitude ranging from maxirevolutions to minirevolutions. There are large-scale upheavals that affect the whole of a science, and even the mode of thought and explanation in other sciences, such as occurred in the Darwinian revolution or the revolutions of relativity and quantum mechanics. Then there are lesser revolutions that may have a very profound effect on only one part of a science and do not affect all thinking in a science or the thinking of other sciences; an example is the revolution that occurred in the foundations of the new experimental psychology, chiefly by Wilhelm Wundt. An attempt to make precise the degree of revolution occurs in a review of earlier opposition to continental drift theory by George Gaylord Simpson (1978, 273), in which he called this change in "physical geology" a "major subrevolution." The reader will find this statement puzzling since Simpson does not explain the fine distinctions that may make a revolution 'major' or 'sub', nor does he indicate what the difference might be between a minor revolution and a major subrevolution. This tendency to rank revolutions began as long ago as the eighteenth century, when the historian of astronomy J.-S. Bailly discussed revolutions on a large scale, such as he perceived to have been produced by Coper-

nicus and by Newton, and minirevolutions attendant on the introduction of new instruments of observation, which may lead to a new way of
thinking or a new basis of knowledge.

A new instrument can also produce large-scale revolutionary effects,
as happened with the invention of the telescope. In his manuscript
records and in his book *The Starry Messenger* (published in 1610), Galileo
reported mountains on the moon, thus confirming — as he said — "the
old Pythagorean opinion that the moon is like another earth." A confirmed Copernican, Galileo unconsciously transformed his observations of spots and bright points of light in the shadow regions of the
moon into conclusions about the surface of the moon, conceiving it to
be like the surface of the earth. As he gazed at the moon through the
newly invented telescope, he 'saw' analogies with terrestrial experience
(see Cohen 1980, 211–215). Galileo found that Jupiter has four moons, a
discovery of enormous consequence for astronomy. In Galileo's day, a
telling argument against the possible orbital motion of the earth was the
problem of how the earth could move around the sun with prodigious
speed (some 20 miles per second) without losing its moon. Galileo could
never solve that knotty problem, but his discovery that Jupiter revolves
without losing four moons destroyed the force of the objection that the
earth could not move without losing one moon. Galileo then found that
the sun has spots and rotates. He observed that Venus exhibits phases, as
the moon does, and deduced from the correlation of the phases of
Venus and its apparent size that Venus revolves around the sun and not
around the earth. He also found that many 'nebulosities' are merely
assemblages of faint stars, invisible to the naked eye, and that the
heavens contain myriads of stars never seen by any human eye before
the telescope.

Astronomy was never the same again. But these revolutionary
changes in astronomy (including the visual demonstrations that the
Ptolemaic system is false) were not 'produced' by the telescope but by
the mind of Galileo drawing Copernican and unorthodox conclusions
from his telescopic observations. The telescope produced a vast change
in kind, magnitude, and scope of the data base of astronomy, providing
the observational materials on which a revolution would eventually
become founded; but these data did not in and of themselves constitute
a revolution in science.

The case is different for the computer, which — like probability and
statistics — has affected the thinking of scientists and the formulation of
theories in a fundamental way, as in the case of the new computer

models for world meteorology. That is, Galileo's telescopic observations changed the data in a way that required an abandonment of traditional theories and the acceptance of a new one, but they did not fundamentally affect the way in which theories are related to experiential data. By contrast, the introduction of probability produced a new kind of theory—and, in fact, a new kind of science—in which the traditional basis of a one-one cause-and-effect is replaced by a statistical foundation. It is the same for the computer, which has also altered the form of scientific theories, in that logically linked propositions and formal mathematical statements have been replaced by complex computer models.

In addition to newness, another feature that revolutions in science share with political and social revolutions is the phenomenon of conversion (discussed in chapter 29). One example will suffice to indicate the revolutionary zeal of the scientific convert. In 1596, in the original preface to *The Secret of the Universe* (1981, 63), Kepler described the stages of his conversion to the Copernican astronomy, a topic which he enlarged upon in the first two chapters. He believed that God had shown him how and why the Copernican system was put together, why there were six planets and "not twenty or a hundred," and why the planets were situated in their respective orbits with the speeds that they there exhibited. His solution was later expressed in what we now call Kepler's third (or harmonic) law, but in 1596 he undertook to prove that God, in creating the universe and regulating the order of the cosmos, had in view "the five regular bodies of geometry, known since the days of Pythagoras and Plato." He later wrote that he owed the Copernican heliostatic system "this duty: that since I have attested it as true in my deepest soul, and since I contemplate its beauty with incredible and ravishing delight, I should also publicly defend it to my readers with all the force at my command."

The comparisons one can make between political and scientific revolutions transcend such inner factors as zeal. For example, every political revolution has as its main feature a series of acts of physical violence related to the takeover of the institutional loci of power. Chalmers Johnson (1964, 6) states categorically that radical changes "not initiated by violent alteration of the system are instances of some other form of social change." Although one might not ordinarily think of a scientific revolution as involving physical violence, many of the great revolutions in science have exhibited a pattern of action similar to the physical overthrow of a government. In a scientific revolution, there is apt to be a series of acts whereby control is gained of the scientific press, the educational system, and the seats of power — in scientific academies and laboratories or on major scientific committees which make policy and apportion resources. One can see this very dramatically in the Lysenko revolution in the Soviet Union, in the course of which the forces of orthodox (Western) genetics were routed. Lysenko and his followers gained control of the genetics section of the Soviet Academy of Sciences and the system of agricultural experiment stations. They rewrote the textbooks to accommodate their new unorthodox ideas and reordered the whole system of teaching and practice of genetics. These revolutionaries drove from their posts all geneticists and even academicians who refused to hew to the new revolutionary line. The leading geneticist of the Soviet Union, N. I. Vavilov, brother of the president of the Soviet Academy of Sciences, dropped out of sight; in fact, at the time of his death in 1943, there was no official obituary telling the date and details of his last years and ultimate death in a concentration camp.

In Nazi Germany in the 1930s, not only were Jews removed from their jobs but the party approved a revolutionary movement to purge German science of the taint of 'non-Aryan' or overly theoretical thinking. Two of the leaders of this movement were Nobel Prize-winning physicists, Philipp Lenard and Johannes Stark. Under Hitler, Stark attempted to reorganize and expand German physics, but he was opposed by brave and decent men, led by Max von Laue, Max Planck, Arnold Sommerfeld, and Werner Heisenberg, whom Stark called "white Jews in science," the "viceroys of the Einsteinian spirit" (see Hermann 1975, 615). Lenard, an old friend and colleague of Stark, was a super-patriot who believed that a "disarmed nation" was a "dishonorable nation" (Hermann 1973, 182). In 1920, at the annual conference of German scientists and doctors, Lenard got into a public debate with Einstein, which was marked by Lenard's "sharp malicious attacks" and "an unconcealed anti-Semitic bias." As early as 1924, Lenard ended his academic lectures on physics with praise for Adolf Hitler as the "true philosopher with a clear mind." He became Hitler's chief authority in physics and published four volumes on experimental physics called *Deutsche Physik* (1936–37), or *German Physics,* which he defined as "Aryan physics" or "physics of the Nordic man." He said that "science . . . is racially determined, determined by blood." The 'Deutsche Physik' group, despite their official Nazi backing, never gained full control of German physics in the way that Lysenko and his followers did in the field of genetics in the Soviet Union. Only a few colleagues came to join Stark and Lenard, and their "efforts remained, despite the support of the Third Reich, fruitless" (Hermann 1973, 182; Beyerchen 1972).

Scientific change through a political kind of power is not limited to twentieth-century totalitarianism in the Soviet Union and Nazi Germany, however. We may see an early example in the various stages by which the Cartesian forces gained intellectual and institutional control of science in France (Sutton 1982). The revolutionary Cartesians fought for power with the forces of orthodoxy — represented by the Jesuits and their schools, the Church and its University of Paris, and the Aristotelians — on every imaginable level. They gained entrée into the influential salons and eventually attracted followers from among the intellectuals. Before long, Cartesians were in control of the schools (the *lycées* and the Jesuit *collèges*) and the universities. Cartesians came to have a strong voice in the Paris Academy of Sciences, where the 'permanent secretary' was Fontenelle, a staunch Cartesian and author of a major work on the Cartesian cosmological system of vortices ('tourbillons'). In the late seventeenth century the general textbook of Jacques Rohault, a noted follower of Descartes, replaced the traditional works and became the standard source of scientific knowledge; it was printed again and again and translated into a number of languages.

When the new revolutionary science of Isaac Newton was set forth in 1687, it was obvious that the real enemies to be routed were not the Aristotelians and scholastics but the Cartesians and their physical cosmology based on vortices. Newton showed, in the conclusion to book 2 of his *Principia,* that the Cartesian hypothesis "is completely in conflict with the phenomena of astronomy" and leads to a "confusion rather than an understanding of the celestial motions." But it was not enough to confute the Cartesians and others; an active campaign had to be mounted on a number of fronts simultaneously. First, there was a definite courting of the forces of the State, initiated when Newton dedicated the first edition of his *Principia* to the Royal Society and its patron, King James II. Edmund Halley, knowing of the King's interest in naval affairs, wrote for him a special account of the portion of the *Principia* dealing with the tides (see Cohen and Schofield 1978, §5). Because the church was such a powerful force in all intellectual matters, the Newtonians wanted control of the new Boyle Lectures (founded under the terms of the will of the chemist and natural philosopher Robert Boyle), which consisted of eight sermons in London churches on the evidences of Christianity (see Guerlac and Jacob 1969). These at once became a major vehicle for expounding the Newtonian science.

The Newtonians followed the path taken by Rohault and introduced popular lectures on the new science, extensively using demonstrations in order to make the subject matter more palatable and easier to understand. The pioneers were William Whiston and J. T. Desaguliers. Newton used his personal influence to replace the scholastic and Cartesian teachers at the major universities with orthodox Newtonians. Before long there was a powerful Newtonian network, comprising Colin Maclaurin in Edinburgh, Roger Cotes in Cambridge, David Gregory in Oxford, and others. To gain control of textbooks, the Newtonian Samuel Clarke added critical notes to his translation of Rohault's book on natural philosophy. This was the same Clarke who spoke for Newton in the famous debate with Leibniz. Eventually Rohault's treatise became the major work for teaching the Newtonian natural philosophy under the guise of a revised Cartesianism. Other Newtonians wrote original textbooks. And finally, when Newton went to London to become head of the Mint, he was elected president of the Royal Society, a position he used to make sure that this institution became a party to the struggle to establish the Newtonian philosophy and to guarantee Newton's priority over Leibniz in the quarrel over the invention of the calculus.[4]

These examples, for the most part, have been taken from successful or partially successful revolutions. But there is, additionally, the category of revolutions that fail. Prominent failures from the realm of politics are the revolutions of 1848 and the abortive Russian revolution of 1905. Scientists and historians of science generally do not talk about failed revolutions. That is, they tend to designate by the name 'revolution' only those movements in science that actually succeed (see chapter 2). No one has as yet written a history of scientific failures. Here, then, is one aspect of revolution in which science notably differs from political and social action.

A final point of difference between political or social revolutions and scientific revolutions is the goal. In one sense, both types of revolution have a specific, narrowly defined goal. For instance, the goal of the Newtonian revolution was to produce a new system of rational mechanics on the basis of which one could retrodict and predict the phenomena observed on earth and in the heavens. It was postulated on concepts of mass, space, time, force, and inertia, and it embraced the concept of universal gravity. This appears similar to the goal of producing a society in which, for example, there may be equality of economic opportunity, political liberty, a system of parliamentary or representative government, and so on. But the real difference is that in most political and social revolutions the goal is alleged to be immediately attainable. For instance, it was certainly the goal of the Russian Revolution to establish a communist state or a classless society. The achievement of this aim was never thought to be simply a prelude to an unending series of political and social revolutions; once the ideal state was achieved, no subsequent revolutions would be necessary. But developments of science, particularly after the revolutionary period of the seventeenth and the eighteenth centuries, have led us to expect that science will provide a series of continual revolutions without end. There is no final particular goal which, once achieved, means that no more revolutions will occur. The Newtonians, for example, were well aware of additional areas in which a scientific revolution was necessary: chemistry, optics, heat, and physiology. And even in terrestrial and celestial dynamics there was still the unsolved problem of the moon's motion under the joint action of the sun and the earth. In science a successful revolution establishes a revolutionary program for future revolutions, whereas a political or social revolution (at least ideally) has a finite program that the revolutionaries hope to achieve.

Revolutionary Science and Society

The scientific revolutionary has a quite different role in society from that of the political or social revolutionary. The social or political radical threatens the established social order or political system by plotting or preaching its overthrow, developing a theory that might be put into practice and that would produce a social or political revolution, or even participating in a revolutionary movement. Thus the social or political radical appears as an immediate or potential danger to our way of life, our mode of government, our value system, and may even seem to put in jeopardy our family system, our homes, our possessions, and our jobs. These considerations obviously apply to the 'haves' more simply than the 'have-nots', but even have-nots may eschew the revolutionary movement in the hope of making it within the established system and becoming haves (even if on a small scale). The scientific radical,[5] on the other hand, threatens directly the current structure of knowledge or the status quo *in science,* but not usually in the society at large. Of course, science does affect the lives of ordinary men and women, but it most often does so at one remove — not directly but indirectly — as a result of practical applications. The basic science of high-polymer chemistry, for example, had — by itself — no effect on society, but its applications to produce manmade fibers have had vast implications for our living style, our economic system, and the reordering of employment possibilities. And it is the same for radar, supersonic flight, nuclear power, conquests of disease, and the exploration of space. The practical fallout from scientific revolutions is technological innovation, with the attendant destruction of old jobs but with the possibility of creating new ones.

A few revolutionary scientific ideas, however, have encountered general opposition because they appear to threaten beliefs that are in some way fundamental to the social order. Darwin's *Origin of Species* (1859) aroused great hostility among laymen, and even among some scientists, on what we would consider to be essentially nonscientific grounds. The world at large was not really concerned with such technical questions as variation, descent, fixity of species, natural selection, struggle for existence, or survival of the fittest — at least insofar as these expressions apply to wild or domesticated plants and animals. But there was real disquiet about the religious implications of Darwinian evolution because it cast doubt on the account of creation in the opening pages of the Book of Genesis. Many people apparently felt a sense of genuine anxiety concerning the dramatic allegation that man shares common ancestry with the ape and does not occupy the unique position in nature which had been accorded to him by all philosophies and religions since the beginning of recorded history. This aspect of scientific revolutions — their effect on the thinking of men and women outside the strict domain of science — has been called the ideological component.

A dramatic example of the ideological component of a revolutionary scientific idea is found in the implications of the Copernican doctrine: the displacement of man and his earthly abode from a central place in the universe. It must have seemed a real blow to man's pride to be told that his planet had been shifted from a fixed place at the center, only to become just "another planet," as Copernicus called it, and physically a rather insignificant planet at that. John Donne (who probably could not have followed the simplest technical astronomical argument for and against the new system) wrote that without a fixed position the earth is lost and no man even knows where to look for it: "all cohaerence [is] gone." Martin Luther, who did not know much (if any) technical astronomy, reacted violently to the Copernican idea before even reading anything Copernicus wrote.

The Newtonian revolution also affected the way in which philosophers, theologians, political and social scientists, and educated men and women thought about the physical universe at large and about nature, "nature's laws," the basis of religion or religious belief, the nature of God, even the forms of government. But perhaps the ideas of Copernicus ultimately had a greater effect beyond the strict limits of the sciences than did those of Newton, since Copernicus's ideas shook the old anthropocentric notions of man's privileged position in the universe and, by implication, his uniqueness. In this respect Copernicus's impact would be more closely akin to Darwin's than Newton's.

Sigmund Freud was writing from bitter personal experience as well as from the long view of history when he compared the hostility to his own innovations with the similar hostile reactions to the ideas of Copernicus and Darwin. But possibly the Einsteinian revolution caused the greatest intellectual stir of the twentieth century in the world at large. Of course, most people had no comprehension of Einstein's theories, but they nevertheless assumed that the new physics of relativity gave grounds for a general relativism, implying that 'everything is relative', that there no longer exist any tenable standards of 'absolute' belief in religion, ethics, and morals.

In a Herbert Spencer Lecture at Oxford in 1973, Karl Popper made a useful distinction between a scientific revolution and an ideological revolution. One is "a rational overthrow of an established scientific theory by a new one," while the other includes "all processes of 'social entrenchment' or perhaps 'social acceptance' of ideologies, including even those ideologies which incorporate some scientific results." The "Copernican and Darwinian revolutions" show how "a scientific revolution gave rise to an ideological revolution," and hence exemplify the way in which a revolution in science may have a distinct "scientific" and "ideological component" (1975, 88). What is perhaps most interesting about these two aspects of revolutions is that a revolution may have profound effects in science and yet have no ideological component. A striking example is the introduction of the physics of fields, largely the work of Faraday and Maxwell, which wholly revolutionized the basis of physical science and produced a radical displacement of the Newtonian basis of physics, which was firmly rooted in the concept of central forces, and opened the way for the physics of relativity. And yet this bold alteration of classical physics had no ideological component, although every physicist from then on was aware that the subject had changed in a very fundamental way.[6] And it is the same for quantum mechanics, "one of the most fundamental scientific revolutions in the history of the theory of matter" (Popper 1975, 90). Physicists have long been puzzled by the fact that the revolution of quantum mechanics has had no ideological component, that Heisenberg's uncertainty principle has never laid hold of the public imagination as relativity had done a few years earlier. And it is noteworthy that — at least as of now — no tremendous ideological component has been associated with the great revolution in molecular biology in our own days.[7]

A second kind of societal hostility to scientific revolution arises — properly speaking — from a reaction to the results or applications of science rather than to the science itself. Because so many rapid advances in both civilian and military technology are sparked by new science or by scientific revolutions, there has been a growing tendency to consider science and technology as one, and even to hold science responsible for the effects of technology. This phenomenon is not wholly new. When, during the Great Depression, a too-rapid rate of science-based innovation was deemed responsible for what was called technological unemployment, there was a call for a "moratorium on science." We have seen objections raised to the enormous costs of the space program in our own day, especially by those who would rather see public moneys spent for improving the conditions of our cities or for achieving other socially beneficent goals and not for revolutionizing our knowledge of the solar system and the rest of the universe. And many people have expressed an obvious deep concern about weapons whose technology is based on the latest discoveries in the biological as well as the physical sciences. Men and women of good will all around us decry the effects of pollution and other aspects of the deterioration of the environment and — rightly or wrongly — attribute such evils to science as the mainspring of techno logical innovation. And so many believe that the revolutions by which science advances are not necessarily benevolent and do not imply real progress for the 'condition of man'.

Apart from considerations of this sort, within the scientific community itself there is a general belief that each revolution in science is a step forward. Of course, some die-hards will always struggle against any major innovation that destroys existing concepts, theories, and general beliefs. Every great revolution in science has engendered an opposition among some scientists; the degree and extent of the antagonism may even be taken as a measure of the profundity of the revolutionary changes. Furthermore, every scientist has a vested interest in the preservation of the status quo to the extent that he does not want the skills and expert knowledge which he has acquired at great cost in time and learning energy to become obsolete.[8] But despite this natural resistance to change, there is no organized conservative party within the scientific system — as is found in the sociopolitical system — that seeks to keep things as they are and to suppress revolutionary movements within the sciences. One will always find radicals and conservatives (even individual counterrevolutionaries) within the sciences, and there will always be those who prefer old-fashioned methods and styles to new ones. But I believe that all scientists would agree with the reply, reported by the late Paul Sears, to a colleague in the humanities who said, "I suppose you will think I am old-fashioned, but I don't think that germs have anything to do with disease." The response was, "No! I don't think you are old-fashioned; I think you are just ignorant."

Because the scientific revolutionary produces an innovation within the sciences that primarily affects other scientists, the radical new science does not need to be comprehensible to nonscientists. It may be beyond the understanding of many other scientists, even most other scientists, especially those in areas other than the specialty to which the innovation pertains. This was certainly the case for Einstein's theory of relativity, the alleged general nonintelligibility of which was enshrined in the popular saying that only eight (or twelve) people understood it. But its nonintelligibility to the public affected neither the acceptance of relativity by the scientific community nor the general lay opinion that Einstein was a genius and that his incomprehensible and revolutionary theory was one of the great intellectual creations of the twentieth century.

Works of art, music, or literature, on the other hand, are not intended primarily (and certainly not exclusively) for the eyes and ears of only other artists, musicians, or writers, in the sense that science is primarily addressed only to other scientists. Literature is meant to be read, art to be seen, and music to be heard by the public. Furthermore, artists, musicians, and writers depend in large measure for their livelihood on the fees and royalties that come from an appreciative audience. This condition all but automatically goes against the true revolutionary in creative areas in which mass tastes may determine canons of acceptability. There are, of course, exceptions, such as Stravinsky and Picasso. In art especially, a kind of mass 'radical chic' seems to have enabled

Picasso to achieve a popular success that by far exceeds the public's comprehension of his oeuvre. No doubt, in the 1920s, James Joyce was read, understood, and fully appreciated by as large a number of writers and critics as there were scientists who truly understood Einstein's general relativity. But Einstein's results were accepted and used by many scientists despite their inability to comprehend the theory fully in all its aspects, or despite their inability to read Einstein's writings with ease or complete understanding. In Joyce's case, on the other hand, there was only a *succès d'estime;* the greater part of the reading public and the writing profession did not adopt and make use of Joyce's radical innovations because of their inability to read and understand *Finnegan's Wake* (called "Work in Progress" when being published serially in the journal *Transition*) and because to have adopted the new style would have separated writers from their readers and so have hindered rather than advanced their professional stature.

It is a curious paradox that conservative societies (and all highly organized and institutionalized societies must be essentially conservative in the sense of being self-preserving) not only tolerate revolutionary activity in the sciences to a degree that is simply not the case for any other form of intellectual or artistic creative effort, but even encourage it. Whereas a man or woman with extremely radical political, social, or economic views may encounter barriers (especially in relation to employment) that impede ordinary career development, and may even — as a dissident — encounter the restraining force of the law or the state, the scientist is especially honored when his or her most radical ideas succeed. Science is an exceptional enterprise in that revolutionary activity has been institutionalized; the system not only recognizes originality and assigns great value to it (as R. K. Merton has taught us) but bestows large cash prizes and social rewards on the successful revolutionary. In art or letters or music an extreme radical is denoted a member of the avant-garde and his or her audiences may be few in number; there are no rewards or prizes or honors for revolutionaries in these creative areas comparable to those in the sciences. It is notable, moreover, that whereas Nobel Prizes are regularly awarded to scientists whose contributions have been radical and truly revolutionary, no such awards in literature have been made to innovative writers of similar radical or revolutionary stature — August Strindberg, Henrik Ibsen, Marcel Proust, James Joyce, or Virginia Woolf.[9]

A major reason why society is usually willing to support and to reward revolutionary science, even an extreme form of ordinarily incomprehensible science, is the constant expectation of practical benefits: healthier and longer lives, better transportation and communication, new and improved manmade fibers, more efficient agriculture and manufacturing processes, greater conveniences for daily life, better instruments for national defense, and so on. The experience of the last half-century has vividly demonstrated again and again that the more innovative and revolutionary the science, the more profound and far-reaching the practical applications.

Predicting Scientific Revolutions

Although every scientist is aware of impending revolutions, no clear universal sign tells even the most astute observer the area of science in which the next revolution will occur or what form it will take. The most brilliant scientists are not able to predict exactly the kind of revolution they themselves will be making. (This is in direct contrast to the political and social revolutionary, who has a program worked out in advance and can, accordingly, direct his revolutionary activity toward a carefully defined goal.)

A major reason why there is no way of predicting precisely where the next revolution will occur, or what it will consist of, is that the sciences are 'arts' to one another. An unpredictable revolutionary innovation in one area may provide the means for effecting a revolutionary break-through in some other area. Because revolutionary advances in one science are apt to depend on revolutions in yet other sciences, the unpredictability rapidly increases exponentially. An example is the rise of molecular biology, notably the enodation of the structure of DNA, which required the use of a technique developed in physics — x-ray crystallography. Since the most rapid changes in technology are apt to come from unpredictable revolutions in the basic sciences, there is also an exponentially rising uncertainty in technological forecasting, nota-bly about coming revolutions in the technological sphere. Computer scientists keep alive the story of the great expert in the nascent specialty of computers who, in the late 1940s or early 50s, is said to have predicted that some six or seven computers would satisfy the future needs of the United States, with a few more for Europe. The number would prove to be small, even for the machines of the giant type then in existence. The anonymous forecaster could hardly have guessed the series of revolutions (such as the one in solid-state physics) that would wholly alter the size, nature, and function of computers in the future.[10]

Revolutions in science are inevitable, in that they cannot be prevented, at least as long as science continues to exist, although they may have to await the arrival of a particular revolutionary genius to ignite the fuse. And scientists, as I have said, do not want the revolutions to be prevented. But the pace of such revolutions, or their frequency of occurrence, may be slowed down or speeded up. That is, the tempo of scientific advance can be accelerated and more areas for revolutionary scientific activity can be opened up by such factors as large-scale financial support, which provides additional manpower for research and enables costly apparatus to be built or obtained. Enormous sums of money are required to mount field work and expeditions, to make observations, to establish better communications within the scientific community, and to give creative men and women in the sciences time to reflect (that is, by relieving them of excessive teaching or administrative obligations). The availability of funds for jobs and training fellowships attracts young men and women with creative potential into science. Contrariwise, a paucity of funds not only limits the possibilities of purchasing and constructing instruments of research and making expeditions but also limits travel and easy communication, the nerve system of scientific intelligence that is so necessary to progress. Even more important, the lack of funds reduces the number of jobs and fellowships and narrows the net of recruiting for the coming generation of scientists. Such a shrinking of manpower directly decreases the rate of scientific revolutions by making smaller the likelihood that a revolutionary genius will be in the right place at the right time.

Changing Concepts of Revolution in Science

Today it is a commonplace to speak of the Scientific Revolution, the Copernican revolution, the Darwinian revolution, the computer revolution, the communications revolution, and so on. Almost every advance in science and technology in recent years is described as a revolution in the daily press. To some extent we can attribute this to a cheapening of our verbal currency, but in part it is a reflection of the simple fact that so many revolutions in science have occurred and continue to occur. As I was writing this chapter, a glance at a single shelf in my study showed almost a dozen books on computers that have 'revolution' in the title. Who would deny that there has been a computer revolution?

Yet even in the twentieth century, scientists and historians of science have not universally held the view that science progresses through a series of revolutions. During the first half of the century, it was generally agreed that revolutions in science were very rare occurrences. Rather, science was thought to develop primarily in an incremental manner, that is, by a cumulative process in which one small step or increment more or less regularly follows another. According to this model, an occasional step of much greater magnitude than usual, corresponding to the activities of a Newton, Lavoisier, Darwin, Rutherford, or Einstein, might be said to constitute a revolution; a revolution could also occur whenever the succession of increments, each one small in itself, adds up to a revolution. But such major restructurings of a scientific field were believed to be rare, if indeed they were believed to occur at all.

George Sarton, one of the primary founders of the academic discipline of the history of science, was not a great believer in scientific revolutions. He went so far as to say that it is only our superficial "first impression of scientific progress" that shows us science advancing by discontinuous giant steps, like a set of "gigantic stairs, each enormous step representing one of those essential discoveries which brought mankind almost suddenly up to a higher level." As we "pursue our analysis," he said, we find "the big steps . . . broken into smaller ones, and these into others still smaller, until finally the steps seem to vanish altogether" (1937, 21–22). Many scientists and historians agreed; Rutherford (1938, 73) stated that "it is not in the nature of things for any one man to make a violent discovery," effectively repeating R. A. Millikan's stern declaration that revolutions in science are very rare occurrences. Sarton's analysis led him to conceive that a primary aspect of science was its cumulative character; in fact, he declared (1936, 5), science is the only "truly cumulative and progressive" activity of mankind — a judgment in which J. B. Conant (1947, 20) and others have concurred. Many analysts considered that revolutions in science — if they did, in fact, occur — were like the great revolutions in the social and political realms, unusual events that occasionally punctuate the otherwise 'normal' mode of regular or incremental progress.[11]

In 1962, Thomas S. Kuhn's *Structure of Scientific Revolutions* radically altered our thinking about scientific change. Few books in the history of science have stimulated so much interest and so continuing a dialogue. Even those who do not follow Kuhn's analysis in all detail have been provoked to consider that the advance of science is not necessarily a cumulative process, that there are successions of great revolutions and intermediate minor ones, and that a revolutionary process is part of the regular pattern in the increase of scientific knowledge.

In his seminal study, Kuhn set forth no ordinary history but rather a social dynamics of scientific change in terms of a sequence of revolutions alternating with what Kuhn calls "normal science." Kuhn's schema has been applied to such diverse areas as historical political theory, science and public policy (with respect to the application of biomedical knowledge), and the nature of the modern university, in addition to the history, philosophy, and sociology of science. One major response to Kuhn's bold presentation has been to challenge some features of his analysis, to show that his schema is not universally applicable but limited to certain sciences or to special periods or selected episodes. Another has been to question the precise meaning (or to explore the ambiguities or multiple meanings) of his technical terms (notably 'paradigm'). Doubts have been raised concerning the propriety of using the concept of revolution in relation to scientific change. These issues and the significance of Kuhn's contribution will be discussed in chapters 2 and 26; here it is only necessary to be aware of the dramatic influence Kuhn has had on the spread of the use of the concept of revolution in discussions of the past, present, and future of science.

How things have changed since the 1950s can be seen by turning over the pages of any book or article on the history of science today and observing the ubiquitous prominence given to scientific revolutions in journals all over the world. Since 1962 a large number of books have appeared that are devoted specifically to *the* Scientific Revolution of the seventeenth century. Five of them[12] (by Basalla, Righini-Bonelli and Shea, Bullough, Kearney, and Rossi) are concerned with historiography and are largely composed of extracts from scholars in various fields who have attempted to define, explain, or analyze the causes of the Scientific Revolution. One of them, edited by George Basalla, deals with "external or internal factors" in the rise of modern science; in this case the editor has "deliberately avoided the term 'Scientific Revolution' and used the less elegant, but more precise, phrase, 'the rise of science in the sixteenth and seventeenth centuries.'" At the Fifteenth International

Congress of the History of Science (Edinburgh, 1977), one out of every six papers in section 11 on philosophy, methodology, and history dealt with aspects of revolution.

In the vast and ever-growing literature on scientific revolutions, containing studies and analyses of almost every conceivable aspect of the subject, the history of the concept is hardly mentioned. An exception is Lewis Feuer's book on *Einstein and the Generations of Science* (1974, 241–252), where some examples are given of the use of revolution in relation to science, chiefly in the late nineteenth and the twentieth centuries. The disregard for this topic on the part of historians of science would be all the more surprising were it not for the fact that historians of science have been generally notorious for neglecting the history of their own discipline and profession (see Thackray and Merton 1972; Thackray 1980).[13]

The purpose of this book is to fill that gap in the literature — to trace the many transformations, throughout four centuries, in the way scientists, philosophers, and historians have conceived of scientific change. In many cases scholars who have used the term 'revolution' may have had in mind nothing other than a historical metaphor for a great change, a truly significant invention. Such usage may be impressionistic and idiosyncratic; I doubt whether scholars have always had in mind the analogy with a particular political or social revolution when referring to a revolution in science. But we will examine many instances in which theories of social and political revolutions have strongly affected scholars' changing concepts of revolution in the sciences. And we will see how these concepts have been further influenced by the actual occurrence of social and political revolutions in the scholars' lifetimes.

For example, the image of revolutionary science in many parts of the world was affected by an abhorrence of bolshevism which grew out of the Russian Revolution of 1917. In the eighteenth century Lavoisier could compare his revolution in chemistry to the political revolution going on in France, which at that stage was a benign alteration of the absolute monarchy of the Bourbons; but the comparison would soon lose this meaning when the revolutionary excesses developed into the Terror, in which Lavoisier himself lost his life on the guillotine. A British historian living in the late eighteenth century, contemplating the Glorious Revolution and even the American Revolution, might very well consider revolutions benign, having the effect of restoring certain natural rights to Englishmen. Yet such a historian could legitimately hold the French Revolution, with its greater social violence and more complete destruction of the established order, to be a pernicious evil. This is hardly a theoretical case, since it is an accurate description of the opinions of Edmund Burke.

A final example of changes wrought by time in the concept of revolution is provided by a current view that the Scientific Revolution may have lasted for a century, or even for three centuries, from 1500 to 1800 (Hall 1954). Not only would this make the Scientific Revolution the longest-lasting revolution in recorded history, but it would imply a wholly different concept of revolution from the models of the Glorious Revolution or the American and French revolutions. That is, current views on the Scientific Revolution consciously or unconsciously invoke a concept of revolution that apparently is not derived by abstracting a set of supposed principles and practices of political and social revolutions and transferring them intact to the consideration of the growth of science.

Whether a given view of scientific change is influenced by social and political theories and events or whether it is influenced by other external considerations, we can safely say that it is always influenced by scientific developments themselves — the actual theories, inventions, or formulations that dramatically alter scientists' thinking about their field and the practice of their profession from day to day. We cannot fully understand the view of scientific change of a historian, philosopher, or scientist from any era without being aware of the nature of the scientific innovations to which he has been witness. Only then can we fully appreciate the ways in which the interpretation of those events has been affected by attitudes and events within the larger society. For this reason, a significant portion of this book concentrates on specific scientific developments — tracing the stages whereby a theory is conceived, discussed, opposed, transformed, and ultimately recognized for the revolutionary new perspective on nature that it makes possible. In short, this book treats not only of the concept of revolution but displays some of the major features of the actual revolutions in science to which the idea of revolution was applied and which exemplify the types of revolution in science in different centuries.[14]

2

The Stages

of Revolutions

in Science

————————

The past decade has witnessed the rise of a variety of analyses of revolutions in science, or of the ways in which science advances, produced by historians and philosophers of science, among them Feyerabend, Kuhn, Lakatos, Laudan, Popper, Shapere, Toulmin, and myself. A vast literature has come into being, much of it a series of arguments about the internal consistency, the universal applicability, or the general usefulness of one or the other of these analyses. A major part of the debate has centered on the ideas of T. S. Kuhn. It is not necessary to agree with Kuhn in every detail in order to appreciate the real worth of his presentation, based originally on the notion of a 'paradigm' (1962; 1970; 1974; 1977), a set of shared methods, standards, modes of explanation, or theories, or a body of shared knowledge. Kuhn sees a revolution in science as a shift from one such paradigm to another, caused — he believes — by a crisis in the state of science that makes a new paradigm necessary. The activity of scientists within one accepted paradigm is called "normal science" and usually consists of "puzzle solving," that is, adding to the accepted stock of knowledge. Such normal science continues until anomalies turn up which eventually cause a crisis, followed by a revolution producing a new paradigm.

There have been serious problems in applying this scheme. One was that the term 'paradigm' was used by Kuhn in a number of different senses (Masterman 1970; Kuhn 1970);[1] another, that all revolutions do not necessarily arise from a crisis; yet another, that the whole scheme seems to work out better for the physical sciences than the biological sciences (Mayr 1976; Greene 1971). But Kuhn's analysis has the solid merit of reminding us that the occurrence of revolutions is a regular feature of scientific change and that a revolution in science has a major social component — the acceptance of a new paradigm by the scientific community. Kuhn has made the notable contribution of shifting the discussion from conflicts among scientific ideas to conflicts among the scientists or groups of scientists who hold those ideas. Additionally, he has stressed certain features of revolutions, such as the occurrence of anomalies that lead to a condition of crisis that in turn precipitates the revolution, the incompatibility between the old and new paradigms (which inhibits meaningful dialogue across the paradigm barrier), the existence of minirevolutions between major revolutions, and much else.

My own research differs from Kuhn's primarily in that I have been exploring the ways in which participating observers and contemporaneous analysts have viewed revolutionary changes in science during the four centuries in which modern science has existed. This line of enquiry sees the concept of revolution as a complex, historically changing entity — affected in turn by revolutionary theory and events in the realm of politics — rather than as a single and simple idea of how scientific change comes about. I have also tried, wherever possible, to juxtapose contemporaneous views of revolutions with the interpretation of later historians and scientists, including those of our own time. I have identified revolutions in science not so much by their fit into a fixed taxonomic scheme as by tests of historical evidence (see chapter 3). A basic step is to examine the mode of genesis and development of revolution-making ideas in science, in the way that I have done for Newton's revolutionary innovations in my book *The Newtonian Revolution* (1980). Another is to examine the fine structure of revolutions in science, as I have done here, taking the genesis of the new ideas or theories or systems (or paradigms) as a starting point, next tracing their public presentation and dissemination, and finally making precise the stages of acceptance by the scientific community that lead to a recognized revolution.

How can we know that a revolution has occurred? There are two types of criteria: one arises from logical analysis in terms of a strict definition, the other from historical analysis. Many of the major revolutions in science turn out to be revolutions on both counts: the Newtonian, Darwinian, Einsteinian revolutions, the Chemical Revolution, and the recent revolutions in molecular biology and the earth sciences. They all pass the tests for revolutions in science that I give in chapter 3. In the present chapter, my aim is to examine the successive stages that I have found to form a characteristic sequence in all revolutions in science and the role of participating observers and contemporaneous analysts in documenting the occurrence of such revolutions. I take it as given that revolutions do occur in the sciences, although I am aware that there are disbelievers and that among believers there is no consensus concerning which events in the development of science constitute revolutions.

From Intellectual Revolution to Revolution on Paper

In the course of studying a large number of revolutions, I have found four major and clearly distinguishable and successive stages in all revolutions in science. The first stage I call the 'intellectual revolution', the 'revolution-in-itself'. This revolution occurs whenever a scientist (or a group of scientists) devises a radical solution to some major problem or problems, finds a new method of using information (sometimes extending the range of information far beyond existing boundaries), sets forth a new framework for knowledge into which existing information can be put in a wholly new way (thus leading to predictions of a kind that no one would have expected), introduces a set of concepts that change the character of existing knowledge, or proposes a revolutionary new theory. In short, this first stage of revolution is what one or more scientists are always found to accomplish at the beginning of all revolutions in science. It consists of an individual or group creative act that is usually independent of interactions with the community of other scientists. It is complete in itself. Of course such an innovation arises from the matrix of existing science and is generally a fundamental transformation of current scientific ideas. Furthermore, it tends to be closely related to certain canons of the received philosophy and modes and standards of the science of the time. But the creative act that expresses itself in new science with a revolutionary potential is apt to be a private or individual experience.

Almost always, the new rules or findings are recorded or written up in the form of an entry in a diary or notebook, a letter, a set of notes, a report, or the draft of a full account which might eventually be published as an article or book. This is the second stage of a revolution — a commitment to the new method, concept, or theory. Very often, this stage consists of writing out a program of research and perhaps, as in the case of Lavoisier, noting that the results are "destined" (see Guerlac 1975, 81) "to bring about a revolution in physics and chemistry." This revolution of commitment is, however, still private.

Every revolution in science begins as a purely intellectual exercise on the part of a scientist or a group of scientists, but a successful revolution — one that influences other scientists and affects the future course of science — must be communicated to colleagues either orally or through the written word. For a revolution in science to occur, the initial intellectual and commitment stages, which are private, must lead to a public stage: dissemination of the ideas to friends, associates, and colleagues and then to the world of science at large. Today this third stage may start with telephone calls, correspondence, conversations with friends and immediate colleagues, or group discussion within one's department or laboratory, followed by more formal presentation at the traditional departmental colloquium or a scientific meeting.[2] If there are no severe adverse reactions of colleagues and if no basic flaws are found by critics or by the author himself, this preliminary communication may lead to a privately circulated preprint or a scientific paper or book submitted for formal publication. The phrase 'revolution on paper' accurately describes this third stage, in which an idea or set of ideas has been entered into general circulation among members of the scientific community.[3]

Very often the intellectual revolution is not completed until the scientist fully works out his ideas on paper. A notable example is Newton's radical construction of celestial dynamics. In 1679, in correspondence with Robert Hooke, Newton learned of a new way of analyzing planetary motion, which he then used to solve the outstanding problem of the cause of planetary motions in ellipses according to the law of areas. He next put his preliminary findings on paper, but he did not (so far as we know) fully write up his ideas and their consequences. He did not even admit publicly that he had made such a breakthrough until Halley's visit (in August 1684) to ask him about the problem of forces and planetary orbits. Then Newton composed a full report of his results and sent them on to be registered at the Royal Society in November 1684 at Halley's suggestion, so that his priority of invention might be secured. As Halley well knew, no one had advanced as far as Newton in the radically new and revolution-making analysis of the forces producing planetary motion. But only after preparing the paper for Halley and the Royal Society, that is, after transforming his private intellectual revolution into a public revolution on paper in the first months of 1685, did Newton go beyond even this extraordinary level of achievement to find that the sun and each planet must act on each other gravitationally and that consequently each planet also acts on and is acted on by every other planet — the essential step that would open the way to the invention of the concept of universal gravity, the basis of the Newtonian revolution in science (see Cohen 1981; 1982).

A revolution in science can fail at any one of these first three stages. Perhaps a private document of an inventor or discoverer will gather dust in the archives and not become known until some later time, when it is far too late for the ideas to produce a revolution. Had the author decided to commit his findings to print or some other form of general circulation, a revolution might have occurred. Two examples of extraordinary scientific advance that never saw the light of day in print until the passage of some three centuries or more occur in the unpublished papers in astronomy, mathematics, and physics of Thomas Harriot (1560–1621) and in the mathematical manuscripts of Isaac Newton (1642–1727). I do not wish to imply that Harriot's discoveries in astronomy and physics (Shirley 1981) or Newton's mathematical innovations (Newton 1967) would necessarily have produced a revolution had they been published. But both examples show how great scientific advances could not realize their revolutionary potentialities simply because they were not made known until the efforts of scholarly research in our own time, some three centuries or more later.

In some cases the fate of the revolution may not depend on a scientist's failure to submit his work for publication, as was the case for both Harriot and Newton. An example is to be found in Evariste Galois's foundational work on algebra (group theory). Galois (1811–1832) did write up his results and submit them for publication to the French Academy of Sciences, but they were not accepted. He was killed in a duel before he had time fully to work up or even to write out all of his mathematical discoveries and his research program. He had life enough only to produce a brief statement of his ideas establishing group theory;

the articles and books that might then have convinced his contemporaries and revolutionized mathematics were never composed.

The career of René Descartes (1596–1650) illustrates another delay in a revolution's advance to the public paper stage. In 1633 he put aside the manuscript of *The World*, a radical text on cosmogony that contains the first complete statement of the general law of inertia. He had just learned of the condemnation of Galileo and of the Copernican doctrines of astronomy, and he did not see how he could publish *The World*, with its Copernican cosmology. He even suppressed the physiological part of the book *Treatise on Man* because he did not see how to disengage this discussion of the life sciences from its Copernican foundation. In this case, the effect was not wholly and permanently to confine the Cartesian revolution to privacy, since both the cosmological and physiological parts of *The World* were published soon after Descartes's death. Furthermore, Descartes went on to write and publish another work, his *Principles of Philosophy*, in which the laws of inertia and some of his cosmogonic ideas were set forth; but the revolution was for a time robbed of a powerful instrument.

From Revolution on Paper to Revolution in Science

Even after publication, no revolution in science will occur until a sufficient number of other scientists become convinced of the theories or findings and begin to do their science in the revolutionary new way. At that time, what had been merely a public communication of an intellectual achievement on the part of a scientist or group of scientists becomes a scientific revolution. This is the fourth or final stage of every revolution in science.[4]

The history of science records the fate of many revolutionary ideas that never got beyond the public paper stage. Mesmerism is a good example. Mesmer proposed a revolutionary system of medical 'science' which was related to his practice of therapy. Although he won a large following among laymen (Darnton 1974) and converted some doctors, Mesmer's concepts and methods were ultimately rejected by the medical and scientific establishment, who found in them no scientific merit. They could not verify the existence of the Mesmeric 'fluid' of animal magnetism.

In our own century, a number of revolutionary areas of 'phenomena' have been similarly rejected when scientific critics could find no real basis for their existence. Among these are the N-rays, discovered in France in 1903. These rays attracted great attention among the scientific community, and their discoverer, René-Prosper Blondlot, achieved great fame and notoriety. But eventually it was shown that N-rays existed only in the mind of their discoverer and of other scientists whose will to believe evidently produced a temporary suspension of their normal scientific disbelief (Rosmorduc 1972; Nye 1980). And it is the same for the mitogenetic radiation discovered in the 1920s in the Soviet Union, comprising rays supposedly given off by growing plants and other living things. They could be transmitted through quartz but not glass. Hundreds of papers were published on this exciting and revolutionary new subject on the borderland of plant physiology and radiation physics. But eventually, careful experiments showed that these rays were nonexistent. In yet another failed revolution of this kind, Paul Kammerer announced in Vienna that he had demonstrated the inheritance of acquired characters. In 1926 his specimens of the mating toad, which presumably were his proof that acquired characters can be inherited, proved to have been doctored by the subcutaneous injection of India ink.

These examples (with the possible exception of Kammerer and his doctored specimens; see Koestler 1971) illustrate how both self-delusion and the excitement of a crowd of followers can lead from a revolution on paper to what may almost become a revolution in science. To some degree these fall within the category of 'fringe' or even 'pathological' science (Langmuir 1968; Rostand 1960), but a failed revolution in science need not be of this sort — although it is often difficult to separate what is extremely radical from what is pathological. Langmuir explained that almost always "there is no dishonesty involved." Scientists are "tricked into false results by a lack of understanding about what human beings can do to themselves in the way of being led astray by subjective effects, wishful thinking or threshhold interactions."

Two abortive revolutions illustrate the difficulty of this problem: one is Velikovsky's radical cosmological physics, the other is polywater. Immanuel Velikovsky attempted to revolutionize physical science with a radical set of ideas concerning the way the solar system came into its present state. One part of his revolutionary theory was that only a few thousand years ago Venus made repeated collisions with the earth and Mars, initiating events recorded in the Bible and other early chronicles; Venus was then a comet. Needless to say Velikovsky's ideas contradicted basic laws of dynamics and gravity. Velikovsky proposed that electrical and magnetic forces overwhelmed the action of gravity on the close encounter of planets. Though widely disseminated, especially in the public press, Velikovsky's radical ideas were not accepted by the scientific community. In fact, they gained serious consideration and even aroused great forces of opposition. In 1973, a debate was held at a meeting of the American Association for the Advancement of Science. Five scientists (Carl Sagan among them) attacked the theory of the collision of planets; only Velikovsky himself defended it (see Goldsmith 1977; Sagan 1979). In a review of this episode in the *New York Times* on 2 December 1979, two weeks after Velikovsky's death, Robert Jastrow listed three predictions of Velikovsky that had been verified and seven major ones that were directly contradicted. He lamented that "the facts" were not "otherwise," since "nothing could be more exciting than to witness a revolution of scientific thought in our own lifetime.""Unfortunately," he concluded, "the evidence does not support this possibility."

Polywater, first known as 'anomalous water', was discovered in 1961 by a Russian chemist working in a small provincial technical institute; the research was almost at once taken over by Boris V. Derjaguin, a well-known Russian physical chemist working at the head of a large group in a prestigious institute of the Soviet Academy of Sciences (see Franks 1981). Produced from ordinary water, this fluid differed from water as we know it in almost all its properties. It had a different boiling point and freezing point. In an article in the leading scientific journal in America, *Science*, on 27 June 1969, spectroscopic evidence was presented to support the judgment that the properties of this substance "are no longer anomalous but rather those of a newly found substance — polymeric water or polywater." The polymerization required a "previously unrecognized type of bonding for a system containing only hydrogen and oxygen atoms." At first, this discovery was not taken seriously by Western scientists, but before long research on polywater began in England and then on a large scale in America, accompanied by many conferences and supported by millions of dollars from the Department of Defense. As a referee for a research proposal wrote to the United States Air Force Office of Scientific Research (Franks 1981, 186), "All of chemistry, including that part of great relevance to the Air Force, will be revolutionized by this type of work." The eminent British crystallographer J. D. Bernal hailed polywater as "the most important physical-chemical discovery of this century" (ibid., p. 49).

Before long there was an avalanche of research papers on polywater published in reputable scientific journals; in November 1970 Derjaguin published an account of this "superdense water" in the prestigious *Scientific American*. Speculations arose on the implications of the new discovery. A professor from Pennsylvania issued a warning, printed in the widely read and authoritative British journal *Nature* (1969, 224: 198), that if "the polymer phase [of water] can grow at the expense of normal water under any conditions found in the environment," all life on earth would become extinct. "The polymerization of Earth's water would turn her into a reasonable facsimile of Venus." Utmost caution was required, he concluded, because "once the polymer nuclei become dispersed in the soil it will be too late to do anything."

Of course there were skeptics, many of them quite vocal. They advised the Air Force, the Office of Naval Research, and the National Science Foundation not to associate themselves by financial support with polywater research, lest they eventually look ridiculous. In a letter to *Science* (1970, *168*: 1397), entitled " 'Polywater' is Hard to Swallow," Joel H. Hildebrand, dean of American physical chemists, expressed the doubts of many members of the scientific community that polywater existed. Eventually it was shown that the properties of polywater came from (Franks 1981, 136) "different types and levels of contamination." An editorial in *Nature* sounded a melancholy note: "The failure of several experimenters to pursue with all the vigour at their command the possibility that contamination might account for most of their observations is nothing to be proud of."

The story of polywater is of special interest in the analysis of revolutions in science, not so much because it is a revolution that failed but because of the way it initially succeeded. Most failed revolutions in science are revolutions that never get beyond the state that I have called a revolution on paper. That is, they do not generate enough support among the scientific community to begin restructuring scientific theory to the extent of constituting a revolution. Other such revolutions fail because they are contradicted by experimental findings. Many others simply do not pass the primary test of being useful. In the case of polywater, however, the revolution was — at least for a time — almost, if not quite, a proper revolution in science. A large number of believers produced and published a considerable amount of research on the subject, much of it sponsored by major prestigious sources of financial support; these publications on the properties of the new substance proliferated in major scientific journals. In one sense, polywater can perhaps be described better as a discovery requiring a revolution (or a revolutionary discovery) than as a proper revolution — in the sense that a revolution would have been required in order to explain how this anomalous polymerization is produced in water. But if polywater represents a revolution in science rather than something that was only revolution-making, then one is tempted to say that for several years the revolution all but succeeded, despite the heavy skepticism of a considerable portion of the scientific community. Such skepticism and even downright hostility is, however, a normal feature of the early stages of any revolution in science.

In the end there was no polywater revolution because of the stern test of experiment that eventually required the abandonment of belief in this polymer of water. One can understand why many scientists must have overcome their fundamental skepticism and joined those doing research on polywater, since there is always a great desire to be active on the forefront of science, to be part of the team working on what is new and challenging. These researchers could not have been engaged in a conspiracy to defraud their fellow scientists, but rather must have suffered a massive self deception, perhaps from too strong a desire to achieve positive results (see Ziman 1970). The history of this mass delusion is a subject worthy of exploration by those studying the sociology and psychology of science and the nature of revolutions in science. The rise and fall of polywater show how men and women actually behave in the laboratory, under the stress of today's extremely competitive scientific system: their behavior does not always conform to the ideal search for abstract truth that has long been the classical image.

The desire to be an active part of a revolutionary movement is often in conflict with the natural reluctance of any scientist to jettison the set of accepted ideas on which he has made his way in the profession. New and revolutionary systems of science tend to be resisted rather than welcomed with open arms, because every successful scientist has a vested intellectual, social, and even financial interest in maintaining the status quo (see Barber 1961). If every revolutionary new idea were welcomed with open arms, utter chaos would be the result.

The rigid and brutal insistence on demonstration which is part of the resistance to change in science actually is a source of strength and stability. Many attempted or proposed revolutions simply do not pass the test. Their predictions may not be verified, the experiential base may prove to be wrong or inadequate, or the theory itself may be shown to have a flaw. If there are no real advantages to a proposed new theory or method, why should anyone adopt it and throw away a lifetime of science? It is this hard test which causes many revolutionary developments in science to be rejected. The scientific enterprise differs from the political and social realms in that it has recognized steps by which other scientists give a revolution legitimation, so that the revolutionary movement — though resisted by conservative forces within science — is not illegal, not outside the accepted canons of scientific change. And in science there is an orderly procedure — not dependent on *force majeure* — for the rejection of revolutions.

Of course the system does not work perfectly. A striking example of a breakdown in the advance of a scientific revolution is seen in the discovery by Gregor Mendel of the founding laws of genetics in the 1860s. Mendel published his work in an admittedly obscure journal, but his paper did enter the bibliographical guides to the subject. Nevertheless, it was ignored for half a century until rediscovered around 1900 almost simultaneously by Carl Correns, Erik Tschermak, and Hugo de Vries (Olby 1966). De Vries, who came upon this work of his illustrious predecessor by chance, brought it to the attention of the world of science. At the time of Mendel's original publication, men of science were searching for variation and blending of inheritance and not for fixities; the scientific world was not yet ready for his discovery and so ignored it. In a sense, Mendel may have been half a century ahead of his time.

Scientists brought up on the dogma that light is emitted, transmitted, and absorbed as a continuous wave phenomenon would obviously have had the greatest difficulty in 1905 in jettisoning this accepted theory of light for Einstein's 'heuristic' concept of discrete light quanta. And it must have been equally difficult for anyone reared in the belief in the fixity of plant and animal species to accept the notion of the evolution of species, as Darwin proposed in 1859. Yet a radical theory may have significant aspects that make it quickly favored over the old one. It may win adherents because it can explain anomalies or predict unsuspected new phenomena; perhaps it can unify branches of science that had been separate or unconnected, or it may bring a new precision to the discussion, even simplifying the kind of assumptions then being made. Sometimes the new theory gains support from a dramatic experiment or observation, as happened when the bending of light in a gravitational field, predicted by Einstein's general relativity theory in 1907, was actually verified during the total eclipse of 1919. But even with verification, general relativity was not at the forefront of the concerns of most scientists during the next four decades or so, being cultivated only by a relatively small number of mathematicians and astronomers interested in cosmological questions. Only after World War II, some forty or more years after the theory had been proposed, did questions of general relativity come to be of primary importance in the active research of a large number of physicists and astronomers. Thus there was a long delay between the revolution on paper — even the confirmation of the theory — and a truly large-scale revolution in physical science.

The gap between a revolution on paper and a revolution in science is documented clearly in the case of Einstein's paper on special relativity of 1905. This paper was entitled "On the Electrodynamics of Moving Bodies," a topic then being studied by the physicist Max Born at the University of Göttingen. Born was a member of a seminar taught by David Hilbert and Hermann Minkowski on "the electrodynamics and optics of moving bodies." Born (1971, 1) records that the seminar students "studied papers by H. A. Lorentz, Henri Poincaré, G. F. Fitzgerald, Larmor and others, but Einstein was not mentioned." After graduation in 1906, Born went to Cambridge, where he attended Joseph Larmor's lectures on the theory of electromagnetism and J. J. Thomson's on the theory of electrons, and "again Einstein's name was not mentioned." It was only later, in 1907–08 in Breslau, that two young physicists — Fritz Reiche and Stanislaus Loria — told Born about Einstein's paper and suggested that he read it. He did, "and was immediately deeply impressed." Born recalls that all anybody knew about Einstein at that time was "that he was a civil servant at the Swiss Patent Office in Berne," all too evidently not a member of the establishment.

In the same year that he published his paper on special relativity, Einstein also proposed his revolutionary transformation of Planck's quantum concept in one of the leading scientific journals, the *Annalen der Physik*. But this remained no more than a revolution on paper until the 1920s. R. A. Millikan, who inaugurated a series of experiments in order to show that Einstein was wrong, found instead that Einstein's bold restatement of quantum theory did predict a law of the photoelectric effect which was verified by experiment. But he nevertheless denied as emphatically as he could that Einstein's revision of quantum theory could have any validity. Although Einstein's new concept was fundamental to Niels Bohr's revolutionary proposal of a new atom model in 1913, its lack of general acceptance is seen in the fact that when Einstein was recommended in 1913 for a post in Berlin, his sponsors (Planck among them) found it necessary to apologize for the candidate's extravagant speculations in the domain of quantum theory.

Sometimes a revolution on paper may not become a revolution in science because the scientist-revolutionary lacks orthodox credentials.

Scientists tend to ignore proposed fundamental revisions of accepted science that come from outside the ranks of the established profession. No doubt, the initial hostility to Velikovsky and his ideas was strongly tinctured by the fact that Velikovsky himself was not a recognized member of the scientific establishment — he was not on the staff of a university, a research institute, or an industrial laboratory; he was an outsider, an amateur. Further, he sinned against orthodox procedure by having the first presentation of his ideas appear in a popular article in *Harper's Magazine* rather than a sober scientific journal. But in the end, the major reason for the rejection of Velikovsky's ideas was that they were wrong, or else that they were so inexact and nonquantitative that they could not really be tested by experiment and observation.

The case was much the same a little over a hundred years ago, in the 1870s, when J. H. van't Hoff published his revolutionary ideas on the asymmetrical carbon atom. Most chemists then were hostile and did not even give serious consideration to this proposed revision of orthodox chemical theory. One of these critics, the great German organic chemist Hermann Kolbe, downgraded van't Hoff's ideas in part because he was only a member "of the Veterinary School at Utrecht." Instead of pursuing sound and "exact chemical research," for which he has "no taste," Kolbe wrote, van't Hoff "has thought it more convenient to mount Pegasus (obviously loaned by the Veterinary School) and to proclaim . . . how during his bold flight to the summit of the chemical Parnassus, the atoms appeared to him to have grouped themselves throughout universal space" (Kolbe 1874, 477; see Snelders 1974, 3). In part, the opposition to van't Hoff's ideas arose also from the fact that he had written of atoms and molecules as if they had physical reality, in direct opposition to the views of most organic chemists, who were willing to use the concept of atom and molecule but were skeptical as to their actuality. Today van't Hoff's revolutionary ideas on the asymmetrical carbon atom are considered to be at the foundation of stereochemistry.

Given these many obstacles in the path to a revolution in science, it is somewhat surprising that any new theory or finding succeeds. In fact, many revolutionary ideas do not survive in a form that would be recognized and accepted by their first proponents; instead, they become transformed in the hands of later revolutionaries. To take an example, the system of the world that was fully elaborated by Copernicus in his *De Revolutionibus* in 1543 had no fundamental impact on astronomy until after 1609, when Kepler published his own radical reconstruction of

Copernican astronomy. From that time on we can begin to discern a revolution in astronomy, culminating in the work of Newton. But this revolution was not merely the Copernican revolution delayed by half a century. Rather, the new astronomy was in a real sense not Copernican at all (though it is often still called 'the Copernican revolution'). Kepler's reconstruction essentially rejected almost all of Copernicus's postulates and methods; what remained was primarily the central idea that the sun is immobile, while the earth moves in an annual circumsolar orbit and has a daily rotation. But this concept was not original with Copernicus, as he was well aware; it came from his ancient predecessor, Aristarchus of Samos.

This same phenomenon of transformation is apparent in the history of continental drift. Here again we have an apparent time-lag between Wegener's revolutionary announcement in the pre-World War I years and the ultimate acceptance of the revolution in the 1960s. But whereas Wegener imagined that the continents had moved in the manner of barges pushing or being pushed apart in the sea, thus making their way through the earth's crust, the eventual revolution was based on the concept of sea-floor spreading, which forces large sections of the earth's crust (plates) to move by a process of accretion at one boundary and disintegration at another. Because these plates may encompass continental land masses, their motion causes a separation of the continents. In this revolution, as in the example of Copernicus, what remains of Wegener's theory is primarily the idea that the continents are not oriented with respect to one another today in the same way they were when the earth was formed.

Scientific revolutions that fail are usually doomed to obscurity. When a political or social revolution (such as the revolutions of 1848 and the abortive Russian revolution of 1905) fails, it may nevertheless be a significant event, serving as an index of political or social conditions and problems that merit the serious attention of historians (Langer 1969; Stearns 1974; Ulam 1981). Some of the goals of failed political revolutions may still be achieved to a limited degree in the post-revolutionary period. But historians of science generally ignore revolutionary failures, unless they are examples of 'pathological' science. Perhaps this is because much of the history of science has been written by scientists themselves, who are more interested in the successive and progressive stages of truth than in the ups and downs of history, with its mixtures of truth and error.

3

Evidence for the

Occurrence of Revolutions

in Science

A discussion of revolutions in science cannot completely avoid a pair of related questions: (1) What *is* a revolution? (2) How can we tell whether or not a revolution has occurred? At first glance, it may seem that these are not wholly distinct, especially if it is believed that all sound definitions must have an 'operational' component. Yet it turns out to be possible to have a working test for the occurrence of revolutions in science, even without a clear-cut definition.

Kuhn's characterization (1962) of a revolution in science as a shift in "paradigms" (to use his original language) that arises when a series of "anomalies" has produced a "crisis" helps us in our attempt to formulate a definition and test. But we face a triple problem in trying to make precise the three notions of anomaly, crisis, and paradigm. Additionally, there is the problem (already mentioned) that not all revolutions in science exactly fit Kuhn's schema.

I have no easy answer to the definitional question of what constitutes a revolution. I repeat that what is of historical significance is that during the four centuries or so in which modern science has existed, scientists and observers of science have tended to call certain events revolutions. These include conceptual changes of a fundamental kind, radical alterations in the standard or accepted norm of explanation, new postulates or axioms, new forms of acceptable knowledge, and new theories that embrace some or all of these features and others. The Newtonian revolution entailed the radical concept of a gravitational force of attraction and achieved the goal of expressing and developing the principles of natural philosophy in mathematical terms; the Cartesian revolution was posited on the 'mechanical philosophy'—the explanation of all phenomena in terms of matter and motion; the kinetic-molecular theory of gases and radioactivity introduced explanations based on probability, while quantum theory denied simple nonprobabilistic causality; the theory of evolution denied the fixity of species and introduced a science that did not permit the prediction of single events; relativity not only sounded the death knell of absolute time and space, but radically altered the apparently simple concept of simultaneity; the Harveyan revolution set forth the idea of a continuous circulation of the blood from the heart out through the arteries and back into the heart through the veins and rejected the ancient and well-established doctrine that blood merely ebbs and flows in the veins, that it is continually being generated in the liver. In all such cases, an event occurred that has been (and is now) generally called a revolution. This is a historical fact, whether we like the word 'revolution' or not, whether we are or are not able to produce a definition that fits all these examples and others.

Since my purpose here is primarily to learn about revolutions that have been acknowledged as having occurred, and not to analyze a concept in abstraction, my method of study has been to examine the ways in which revolutions in science have been perceived. And this leads at once to a series of four tests that may be universally applied to all major scientific events that have occurred during the past four centuries. The basis of these tests is purely historical and factual. It consists, in its first part, of the testimony of witnesses: the judgment of scientists and nonscientists of that time. Among such witnesses I would include philosophers, political scientists, people active in political affairs, social scientists, journalists, literary figures, and even educated laymen. Both Newton and Leibniz were still alive and working on the development of the calculus when Fontenelle recorded his contemporaneous impression that their creation had produced a revolution in mathematics. Within a decade of Newton's death, Clairaut hailed Newton's *Principia* as the "epoch" of a revolution in the science of mechanics. Lavoisier's radical reform of chemistry was seen to be a revolution in chemistry by many of the scientists of his day. A number of Darwin's contemporaries wrote of the theory of evolution as a revolution in biology. To many earth scientists of the 1920s and 30s it was obvious that Wegener's ideas concerning the motion of continents were revolution-making, long before continental drift had changed its status from revolution on paper to revolution in science. These revolutions all pass the first test — the testimony of contemporaneous witnesses.

In three of the above examples, the scientist chiefly responsible for the revolution (Lavoisier, Darwin, Wegener) said expressly that his own work would create a revolution. This concurrence gives added strength to the testimony of other witnesses. It is obvious, however, that one should not make too much of the lack of such special testimony, since most scientists are usually too modest or too restrained by the conventions of the scientific enterprise to make such judgments about their own creations.[1] On the other hand, I would not put much trust in a later historical judgment that a revolution in science had actually occurred if there were no witnesses to testify to the event (such as a Mendelian or Babbagian revolution in science in the nineteenth century).

A scientist may believe he is making or has made a revolution even though later events show that such a revolution never occurred. Two examples are Symmer's theory of electricity and Marat's theory of optics. And as we saw in chapter 2, there are many examples of revolutionary movements in science that simply do not go on to develop into full-scale revolutions — Mesmerism, N-rays, and polywater, to name only a few. And so we need additional tests to supplement the testimony of witnesses.

A second test is an examination of the later documentary history of the subject in which the revolution is said to have occurred. A study of the treatises and textbooks of astronomy written between 1543 and 1609 does not show the adoption of Copernicus's ideas or methods. Hence this test indicates the nonexistence in those years of a Copernican revolution. By contrast, most of the mathematical writings of the eighteenth century — whether treatises, articles, or textbooks — are written in terms of the new calculus (whether the Leibnizian or the Newtonian algorithm), thus giving confirmatory evidence to Fontenelle's statement that the invention of the calculus was the epoch of a revolution in mathematics. Similarly, we have evidence of the Newtonian revolution if we compare and contrast post-1687 mathematical astronomy (with its strong component of gravitational celestial dynamics) with astronomy before the *Principia*. Obviously, by itself this test leads to what can be at best a subjective judgment on the degree of restructuring: whether it is of sufficient magnitude to constitute a revolution in science. But this test is conclusive in a negative judgment that no such influence is to be found in the major writings of a particular science. In many cases the evidence is overwhelmingly positive (as for the calculus) or at least strongly confirmatory. A confluence of results of these first two tests gives us a powerful indication of the occurrence of a revolution in science.

A third test is the judgment of competent historians, notably historians of science and historians of philosophy. This would include not only historians of the present and recent past but also historians who lived long ago. An example is the eighteenth-century historian of astronomy, J.-S. Bailly, writing about events associated with Copernicus in the sixteenth century. There is no want of historians or historically minded scholars (philosophers, sociologists, and other social scientists) to testify for a Newtonian revolution, a Chemical Revolution, a Darwinian revolution. The conjunction of affirmative answers to all three of these tests leads to a very strong conviction that these events were revolutions. But there are some episodes that may generally be considered by historians to be revolutions but for which there is no contemporaneous body of opinion that this was the case. The primary example, once again, is the so-called Copernican revolution. We shall see that the idea of a Copernican revolution in astronomy's having occurred in the sixteenth century is a fiction invented and perpetuated by later historians, apparently first by Montucla and Bailly in the eighteenth century. The discrepancy — between the testimony of ancient witnesses and the opinions of later historians — should have warned historians to treat this alleged revolution with skepticism. A close analysis of the events in this case shows how the error arose and how it has depended on events connected with Galileo and Kepler a half-century or more after the publication of Copernicus's treatise (1543). It is nevertheless a fact of history that for some two centuries historians and scientists have believed that there was a Copernican revolution. Such judgments made long after the events must be examined critically, especially when the people making those judgments lived before present-day standards of historical evidence.

I believe it is a sound historical judgment that there was a great revolution in science in the nineteenth century in the domain of statistics and of statistical thinking. Some glimmerings of this revolution's occurrence may be found in the writings of Adolphe Quetelet, J. Clerk Maxwell, Ludwig Boltzmann, and John Herschel. But I do not know of a body of clear-cut statements by contemporaries about this revolution, such as exist for the Chemical Revolution and the Darwinian revolution (although remarks by Herschel come close). This may perhaps be no more than a sign of our ignorance, a reflection of the rather primitive state of our knowledge of the history of this subject. Since there are so few serious historians who are or have been concerned with the development of probability and statistics, the third criterion for a revolution is just not fully applicable to this case. But there is a fourth and final test that may be applied to the statistical revolution: the general opinion of working scientists in the field today. In this case, the physical, biological, and social scientists of the twentieth century have tended to be aware that the establishment in their own times of a statistically based physics (radioactivity and quantum physics), biology (genetics), and social science has constituted a sharp break with the past and that there has been a statistical revolution.

In this fourth test, I give considerable importance to the living scientific tradition, to the mythology that is part of the accepted heritage of practicing scientists. Myths play a significant and as yet not adequately appreciated role in science, somewhat analogous — I am sure — to that of myths in society at large. While myths about heroes of science and the revolutions they are believed to have made do not constitute historical evidence concerning past events, they do give us clues to the existence of major episodes that have had a formative significance in the development of science. The general beliefs of scientists about their past reinforce the kinds of evidence supplied by the other three tests.

The fourth test is not wholly independent of the third, however. It is obvious that scientists may be influenced by historians and historians by scientists. And both may be under the spell of a long tradition, as in the case of the Chemical Revolution. Even a tradition based on error may strongly affect the thinking of later historians and of scientists, as references to a Copernican revolution show plainly.

An instructive example, in which all four tests yield the same result, is the revolution in the earth sciences of our own times. The basic concept of this revolution is that there has been and still is a relative motion of the continents on the surface of the earth, a continental drift. When first put forward by Alfred Wegener just before the 1914 war, the theory of continental drift was generally recognized by earth scientists to be revolutionary, and it was widely debated (although not really accepted by the community of geologists and geophysicists) during the 1920s and 30s — thus passing the first test: the contemporaneous opinion of scientists. Furthermore, Wegener himself was fully aware of the revolutionary character of his new idea. When in the 1960s and 70s a new version of continental drift, based on the idea of plate tectonics, came to be part of the belief of earth scientists, they tended to speak of this change as a revolution. The literature of the earth sciences documents the dramatic changes that have occurred in this subject, consistent with a revolution. So continental drift passes the second and fourth tests. Finally, in the third test, we may note that historians have produced works in which the rise and acceptance of the ideas of continental drift have been described in terms of a revolution in science. A number of the discussions of continental drift, by both historians and scientists, even invoke the ideas of Kuhn and present the subject in terms of paradigm and paradigm-shift. Since, in this example, all our tests concur, can there be any doubt that a revolution has taken place? The theory of continental drift passes all tests for being a revolution.

For me, the testimony of contemporaneous witnesses weighs very strongly. Unlike later judgments, which are reflections less on the events of the revolution than on the revolution's long-term effects or on the postrevolutionary history of the science, these evaluations provide a direct insight into what was going on. There is a real significance, for example, to the fact that Charles Darwin not only believed his new ideas would create a revolution but actually said so in the conclusion to the *Origin of Species* in 1859. His prediction of a "considerable revolution in natural history" is a rare instance when a scientist was so bold as to make such a declaration in print — in this case, in the major publication announcing the discovery.[2] Darwin's judgment was echoed by a number of his correspondents. The statements by Lavoisier and Darwin about the revolution implicit in their own ideas are reinforced by the corroborative judgment of their contemporaries and are sustained by the evaluations of later historians and scientists. But self-evaluation may be unreliable. Few scientists or historians have ever heard of Robert Symmer and those who have would hardly agree with his own opinion that his contribution to electricity was "revolutionary." And our judgment is even stronger that Jean-Paul Marat, despite his self-evaluation, never did create a revolution in science.

Very few scientists appear to have described their own work in terms of revolution. Some fifteen years of research on this subject, aided by the contributions of many students and friends and the fruits of the investigation of several research assistants, have uncovered only some dozen or so instances of a scientist who said explicitly that his contribution was revolutionary or revolution-making or part of a revolution. These are, in chronological order, Robert Symmer, J.-P. Marat, A.-L. Lavoisier, Justus von Liebig, William Rowan Hamilton, Charles Darwin, Rudolf Virchow, Georg Cantor, Albert Einstein, Hermann Minkowski, Max von Laue, Alfred Wegener, Arthur H. Compton, Ernest Everett Just, James D. Watson, and Benoit Mandelbrot.[3]

Of course, there have been others who have said dramatically that they have produced a new science (Tartaglia, Galileo) or a new astronomy (Kepler) or a "new way of philosophizing" (Gilbert). We would not expect to find many explicit references to a revolution in science prior to the late 1600s. Of the three eighteenth-century scientists who claimed to be producing a revolution, only Lavoisier succeeded in eliciting the same judgment of his work from his contemporaries and from later historians and scientists.

Evidence from contemporaneous observers or participants concerning a revolution in science is subject to certain obvious uncertainties. The survival of evidence from an earlier period may be a matter of chance; even if it exists in some physical form (published reports or diaries, notes, correspondence, and so on), it might not be known to historians today. So the lack of contemporaneous documents stating explicitly that a revolution has occurred (or is about to occur) cannot always be used as positive proof of the nonoccurrence of a revolution. In other words, such contemporaneous evidence is one of the sufficient conditions for our judgment that a revolution has occurred, but is not always a necessary condition.

Information arising from the period under discussion may be extraordinarily valuable. A case in point is the annual report of the president of the Linnean Society of London for 1858, the year in which Darwin and Wallace published their first joint communication on the evolution of species by natural selection. The president said that the past year had not been noted for one of those revolutions that change the face of a science. Are we to assume that he was merely insensitive to the revolutionary implications of evolution? Not necessarily. As we shall see, his report shows that he believed revolutions occur in science, and that he supposed the time was ripe for a significant revolution in the life sciences. His statement thus proves that the great Darwinian revolution was not produced merely by the enunciation of bold ideas concerning evolution and natural selection. For a revolution to occur, there was needed the careful and complete body of documentary evidence and the fully worked-up theory that Darwin provided in his book a year later. The Darwinian revolution was not produced by the mere statement of radical ideas but by an interplay between an overwhelming mass of factual data and theoretical inferences on a high level.

Admittedly, these four criteria are in the end somewhat subjective. And obviously they do not cover every possible contingency. But at the very least they do provide the conditions sufficient for our judgment that a revolution has occurred, which may be buttressed by further research and critical reflection.

Historical Perspective

on 'Revolution' and

'Revolution in Science'

II

4

Transformations
in the Concept
of Revolution

A political revolution is commonly conceived to be a change that is sudden, radical, and complete, often accompanied by violence or at least the exercise of force. Such a fundamental change has a dramatic character that usually enables observers to discern that a revolution is taking place or has just done so. The classic revolutions of early modern times — the American and the French revolutions — were notable for their alteration of the system of political organization, more drastic for the French than for the American Revolution. In both cases, the government or the ruler was renounced and overthrown and a new form of government was substituted for the old one by the action of the people being governed or their representatives. To some extent this had also been true of the Glorious Revolution.

In the nineteenth century, revolution and revolutionary activities began to transcend such purely political considerations as the form of government, and came to embrace fundamental socioeconomic policies. The word 'revolution' eventually came to connote not merely events leading to radical political or socioeconomic changes, but the activities (either unsuccessful or not as yet successful) intended to effect such changes. Thus in 1848, Marx and Engels set forth in *The Communist Manifesto* a blueprint for a revolution and a call for "a Communist Revolution." A year later, Marx announced among the "auguries of the year 1849" (1971, 44): "revolutionary uprising of the French working class, and world war."

From the eighteenth century onward, a revolution was more than an armed uprising, more than a defiance of established authority, more than a revolt or active renunciation of allegiance or subjection to a government. That is, a revolution transcended acts of revolt or rebellion that would not necessarily produce a new form of government or a new socioeconomic system.

A shift from one ruling house to another or a dynastic change was no longer deemed to be a revolution. Generally, a mere opposition to authority, especially if open and armed, has tended to be called a rebellion — this is especially the case when the act of rising up against authority has proved to be unsuccessful in either the short or the long term. For example, what we know today as the American Civil War was formerly called The War of Secession or The Rebellion, and a Confederate soldier was colloquially referred to by Northerners as Johnny Reb. (The rebel yell was the name given to the prolonged high-pitched scream or shout of the Confederate soldiers.) The events associated with the English Civil War — the military encounters between Cavaliers and Roundheads, the execution of Charles I, and the establishment of the Commonwealth — were referred to by the seventeenth-century historian and chronicler Clarendon as "The Rebellion and Civil Wars in England."

The history of the concept of revolution cannot be separated from the history of the ways in which the word itself has been used. This history has a number of closely related themes that are relevant to the subject of revolution in science. First: the origins of the word itself in late Latin, as a substantive deriving from the verb 're-volvere' as 'to roll back' and hence also 'to unroll', 'to read over', 'to repeat', and 'to think over'; whence the further meanings of 'to return', 'to recur'. Second: the employment of the substantive 'revolutio' as a technical term in astronomy (and in mathematics), beginning in the Latin Middle Ages. Third: the gradual introduction of 'revolution' in a political sense, to signify a cyclical process or an ebb and flow, implying a return to some antecedent condition, and eventually to indicate an 'overturning'. Fourth: the association of 'revolution' with the process of overturning in the realm of political affairs, and the subsequent separation of the sense of an 'overturning' from the cyclical connotations of 'revolution'; at the same time the word 'revolution' came into use to indicate a more than ordinarily significant event. Of considerable importance in the development of thinking about revolutions was the rather early recognition that a revolution had occurred in England (the Glorious Revolution in 1688) and that a revolution was going on in the sciences. By the beginning of the eighteenth century, revolutions (in a sense very much like that which we would use today) were thought to occur not only in relation to the state but also in the realm of intellectual and cultural affairs, specifically in the growth of science; there was an awareness that by the time of Newton a revolution had occurred in science. This period is notable for the recognition by at least three different scientists that their individual research could lead (or was leading) to a revolution in science.

In the last quarter of the eighteenth century, the American and French revolutions gave a factual demonstration that revolutions are part of a continuing political and social process, and at the same time Lavoisier announced a new revolution in science, the Chemical Revolution. By this time it had also become generally agreed that there had been a Copernican revolution as well as a Newtonian revolution, plus a succession of minor revolutions in the sciences.

During the nineteenth and twentieth centuries the name 'revolution' was applied to a series of social and political revolutionary events, unsuccessful as well as successful ones. A body of revolutionary theory was also developed, accompanied by the formation of a revolutionary movement dedicated to putting the theory into practice through the activities of organized groups of committed revolutionaries. Above all there arose the concept of 'permanent' (or continuing or ongoing) revolution, rather than a revolution consisting of a series of events closely packed within a relatively short interval of time. In the twentieth century, a succession of major and minor revolutions has made everyone keenly aware of revolutions as a regular feature of political, social, and economic change, and today they have been generally accepted as an equally regular feature of scientific change.

Revolution in Antiquity

Students of political theory trace a history of the analysis of revolutions going back at least as far as the philosophers Plato and Aristotle and the historians Herodotus and Thucydides. Although a number of events in ancient times might be called revolutions, the Greeks did not have a single agreed-upon word to describe them. Greek philosophers and historians tended to use a number of different words for what we would call revolutionary uprisings and changes. Hence, "though the Greeks had their fill of revolution they had no single word for it" (Hatto 1949, 498). In short, 'revolution' was not a clear and fully developed concept in the sense that we have known it since 1789. Arthur Hatto, who has made the fundamental study of the early history of the word and concept, has pointed out that Plato's 'revolution' is more properly an evolution, to the degree that "his ideal state tends to deteriorate into a timocracy and a timocracy into an oligarchy and so on through democracy into tyranny" (ibid.). Apparently Plato himself did not actually complete the cycle and conceive that this sequence of events would be repeated again and again, which would require that tyranny give way once again to an ideal state. That step was taken by Polybius, who alleged that he was making a summary of what Plato had said. It was Polybius and not Plato who conceived that kingship passes "into tyranny, tyranny into aristocracy, aristocracy into oligarchy, oligarchy into democracy"; then "democracy into mob-rule and mob-rule into that state of nature which . . . must inevitably produce kingship and a new cycle" (p. 499). In Polybius' own words, "Such is the cycle of political revolutions, the course appointed by nature in which constitutions change, disappear and finally return to the point from which they started." This cyclical view was expressed by Polybius (*Hist.* VI, 9, x) in the word 'anakyklosis' (from the stem 'kyklos', circle or wheel, the root of our word 'cycle') as in the turning of a wheel; "the force behind its turning is Fortune" (or 'tyche').

Book 5 of Aristotle's *Politics* is a discussion of revolutions, including a refutation and rejection of the cyclical theory of revolutions (V, 12, vii). Aristotle's 'regular term' for 'revolution' is 'metabole kai stasis' (change with uprising); if there is no violence, the 'metabole' is used by itself. Hatto (p. 500) concludes that the Greeks obviously had a sense of the concept of revolution and had experienced revolutions. But although a word could always be found to express this concept, or some phase of it, the Greek writers "did not always choose the same word and sometimes chose two or more." Perhaps the reason may be that although they experienced many revolutions and near-revolutions or proto-revolutions, the Greeks were not witness to a "classic revolution" in the sense that Europe was "in *the* Revolution of 1789" (ibid.).

The Romans also had no single word for 'revolution' (Hatto 1949, 500). The major expression in Latin that comes close to our 'revolution' is 'novae res' (new things, innovations), which really was a designation for what we would call the result of a revolution. Among the phrases used to express the action of revolution are 'novis rebus studere' (to strive for innovations) or 'res novare' (to innovate). Two other expressions from Classical times are 'mutatio rerum' (change of affairs) and 'commutatio rei publicae' (alteration of government); these expressions survived in various Renaissance translations into Latin of Aristotle's *Politics*.

Cicero adopted and popularized the Plato-Polybius theory of cyclical changes in constitutions (*Rep.* 1.45): "Wonderful are the cycles and what might be called the revolutions of the upheavals and vicissitudes in governments [Mirique sunt orbes et quasi circuitus in rebus publicis commutationum et vicissitudinum]." In this place as elsewhere, Cicero expresses the concept of change that occurs in cycles by using the word 'orbis' (wheel, circle, sphere, cycle). According to M. L. Clarke (quoted in Hatto 1949, 501), Cicero considered these cyclical changes as "natural, but not inevitable"; that is, "the wise statesman can foresee and prevent them." This concept of cyclical changes was applied by Cicero both to events of the past and to political changes occurring in his own day: "Soon you'll see the wheel turning [Hic ille iam vertetur orbis]" (*Rep.* 2.45), or "the wheel of the political situation has turned [orbis hic in republica est conversus]" (*Att.* 2.9.1: cf. 2.21.2). In a work written late in life (*De divinatione* 2.6), Cicero spoke of "what might be called political revolutions [quasdam conversiones rerum publicarum]." Here Cicero is using the noun 'conversio', which means 'a turning around' and hence 'revolving' or 'revolution' (as in the periodical return of the seasons),[1] in the sense of a complete change or even a violent change somewhat akin to our 'political revolution'; he uses 'conversio' this way also in combination with 'motus' (*Sest.* 99) or 'perturbatio' (*Phil.* 11.27). In the preface to his treatise *De Revolutionibus*, Copernicus refers to his having found in Cicero a statement that "Hicetas supposed the earth to move" (1978, 4). The reference is to Cicero's *Academica* (*Acad. prior.* 2.123), in which Cicero records Theophrastus' saying that according to Hicetas the earth "turns and twists about its axis with extreme swiftness," so that it appears to a terrestrial observer that the heavens move. Cicero's words are "quae [terra] cum circum axem se summa celeritate convertat et torqueat," where the verb 'convertere' (used reflexively) has the sense of turning on an axis or of rotating and so is akin to revolving.[2]

In later Latin, the noun 'revolutio' had the sense of 'conversio' in classical Latin. Two instances may be cited from the fifth century: Martianus Capella (9.22) writes of "the courses of sidereal revolution [sidereae revolutionis excursus]," and Augustine (*City of God* 22.12) describes metempsychosis as many "revolutions through different bodies [per diversa corpora revolutiones]."

The Middle Ages and the Renaissance

The Middle Ages were not witness to revolutions in the sense of a dramatic and complete overturning of the political and social hierarchical system, although changes in government were sometimes brought about by revolts or forced removals of dynastic rulers. The Peasants' Revolt in England in 1381 had many features of an incipient revolution, including "the burning of manors, destruction of records of tenures, game parks, etc., assassination of landlords and lawyers, and a march (100,000 [?] men) . . . on London," where "lawyers and officials were murdered, their houses sacked, the Savoy (John of Gaunt's palace) burned" (Langer 1968, 290). But this was not a revolution in the present sense of the word because it did not have an organized program, did not even envisage the end of the monarchy or the abolition of the aristocracy, and was limited, insofar as there was a program at all, to the correction of particular grievances or excesses. Some scholars (Rosenstock 1931, 95; Hatto 1949, 502) have said that the beginnings of the now-current usage of 'revolution' can be seen in the early Italian Renaissance, for example, in the fourteenth-century *Cronica* of Matteo Villani (4.89 = Villani 1848, 5: 390), who wrote about "la subita revoluzione fatta per i cittadini di Siena" ("the sudden revolution made by the citizens of Siena") in 1355. Here, apparently, was a political event that was produced by men and that did not occur as a result of forces beyond human control, but, as Hatto warns us, we must be careful not to make too much of this single revolution that allegedly resulted from man's actions, since in another passage (4.82 = 5: 384) Villani refers to this very same event in terms of "le novità fatte nella città di Siena" (the changes made in the city of Sienna), and he also uses 'rivoluzione' (9.34 = 6: 223) and "revoluzioni" (5.19 = 5: 413) for general political unrest.

Scholars have found a few other early examples of use of the term 'rivoluzione', but it was not in general circulation as either a political name or a concept. Macchiavelli, whose writings show that he did begin to approach our idea of political revolution, tended to use the Italianate form 'mutazione di stato' of the conventional Latin 'commutatio rei publicae' or 'mutatio rerum', although at least once (*The Prince,* ch. 26) he wrote of 'revoluzioni' in the more general sense of changes (Hatto 1949, 503). By the early sixteenth century, the Florentine historian Guicciardini (1970, 81) was writing of a change in government as a 'rivoluzione'. There seems to be general agreement that the new political sense of revolution as a political change arose in Italy during the late Middle Ages and early Renaissance and then spread northward.[3]

The primary signification of 'revolution' in the Middle Ages and Renaissance was astronomical and thus—either associatively or derivatively—astrological. The daily revolutions observed in the stars and in the sun, moon, and planets, and the orbital apparent motions of the planets (or of the spheres to which it was thought they were attached) are recorded in an unambiguous use of the term by Dante in Italian and Latin, by Chaucer in English, by Alfraganus (who was a major source of astronomical information for Dante) and Messahala in Latin translations, by Sacrobosco and others. At the beginning of the Scientific Revolution, this word appears boldy in the title of Copernicus's celebrated book *De Revolutionibus Orbium Coelestium (On the Revolutions of the Celestial Spheres,* 1543) and it occurs not infrequently in Galileo's dialogue *On the Two Chief World Systems,* 1632. It may be found in almanacs; in the many editions and translations of Leurechon's *Récréation mathématique* (which was translated into English by William Oughtred as *Mathematicall Recreations*) and similar works in Vincent Wing's popular compendiums of astronomy and astrology and in Streete's *Astronomia Carolina* (1661), from which the youthful Newton recorded Kepler's third law. In other words, from the twelfth to the seventeenth century and later, revolution occurs frequently and prominently in technical treatises on astronomy and astrology (both Latin and vernacular) and in general works such as the *Divine Comedy,* to indicate the motion of celestial bodies (or their spheres) turning through 360° and completing a circuit or the quantitative measure of such revolutions (the period of revolution). But 'revolution' also had the extended meaning of any turning or rolling around—ranging from such physical events as the turning of a wheel to the figurative notion of turning something over in one's mind.

By the time of the Renaissance, and in the seventeenth century, 'revolution' began to acquire an even wider range of meanings than the primary astronomical and astrological signification, well beyond simple mathematical and physical exemplifications.[4] A revolution could be any periodic (or quasi-periodic) occurrence, and eventually any group of phenomena that go through an ordered set of stages — a cycle (in the sense of 'coming full circle'). Even the rise and fall of civilizations, or of culture, as a kind of tidal ebb and flow, was called a revolution. All of these senses are obviously linked to the primary astronomical meaning.

A similar word is 'rotation', with which 'revolution' is sometimes confused. Today we like to make a clear distinction between the turning of a body on its axis (rotation) and the motion in a circuit, as along a closed path or orbit (revolution); we thus speak of the daily rotation of the earth about its axis and the annual revolution of the earth in its orbit about the sun. But as late as the end of the seventeenth century, these two words were apt to be used interchangeably, as they are in Newton's *Principia* (1687). 'Rotate' comes from the Latin verb 'rotare' (to turn, or to swing around); the Latin noun 'rota' means wheel (and hence also chariot) and can even have the figurative sense of alternation or fickleness. This word 'rota' survives in current English usage to denote a fixed order of rotation of individuals or of duties, or even a roll or list of persons. A late Latin noun 'rotatio' has given us our word 'rotation'.

In the fortune-telling 'tarocchi' (or tarot cards) of the late Middle Ages and Renaissance, as in those of today, a major card is the 'rota di fortuna' or Fortune's Wheel. The fate of men was supposed to be determined by this wheel or 'rota' and its rotation. There would thus be two major sources of 'turning' that were believed to affect or even determine the course of men's lives and of the state: the rotation or turning or spinning of Fortune's Wheel and the revolution of the celestial spheres. Possibly, the rise of the word 'revolution' may be associated with Fortune's Wheel as well as with the celestial spheres (this notion has been advanced by Henry Guerlac). Evidence for this association could be found in the frequency of the occurrence in a political context of 'revolution' or 'rivoluzione' in association with Fortune's Wheel or the 'rota di fortuna'. In Dante, 'revoluzione' occurs in the *Convivio* as a term for the circular motions of the heavens; but there he does not invoke the image of the 'rota di fortuna'. Although the turning of a wheel is cyclical, there is no implication that the wheel will end up at the end of the spin at the same place where it began. Hence the sense of return or going back or completing a cycle is not necessarily implied by Fortune's Wheel in the same sense that is true of the revolutions of the celestial spheres.

There is abundant evidence of a widespread belief during the late Middle Ages and Renaissance that the affairs of state were controlled by the planets in their revolutions. Eugen Rosenstock-Huessy (Rosenstock 1931, 86–87; cf. Hatto 1949, 511) turned up a sixteenth-century example from Germany, in which events of human history are linked with planets in relation to zodiacal signs "in the first revolution" ("in der ersten Revolution"). Villani (Hatto 1949, 510) has an entry for the year 1362, in which astrology provides the exact hour when the Florentines should set out against the Pisans. Both Kepler and Galileo cast horoscopes for the rulers as part of their professional employment. Kepler (1937, *4*: 67; cf. Griewank 1973, 144) held that the occurrence of comets was associated with prolonged evils that cause sufferings "not only because of the departure of a potentate and the consequent change in government [nicht eben durch Abgang eines Potentatens und darauf erfolgende Neuerung im Regimen]." In a letter of 1606 Kepler (1937, *1* 5: 295–296) criticized superficial astrological predictions about human history made "on the basis of the revolution of the universe [ex revolutione mundi]." There is pictorial evidence that the royal power and the basis of monarchical government of both Queen Elizabeth and Louis XIV were associated with astrology (see figures 1, 2, and 3).

Anyone who lived in the Renaissance, or in the sixteenth or seventeenth century, would at once associate the word 'revolution' with the idea of the unrolling of the great wheel of time. The notion of the wheel of time and its revolutions was not only used as a purely intellectual metaphor but was exemplified in definite physical images and objects. For instance, on the clock towers of Renaissance buildings everyone could see the continual revolution of the hand to mark the passage of time.[5] (There was only one hand, to mark the hours.) Another image of the passing of time would be the daily apparent motion in revolution of the celestial sphere with sun, stars, and moon. The wheel of time could also invoke the image of the motion of the sun in its annual apparent orbit among the fixed stars. The daily revolution (which we today call rotation) of the celestial sphere brings with it the change from morn to noon to evening to night and marks a twenty-four-hour cycle of days. The revolution of the sun in its orbit during the course of a year brings with it a change in the rising and setting point of the sun, the relative length of hours of daylight and darkness, and the seasons.

The significant quality of these revolutions is not merely that they are cyclical or repeat a succession of phenomena in the sense that the word 'revolution' itself means to roll back, but rather that during the course

Figure 1. The traditional finite geocentric universe. A diagram published in Petrus Apianus's *Cosmographia* (Antwerp, 1539). At the center is the immobile earth (with its water and the spheres of air and fire), surrounded by spheres of the seven traditional 'planets': the moon, Mercury, Venus, the sun, Mars, Jupiter, and Saturn. Then come the sphere of the firmament or fixed stars, the ninth crystalline sphere, and the tenth sphere of the prime mover. Surrounding all is the EMPYREAL HEAVEN, ABODE OF GOD AND OF ALL THE ELECT. (Courtesy of Houghton Library, Harvard University.)

Figure 2.   Political power and royal virtue linked with the revolutions of the celestial spheres in John Case's *Sphaera Civitatis* (Oxford, 1588), an "Aristotelian treatise on political moral philosophy," according to Frances Yates, *Astraea* (London, 1975, 64–65). As is partially explained in the facing poem and amply clear in the diagram itself, IMMOVABLE JUSTICE is shown as the immovable earth in the center. The inner spheres are those of the seven planets of the civic universe, with ABUNDANCE as the moon, ELOQUENCE as Mercury, CLEMENCY as Venus, RELIGION as the sun, FORTITUDE as Mars, PRUDENCE as Jupiter, and MAJESTY as Saturn. The sphere of the fixed stars consists of the *Star Chamber, Nobles, Lords, Counsellors,* while the outermost sphere is ELIZABETH BY THE GRACE OF GOD QUEEN OF ENGLAND, FRANCE, AND IRELAND, DEFENDER OF THE FAITH. Queen Elizabeth, therefore, as representative of the Deity, is the "prime mover" of the "Sphere of the State." (Reproduced by permission of The Huntington Library, San Marino, California.)

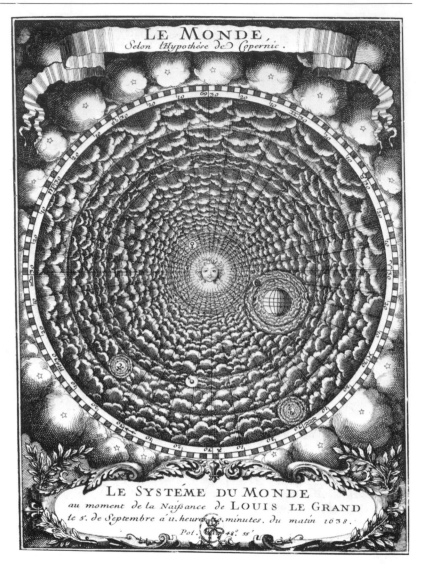

Figure 3. The Cartesian system of the world and royal power. This very realistic seventeenth-century engraving depicts "THE WORLD, According to the Copernican Hypothesis" as a system of swirling Cartesian vortices, surrounded by external systems of stellar vortices. The political significance of this representation is that it shows the configuration of the system of the world "at the moment of the Birth of LOUIS THE GREAT" and hence links Copernican astronomy and Cartesian physics and cosmology with political astrology. (Cabinet des Estampes, Bibliothèque Nationale, Paris.)

of each of these revolutions of time there are dramatic and significant changes. What could be more different than day and night or winter and summer! They are as different as birth, life in a mature state, death, decay, eventual resurrection — the cycle of life on earth and life ever-lasting. The cyclical revolution of astronomical time comprises a sequence of changes so dramatic that they were quite properly defined by the word mutation, which was used by Montaigne and other writers of the Renaissance to signify a great change, similar to what we would call a revolution. Until the seventeenth century, a revolution was a sequence of events, a cycle, an ebb and flow in the tides of human affairs and the fate of nations, or a return (more or less) to some antecedent state, while the individual occurrences or particular events in the sequence were apt to be called mutations. But even if a great event or change was not necessarily part of a regular sequence, it could be described in terms of revolution because it occurred in time and came into being with the unrolling of the great wheel of time. A revolution could also be an event that altered the normal course of history, one that — so to speak — nudged forward the wheel of time with a slight acceleration, an event that marked an epoch (or 'epoca'), the beginning of an epoch. In the sixteenth and seventeenth centuries, and even in the eighteenth, great changes are called revolutions in ways that reflect the background of thinking about astrological causes, the Wheel of Fortune, the ebb and flow or cyclical aspect of events, and the unrolling of the wheel of time.[6]

What is perhaps of the greatest interest in this emergence of 'revolution' is that it implied a determining of events beyond human will and human forces — whether by astrological causes or simply by the laws of cyclical succession, which come from the revolution of the wheel of time. Thus human events and the course of history would follow the same inexorable and fixed schedules as the motions of the stars, sun, moon, and planets, alterable by the direct intervention of God, much as in the occurrence of a miracle. Possibly a revolution could arise by man's intervention, transcending or momentarily replacing the inexorable sequence determined by the revolutions of the stars.

The Seventeenth Century

Out of this set of usages and implications, there gradually emerged the concept of revolution as a noncyclical event of enormous change. In this development we must keep in mind that in the sixteenth and early seventeenth centuries, the word 'revolution' occurs in two apparently opposed general senses. One is the act of passing through the stages of a cycle that can ultimately lead to a condition that is identical or similar to some antecedent one, or a continuation of such a cycle, or an ebb and flow that need not be strictly periodic. The other is a turning over, an overturning, a 'mutatio rerum', a change of considerable magnitude in the affairs of the state, in dynastic succession, or in a constitution. The first invokes the notion of a full cycle or a shift through 360°, the second a topsy-turvy shift of 180°which, being in effect nothing more than a radical change in a short time, sounds more like our post-1789 concept of revolution (that is, political revolution). But the difference between the two will not necessarily be so great as might at first seem to be the case. For it was widely believed at this time, as throughout much of recorded history, that the way of improvement was to return to the better days of an earlier age.

Ever since antiquity, men and women have conceived of a radical improvement as a return to some previous condition, some Golden Age. The belief that progress consists of turning back the clock or calendar is associated with the concept of a continuous decay of the world itself or of the conditions of life, a decline — according to Western religious thought — that can be traced back to the Fall of man, to the expulsion from Eden. Who among us was not told by his parents that things were better in the 'old days'? Our parents were right. Food certainly tasted better and was more nourishing when fresh and not frozen, not adulterated with chemical colorants and preservatives, and not wrapped in airless plastic packages. It was plainly more comfortable to cross the ocean in the peace and quiet of a stateroom, with a steward and stewardess in attendance, than to be packed like sardines eight or ten abreast in a wide-bodied jet. And no doubt, our parents were also right when they said that children were more respectful of their elders and were better mannered when they were young. No one today, living as we do under the continual threat of chemical and biological warfare and possible nuclear annihilation, can help but look back at some of the blackest days of the past as being in some sense better times than our own. And in the same way the political and social reformers of the sixteenth, seventeenth, and eighteenth centuries hoped for a return to some better or simpler times, to Biblical conditions, and to a world governed by the principles of justice announced in the Sermon on the Mount. Thus radical change was thought of as a 'revolution' in the sense of a return to a better time, a building of "a new Jerusalem," a renewal

—as a 1649 *Manifestation* of the Levellers put it (Aylmer 1975, 153)—of the principles of the "[Voluntary] Community [which existed] amongst the primitive Christians." As late as the American Revolution, the word 'revolution' still had decided overtones of return, in this case a return to the principles of the Bill of Rights (1689), which had been denied Englishmen in the American Colonies.

In the sixteenth and seventeenth centuries, and even in the eighteenth, it is not always easy (and sometimes not even possible) to tell which sense of 'revolution' a particular author may have in mind: a definite return (a cyclical phenomenon or an ebb and flow), or an event of major proportions (which may entail the establishment of something new), or merely an event in a sequence. For instance, in 1603, in John Florio's translation of Montaigne's *Essays* there occurs (p. 74) a modern-sounding passage: "In viewing these intestine and civill broiles of ours, who doth not exclaime, that this worldes vast-frame is neere vnto a dissolution, and that the day of judgement is ready to fall on vs? never remembring that many worse revolutions have been seene. . . . ?" In isolation this sounds much like a comment made in the post-1789 sense, and it is thus interpreted by Vernon F. Snow (1962, 169), but the presence of the modifiers "many worse" suggests that Florio had in mind nothing more than events in earlier cycles or perhaps merely earlier events; this interpretation is supported by the fact—not noticed by Snow—that Florio's "revolutions" renders Montaigne's "choses" (1595, "97" = 88; 1906, 204). To indicate events which are like those that we would call 'revolutions', Montaigne used the expression 'mutation d'estat', from the Latin 'commutatio rei publicae'.

The cyclical sense certainly predominates in another example brought forward by Snow (and also involving a translation to which he does not make reference). The 1614 edition of Camden's *Remaines* includes a chapter on "Apparell" which is not contained in the first edition of 1605. Towards the end of this chapter (p. 237) Camden says: "They which mislike most our present vanity herein, let them remember that of *Tacitus*. All things runne round, and as the seasons of the yeare, so mens maners have their revolutions." This obviously includes a rendering of what Tacitus says in a similar context and with similar meaning, though without, of course, the word 'revolution' (*Annals* 3.55.5): "Nisi forte rebus cunctis inest quidam velut orbis, ut quem ad modum temporum vices, ita morum vertantur . . . "

A striking instance of the cyclical use of the word 'revolution' in relation to human affairs and life occurs in the famous grave scene in

Hamlet (5.1.98). Shakespeare has Hamlet talk to the skull turned up by the clown: "Here's fine revolution, an we had the trick to see't. Did those bones cost no more the breeding, but to play at loggats with 'em? Mine ache to think on't . . . " Has Shakespeare (as suggested by Snow 1962, 168) equated "*revolution* with a restoration to one's former status, or a return to a previous point in the life-death cycle"? That is, is there here the idea of an ebb and flow, a sense of that 'reversal of fortune' to which some authors refer? It was in this vein that Molière wrote of "all the revolutions to which inhuman fortune can expose us" (*Psyché*, lines 611–612).

During the first part of the seventeenth century, the general or nonscientific use of the word 'revolution' most often implied some cyclical or quasi-cyclical phenomenon, somewhat similar to the astronomical meaning. Thus in a dictionary of 1611, 'revolution' was defined only as "a full compassing, rounding, turning backe to its first place, or point; the acommplishment of a circular course." But revolution was also coming to mean a great event, a change. We may perhaps see how these two senses of the word 'revolution' occur together, in the following extract from a letter written in 1646 by James Howell: "I think God Almighty hath a quarrel lately with all Mankind . . . for within these twelve years there have the strangest Revolutions and horridest Things happen'd not only in *Europe,* but all the World over, that have befallen mankind, I dare boldly say, since *Adam* fell, in so short a revolution of time." In the phrase "so short a revolution of time" Howell (1890, *1*: 512) has used the word in the traditional and literal sense; but in "the strangest Revolutions" he may or may not have had in mind the political events of those turbulent days.[7]

The sixteenth century knew no major or large-scale revolutions in the social and political realms in any sense in which we use the word today. There were thus no political or social events of the sixteenth century or of the early seventeenth century that could serve as concrete examples of revolutionary theory, or that could provide examples or conceptual models for revolutions (in the sense of a drastic and even sudden secular change) in the areas of human creative effort. But by the middle of the seventeenth century the development of the theory and concept of revolution was given a tincture of reality by political upheavals, notably the Glorious Revolution of 1688 — the first recognized political revolution of modern times. (On the Reformation, see §4.1.)

The historical significance of the Glorious Revolution (see below) is somewhat obscured today by considerations of the series of events that occurred a few decades earlier, in the mid-seventeenth century, and which today are sometimes known collectively as the English Revolution—a name that is far from being in universal usage among historians, many of whom do not consider these events to have been a revolution at all. There is further confusion arising from the fact that a historian such as Acton (1906, 219) refers to the later Glorious Revolution as the English Revolution. This English Revolution is almost never defined, even by those historians who believe that a revolution occurred. A central feature is a constitutional and religious upheaval punctuated by dramatic events: the Civil War (1642–1646), the trial and execution of King Charles I (1649), the interregnum of the Commonwealth and Protectorate under Oliver Cromwell. Samuel Rawsun Gardiner, an eminent constitutional historian of the nineteenth century, saw this episode as "the Puritan Revolution 1625–1660,"[8] and so titled his great documentary history (1906); but in his text (for example, pp. x, xi) he referred to "the English Revolution." Although this English Revolution was marked by violence (civil war, regicide) and temporarily produced a change in the outward form of government (commonwealth rather than monarchy), no basic political or social changes "of lasting value" occurred. Even the fundamental issue of the king's divine right versus the power of Parliament (with real sovereignty based on the electorate) was not fully resolved until the Glorious Revolution.

The name Puritan Revolution, put forth by Gardiner (1886, and other works), was based on the plain fact that the opposition to the king was primarily Puritan, but the issues were economic as well as religious (the anti-royal party comprised many of the rising class of merchants and artisans who wanted to have a greater role in government and to effect a lessening of the restrictions imposed by government on finance and commerce). Within the Puritan movement, there were true revolutionaries, of whom a most extreme sect were called the Levellers (a pejorative expression applied to them because of their beliefs in democracy and equality). The Levellers were twice defeated by Cromwell and "the 'revolution' which *they* wanted never took place" (Aylmer 1975, 9). They wanted to abolish monopolies and special privileges (but not private property), to establish universal "male household suffrage" but not "unqualified manhood suffrage" (p. 50). Their aim was to revolutionize the mode of government by extreme parliamentary reforms, election of local magistrates and other officials, rotation in office, and decentralization of government plus a severe limitation of its powers, together with the abolition of the monarchy and the House of Lords.

Christopher Hill, the most important writer on the English Revolution in our day, asserts in *The Century of Revolution* (1972, ch. 11, pp. 165 ff.) that "a great revolution took place" during "the decades 1640–60 . . . comparable in many respects with the French Revolution of 1789." It was a "great revolution" because "absolute monarchy on the French model was never again possible." The "instruments of despotism, Star Chamber and High Commission, were abolished forever" and "Parliamentary control of taxation was established." But here again Hill asserts that this "was a very incomplete revolution," that "between 1640 and 1660 there had been two revolutions, of which only one was successful." Hill also insists that there had been a "great revolution in human thought" — a "general realisation . . . that solutions to political problems might be reached by discussion and argument," that "questions of utility and expediency were more important than theology or history," and that "neither antiquarian research nor searching the Scriptures was the best way to bring peace, order, and prosperity to the commonwealth." This being so, we would agree with Hill that it constituted "so great an intellectual revolution that it is difficult for us to conceive how men thought before it was made." In this same work, Hill sums up the effects of the decades from 1640 to 1660 by making a contrast between "the Puritan revolution" which "was defeated" and "the revolution in thought [which] could not be unmade." The latter included "the revolution in science led by the men who were to form the Royal Society after the Restoration" and "the revolution in prose which the same Royal Society was to consecrate."

This so-called English Revolution was not generally called a revolution until the nineteenth century; in its own century it was referred to as "The Great Rebellion" and "The Civil War." The nineteenth-century historian and statesman François Guizot produced a very influential six-volume *Histoire de la révolution d'Angleterre* (1826–56), in which he drew parallels between the French and English revolutions (both marked by regicide) and expressed his great admiration for the relative moderation of English revolutionism. This work particularly aroused the ire of Karl Marx, who wrote a major essay attacking Guizot in 1850. Marx and Engels discussed the English Revolution (and also the Glorious Revolution) in many writings. By the twentieth century, many books on English history referred to the English (or Puritan) Revolution along with the Glorious Revolution.[9]

The Glorious Revolution

Although a number of historical and political writers of the seventeenth and eighteenth centuries referred to the English Revolution as a revolution, it was not then generally considered to be an embodiment of the growing concept of political revolution, whose history we are tracing here. In the mainstream of thought, the first revolution in modern times was rather the Glorious Revolution, perhaps because the changes it produced were so permanent. In the mid-eighteenth century, in the article on 'révolution', the *Encyclopédie* listed the Glorious Revolution as prototypical and did not even mention the English Revolution. In the first edition of the *Encyclopaedia Britannica* (1771, *3*: 550), a revolution "in politics" is said to signify "a grand change or turn in government." In this sense, it is said, the word is used "by way of eminence" for the "great turn of affairs in England, in the year 1688, when king James II abdicating the throne, the prince and princess of Orange were declared king and queen of England, &c." Four decades later, by the time of the fourth edition (1810, *17*: 789), the *Britannica* listed four political revolutions: "that which is termed *the revolution* in Britain" (the Glorious Revolution, 1688), the "American revolution," "the revolution which took place in Poland about the end of the 18th century" (by which Poland was partitioned among Austria, Prussia, and Russia), and "the French revolution" — "the most extraordinary of all, whether considered with regard to the events which accompanied, or the consequences which followed it."

The Glorious Revolution consisted of two major events and the steps that led up to them: the abdication of James II and the ascent of William and Mary to the throne. Unlike most subsequent revolutions in history, this one was relatively peaceful and bloodless, although accompanied by a great show of arms. The revolution turned the monarchy from a Catholic to a Protestant line and guaranteed a Protestant succession for all times. The significant radical step, however, was the affirmation that the king's power was not his absolutely by divine right but rather by and with the consent of the governed, at least insofar as the Parliament represented the governed. This occurred when the throne was said to be 'vacant' because James had — so it was declared — 'abdicated' the government; he "was not declared to have been 'deposed', nor to have 'forfaulted', that is 'forfeited', the Crown" (Trevelyan 1939, 145); the "word 'vacant' destroyed divine hereditary right in the realm of theory," while the Act of Settlement, in giving the throne jointly to

William and Mary, destroyed that right in practice. Within a year, in 1689, certain rights and privileges of Englishmen were spelled out in a series of 'articles' comprising the Declaration of Right, the instrument setting forth the conditions that had to be accepted by William and Mary in order to be King and Queen. They could not be elevated to the throne unless they accepted the declared limitations of the royal power. When William and Mary simultaneously accepted the Crown and the Declaration of Right, they formally agreed to a contract which has not needed fundamental alteration for three centuries. England had "acquired the outline of a Constitution" which has worked and worked well. But we must note that the Declaration of Right "introduced no new principle of law, not even Toleration for Dissenters or irremovability of Judges, though there was entire agreement on the immediate necessity of those two reforms" (Trevelyan 1939, 150).

Today the revolutionary aspects of the Glorious Revolution may seem minimal, especially in contrast to the French and Russian revolutions. But in the succeeding century, men of as different political views as the conservative David Hume and the radical Joseph Priestley were as one with respect to the significance of the principle that a monarch rules by and with the consent of the governed. According to Priestley (1826, 286–7),

> the most important period in our history is that of the revolution under king William. Then it was that our constitution, after many fluctuations, and frequent struggles for power by the different members of it (several of them attended with vast effusion of blood), was finally settled. A revolution so remarkable, and attended with such happy consequences, had perhaps no parallel in the history of the world, till the still more remarkable revolutions that have lately taken place in America and France. This it was, as Mr. Hume says, that cut off all pretensions to power founded on hereditary right; when a prince was chosen who received the crown on express conditions, and found his authority established on the same bottom with the privileges of the people.

For most Englishmen this had been a beneficent revolution. No doubt the Glorious Revolution thus contributed to the intellectual linking of revolutions with the idea of progress.

The contrasting progressive and conservative aspects of the Glorious Revolution were presented in a dramatic way in an essay by Lord Acton

(1906) entitled "The English Revolution." Acton there presented the point of view of Burke and Macaulay, who—he said—"have taken pains to show that the Revolution of 1688 was not revolutionary but conservative, that it was little more than a rectification of recent error, and a return to ancient principles." The revolution was "essentially monarchical," and "no change took place in the governing class," that is, "there was no transfer of force from the aristocratic element of society to the democratic." Neither in the Convention nor in the subsequent Bill of Rights was there any mention of "free government, religious liberty, national education, emancipation of slaves, freedom of trade, relief of poverty, freedom of the press, solidarity of ministers, publicity of debates." Even so, according to Acton, the revolution "is the greatest thing done by the English nation." The reason? Because "it established the State upon a contract, and set up the doctrine that a breach of contract forfeited the crown." Since "Parliament gave the crown, and gave it under conditions," Parliament "became supreme in administration as well as in legislation": "All this was not restitution, but inversion" (p. 231).

The Glorious Revolution, in linking the two main aspects of revolution—a significant alteration in the form of government and a return to older principles or conditions—set the stage for the growing use of the originally cyclical word 'revolution' to express the singularity of change. Eventually, as the century wore on, revolution came to mean primarily the introduction of something wholly new, as in the American and French revolutions, and ceased to signify a reaffirmation or restoration.

The sense in which the Glorious Revolution partook of a return is plainly stated in the general article on political revolution in the sixth edition of the *Encyclopaedia Britannica* (1823, *17:* 789). Here it is said that not only did the revolution establish (reestablish) the Protestant succession, but the constitution was "restored to its primitive purity." This "important transaction," furthermore, "confirmed"—rather than established or set forth for the first time—"the rights and liberties of Britons." This is similar to the use of the word 'revolution' by Clarendon (d. 1674) in his *History of the Rebellion and Civil Wars in England* (bk. 11, §207). Clarendon referred to the situation following the Restoration in 1660 as a time when "many of those excluded members, out of conscience or indignation, forbore coming any more to the house for many years, and not before the revolution."

The almost identical political sense of a return or cycle is also to be found, expressed in a forceful way, in Thomas Hobbes's history of the Long Parliament (1969, 204): "I have seen in this revolution a circular motion of the sovereign power through two usurpers, father and son, from the late King" to his son. It "moved from King Charles I. to the Long Parliament; from thence to the Rump; from the Rump to Oliver Cromwell; and then back again from Richard Cromwell to the Rump; thence to the Long Parliament; and thence to King Charles II., where long may it remain." Another kind of cycle, the revolution of stars, was invoked by the Earl of Clarendon in a "Speech about Disbanding the Army," September 13, 1660: "The *Astrologers* have made us a fair excuse, and truly I hope a true one; all the motions [!] of these last twenty Years have been unnatural, and have proceeded from the evil Influence of a malignant Star; and let us not too much despise the Influence of the Stars. And the same *Astrologers* assure us, that the Malignity of that Star is expired; The good *Genius* of this kingdom is become superiour and hath mastered that Malignity, and our own good old Stars govern us again" (*State Tracts*, 1692, 3).

I do not know when the revolution of 1688 was first called 'glorious', but in that year John Evelyn wrote to Samuel Pepys to ask how "I may serve you in this prodigious Revolution."[10] A text of the following year mentions "this great revolution." As early as 1695, the word 'revolutioneer' was being applied to a supporter of the revolutionary settlement of 1688. A volume of *State Tracts* (1692) from 1660 to 1669 was stated to have the purpose "to shew the necessity, and clear the legality of the late Revolution." The opening decades of the eighteenth century also produced a number of references to the revolution of 1688; in Dr. Johnson's *Dictionary of the English Language* (1755), the third definition of 'revolution' reads: "change in the state of a government or country. It is used among us . . . for the change produced by the admission of king William and queen Mary."

Conservative Catholic opinion in France did not view the revolution settlement as a beneficent or glorious event. Rather, there was seen to be a cycle, a parallelism between the execution of Charles I and the flight of James II, both having been Catholic monarchs who lost their thrones, both having been replaced by Protestants: Cromwell and William of Orange. There was a quite natural fear that a similar cycle would unfold its revolution in France. A main theme of *The History of the Revolutions in England*, written by the French Jesuit P. J. d'Orléans, is that there was no inexorability in these events. As he says in dedicating the book to Louis XIV (translated in 1711 into English and reissued in a second edition in 1722), "It was no Failure in Your Majesty, that the last of [the revolutions of England] . . . was not prevented." Had Louis's "Advice been followed," and his "Succours accepted of, the King of *England* had been still on his Throne."

French Protestants, however, took new hope from the revolution of 1688. At the end of that year Pierre Jurieu, in one of his *Lettres pastorales, adressées aux fidèles de France qui gémissent sous la captivité de Babylone*, expressed his Protestant hope that this "great and surprising revolution will doubtless lead to others which will be not less considerable" (quoted in Goulemot 1975) than the revolutionary succession of William and Mary. Jurieu could find hope that "the tyranny of the anti-Christ [that is, Louis XIV] will fall without bloodshed, fire, and sword." In 1691, in a discussion of the execution of Charles I and the rise of Cromwell, the Catholic Raguenet invoked the image of "those idle and uneasy souls who are disgusted with a constantly even life and who delight in revolutions; in a word all those who hoped to gain some advantage in change or in general confusion entered with pleasure into this Cabal and spared nothing to make it succeed."

The Spread of a Concept

Jean-Marie Goulemot, in his *Discours, révolutions, et histoire* (1975), has shown that in the last decade of the seventeenth century the word and concept of 'revolution' was used rather extensively in France in relation to the English revolution of 1688, not considered in any way 'glorious' but rather as a Protestant threat to the established monarchy. In particular, Goulemot has traced the revolutionary idea during the closing years of the seventeenth century and opening years of the eighteenth in literature (tragic drama and romantic fiction) and in historical writing. The wealth of example he has found, showing the growing popular currency of the concept and term 'revolution', helps to explain the introduction during these years of the idea that there are 'revolutions' in mathematics and the sciences.[11] Unfortunately, this extraordinary book, while developing at great length the author's theme concerning the ideas of revolution in the seventeenth century, does not make a constant and consistently clear distinction between the seventeenth-century view and his own interpretation, strongly and admittedly influenced by the political events of the 1950s, 60s, and 70s. This is especially the case with respect to the actual occurrence of the word 'revolution', as opposed to some kind of idea of revolution (as seen by a twentieth-century thinker) in the works under analysis. And even in the examples given, there is not always a really careful attempt to discover whether 'revolution' in its actual occurrences means a cyclical phenomenon or a single event of major proportions.

Nevertheless, the number of instances in which 'revolution' does occur provides convincing evidence of the growing currency of this word and of the concept of radical change it carries.[12] An example is Fénelon's *Aventures de Télémaque* (published in April 1699); the "éditions commentées," published in 1719 and afterwards, "connect numerous episodes of the romance with the death of Charles I, the restoration of Charles II, the dictatorship of Cromwell, and the fall of James II" (see Goulemot 1975). Fénelon has his characters discuss "the revolts" and "the cause of the revolts" (particularly "the ambition and the restlessness of the great personages in the state"). There are three "fictional revolutions," each occurring in a monarchy in which a prince has become a tyrant; in two of them the tyrant is killed, but in a third he is exiled. As Goulemot observes, in two cases there is an uprising ('révolte'), in which the people rise up in order to gain their liberty, but they do not get rid of the monarchy and establish a republic. They chose a new king by legitimacy of succession or by elective vote; hence, it has been said that this kind of "revolution is not at all the creation of a new order, nor even a radical modification of the mode of exercising sovereignty, but a return to the old political order which tyranny has perverted." Fénelon says, "There is only a sudden and violent revolution that can bring back into its natural path this overflowing power" (quoted in Goulemot 1975). In 1697 Le Noble gave a Jacobite view of the revolution in England in a novel entitled *Milord Courtenay, ou Histoire secrette des premières amours d'Elisabeth d'Angleterre*, in which he wrote: "England is a perpetual Theatre of revolutions, in an instant the calm is changed into the most furious tempest, and this tempest changes in a moment into calm." In many late-seventeenth-century French novels, which turn out to be "oeuvres historico-galantes," revolutions abound. Thus, in *Abra Mulé, ou l'histoire du détrônement de Mahomet V*, Le Noble tells the story of "the revolution that occurred in the Ottoman Empire in the month of November 1687, by the deposing of Sultan Mahomet and by the elevation of his brother Soliman to the throne."

The new usage of 'revolution' as a transfer of an astronomical concept to the world of political affairs and even to conditions of life is illustrated by a seventeenth-century French-Latin dictionary by François Pomey. His *Royal Dictionary* (3rd ed., Lyon, 1691) has two separate entries for 'revolution'. First the technical sense, the traditional motion of circularity and the going around of the heavenly bodies: "*tour, cours des Astres.* Astrorum circumactus, circuitus, circuitio, conversio." But there is a second entry for 'revolution', devoted to political change, change in general, and even the advance of time and the vicissitude of fortune: "*changement d'état.* Publicae rei commutatio, conversio, mutatio. Temporum varietas, fortunaeque vicissitudo."

The spread of the new sense of revolution is seen in a book by John Ovington, *A Voyage to Suratt, in the Year 1689* (London, 1696). Of the four appendixes, the first was announced as "The History of a Late Revolution in the Kingdom of *Golconda.*" The revolution in question appears to have been a change in government in which a puppet king rather peacefully became a true monarch by taking power from his ministers. In the introduction, Ovington describes how he set sail from Gravesend on "*April* the 11th, 1689, the Memorable Day whereon their Majesties, King *William* and Queen *Mary* were Crown'd." The ship was sent to the East Indies, he says, "as an Advice-Ship of that wonderful Revolution, whereby their Sacred Majesties were peaceably setled in the Throne, and had been receiv'd with the Universal Joy of all the Nation." Ovington also used the word 'revolution' in an implied sense of a return when he talks of the "Rebellion of Cha-Egber against his Father" (new ed., Oxford, 1929, pp. 108–109). He "daily waits for some favourable Revolution," Ovington says, "when he may return to *India* again, whither he hopes to be recall'd by his Father's death."

In this new age of revolution, an earlier work on the English Revolution had a new relevance. Anthony Ascham's *Of the Confusions and Revolutions of Governments* (London, 1649; for which see Zagorin 1954, ch. 5) was an enlarged version of a work published in 1648. He used the phrase "confusions and revolutions" in a general, rather than a specific sense, but what made his book seem important after the Glorious Revolution was his political discussions of lawful and unlawful monarchical powers.

We may conclude this discussion with a presentation of the use of 'revolution' by Thomas Hobbes and John Locke.[13] Hobbes was perfectly familiar with the traditional scientific sense of the word 'revolution' and he used this expression in his writings on geometry and on natural philosophy. He wrote of "a contrary revolution," "epicycles," and of revolutions in the sense of completed circular motions. In his study of "the civil wars of England," or *Behemoth* (pt. 4, concl.), Hobbes transferred this scientific term to politics, writing (as we have seen) of "this revolution" as "a circular motion of the sovereign power through two usurpers, from the late King to his son."

Nevertheless, when Hobbes came "to describe a sudden political change" (Snow 1962, 169), he — like Bacon, Coke, Greville and Selden —"used such words as 'revolt,' 'rebellion,' and 'overturning.'" Locke, in both his *Elements of Natural Philosophy* and his *Essay on Human Understanding,* used the word 'revolution' in reference to the earth's annual motion about the sun (her "annual revolutions") and referred to the sun as the "Center" of the planets' "Revolutions" (Snow 1962, 172; Laslett 1965, 55). In the political sphere, Locke followed François Bernier (whose *Histoire de la dernière révolution des états du Grand Mingol* he had studied in close detail) in his use of the term 'revolution' in the sense of a completed dynastic change. In his famous *Second Treatise,* notable for its defense of the Glorious Revolution and for its presentation of the theory of government based on compact, Locke used the word 'revolution' only twice (bk. 2, § §223, 225), each time referring to a political cycle in which there was a return to some previous state with regard to some constitutional points. Thus he mentioned the "slowness and aversion in the People to quit their old Constitutions," which "has, in the many Revolutions which have been seen in this Kingdom, in this and former Ages, still kept us to, or after some interval of fruitless attempts, still brought us back again to, our old Legislative of King, Lords, and Commons."[14]

5

The Scientific Revolution:

The First Recognition

of Revolution in Science

A number of historians— among them Roger B. Merriman (1938), H. R. Trevor-Roper (1959), E. Hobsbawm (1954), and J. M. Goulemot (1975) — have called attention to the almost simultaneous occurrence of revolts, uprisings, or revolutions in different parts of Europe in the middle of the seventeenth century — in England, France, the Netherlands, Catalonia, Portugal, Naples, and elsewhere. This was obviously a time of crisis and instability, and it would almost seem that there was a general revolution, of which the geographically separate events were but individual manifestations. That there was a "general crisis," as Trevor-Roper put it, was apparent to sensitive men of that day. In a sermon preached before the House of Commons on 25 January 1643, Jeremiah Whittaker declared that "these days are days of shaking," and the "shaking is universal: the Palatinate, Bohemia, Germania, Catalonia, Portugal, Ireland, England" (see Trevor-Roper 1959, 31, 62 n.1).

The seventeenth century was also the time of the Scientific Revolution. The first Civil War in England, in 1642, began just four years after the publication of Galileo's *Two New Sciences*, the founding work in the science of motion, and five years after the publication of Descartes's *Discourse on Method* and *Geometry*. Newton's *Principia* — the most significant and influential book of the Scientific Revolution — appeared in 1687, one year before the Glorious Revolution; it was, in fact, dedicated to James II and the Royal Society. In many ways the Scientific Revolution was more radical and innovative than any of the political revolutions of the seventeenth century and its effects have proved to be more profound and lasting. But so far as I know, no one has linked the Scientific Revolution to the other revolutions that occurred in that same century, or speculated that the revolutionary spirit which moved in the realm of politics might have been the same as that which caused upheavals in the sciences.

The best way to assess the depth and scope of the Scientific Revolution is to compare and contrast the science that came into fruition in the seventeenth century with its nearest equivalent in the late Middle Ages. Consider the central problem of motion (since "to be ignorant of motion is to be ignorant of nature"). Medieval scholars understood motion in the general Aristotelian sense of any change from potentiality to actuality. The laws of motion were thus not limited to mere local motion (a change of place) but embraced any change that could be quantified as a function of time, including the acquisition or loss of weight with age, or of grace. When scholars considered specifically local motion, in the fourteenth century, they became well aware that motion could be uniformly accelerated or nonuniformly accelerated, and they were able to prove mathematically that the effect of a uniformly accelerated motion in a given time is exactly equivalent to a uniform motion during that same time if the magnitude of the uniform motion is the mean of the accelerated motion. But the mathematical philosophers of the fourteenth century and those who discussed their work in the fifteenth century never put these mathematical principles to the test by applying them to physical events, say falling bodies. Galileo's approach to this same problem, on the other hand, was to view these principles, and others, not as pure mathematical abstractions but as laws that governed actual physical processes and occurrences in the world of experience. Galileo even tested and confirmed the law of freely falling bodies by the famous experiment of the inclined plane, described in his *Two New Sciences*. Galileo's development of such laws was no less mathematical than that of his fourteenth-century predecessors, but his mathematics was conceived in a physical context and was tested by physical experiment. Some otherwise incomprehensible manuscript notes of Galileo were found by Stillman Drake (1978) to be a set of actual experiments that led him to the discovery of these laws.

This example shows us how novel and revolutionary it was to discover principles by experiment combined with mathematical analysis, to set scientific laws in the context of experience, and to test the validity of knowledge by making an experimental test. Traditionally, knowledge had been based on faith and insight, on reason and revelation. The new science discarded all of these as ways of understanding nature and set up experience — experiment and critical observation — as the foundation and ultimate test of knowledge. The consequences were as revolutionary as the doctrine itself. For not only did the new method found knowledge on a wholly new basis, but it implied that men and women no longer had to believe what was said by eminent authorities; they could put any statement and theory to the test of controlled experience. What counted, therefore, in the new science of the seventeenth century was not the qualifications or learning of any author or reporter but rather his veracity in reporting, his true understanding of the method of science, and his skill in experiment and observation. The simplest and humblest student could now test (and even show the errors in) the theory or laws put forth by the greatest scientist. Knowledge thus took on a democratic rather than a hierarchical character and no longer depended so much on the insight of a chosen few as on the application of a proper method, accessible to anyone with sufficient wit to grasp the new principles of experiment and observation and the way to draw proper conclusions from the data. It is not surprising, therefore, that so much attention should be given during the Scientific Revolution to codifiers of the method — men like Bacon, Descartes, Galileo, Harvey, and Newton, who wrote about the way to proceed in scientific inquiry.

The scientists of the seventeenth century and of the late sixteenth century were fully aware of the newness of their approach in directly appealing to nature. This approach is apparent in the books on plants and animals of the late sixteenth century. Not only do they show a new sense of realism derived from the use of perspective, but they state explicitly that the illustrations were made from living specimens. Thus, Fuchs's herbal of 1542 has a plate showing the artist and the woodcutter working from a plant held in front of them. In Vesalius's great book on *The Construction [or Fabric] of the Human Body* (1543), a plate shows all the tools necessary to make dissections. The message is plain: "Do it yourself." Vesalius not only wanted his student-readers to duplicate his results and confirm his findings, and then to go on to add to our knowledge; he was also making it evident that his revolutionary book was based on experiential and testable facts.

This sixteenth-century fascination with nature was evident in the response of men and women to the discovery of new worlds, especially North and South America. What was of interest was not just the land forms and geological deposits but the forms of life, plant and animal. Were these animals spared the Noachian flood and so different from the animals of Europe? Or were they separate and special post-diluvian creations? Both questions were disturbing because they seemed to have answers that ran counter to Scripture. And the question of the men and women native to the New World was more disturbing still.

In the opening decade of the seventeenth century, when Galileo's telescope made known for the first time what the heavens are like, the excitement was worldwide. Marjorie Nicolson has chronicled for us the eagerness with which people all over Europe waited for each new revelation of Galileo's telescope and the way in which his discoveries quickly took their place in the images used by poets. A masque of 1620 by Ben Jonson entitled *Newes from the New World* is not about America but about the heavens, the moon in particular, containing mention of the telescope — in keeping with Galileo's name for the account of his discoveries, *The Message of the Stars,* or *The Messenger* (both are suitable translations of the Latin *Sidereus Nuncius*). As an announcement of novelty, Jonson's work is the humorous equivalent of Monardes's description of the medicinal flora of America called *Joyfull Newes out of the Newe Founde Worlde.* Here was a symbolic beginning of the revolutionary newness of the sciences. For Galileo announced not only new facts, new information, but he speedily concluded that the new data of telescopic observations falsified the Ptolemaic system (which it did) and proved the Copernican (which it did not).[1]

Many of the seminal books of the Scientific Revolution bear the word 'new' in their titles. Kepler published (1609) a *New Astronomy,* based on physical principles. Galileo's last book (1638) bears the title *Two New Sciences;* though this may not have been the title he chose, he does introduce the third book, on motion, by stating that he had discovered many new things worthy of note. Tartaglia called his work a *New Science* (1537). An account of revolutionary experiments made with the newly invented vacuum pump was called *New Experiments Made at Magdeburg* (1672) by von Guericke. Boyle used the word 'new' in the titles of many of his books. In 1600 William Gilbert published a work significantly entitled *On the Magnet . . . a New Physiology, Demonstrated by Many Arguments and Experiments*. He dedicated this "nature-knowledge" which is almost "entirely new and unheard-of" to "you alone, true philosophizers, honest men, who seek knowledge not from books only but from things themselves." Gilbert was aware that only a small company as yet were devoted to "this new sort of Philosophizing."

The Scientific Revolution, which produced a new kind of knowledge and a new method of obtaining it, also produced new institutions for the advancement, recording, and dissemination of that knowledge. These were societies or academies of like-minded scientists (and some others who were just interested in science) who met to do experiments in concert, to see performances and tests of experiments done elsewhere, to hear reports on scientific work done by members, and to learn what was going on in other scientific groups and in other countries. The emergence of a scientific community is one of the distinguishing marks of the Scientific Revolution. By the 1660s permanent national academies arose in France and England, and both had official journals for the publication of research done by their members.

The example of Isaac Newton shows how significant election to such membership could be. In 1671 Isaac Barrow (Newton's predecessor as Lucasian Professor) took an example of Newton's newly invented reflecting telescope to London to show to the Royal Society. Newton's invention was "applauded" and Newton was shortly afterwards elected a Fellow of the Royal Society. Pleased at being so recognized by his fellow scientists in London, Newton wrote then to ask when the Society met so that he could present a report on his experiments on light and color, the ones that were the basis of the invention of the new telescope. With the pride of youth, writing to the secretary of the society that had reached out to make him a member, Newton said that his discovery was "the oddest" detection yet made in the operations of nature. Newton's eagerness to share his discovery at once with his new associates in science is in such marked contrast to his later reluctance to print (or to have printed) any of his findings that it gives us an indication of how significant formal admission to the established scientific community can be to a scientist.

Newton's paper on light and colors carried a number of firsts: the first scientific publication of Isaac Newton; the first or founding paper in the physics of colors; the first major scientific discovery to be published as an article in a scientific journal. And it is remarkable, furthermore, because it describes Newton's experiments and the theoretical conclusions he drew from them without expounding a system of cosmology or a doctrine of theology; it is science, pure and simple, as we have understood that term ever since.

A revolutionary feature of the emerging scientific community was the establishment of a formal information network. In part, this was accomplished by personal visits and correspondence, but chiefly by scientific journals and reports. The short-lived Galilean Accademia del Cimento (Academy of Experiment) published the result of its labors in a volume of *Saggi* (1667) in Italian. These were made available in English in 1684, in a volume that had a symbolic frontispiece showing how the Italian academy had handed on its tradition to the Royal Society of London. The *Philosophical Transactions* of the Royal Society contained articles in English and Latin. Soon there appeared volumes intended for Continental readers which translated the English articles into Latin. The summary volumes, or abridgements, of the *Philosophical Transactions* were wholly in English but were rapidly translated into French, while the findings of the French Academy of Sciences were made available in English. A surprising number of the great works of science in the seventeenth century were not published in Latin, as is often supposed, but in the vernacular languages. Some examples are Galileo's *Dialogue Concerning the Two Chief World Systems* (Italian, 1632; Engl. trans. 1661; Latin trans. 1635), Descartes's *Geometry* (French, 1637; Latin trans. 1649, 1659), and Newton's *Opticks* (Engl. 1704; Latin trans. 1706). Other examples are Descartes's *Dioptrique* (1637), Huygens's *Traité de la lumière* (1690), and Hooke's *Micrographia, or Some Physiological Descriptions of Minute Bodies* (1665).

We can see the information network in operation through the extensive correspondence of Henry Oldenburg, the inaugural Secretary of the Royal Society. In 1668 Oldenburg wrote a letter to Huygens, who was in Paris, expressing the desire of the Society to have him communicate to them "what he had discovered on the subject of motion," even though he "did not yet think fit to print [it]." He was asked whether he "would impart to them his theory of it, together with such experiments, as he grounded his theory upon." Huygens agreed, "not doubting but the society would secure to him the honour of that discovery, by giving it place in their Register-book, as coming from him." A few months later, Huygens's text was sent over and studied, in part by Christopher Wren. Then "several experiments were tried" as a means of testing Huygens's theory and also Wren's, but the apparatus did not work perfectly, and the experiments were ordered to be performed again at a meeting the next week. A priority question between Huygens and Wren would shortly become apparent. Huygens sent on to the Royal Society a statement of new results in a "cypher or anagram," to be entered into the Register-book as a "way of securing his discoveries or inventions for the future" until "he should think it convenient to explain them in a common language." Some two decades later, Edmond Halley would urge Newton to send an account of his discoveries to be registered in the Royal Society, to ensure his priority. The tract *De Motu*, which Newton wrote in autumn 1684, is still to be seen in the Register-book; he later expanded this tract into the famous *Principia*.

The role of scientific societies and academies in establishing the record of priorities in discovery and invention underlines another major aspect of the Scientific Revolution. This was the first revolution in history dedicated to a continuing process rather than a goal. It was mentioned earlier that political and social revolutions tend to have a well-defined end, the establishment of a certain kind of state or social system, even though that state may not be envisioned as capable of being achieved in the immediate future. But almost at once the new science was conceived as a process of finding out, a never-ending search. Provision was made to publish and disseminate discoveries, to establish laboratories and observatories, gardens and menageries, where discoveries could be made. The process of continual change was institutionalized in the form of journals for the publication of new results, depositories for the registration of discoveries to ensure priority, and prizes for the most revolutionary advances. I know of no other revolution, or revolutionary movement, which so institutionalized the continuing process of revolutions to come. This was indeed something new under the sun.

But despite the expectation that science was an unending search for truth, it was widely hoped that advances in science would lead to practical inventions and improvements in medicine of value to mankind. Such statements appear early in the seventeenth century, as part of the methodological treatises of Bacon and Descartes. If only there were some wealthy man to support him, wrote Descartes in his *Discourse on Method* (1637), what improvements could be made in such practical arts as mechanized agriculture, in medicine and health care. The same theme occurs again and again in Bacon, who argues that science — knowledge of nature — will lead to control of our environment, giving us new powers. Bacon adds, very wisely, that such practical applications are of more value as "earnests and guarantees of truth" than as means of adding to the comforts of life. What he meant was that because the new science was empirically based, its principles could be embodied in real devices. Functioning machines embodying or based on new principles provide tangible evidence of the truth of the principles.

All of these revolutionary features aside, what did the Scientific Revolution actually achieve in the way of basic scientific advance? We have already seen that abstract laws of motion were replaced by Galileo's laws of freely falling bodies. Furthermore, free fall—a type of accelerated motion—could be combined with uniform horizontal motion to yield, as Galileo showed, the parabolic path of projectiles. The seventeenth century also saw the beginnings of the science of magnetism. Kepler found the three laws of planetary motion that bear his name and devised the modern heliocentric system of the universe that we often call Copernican. Newton not only founded the science of colors but produced a mathematical system that simultaneously embraced the new terrestrial and celestial physics. His principle of universal gravity accounted for Kepler's laws and the laws of falling bodies and could explain the occurrence of tides in the sea and the shape of the earth. It even set the ground for the successful prediction of a comet some four or five decades ahead. In the simplicity of its explanations and in the depth and scale of its applications, Newtonian physics was without a doubt a revolutionary force.

But the physical sciences were not the only ones that saw a revolution in our understanding of nature. The life sciences were also active, culminating in Harvey's discovery of the circulation of the blood, which revolutionized physiology. Here, as in the science of motion, the revolution had definite nonarguable aspects of falsification. Just as the prediction of Aristotelians (if not of Aristotle himself) that heavy bodies fall in air more swiftly than light bodies in proportion to their weight is demonstrably false—as is easily seen by making the experiment—likewise, Galen was simply incorrect in saying that the blood ebbs and flows in the veins and can ooze from one side of the heart to the other through pores in the septum, or inner wall, of the heart.

Contemporaneous Views on Revolution in Science

Although it was hard to deny the tremendous progress science had made in the sixteenth and seventeenth centuries, some observers nevertheless preferred to think of the advances as improvements rather than as revolutions, and there were some who denied that such truly great progress had been made at all. For example, the works written at the end of the seventeenth century and the beginning of the eighteenth century in the controversy known as the Battle of the Books or the Quarrel between the Ancients and the Moderns—by Fontenelle, Glanvill, Perrault, Swift, Temple, and Wottan—tend to invoke the concept of 'improvement' of knowledge, even in science and medicine, rather than 'revolution'. This fact is all the more surprising in that Fontenelle and Swift wrote of revolutions in other contexts, and Fontenelle applied this very word and concept to the new mathematics. In presenting the superiority of the 'moderns' over the 'ancients', and the great achievements of what we would call the Scientific Revolution, these authors (with one exception) seem to have eschewed the word 'revolution' altogether. And it is much the same with regard to Thomas Sprat's defense of the Royal Society (1667), a work dedicated to a display of the achievements of the new science and the changes that science would bring—even to language.[2] The themes are primarily newness and improvement, not revolution.

By the end of the seventeenth century revolutions in science began to be recognized. I have not found any clear-cut unambiguous statement that revolutions occur in science which was written before the century's end, although Gilbert, Galileo, Kepler, Harvey, and others stressed the newness of their work. But a letter written in Italian in 1637 contains a striking reference to the revolutionary character of Harvey's work.

For the study of the history of revolution in science this letter is a truly extraordinary document. It shows unambiguously how new discoveries in the sciences were perceived to be revolutionary, but it also indicates how difficult it was to express this revolutionary character by a single word. The letter was written in the year of publication of Descartes's *Discourse on Method* and *Geometry*. The writer was Raffaello Magiotti, a priest and scientist in Rome. He sent the letter to a fellow priest, Famiano Michelini in Florence, to inform his friends, including the aged Galileo, about the new physiological discovery made by Harvey and published in 1628. "This is the circulation which the blood makes in us," he wrote. It is "enough to overturn all of medicine, just as the invention of the telescope has turned all astronomy upside down, and as the compass [has done to] commerce, and artillery to the whole military art" (Galileo 1890, *1 7*: 65).

The year 1637 was too soon to have a single word or concept 'revolution' to express the radical character of Harvey's discovery. Half a century or more would be needed before it could have been said that the discovery of the circulation of the blood would inaugurate a 'revolution in medicine'. The verb Magiotti used was 'rivolgere' ("bastante a rivolger tutta la medicina"), which means 'to turn', 'to turn over' (as in 'to turn over in one's mind'), and sometimes 'to overturn'. But in order to make sure that his readers got the message, he explained what he meant by the use of this word, since at that time it was not a common occurrence for discoveries to have such an overturning (that is, revolutionary) effect on a science. And so Magiotti compared its effects with two major breakthroughs in technology: gunpowder and the magnetic compass. This pair of technological innovations along with printing from movable type were said by Bacon to have produced the most radical changes in the modern world. (Bacon, we may observe, also did not have at hand the word 'revolution', nor the concept that this word implies in its current sense.) Magiotti was saying in effect that this new phenomenon of turning a subject of science upside down, for which there was no name or clear concept and which was not as yet a well-established kind of event, was like those extraordinary inventions that had changed the nature of world trade, exploration, and warfare. Further, in order to make his point effectively, Magiotti compared Harvey's discovery with what was in 1637 the single most dramatic and most subversively revolutionary discovery as yet made in any branch of science: the new phenomena of the heavens revealed by Galileo who demonstrated with a mighty blow that the Ptolemaic system was false and that for thousands of years astronomers had written about the heavens without any true conception of what the heavenly bodies were like. Equally, Harvey had shown that the Galenic system was false and that therefore the whole medical system based upon Galenic physiology would have to be replaced. And so Magiotti said that the effect of the discovery of the circulation of the blood was to be compared to "the invention of the telescope," which has turned "astronomy upside down." In this case, Magiotti did not use the verb 'rivolgere' (as he had done a moment earlier), but rather 'rivoltare', which means not only 'to revolt' but also 'to turn upside down' and 'to turn inside out', and so 'to turn over', 'to overthrow'.

The word 'revolution' was actually coupled with Harvey's discovery in an essay written later in the seventeenth century by Sir William Temple. We can also see the first stages in the emergence of the modern concept of revolution in the ways in which the author used this word. In the essay, entitled "Of Health and Long Life" and probably written

before 1686 (see Woodbridge 1940, 212), Temple referred to the establishment of the ancient systems of medicine by Hippocrates and Galen,
the attempts of Paracelsus "to overthrow the whole scheme of Galen"
and his introduction of "the use of chymical medicines," and then
discussed Harvey and the circulation of the blood. This sequence of
events Temple (1821, *1*: 73) called "great changes or revolutions in the
physical empire," that is, in the empire of 'physic' or medicine. The
word 'empire' suggests that Temple was not referring here to the new
sense of a single dramatic event but rather to the traditional use of the
word 'revolution' in the phrase 'revolutions of empires'. And this is all
the more likely in that Temple elsewhere ("Heroic Virtue," 1821, *1*: 104)
invoked the image of revolution of empires as an unfolding or sequence
of events. Furthermore, Temple himself did not really believe in a
Harveyan revolution and said that the doctrine of the circulation "was
expected to bring in great and general innovations into the whole practice of physic," but actually "has had no such effect."

In his *Ancient and Modern Learning* (1690 [1963], 71) Temple generally
came out in favor of the ancients, arguing that the older books were
best, that—in the words of Alfonso el Sabio—the only things in life
worth pursuing are "old wood to burn, old wine to drink, old friends to
converse with, and old books to read." He asked, "What are the sciences
wherein we pretend to excel?" For 1500 years there have been no new
philosophers of note, "unless Des Cartes and Hobbs should pretend to
it." In astronomy he found "nothing new . . . to vie with the ancients,
unless it be the Copernican system, nor in physic, unless Harvey's circulation of the blood." But Temple had no doubt that even "if they are
true," "these two great discoveries have made no change in the conclusions of astronomy, nor in the practice of physic." So, although these
discoveries have been the source "of much honour to the authors," they
have been "of little use to the world. (pp. 56–57, 71)."

The subject of a revolution in medicine also appears in Fontenelle's
Nouveaux dialogues des morts, issued in 1683, which contains a dialogue
between the Alexandrian physician and physiologist Erasistratus and
William Harvey (called Hervé). Erasistratus opens the dialogue with a
summary of the marvels ('choses merveilleuses') reported by Harvey:
the blood circulates in the body, the veins carry the blood from the
extremities to the heart, and the blood then leaves the heart and enters
the arteries which take it toward the extremities. He admits how wrong
were the doctors of antiquity who thought the blood had only a very

slow movement from the heart to the extremities of the body and he
records how grateful the world is to Harvey for "having abolished that
old error." But in the ensuing dialogue, Erasistratus admits that the
moderns are better scientists than the ancients and that they have a
better knowledge of nature; yet he avows that they are "not better
doctors," since the doctors of antiquity cured sick people just as well as
the doctors of the newer age.

Harvey counters with the observation that ignorance of the blood's
circulation had been responsible for the death of many patients.
"What," replies Erasistratus, "you believe your new discoveries to be
truly useful?" When Harvey answers affirmatively, Erasistratus asks
him why there are now as large a number of the dead coming to the
Elysian Fields as formerly. "Oh!" says Harvey, "if they die, it's their
fault, not that of the doctors." Harvey ends his reply on an optimistic
note for the future, when the world will have had "the leisure to make
useful application of discoveries made only recently," since "very great
effects" will be seen with the passage of time. In the English translation
by John Hughes (Fontenelle 1708), Erasistratus's acerb comment is that
there will be "no such Revolutions, take my Word for it." That is, man
early attained "a certain Measure of useful Knowledge," to which some
small additions have been made, but which can never be surpassed.
Fontenelle ends the dialogue on a pessimistic note: that whatever scientists may discover about the human body will be in vain, since "Nature
will not be baffled" and men will continue to die at the appointed time.

This dialogue is of extraordinary interest in the present context. First
of all, Fontenelle compares a discovery like Harvey's ("to find out a new
Conduit in Man's Body") with that of an astronomer discovering a "new
Star in the Heavens" — both are of little or no practical use. Second,
Fontenelle, despite his rather strict adherence to the Cartesian philosophy, takes direct issue with Descartes's boast in his *Discourse on Method*
that sponsored medical research would produce an indefinite prolon-
gation of the life-span. Finally, we would note that Fontenelle's assertion (through Erasistratus) that there will be no revolutions in medicine
is the direct antithesis of Fontenelle's own recognition of the occurrence of a revolution in mathematics. In this case, the denial of a possible revolution may be taken as an index of the general rejection of
Harvey's great discovery by French doctors (see Roger 1971, 13, 169).
Despite Descartes's own warm espousal of the circulation of the blood,
Fontenelle could not conceive that this was a discovery of any great
consequence for medicine. In fact, Fontenelle does not seem to have

believed that revolutions in medicine would ever occur. The words "no such Revolutions," in the speech of Erasistratus, no doubt would have expressed Fontenelle's own convictions, but his own words were somewhat different. Where John Hughes has Erasistratus say, "No such Revolutions, take my Word for it," Fontenelle wrote, "Sur ma parole, rien ne changera" (Take my word for it, nothing will change).

Revolution appears in a letter of November 1656 by the chemist and physicist Robert Boyle, but in the context of intellectual endeavor in relation to divinity:

> I tell you so triviall a Story, to Lett you see to what a Height of Madnesse the giddy presumption of fond Man may carry him: & what strange Absurditys the Impudence of some, blusheth not to entitle the Spirit to. As for Publicke Intelligence, the Compleatnesse & entireness of the Late Successes have so confin'd Newes to what passeth within the Walls of Westminster, that I could at present but Transcribe, or at best, Anticipate, Diurnalls. What our new Representative will prove, or whither we shall have any, I dare not pretend to conjecture: especially in Blacke & White; only I shall not scruple to confesse that my Hopes and Feares have very peculiar Motives; & that the Clouds from which I expect either fertile Showres or Boisterous Storms, are not yet in their invisible & uncondensed Vapors. And as for our Intellectual Concernes; I do with some confidence expect a Revolution, whereby Divinity will be as much a Losser, and Reall Philosophy flourish, perhaps beyond men's Hopes.
>
> [British Library Harley MS 7003, fols. 179/80]

I have not found any similar statement by Boyle in a scientific context (nor is any mentioned by James Jacob in his book on Boyle as a revolutionary, 1979). But, considering the prolixity of Boyle's writings, it would be a bold scholar indeed who would assert that no such reference exists.

I have mentioned that many of the scientists of the seventeenth century were aware of the newness of their achievement and expressed this in the titles of their books, and that some of the greatest scientists of the seventeenth century (Gilbert, Kepler, Galileo, Descartes, Harvey, Newton) made explicit statements concerning the nontraditional quality of their work, pointing to the errors of ancient and medieval writers and adopting a revolutionary stance. A magnificent statement of the new science-in-the-making appears in the conclusion of Henry Power's *Experimental Philosophy* (1664). "This is the Age," he wrote, when "Philosophy comes in with a spring-tide." The "Peripateticks may as well hope to stop the Current of the Tide" as "hinder the overflowing of free Philosophy." He declared that "all the old Rubbish must be thrown away, and the rotten Buildings be overthrown," because these "are the days that must lay a new Foundation of a more magnificent Philosophy, never to be overthrown." This new philosophy, he said, "will Empirically and Sensibly canvass the *Phaenomena* of Nature, deducing the Causes of things from such Originals in Nature, as we observe are producible by Art, and the infallible demonstration of Mechanicks." This "is the way, and no other, to build a true and permanent Philosophy."

I have found that a rather early, fully modern and explicit statement concerning a revolution in mathematics occurs in the writings of Fontenelle, in the first years of the eighteenth century. Fontenelle was writing about the calculus, which had been invented by Newton and Leibniz and which was by all odds the most truly revolutionary intellectual achievement of the seventeenth century. Again and again in his writings Fontenelle invoked the new concept of revolution to exclaim how extraordinary this mathematics was. It gave scientists powers that by far surpassed what previously one would not have "dared to hope" might be achieved. The revolution had only begun, but already it had made mere beginners more skilled in solving mathematical problems than the wisest and most experienced mathematicians of the very recent past.

In the field of medicine, we find a reference to Paracelsus by W. Cockburn, M.D., which uses the term 'revolution' explicitly in the new sense and even implies that the occurrence of revolutions is a feature of the growth of medical systems. The date was 1728, shortly after Newton's death.

Three decades later, the mathematician Clairaut hailed Newton for inaugurating a revolution in the science of rational mechanics, a borderline subject that embraced both mathematics and physics. It is significant that Newton's great contributions to pure mathematics and to mathematical physics were so plainly recognized in their revolutionary dimensions, for Newton's achievements represent the high point of the Scientific Revolution. Contemporaneous testimony thus confirms our own judgment and stresses that the areas of most revolutionary achievement in the seventeenth century were pure mathematics and rational mechanics.[3]

6

A Second

Scientific Revolution

and Others?

The Scientific Revolution differs from other revolutions discussed in this book, and in most works on the history of science, in that it affected all of scientific knowledge. It radically altered the groundwork of science, giving prominence to experiment and observation, proclaiming a new ideal of mathematical theory, emphasizing prediction, and trumpeting the making of new discoveries in the future that would both advance knowledge of ourselves and our universe and increase our dominion over the forces of nature. There was also a concomitant revolution in the institutional structure. The recognition that such a large-scale intellectual and institutional revolution occurred naturally has led historians of science and other scholars interested in history to explore the possibility that there have been (and will be) yet other Scientific Revolutions of this kind.

Revolutions in the Institutions of Science

We saw in chapter 5 that a major revolutionary feature of the Scientific Revolution was the rise of a scientific community, as embodied in the academies. By the early decades of the nineteenth century, these older academies — the Royal Society, the Paris Académie des Sciences, with their younger cousins in Berlin, Stockholm, St. Petersburg, and elsewhere — could no longer contain the enlarged numbers of active scientists. Provincial scientific academies and journals devoted to scientific specialties, such as the French *Journal de Physique* and the British *Philosophical Magazine* for the physical sciences, proliferated. The explosion in numbers of scientists and supporters of science was accompanied by the rise of specialist scientific organizations, such as the British Association of Geologists. This enormous increase in science professionals and the institutions that support them has been described by Roger Hahn (1971, 275) as a " 'second' scientific revolution in the early nineteenth century."

The British Association for the Advancement of Science, established in 1831, along with its counterpart organizations in France, America, Germany, and elsewhere, was open to membership at large and was even a proselytizing organization. These institutions pushed for "advancement of science" by working with local groups, meeting each year in a different city, so that eventually the whole nation could become part of the scientific movement. At its meetings, the British AAS (traditionally pronounced 'British Ass'), the prototype organization, was divided into scientific sections (mathematics, physics, chemistry, astronomy, and so on), a feature that appeared also in the annual published record of meetings. But there were always a few general talks and major addresses and even sessions which could arouse interest among the public at large. The most famous example of the latter occurred at the Oxford meeting of the BAAS in 1860, when Bishop Wilberforce and T. H. Huxley held their debate on Darwinian evolution.

I believe that a good case can be made for a third Scientific Revolution during the latter part of the nineteenth and first decades of the twentieth century. This revolution also had a number of institutional aspects. First, it was during this time that universities truly became the centers for research and graduate training on a large scale that has been the pattern for the last hundred years or so. Self-trained scientists — amateurs like Faraday and Darwin — tended to be replaced by scientists with specialist and advanced disciplinary training and initiation into research, often accompanied by travel and graduate degrees (the M.A., Ph.D., D.Sc.). New universities like Johns Hopkins were founded with an express commitment to graduate study and research, while older universities established research institutes. A primary example of the latter is the Cavendish Laboratory at Cambridge University; others are the Yerkes Observatory of Chicago and the Museum of Comparative Zoology at Harvard. Many such institutes had no direct connection with universities—the Cold Spring Harbor Laboratory for genetics, the Carnegie Institution of Washington, and the Rockefeller Institute in the United States, the Institut Pasteur in France, and the institutes of the Kaiser Wilhelm Gesellschaft in Germany, where Nernst, Planck, and Einstein worked.

The third Scientific Revolution was also the time when scientific bureaus and institutes were either established or enlarged within government. But perhaps most important of all, this era witnessed the emergence of industrial laboratories and the large-scale harnessing of scientific research for the purpose of inventing new products, introducing improvements in the manufacture of existing products, and establishing standards. The first industry to produce awesome economic and social effects from the large-scale coupling of science and technology was the chemistry of dye-stuffs. One of the most significant aspects of the revolution in dye chemistry in Germany in the late nineteenth century was the mutual intellectual endeavor of universities, industrial concerns, and government for a practical end product. Science-based technological advance involving the cooperation of several institutions has been a feature of our world ever since.

Mention of government leads us directly into what I see as a possible fourth Scientific Revolution, one that has occurred during the decades since World War II.[1] The two main institutional features of this revolution are the expenditure of large sums of money by government (as much as three per cent of the gross national product of the United States in the 1960s) and research in groups. Both of these features of the fourth Scientific Revolution may be traced back to World War II, with the tremendous expenditure for the invention and production of atom bombs (along with large-scale but lesser expenditures for such devices as radar and the proximity fuse) and the development and production of antibiotics. In some branches of science, most notably high-energy physics and space research, the state of knowledge today is directly linked to the sum of money that government is willing to spend on particular projects. The nineteenth-century image of a Charles Darwin, who could live for decades in the country outside of London, in Down House, and there study and think in privacy while occasionally conducting significant experiments at miniscule cost, must seem as foreign and strange to today's scientists as the concept of scientific research done by

Martians. A measure of this difference is that an enormous part of the time and intellectual effort of today's scientists is spent not on conducting direct research at all but on preparing grant proposals, refereeing scientific papers and grant proposals of other scientists, writing status reports, attending committee meetings, and traveling to colloquia, conferences, and other scientific meetings.

The third Scientific Revolution was the time of flowering of specialized scientific societies, not merely disciplinary organizations such as the American Physical Society or the American Chemical Society but specialist groups within disciplines. Examples are the Optical Society of America, the American Rheological Society, and Society of Plant Physiologists. These sponsored general disciplinary journals (*The Physical Review, Reviews of Modern Physics*) and specialist publications of all sorts. The fourth Scientific Revolution has been marked by yet new forms of scientific communication. These include xeroxed preprints distributed on a large scale, sometimes even before an article has been accepted for publication by a journal, and the publication of short notes (*Physical Review Letters* publishes such communications far more rapidly than the older, more classical parent journal, *Physical Review*). A smoothly functioning communications network among those doing research on the same topic or allied topics—a group to which the late Derek de Solla Price gave the name of Invisible College—has also grown up. Because of the importance of financial support for today's 'big science', new institutions within government have been created (or altered) to serve as organizers, evaluators, and distributors of public funds for research. In the United States these have included not only the specially created National Science Foundation (NSF) and National Institutes of Health (NIH) but granting divisions in the armed forces, the National Aeronautics and Space Administration (NASA), and the Atomic Energy Commission.

Conceptual Revolutions in the Sciences

The four Scientific Revolutions have been described thus far almost wholly in terms of their institutional features. But intellectual changes also occurred in science more or less simultaneously with these four revolutions.[2] *The* Scientific Revolution established experiment and observation as the basis of our knowledge of nature and set forth the goal of mathematics as the key to science and the form of its highest expression. This revolution reached its peak with the appearance of Newton's *Principia*, whose very title expressed the aim of Copernicus, Galileo, Kepler, and others: to produce *Mathematical Principles of Natural Philosophy*. In the succeeding century and a half, the mathematization of nature continued to be most successful in the areas of rational mechanics and astronomy; the great eighteenth-century revolution in chemistry, for example, did not conclude in the mathematical style of Newton. The next area of physics to become Newtonian in this sense appears to have been the undulatory theory of light as developed by Augustin Fresnel in the 1820s. The Newtonian style, the acme of the first Scientific Revolution, evidently was not simply transferable to other branches of science.

In an incisive discussion of this topic, T. S. Kuhn (1977, 220) directs our attention to "an important change in the character of research in many of the physical sciences" that occurred some time between 1800 and 1850, "particularly in the cluster of research fields known as physics." This change, "the mathematization of Baconian physical science," is said by Kuhn to be "one facet of a second scientific revolution." Kuhn stresses the fact that "mathematization" was no "more than a facet" of this second Scientific Revolution: "The first half of the nineteenth century also witnessed a vast increase in the scale of the scientific enterprise, major changes in patterns of scientific organization, and a total reconstruction of scientific education." Kuhn rightly stresses the fact that "these changes affected all the sciences in much the same way." Accordingly, some other factors are needed in order to "explain the characteristics that differentiate the newly mathematized sciences of the nineteenth century from other sciences of the same period."[3]

Kuhn's implied suggestion of an intellectual revolution and an institutional change has been generalized in a very attractive manner by Ian Hacking (1983, 493). Referring to *the* Scientific Revolution and Kuhn's proposed second Scientific Revolution as 'big revolutions', Hacking has advanced an "initial rule of thumb," which is that every big revolution must have a concomitant "new kind of institution that epitomizes the new directions." In this analysis, the second Scientific Revolution embraces not only the Kuhnian mathematization of the Baconian sciences but the advent of Darwinian natural history as a new biology. Darwinian biology carried with it both institutional and intellectual novelties. It drew heavily on information amassed by nonscientists for nonscientific purposes, namely the records and experiences of plant and animal breeders, and it essentially produced a non-Newtonian science. It was the first major scientific theory of modern times that was causal, but not predictive. Whereas biologists or naturalists had been longing for their Newton, the fact is that when the 'Newton' appeared in the guise of Charles Robert Darwin, his theory lacked an essential feature of the science of the *Principia.* Darwin showed that the way of progress in all of the sciences was not necessarily mathematical in the Newtonian style, that great advances in science could still be made in a nonmathematical Baconian way. I believe, furthermore, that the form of discussion after the publication of the *Origin of Species* in 1859 was an aspect of the large-scale public participation in science that was a planned institutional feature of the British Association for the Advancement of Science.

Were there also intellectual changes in science that accompanied the third and fourth Scientific Revolutions and are characteristic of them? This is a difficult question. The third Scientific Revolution spans three great revolutions in physics (the Maxwellian revolution and the great revolutions of relativity and quantum mechanics), numerous revolutions in chemistry, and revolutions in the life sciences, of which perhaps the most significant is the foundation of the science of genetics. If I had to choose a single intellectual characteristic that would apply to the contributions of Maxwell (although not directly to his revolutionary field theory), Einstein (but not the revolution of relativity), quantum mechanics, and also genetics, that feature would be the introduction of probability. In this sense, then, just as the first Scientific Revolution is dominated by simple Newtonian one–one causality of physical events, the third Scientific Revolution was a time in which there was introduced into many fields of science (and also into the social sciences) a set of theories and explanations that were based on probability rather than simple causality.

It is difficult to think of any such single intellectual feature that marks the fourth Scientific Revolution. But of major significance is the fact that a considerable part (though by no means the whole) of the biological sciences can be construed as almost a branch of applied physics and chemistry. At the same time, in the world of physics, the most revolutionary general intellectual feature would be the abandonment of the vision of a world of simple elementary particles with only electrical interacting forces between them.

Although there is a danger of overworking the idea of a simultaneity of four institutional revolutions and four conceptual revolutions within science, one is tempted to hope that some causal relationships will yet be discerned between changes in the intellectual content and style of science and changes in the institutions and style of *doing* science.

Historians' Views on Other Great Scientific Revolutions

So far as I can discover, the term 'second Scientific Revolution' was introduced into the literature of the history of science by T. S. Kuhn in an article in *Isis* in 1961 on the role of measurement in physics. Kuhn's paper (1977, 178ff.) was presented at an A.C.L.S. symposium on measurement in 1960.[4] Other authors may have referred to a second Scientific Revolution before Kuhn, although in a different sense; yet it was through Kuhn's discussion, so far as I can determine, that this expression formally entered the discourse of history, philosophy, and sociology of science.

Roger Hahn's concept of the second Revolution, presented earlier, differs notably from Kuhn's. For Hahn, whose views appear in his celebrated study of the Paris Academy of Sciences (1971, 275ff.), the second scientific revolution is "the crucial social transformation that ushered science into its more mature state and, like the first revolution in the seventeenth century, cut across national boundaries." In this presentation, Hahn does not discuss the actual development of science during the second Scientific Revolution but focuses on the institutional changes that were features of the revolution: "the eclipse of the generalized learned society and the rise of more specialized institutions" and "the concurrent establishment of professional standards for individual scientific disciplines." This second Scientific Revolution was accompanied by the rise of universities and research institutes, and particularly the cultivation of "professionalized science" in "institutions of higher learning." It was an age when "specialized laboratories" were replacing the "academies that had dominated the scene since the middle of the seventeenth century."

Hahn particularly directs our attention to the enormous increase in the size of the scientific community — a factor of bulk which, by itself, "required institutional differentiation." He finds that the growth of professionalization was a necessary consequence of an "increased technicality of disciplinary problems" within each of the sciences, as well as "experimental requirements peculiar to each subject." Finally, Hahn would link the rise of specialization to a "narrowing of the gap between science and its direct applications," a factor which would tend "to reduce the usefulness of general [as opposed to specialized] science in the face of the specific demands of technology." Hahn sees that a severe problem arose concerning training; in order to function efficiently, a "fully educated engineer or doctor" would need specialized knowledge of the highest degree available, and hence "could not at the same time be expected to have a deep grasp of the old generalized science, natural philosophy."

Another historian who has speculated about other Scientific Revolutions is Hugh Kearney (1964, 151–155). He suggests that the "scientific movements" of ancient China and ancient Greece "may be regarded not unfairly as revolutions" and that since Newton's day "other Scientific Revolutions have occurred." He finds a great revolution — comparable to that of Copernicus, Galileo, and Newton — to have occurred at the end of the nineteenth and beginning of the twentieth century: "The Galileo of this scientific revolution was a Scotsman, Clerk Maxwell, its Padua was the Cavendish laboratory at Cambridge and its Kepler was Albert Einstein. Among the others who were associated with the revolution were Rayleigh, Rutherford, Bohr, Schrödinger and Heisenberg." Of particular interest in the present context is Kearney's statement that "whatever view one takes of the importance of the universities in the first scientific revolution, the pre-eminence of the universities in the second seems unmistakable." He also suggests that the "relationship between government patronage of science and the second scientific revolution may also claim our attention." Finally, as "postscript," he advances the notion "that a Third Scientific Revolution also took place in the nineteenth century, distinct in every feature from what was taking place in the world of Faraday and Clerk Maxwell." He explains this as follows: "The nineteenth century also witnessed an equally radical revolution in its approach to time . . . First the age of the world, then of man and finally the age of the universe came to be regarded in a new historical dimension. This revolution in men's approach to the universe was as significant in its own way as the Mathematical Revolution of the seventeenth century." But this third Scientific Revolution, unlike Kearney's second revolution, did not involve specific institutional innovations. And he does not include the great Darwinian revolution in his presentation, limiting himself to the physical sciences. But he does make the important point that by "the mid-twentieth century" the "achievement of Copernicus, Galileo and Newton" no longer appear to historians "to constitute a unique Scientific Revolution, without parallel in human history."

Yet another statement concerning a second Scientific Revolution occurs in an article by Everett Mendelsohn on "The Context of 19th-Century Science" (intro. to Jones 1966). In this presentation, Mendelsohn stresses changes in the "social structure of nineteenth-century science," focusing on new periodicals, new scientific societies, and the growth of two kinds of organizations: the broadly-based scientific organizations, such as the British Association, and the new specialized organizations devoted to specific subdisciplines within the sciences. In referring to the "changes in the social institution within which science was practiced," he suggested that they may well be called a 'second scientific revolution'. For him this revolution entailed a radical alteration in the type of person who was a scientist. In the seventeenth and eighteenth centuries, Mendelsohn points out, scientists tended to be amateurs. That is, they did not earn their livelihood through the practice of science, but either were men of independent wealth or made a living in quite a different field of activity, such as medicine, commerce, trading, and ship building. In the nineteenth century, scientists began to come from the middle or even the lower-middle class, with the result that "the nineteenth-century scientist had to look for support for his scientific activity in the practice of science itself." One notable feature of this change was that the scientific community became "conscious of the vocational needs of its members," with a consequence that "a good deal of time was spent seeking recognition of and support for scientists."

The historian Stephen Brush (1982) has also advanced the idea of two Scientific Revolutions. He believes that the first Scientific Revolution "occurred in the period 1500–1800 as the result of the work by Copernicus, Galileo, Descartes, Newton and Lavoisier;" the second—during the period 1800–1950—was "brought about by Dalton, Darwin, Einstein, Bohr, Freud and many others." He claims that "our civilization has seen only two complete Scientific Revolutions of this magnitude." Brush's second Scientific Revolution is, I believe, the second longest revolution of any kind that has been suggested to have occurred in historical times; it is just half of the time-span of the longest such revolution, the 300-year event from 1500 to 1800 which was first proposed by Rupert Hall.[5] Brush's comparison of Darwin and Darwinism to "the revolution in physics" of the twentieth century is provoking, as in his insightful observation concerning the similar reasons why Copernicus opted for a geostatic system and Einstein for special relativity. But these topics, along with Brush's concluding remarks on a possible future third Scientific Revolution, would take us far afield. For me, however, too much seems lumped together from the years 1500 to 1800, without differentiation, to constitute a meaningful single Scientific Revolution.

Enrico Bellone has written a volume "Studies on the Second Scientific Revolution," under the general title of *A World on Paper* (Ital. ed. 1976; Engl. trans. 1980). It is difficult to make clear in a few words what Bellone conceives to have been the second Scientific Revolution. He places its origins at a period somewhere between the closing decades of the eighteenth century and the opening decades of the nineteenth. The revolution consists, in part, in "a growing awareness of the need for a radical change in the mechanistic view of the world." He finds that the "premises for the overturn of the scientific view of the world" come from a series of rational investigaions "of natural phenomena" which call into question "the belief in a universe understood as a cosmic clock without history." From "this revolution" there arises a "new view of the world, wherein events no longer repeat in accordance with cyclic models and are no longer governed by unalterable rules." This new world is, by contrast, "subject to an evolutionary process that affects both organic and inorganic forms of matter." The foundations "for the second scientific revolution" come from an "intense effort" to elucidate problems and contradictions "within the mechanistic legacy" which are revealed by this new vision, "together with the reflections on scientific explanation inspired by such efforts."

The revolution begins "with the new theories of thermodynamics, radiation, the electromagnetic field, and statistical mechanics." What Bellone finds in common in all of these theories is that they "raised questions about the structure of matter and the very meaning of physical law," and in this way influenced the Galilean – Newtonian tradition. While this is essentially a revolution within the physical sciences, involving a "complete rethinking on the foundations of mechanics," the history of the nineteenth century shows that the "new physical view of the world" had a profound effect "on other disciplines, such as biology, chemistry, and geology."

Bellone says his "intent" is "to vindicate the revolutionary character of nineteenth-century classical physics," although he insists that this "does not necessarily entail a devaluation of the innovative character normally ascribed to the theory of relativity and to quantum mechanics." He rather considers that the "physical sciences of our century" ought to be seen as "the most problematic products of the revolution that began between the end of the eighteenth century and the first decades of the nineteenth." Bellone concludes "that the second scientific revolution is still going on today."

In a perceptive review of Bellone's book,[6] Stephen Brush begins by setting forth his own views on the definition of "the 'second scientific revolution'" — "the historical events that replaced Newtonian physics with quantum mechanics and relativity as the foundation of physical science." Most scientists and historians place these events within a time period beginning in 1887 and ending in 1927 (but not necessarily referring to them as a 'second scientific revolution' or even necessarily as a continuous 'scientific revolution') — these being, respectively, the dates of the Michelson – Morley experiment and Heisenberg's indeterminancy principle. Brush's presentation contrasts Bellone's interpretation with the more customary analysis. Usually, emphasis is placed on "the collapse of the mechanistic/deterministic worldview and the proliferation of startling experimental results that force the abandonment of classical concepts of space, time, matter, and energy." But, as Brush indicates, Bellone argues that "the second scientific revolution really began much earlier in the nineteenth century." Furthermore, this revolution "did not result from the decline of mechanism or a specific set of experiments, but rather involved the emergence of mathematical theory as a source of scientific problems and objective knowledge."

Whereas Kuhn and Bellone see a second Scientific Revolution (obviously it is not quite the same one) in terms of the relation of mathematics to the physical sciences, without indicating any component of revolutionary institutional change, Hahn has stressed the institutional change as a main feature of a second Scientific Revolution. Mendelsohn has also stressed the institutional or sociological features of the second Scientific Revolution. Kearney's main concern is a change in the science of physics, but he notes that in the nineteenth century there were quite distinct national traditions of science and that the support of science by government differed from country to country. Only Ian Hacking has made the brilliant and daring leap of suggesting a link between a conceptual second Scientific Revolution and an institutional second Scientific Revolution.

Scientific Revolutionaries

of the

Seventeenth Century

III

7

The

Copernican

Revolution

———◆———

Whenever historians write about dramatic changes in the sciences, one of the first images that leaps to mind is the radical shift of center that translated the stationary midpoint of the universe from the earth to the sun. This change, commonly known as the Copernican revolution, is usually presented as a complete alteration of our frame of reference which had repercussions on many levels. It is this cosmological shift that is considered to have been revolutionary; thus Copernicus was a "rebel cosmic architect" who produced a "revolution in the conceptual structure of the universe" (Rosen 1971, pref.). For Thomas Kuhn (1957) the Copernican revolution was a "plural" event (despite its "singular" name), being a "revolution in ideas, a transformation in man's conception of the universe and of his own relation to it." This "epochal turning point in the intellectual development of Western man" is said to require consideration on three distinct levels of meaning, since it was, first of all, a "reform in the fundamental concepts of astronomy"; second, a "radical" alteration "in man's understanding of nature" (culminating "a century and a half later" in such "unanticipated by-products" as "the Newtonian conception of the universe"); and third, a "part of a transition in Western man's sense of values" (pp. vii, 1, 2). According to Kuhn, therefore, the alleged Copernican revolution was not merely a revolution in the sciences but a revolution in man's intellectual development and value system. But others (for example, Crombie 1969, 2: 176–177) assert merely that "the Copernican revolution was no more than to assign the daily motion of the heavenly bodies to the rotation of the earth on its axis and their annual motion to the earth's revolution about the sun."

The Copernican revolution is of particular interest to the critical analyst of the concept of revolutions in science because Copernicus's writings and doctrines did not in their own day create any immediate radical change in the basic system of accepted astronomical theory, and only very slightly affected the practice of working astronomers. Those historians and philosophers who have accepted the notion of a Copernican revolution have not been concerned with principles or details of Copernican planetary theory, with the theory of the moon, or with the day-to-day assignments of practicing astronomers — that is, the actual work of computing planetary and lunar positions and the making of ephemerides, all of which were needed for the casting of horoscopes. Had they been primarily interested in the 'hard' science of astronomy, and centered their research on the possible ways in which Copernican ideas might have affected astronomical work, these historians and philosophers would never have alleged that there had been an astronomical revolution in the sixteenth century, much less a Copernican revolution in the sciences at large. For the sciences, the real impact of Copernican astronomy did not even begin to occur until some half to three-quarters of a century after the publication of Copernicus's treatise (1543), when in the early seventeenth century considerations of the physics of a moving earth posed problems in the science of motion. These problems were not solved until a radical new inertial physics arose that was in no way Copernican but rather was associated with Galileo, Descartes, Kepler, Gassendi, and Newton. During the seventeenth century, furthermore, the Copernican astronomical system became completely outmoded and was replaced by the Keplerian system. In short, the idea that a Copernican revolution in science occurred goes counter to the evidence, as this chapter will show, and is an invention of later historians. (The earliest references I have found to a Copernican revolution — by J.-S. Bailly and J.-E. Montucla — are analyzed in §7.4.) There is an obvious parallel here with the so-called English revolution of the mid-seventeenth century, which—as we have seen—was not generally conceived to have been a revolution until after the French Revolution, a century and a half later.

The Copernican System

All too many presentations of the Copernican system by philosophers and historians (and historians of science) are limited to the opening pages of Copernicus's treatise *De Revolutionibus Orbium Coelestium*. Here Copernicus describes what usually goes under the name of 'the Copernican system', pictured in an oft-reprinted diagram (see Figure 4, left) as a set of concentric circles. Simple as this diagram appears, its interpretation is far from simple. The manuscript shows a set of eight concentric circles, but does not explain fully what they represent. The central circle contains the word 'Sol' for the sun, at rest. Reading inward from the outermost circle, the spaces between circles are numbered from 1 to 7: a circular band for the fixed stars and then a circular band for each of the planets, 2, Saturn; 3, Jupiter; 4, Mars; 5, Earth; 6, Venus; 7, Mercury. Each planetary circular band not only carries the name of a planet but also the planet's sidereal period of revolution. For example, the third outermost band is captioned "3 Iovis xii annorum revolutio" (3. The revolution of Jupiter of 12 years). The band for the earth carries the inscription "5. Telluris cū Luna an. re." (Telluris cum Luna annua revolutio: The annual revolution of the earth, with the moon).

What are these circles and bands? To the eye of an unschooled reader, they may seem to be circular orbits, but the Copernican scholar Edward Rosen (1971, 11–21) has alerted us to the fact that these are not planetary orbits. They are some kind of physicalist planetary spheres. Copernicus is harking back to a concept of spheres in which planets are embedded — a concept going back to the ancient doctrine of Eudoxus, Aristotle, and Callippus, who held that the planets are carried around (the earth) in huge rotating spheres. Hence, the title of Copernicus's book, *De Revolutionibus Orbium Coelestium*, should be rendered as *On the Revolutions of the Celestial Spheres* in the sense of the spheres introduced into cosmology by Eudoxus and popularized by Aristotle. But we note that Copernicus has transformed the old Greek idea of earth-centered spheres into new sun-centered spheres. The title of the book is hardly revolutionary, but rather suggests a kinship with ancient ideas about the

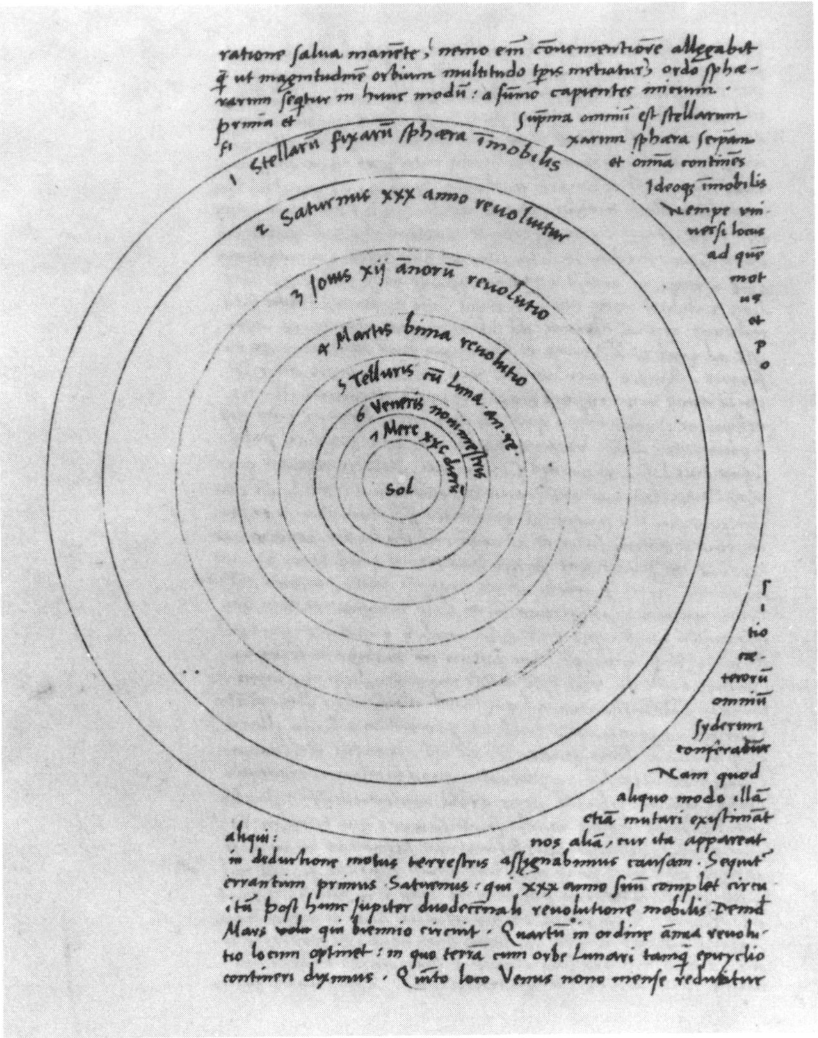

Figure 4.   Two versions of the simplified Copernican system. The diagram on the left is from Copernicus's manuscript, now preserved in the Jagiellonian Library, Cracow; the one on the right is from the editio princeps of *De Revolutionibus* (1543).

NICOLAI COPERNICI

net, in quo terram cum orbe lunari tanquam epicyclo contineri
diximus . Quinto loco Venus nono menſe reducitur. Sextum
deniq̃ locum Mercurius tenet, octuaginta dierum ſpacio circũ
currens. In medio uero omnium reſidet Sol. Quis enim in hoc

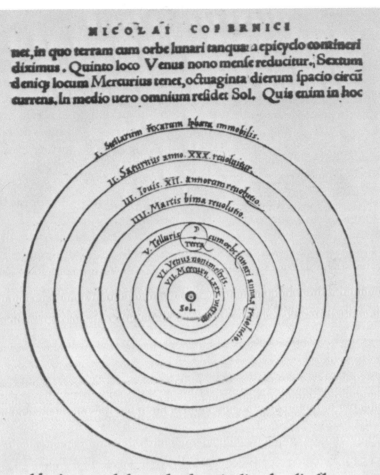

pulcherimo templo lampadem hanc in alio uel meliori loco po
neret, quàm unde totum ſimul poſsit illuminare: Siquidem non
inepte quidam lucernam mundi, alij mentem, alij rectorem uo-
cant. Trimegiſtus uiſibilem Deum, Sophoclis Electra intuentẽ
omnia. Ita profecto tanquam in ſolio regali Sol reſidens circum
agentem gubernat Aſtrorum familiam. Tellus quoq̃ minime
fraudatur lunari miniſterio, ſed ut Ariſtoteles de animalibus
ait, maximã Luna cũ terra cognationẽ habet. Concipit interea à
Sole terra, & impregnatur annuo partu. Inuenimus igitur ſub
hac

universe.[1] The use by Copernicus of the doctrine of the spheres suggests additionally that Copernicus may have conceived his work to be an improvement on ancient astronomy rather than a revolutionary replacement. This is further borne out by the way Copernicus's order and mode of presentation closely follow the plan of Ptolemy's *Almagest* (see below).

Now there has been some rather acrimonious debate in recent years concerning the actual nature of the Copernican celestial spheres. Noel Swerdlow (1976, 127–129) has marshaled some rather convincing arguments that Copernicus may have had in mind a series of contiguous spheres. Swerdlow points out that in his manuscript Copernicus has seven numbered captions and eight circles, so that it would appear that the captions do refer to the seven spaces between the circles. He concludes that these spaces would correspond to "the spheres themselves, each being of a certain thickness (not drawn to scale) and everywhere contiguous to the sphere above and below it." Our problem is made complex by a woodcut appearing in the printed version (Nuremberg, 1543), but not checked and approved by Copernicus (see Figure 4, right). Here there are nine circles, plus an additional small circle for the moon's orbit around the earth. The woodcutter may simply have placed the captions on the wrong sides of the circles, but there are then too many circles, because of the two uncaptioned circles on either side of the circle marked for the earth and its moon. Without getting too deep into the controversy concerning the actual nature of the spheres, their degree of solidity and contiguity, we may nevertheless consider the drawing in Copernicus's manuscript as more authentic than the one made by a far-off woodcutter, and conclude that these are indeed spheres and not free circular orbits in an empty space as in more modern conceptions.

The outermost sphere is "1. Stellarum fixarum sphaera immobilis" (The immobile sphere of the fixed stars); here again is an ancient concept, a celestial sphere of stars. But once more Copernicus has had to introduce a transformation, since the traditional sphere of the fixed stars had to have a daily rotation in order to account for day and night, whereas in the Copernican scheme this sphere is immobile. In the Copernican system, the phenomena of day and night result from the diurnal rotation of the earth on its axis.[2] These stars are 'fixed', because they do not wander with respect to one another in their sphere — as opposed to the planets (or 'wandering' stars), which move both with respect to one another and with respect to the fixed stars.

Copernicus assumed that the fixed stars have to be very far away since they exhibit to the naked eye no observable annual parallax. They could not really be infinitely far away since the sun is supposed to be at their center—all right for a sphere but impossible for an infinite stellar region, which can have no geometric center. Copernicus writes: "Stellarum fixarum sphaera, seipsam et omnia continens, ideoque immobilis, nempe universi locus" (The sphere of the fixed stars, containing itself and everything else, and therefore immovable, since it is the place of the universe). But as Clark (1959, 125) has pointed out, this contradicts a statement made a few pages earlier: "Mobilitas . . . sphaerae est in circulum volvi, ipso actu formam suam exprimentis" (Rotation is natural to a sphere and by that very act is its shape expressed).

Copernicus's diagram of the spheres has been mistakenly interpreted as a version of the Copernican system of the universe (for example, by Wolf 1935, 16), with the circles actually labeled "II. Orbit of Saturn,""III. Orbit of Jupiter," and so on. But of course Copernicus was fully aware that no set of simple circular motions could give an accurate representation of the heavenly world. Hence he was led to construct a complex system, first set forth in the preliminary tract called *Commentariolus* (written in 1514 but not printed until the nineteenth century) and then fully developed in *De Revolutionibus.* Anyone conversant with astronomy would be aware that the diagram in book 1 of *De Revolutionibus* was at best schematic, a greatly oversimplified model of the system. In order to account for a variety of phenomena, Copernicus introduced not only a certain number of epicycles (which serve a very different function than those in the Ptolemaic system) but even epicycles on epicycles (or secondary epicycles, that is, epicyclets). We shall see below that the claim for a great simplicity of the Copernican system, as opposed to a great complexity of the Ptolemaic system, must therefore— insofar as the number of circles is concerned—be taken *cum grano salis,* in fact, with the whole saltcellar. Even Copernicus himself, in the *Commentariolus,* admitted that "thirty-four circles" are needed in order to "represent the entire structure of the heavens and the entire choric dance of the planets" (Swerdlow 1973, 510).

In considering the possible revolutionary impact of *De Revolutionibus,* we must take into account the difference between the opening book 1 and the remaining five books. This difference has been summarized very clearly by E. J. Dijksterhuis (1961, 289), who reminds us that "*De Revolutionibus* consists of two parts, which differ widely in aim, character, and importance."

The first part is formed exclusively by the first of the six books into which the whole work is divided. It . . . gives a very lucid and greatly simplified exposition of the new world-system.

The second part, comprising Books II – VI, . . . in a rigorously scientific form . . . gives the highly complicated details of the system, and thus constitutes a text of the same grade of difficulty as the Almagest.

Book 1 is where the arguments favoring the motion of the earth and the fixity of the sun are found.

Copernicus's Differences with Ptolemy

In both *De Revolutionibus* and the *Commentariolus* Copernicus attacks the Ptolemaic astronomy not because in it the sun moves rather than the earth, but because Ptolemy has not strictly adhered to the precept that all celestial motions must be explained only by uniform circular motions or combinations of such circular motions. Ptolemy had recognized that an accurate representation of planetary motion necessitated the abandoning of uniform circular motion, and he boldly introduced what was later called an "equant," from which nonuniform motion along an arc would appear uniform. From the point of view of accuracy, this was a great step forward (see Figure 5), indeed, the best representation of planetary motion before Kepler. But Copernicus considered the use of an equant to be a violation of fundamental principles and devoted his original astronomical research to devising a system of sun, planets, moon, and stars in which the planets and the moon glide with uniform motion along a circle or with some combination of such motions.[3]

Copernicus set two goals for his astronomy. He wanted agreement with the motions known to be produced by Ptolemy's models (and not really with observations); and he insisted on the physical principle that all celestial motions must be circular and uniform. Copernicus mentioned with approval in both the *Commentariolus* and *De Revolutionibus* the ancient doctrine of Callippus and Eudoxus, in which combinations of circular motions (or rotations of spheres) had been used to account for the phenomena, but he recognized that this particular system had deficiencies. Far better, so far as numerical results are concerned, Copernicus wrote in the *Commentariolus,* was the planetary theory of Ptol-

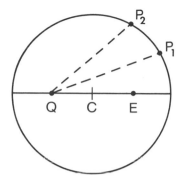

Figure 5. In Ptolemaic astronomy, a 'planet' can move with a nonuniform speed along a circle (center C) while its motion is uniform with respect to a point (Q) called the 'equant'. This point is chosen along a diameter of the circle containing E and Q in such a way that EC = CQ. E is the earth. As a point moves along the circle (say from P_1 to P_2) the angle (or angle P_1QP_2) increases uniformly. The term 'equant' was not used by Ptolemy, and Copernicus evidently considered it to be of 'recent' origin, as something "quem recentiores appellant aequantem" (*De Rev.*, 1543, 164v = bk 5, ch. 25). By 'recentiores' Copernicus means medieval and so not ancient.

emy and "most other" astronomers, which made use of epicycles (see Figure 6); but as Copernicus lamented (in the introduction to the *Commentariolus*), the fact that equants are introduced means that "a planet never moves with uniform velocity either in its deferent sphere or with respect to its proper center." As Noel Swerdlow (1973, 434) has remarked, Copernicus, "in his comment on the Ptolemaic model . . . concedes that the representation of planetary motion is accurate for purposes of computation," but he "objects on principle to the violation of uniform circular motion." It has been generally believed that Copernicus's insistence on uniform circular motion is part of a philosophical or metaphysical dogma going back to Plato, but Swerdlow (p. 435) has suggested a physical basis for Copernicus's position (at least in the *Commentariolus*),[4] and he concludes that "speculations about such things [as philosophical or metaphysical principles about the motion proper to the substance of the heavens] do not belong to the domain of mathematical astronomy."

Copernicus apparently believed that one of his major achievements in astronomy was to restore the principle of uniform circular motion. His follower, Reinhold, alleged that Copernicus thought it was a more

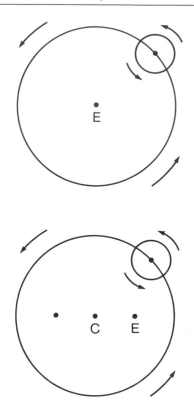

Figure 6. In the Ptolemaic system, the construction of planetary orbits made use of a geometric device in which a planet moves on the circumference of a circle (the epicycle), whose center moves in the circumference of another circle (the deferent). If the earth was at the center (C) of the deferent circle, the system was concentric *(top)* but if the earth (or its center) did not coincide, then the system was eccentric *(bottom)*. An additional complication was to have the motion be nonuniform along the circle, but 'equalized' with respect to an 'equant' (see Figure 5).

significant contribution to have eliminated the equant and to have gone back to pure uniform circular motion than to have dethroned the earth from the center of the universe and to have placed the sun there (Gingerich 1973, 515). Erasmus Reinhold, who composed the *Prutenic Tables* (1551), wrote out (in Latin) on the title page of his personal copy of *De Revolutionibus:* "The axiom of astronomy: celestial motion is circular and uniform or made up of circular and uniform parts" (Gingerich 1973, 515).

If this return to Greek canons of circularity and uniformity was the revolution, there would have been a Copernican revolution only in the old sense of a return to the ideals of the past, a ritual of purification in which later innovations would be eliminated; this would not be a revolution in the new sense of a radical break with the past, which is usually implied by the name 'Copernican revolution'. Copernicus's treatise may be considered as a kind of valedictory for uniform motion; at least, that is how he wished it to be understood. But in that case, it was more a philosophical than an astronomical success, as Neugebauer has suggested, since, as Kepler would prove less than a century later, the planetary motions are not uniform and cannot be represented with sufficient accuracy by simple combinations of uniform circular motions.

Copernicus's Impact on Astronomy

De Revolutionibus was written primarily as an astronomical treatise and not as a philosophical discourse on the motion of the earth. The burden of *De Revolutionibus* was to produce a "mathematical construction" of the universe, as Ptolemy had done and indicated in the title of his great treatise. The mathematical aspect of Copernicus's treatise was stressed in the preface, which states that "mathematics is for mathematicians"; this was further emphasized for readers in the Platonic motto printed in Greek on the title page, "Let no one ignorant of geometry enter here." *De Revolutionibus* is 196 folios (391 pages) long in the first edition, and of these only 7 folios (14 pages) are devoted to generalities, physical principles, his philosophical point of view, and his reasons for having the earth move rather than the sun. This includes Copernicus's arguments that the apparent motions of the planets arise from their own orbital motions about the sun, modified by the shift in the viewing position caused by the earth's annual orbital motion. Almost all of the treatise is devoted to 'hard' mathematical astronomy. Copernicus showed how to deal with planetary and lunar latitudes and longitudes and with a whole range of planetary and lunar phenomena. Copernicus devised sets of circles for the motions of the outer planets Mars, Jupiter, and Saturn and for Venus, an inner planet; Mercury required a special and quite different scheme of its own. The moon was a case apart and unto itself (see below). Because, unlike Ptolemy, Copernicus disdained to use an equant directly, he had to introduce a cumbersome machinery of circles on circles: an epicycle with center on a deferent plus an epicyclet with center on the epicycle. Because Copernicus's models were direct transformations of Ptolemy's so as to fit a heliocentric arrangement, Copernicus located the center of the planetary spheres at an empty point in space — the center of the earth's sphere, a kind of 'mean-sun' — rather than centering the planetary universe on the sun itself. Hence, in actual fact, the doctrine of Copernicus's *De Revolutionibus* is not truly heliocentric (or sun-centered), as it is commonly described, but rather only heliostatic (with the sun immobile). The true heliocentric system of modern astronomy was introduced by Kepler, not by Copernicus, in his treatise on Mars in 1609.

But for astronomers, the main question was not whether the arguments favoring a moving earth and a stationary sun seemed more convincing or less convincing than those favoring a stationary earth and a moving sun (as in the beginning of book 1). Rather, astronomers had to decide whether the mathematical theory of the motions of the planets, the earth (equivalent to the apparent motion of the sun), and the moon were or were not superior to those to be found in Ptolemy's *Almagest* and later tables. This question has two parts: (1) Did Copernicus's methods of calculation yield results more in harmony with observation than Ptolemy's? (The answer, as we will see, is no.) (2) Were Copernicus's methods of calculations easier to perform (that is, simpler) than Ptolemy's? (There is no evidence that in the late 1500s this question was discussed.)

These two fundamental questions could be posed without direct relation to either the philosophical issue (whether uniform circular motion is a requirement) or the cosmological issue (whether it is the earth or the sun that 'really' moves). To us it may seem that the methods of calculation could not be evaluated without reference to the philosophical and cosmological debate about the motion of the earth, but in the sixteenth century these two topics were considered independently. That is, the Copernican mathematical astronomy, independent of its cosmology, was conceived as a hypothetical basis for computation. In fact, *De Revolutionibus* was published with what — to all intents and purposes — was an opening statement by Copernicus himself which gave sanction to this position. By the seventeenth century it became recognized that Copernicus was not the author of the preliminary statement that the Copernican system may be considered only as a computing hypothesis.[5] But as late as the early nineteenth century the scholarly astronomer-historian J.-B. Delambre still thought this declaration concerning hypotheses to have been written by Copernicus himself.

In considering a possible Copernican revolution in astronomy (not in cosmology or the philosophy of motion in circles), we must compare and contrast the Copernican and Ptolemaic schemes for computing the motion of the earth (or apparent motion of the sun), of the planets, and of the moon. Did the methods of Copernicus provide astronomers with more accurate results? Owen Gingerich has used the computer to find out where the planets actually were in the sixteenth century and has compared these results with those given by sixteenth-century Ptolemaic tablemakers. He finds that errors in the longitude of Mars could be as great as 5°. But he points out that "in 1625, the Copernican errors for Mars reached nearly 5°, as Kepler complained in the preface to his Rudolphine Tables" (Gingerich 1975, 86). In short, the Copernican results were numerically no better than the Ptolemaic ones they were supposed to supersede. Yet if Copernicus had used the observations of Walther rather than his own (see Kremer 1981), he could have reduced these errors considerably.

How accurate did Copernicus himself believe his planetary astronomy to be? Copernicus is reported by Rheticus (*Ephemerides Novae . . . MDLI*, p. 6; see Armitage 1957, 159) to have said to him that if his planetary theory agreed with the observed positions of the planets (that is, to within ten minutes of arc), he would be as well pleased with himself as Pythagoras had been when he discovered the famous theorem associated with his name. In fact, however, Copernicus never attained this accuracy. To see how large or small this value is, it may be pointed out that the average naked-eye observer can just distinguish as two a pair of near-by stars four minutes of arc apart. According to Neugebauer (1968, 90), ten minutes was considered adequate agreement of observation and theory prior to Tycho Brahe at the end of the sixteenth century. Before long, ten minutes of arc was considered to be so far off the mark that a difference of approximately this magnitude between a theory and the observed positions of Mars determined by Tycho Brahe could decide that a theory was worthless and should be cast aside. For Kepler it was unthinkable that there could be an error of even eight minutes of arc in Tycho's planetary observations. The positions Tycho assigned to certain fundamental stars were generally less than one minute of arc from the true positions (Berry 1898, 142), and it can be assumed that errors in his planetary positions, save for certain exceptions, may not have exceeded one or two minutes. In the *Astronomia Nova* (1609), Kepler, who was heir to Tycho Brahe's observations, wrote (pt. 2, end of ch. 19: trans. Berry 1898, 184):

Since the divine goodness has given to us in Tycho Brahe a most careful observer, from whose observations the error of 8′ is shewn in this calculation . . . it is right that we should with gratitude recognise and make use of this gift of God . . . For if I could have treated 8′ of longitude as negligible I should have already corrected sufficiently the hypothesis . . . discovered in chapter XVI. But as they could not be neglected, these 8′ alone have led the way towards the complete reformation of astronomy, and have been made the subject-matter of a great part of this work.

Historians who believe in a Copernican revolution in astronomy are apt to cite Erasmus Reinhold's *Tabulae Prutenicae* (*Prutenic* or *Prussian Tables*), named in honor of two 'Prussians': Copernicus[6] and Reinhold's patron, Duke Albrecht of Prussia. Published in 1551, only eight years after *De Revolutionibus,* and admittedly Copernican, the general arrangement followed the model of *De Revolutionibus,* although the tables were given to seconds of arc "where Copernicus had only given minutes" (Dreyer 1906, 345). These tables enjoyed a real success which no doubt "enhanced Copernicus' reputation" (Gingerich 1975a, 366), but his method of making "small changes in the planetary parameters in order to have them conform more accurately with the observations recorded by Copernicus" was "an exercise in futility because of serious errors in the Copernican planetary positions" (p. 366). Dreyer (1906, 345) concluded that because of "the extreme scantiness of recent observations," Reinhold's tables "were not very much better than those they superseded . . . and nothing better could be done until the work of Tycho and Kepler had borne fruit."

One final point (suggested to me by Owen Gingerich) is that in the late sixteenth century virtually no one computed planetary positions by following Copernicus's system of epicyclets, or small circles, with centers located in epicycles, whose centers lay in the deferent or circle of reference. Rather they used the tables given by Copernicus in *De Revolutionibus* or by Reinhold in his *Prutenic Tables.* Furthermore, Copernicus used extreme rather than mean positions, so that there never would be the ambiguity as to whether a given correction was to be added or subtracted, a feature in the older tables (based on mean positions) that was a serious problem and source of error. Accordingly, the tables in Copernicus's *De Revolutionibus* had a real (and positive) influence on computational astronomy, even if the basic features of Copernican heliostatic astronomy did not.[7] But it is the set of 'Copernican' astronomical concepts and the system of the universe that are held to constitute the Copernican revolution and not the tables he computed.

Even if the Copernican system did not produce more accurate results, it has often been alleged that the system "was simpler and more elegant than the Ptolemaic scheme" (Mason 1953, 102), that "astronomical computations were rendered easier by the Copernican scheme, owing to the smaller number of circles involved in the calculations." A biography of Copernicus, subtitled "The Founder of Modern Astronomy," would have us believe that "by making the Earth rotate on an axis and revolve in an orbit, Copernicus reduced by more than half the number of circular motions which Ptolemy had found it necessary to postulate" (Armitage 1957, 159). Many accounts of this subject display what Robert Palter (1970, 114) has called the "80 – 34 syndrome," a dogma going back at least to Arthur Berry's *Short History of Astronomy* of 1898, according to which the Copernican universe required only 34 circles whereas Ptolemy or his followers needed 80. In fact, it is not easy to say exactly how many circles each system required; the number depends on the mode of reckoning and the state of development of the system. We have seen that Copernicus said at the close of his *Commentariolus* that he needed only 34 circles,[8] but the German historian of astronomy Ernst Zinner (1943, 186) said Copernicus really needed 38. Arthur Koestler (1959, 572 – 573) reckoned the number of circles used in *De Revolutionibus* to be 48. Neugebauer (1975, 926) put the number of spheres needed by Ptolemy at 43 — five less than in *De Revolutionibus*. Owen Gingerich found the "comparison between the Copernican and the classical Ptolemaic system" to be "more precise if we limit the count of circles to the longitude mechanisms for the (Sun), Moon, and planets: Copernicus requires 18, Ptolemy 15." Thus he concluded that "the Copernican system is slightly more complicated than the original Ptolemaic system" (Gingerich 1975, 87).[9]

Evidently, there was no Copernican revolution in simplifying the astronomical system. In any event, it is not merely the gross number of circles that determines which of the two astronomical systems is the simpler. However many circles Copernicus may actually have required (or supposed that he required), the fact is that it takes only the most cursory leafing through the pages of *De Revolutionibus* (in any of the three English translations, in either of the two facsimile editions of the holograph manuscript, in any of the original printings or the facsimile reproductions, or any later Latin reprint) to be struck by Copernicus's use of epicycles page after page. Even a neophyte will recognize in the diagrams of *De Revolutionibus* and the *Almagest* a kinship of geometrical methods and constructions that belies any simple claim that Copernicus's book is in any obvious sense a more modern or a simpler work than Ptolemy's.

Copernicus was able to explain (or explain away) certain features of the received system of Ptolemy. For example, in order to explain why Venus is never seen far from the sun, Ptolemy had assumed that the center of Venus's epicycle is always on a line from the earth to the sun (see Figure 7). For Mercury this particular feature is the same, although there are further complications. But Copernicus accounted for this same phenomenon by the simple fact that Venus and Mercury have smaller orbits around the sun than the earth has. For the three superior or outer planets, the Ptolemaic theory contains the condition that the radius of each of the three planetary epicycles is always parallel to a line drawn from the terrestrial observer to the (mean) sun. In the Copernican explanation, these two lines converged, as it were, or — to put it differently — "the perpetually parallel orientations of the epicycle's radius directed to the planet and of the line earth-sun were no longer an unexplained coincidence but rather an indication of a physical phenomenon, the earth's orbital revolution around the sun" (Rosen 1971a, 408).

It is often said that a major feature of Copernicus's system in comparison to Ptolemy's lay in this 'natural' explanation of planetary motions. In the Ptolemaic system, in which the sun moves around the earth and is just another planet or 'wandering star', there is no reason why the motion of Mercury and Venus and of Mars, Jupiter, and Saturn should exhibit features that are related to the sun. This oddity, it is alleged, becomes reasonable or understandable when the reference center of the system is shifted from the earth to the sun. But it must be noted, in this context, that in Copernicus's system there are features of the motions of these same five planets that are related to the earth, even though for Copernicus the earth is a planet just like them (see Neugebauer 1968, 102–103).

Copernicus was proud of his own theory of the moon's motion. The Ptolemaic explanation of the moon's motion not only violated the principle of uniform motion but also achieved a tolerable accuracy in lunar position only at the expense of greatly exaggerating the variation of the distance of the moon, although there was no corresponding variation in the apparent size and parallax of the moon. In the *Commentariolus* Copernicus (Rosen 1971, 72) explicitly criticized Ptolemy's lunar theory

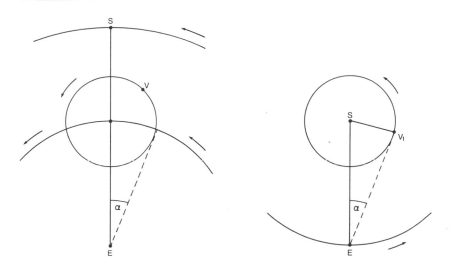

Figure 7. Two explanations of why Venus is never seen far from the sun. In the Ptolemaic system *(left)* the center of Venus's epicycle is always to be found on a line drawn from the earth to the sun. Hence the angular distance from Venus to the sun can never exceed a small angle α. In the Copernican system *(right)* the orbit of Venus around the sun has a smaller radius than the earth's orbit. Hence there is a maximum angular distance (α) of Venus from the sun. In order to determine the distance of Venus from the sun, Copernicus observed the angular separation (elongation) until at some position V, this value became maximum. Under these conditions the line of sight EV from the earth to Venus is tangent to the orbit of Venus. From simple geometry, it follows that the tangent EV_1 is perpendicular to the radius SV_1 of Venus's orbit (presumed to be a circle). Hence, triangle ESV_1 is a right triangle in which $SV_1 = ES \times \sin \alpha$. Since angle α has been found by observation, this equation yields the distance SV_1 from Venus to the sun in terms of ES, the earth-sun distance, or the fundamental astronomical unit. A variation of this method yields similar values for the three superior planets Mars, Jupiter, and Saturn. This simple and straightforward method yielded very accurate results.

because it predicts that "when the moon is in quadrature and at the same time in the lowest part of its epicycle it should appear nearly four times greater . . . than when new and full." In the same way, "the lunar parallax should increase very greatly at the quadratures." But anyone who makes careful observations, Copernicus concluded, "will find that in both respects the quadratures differ very little from new and full moon." Copernicus's own theory of the moon, developed fully in book 4, ch. 3, of *De Revolutionibus,* was long considered perhaps the most original part of that treatise; it makes use of a second epicycle, or epicyclet, a small circle that is centered on the epicycle. Having the moon thus travel on an epicyclet eliminates the nonuniform motion and the obviously incorrect, and unobserved, great changes in the apparent size of the moon. Recent scholarship has shown that this kind of lunar theory had been developed about a century and a half earlier by the astronomer Ibn al-Shātir of Damascus (see the series of articles by E. S. Kennedy, V. Roberts, F. Abbud, and Willy Hartner), but we do not have any evidence as to how Copernicus was influenced by his Muslim predecessor. (See Copernicus 1978, pp. 358, 385; *De rev.,* bk. 3, ch. 4.)

That *De Revolutionibus* has a close affinity with Ptolemy's *Almagest* and does not really constitute a radical new departure may be seen, furthermore, in the fact that in these two works, as in the medieval *Opus Astronomicum* of al-Battānī, there is a parallelism that occurs "chapter by chapter, theorem by theorem, table by table" (Neugebauer, 1957, 206). It was only with Kepler and also with Tycho Brahe that "the spell of tradition was broken"; we may agree with Neugebauer that "never has a more significant title been given to an astronomical work than to Kepler's book on Mars: *Astronomia nova.*"

J. L. E. Dreyer, who was generally laudatory of Copernicus's achievement, could not help but conclude that Copernicus's work had "a serious defect" (1909, 342). It was not just that Copernicus himself had made few actual observations and that the book suffered from a "want of new observations." Rather, the fault stemmed partly from Copernicus's "excessive confidence in the accuracy of Ptolemy's observations" and partly from the feature "that Copernicus in many cases had kept too close to his great predecessor."[10] Kepler was apparently the first astronomer to make this criticism when, in his *Astronomia Nova,* he criticized Copernicus for trying "more to interpret Ptolemy than nature." Almost all commentators make the point that Copernicus and Ptolemy use the same data. Neugebauer (1957, 202–206) has compared "Ptolemy's model of the motion of Mercury with the Copernican theory" to conclude that "cinematically the two models are hardly different except for Copernicus's insistence on using circles for every partial motion where Ptolemy had already reached much greater freedom of approach" (p. 204).

Was There a Copernican Revolution?

What, then, can be concluded about a revolution in relation to Copernicus and his *De Revolutionibus?* So far as practical or computational astronomy was concerned, the innovations that Copernicus introduced were hardly revolutionary and in some cases were even retrograde steps. But Copernicus may have been revolutionary in advocating a philosophy of realism in place of the current instrumentalism (for which see §7.1). We have seen that the alleged claims of greater simplicity represented by a severe decrease in the number of required circles proves on close examination to be spurious. The introduction of uniform motions in circles, a conspicuous feature of the Copernican system, may have been more satisfying from a particular physical or philosophical point of view than the Ptolemaic equant, but it did not improve the case of making astronomical observations and was abandoned by Kepler, who began by returning to a form of the Ptolemaic equant in his successful construction of a new astronomical system based on elliptical orbits.

There was some debate during the second half of the sixteenth century about the Copernican system, insofar as the motions of the earth are concerned (for which see the writings of J. L. E. Dreyer, T. S. Kuhn, Dorothy Stimson, Ernst Zinner). Yet it is significant, I believe, that the only major sixteenth-century instance of an advance in planetary astronomy dependent on Copernicus was the production of Reinhold's *Prutenic Tables.* For these, it was Copernicus who provided the observations, the models, the mode of calculation, and the original derivations and values that Reinhold reworked. But the production of these tables — as we have seen — provided "no occasion for Reinhold to make a confession of scientific faith, and he gave no hint as to whether the system of Copernicus was the physically true one or not" (Dreyer 1906, 346). In short, despite the use of Copernican tables and some of his computational methods, the literature of astronomy from 1543 to 1600 does not show any signs of a revolution. According to the tests presented in chapter three, we must conclude that if there was a Copernican revolution, it occurred in the seventeenth and not in the sixteenth century and was associated with the names of Kepler, Galileo, Descartes, and Newton. The transformations made by these scientists so altered the astronomical system that it was no longer strictly Copernican, although Kepler honored Copernicus by embodying the final statement of his own innovations in a mammoth work which he called *Epitome of Copernican Astronomy.* Many of the writers on science of the seventeenth century did not give much prominence to Copernicus (see §7.2), which is yet another indication that there had not been a Copernican revolution in astronomy.

From the strictly astronomical rather than cosmological (metaphysical) point of view, O. Neugebauer (1968, 103), the outstanding scholar in our times in the field of early astronomy, could conclude:

> Modern historians, making ample use of the advantage of hindsight, stress the revolutionary significance of the heliocentric system and the simplifications it had introduced. In fact, the actual computation of planetary positions follows exactly the ancient patterns and the results are the same. The Copernican solar theory is definitely a step in the wrong direction for the actual computation as well as for the underlying cinematic concepts. The cinematically elegant idea of secondary epicycles for the lunar theory and as substitute for the equant — as we now know, methods familiar to a school of Islamic astronomers — does not contribute to make the planetary phenomena easier to visualize. Had it not been for Tycho Brahe and Kepler, the Copernican system would have contributed to the perpetuation of the Ptolemaic system in a slightly more complicated form but more pleasing to philosophical minds.

According to Neugebauer (1957), Copernicus made three main contributions to astronomy. He made clear the steps from observations to the determining of the values of parameters, an improvement in methodology. He had the insight to see that the planetary distances from the sun can be found by simple calculations without additional and arbitrary assumptions. And his postulate of a single center for all the planetary orbits suggested the solution to the problem of planetary latitudes.

But looking at the situation as of 1600, there would have been no revolution discernible in astronomy, save possibly a then-current one being carried on by Tycho Brahe, who was reforming astronomy completely by his new methods. These consisted of using ingeniously contrived and well-made astronomical instruments (designed on a large scale and with a system of "pinnules," so as to give clear readings of fine divisions of the arc), new tables of atmospheric refraction, a new system of observation, and—perhaps above all—a new practice of continual observation of a planet, night after night, during its whole time of visibility. Tycho's innovations, like Galileo's telescopic observations of the surface of the moon, did not of themselves constitute a revolution in science, but they did provide new and accurate data for the new astronomy of Kepler, that would eventually lead to the Newtonian revolution.

The Copernican doctrine achieved a kind of revolutionary notoriety in 1616, when *De Revolutionibus* was placed on the *Index Librorum Prohibitorum;* Galileo's Copernican *Dialogue Concerning the Two Chief World Systems* was similarly prohibited in 1633. But *De Revolutionibus* was proscribed only, as it was said, "donec corrigatur" (until it is corrected), whereas Galileo's *Dialogue* was entered in the *Index* unconditionally; and so remained until well into the nineteenth century. The nonrevolutionary character and style of *De Revolutionibus* is evident in the list of corrections demanded by the Sacred Congregation of the Index in 1620. Almost all the corrections demanded were changes from statements of reality or certainty to conditions or hypotheses. For instance, the title of bk. 1, ch. 11 was to be corrected by a stroke of the pen from "On the Demonstration of the Triple Motion of the Earth" to "On the Hypothesis of the Triple Motion of the Earth and Its Demonstration."

The great advances in physical science made in the seventeenth century that culminated in Newton's *Principia* (1687) did not stem from the complex system of circles on circles of Copernicus but rather from the new system of Kepler that was sun-centered and that had a single simple curve (an ellipse) for each planetary orbit, and from the physical ideas of Galileo and Descartes, which were decidedly un-Copernican. As we shall see in chapter 8, the Keplerian system contradicts Copernicus on almost every fundamental principle. In much of the seventeenth century and afterward, when scientists discussed the Copernican system, they almost always meant the Keplerian system. As Dreyer (1909, 344) put it simply and boldly, "Copernicus did not produce what is now-a-days meant by 'the Copernican system.'" If there was a revolution in astronomy, that revolution was Keplerian and Newtonian, and not in any simple or valid sense Copernican.[11]

8

Kepler, Gilbert, and Galileo:

A Revolution

in the Physical Sciences?

———◆———

Scholars who have written about a Copernican revolution usually conclude that this revolution did not happen until the innovations of Kepler and Galileo. In fact, the bold new ideas of these two scientists go far beyond any simple Copernican fulfillment. Galileo was an ardent advocate of Copernicanism, which he saw vindicated by his own telescopic discoveries. But his contribution to the science of motion by means of mathematical analysis and experiment was far more revolutionary than the work of his predecessor Copernicus. Kepler, too, was allegedly a Copernican, though in the end he jettisoned all but the two most general Copernican axioms: that the sun stands still and that the earth rotates and revolves. In place of the complex machinery of *De Revolutionibus*, Kepler set forth a new and wholly different astronomical system of the universe that is essentially accepted today. He also proposed a new basis in dynamics for all astronomy.

Kepler's double-pronged reformulation of astronomical science was obviously 'revolutionary' to the highest degree. But we must ask whether this was a self-contained or private revolution or a public one. And if the latter, did it produce in its time, and in and of itself, a revolution in science? Or did it remain a revolution on paper until Newton or some later scientist realized its revolutionary potential? These same questions must be asked in relation to Galileo. We will also look briefly at the work of William Gilbert, an older contemporary who was revolutionary not only in his advocacy of the art of experiment but in his idea that the earth is a huge, spherical magnet. This notion furnished Kepler with a hint that planetary magnetic forces might be the dynamical cause of planetary motions.

Kepler, the Enigmatic Revolutionary

Committed to planetary dynamics (the analysis of the forces that produce the planetary motions) and an astronomy based on physical causes rather than on kinematical dogmas, Johannes Kepler was in part a true modern. And yet he was deeply mired in the traditional past. He was a true believer in astrology (the last major astronomer, in fact, to be in any degree a convinced astrologer), his scientific thought is suffused with what has been called number mysticism, and he argued from first principles of cosmological necessity. He was particularly proud of his early 'discovery' of a direct relationship between the number, size, and arrangement of the planetary orbits and the existence of five (and only five) regular geometric solids. He made one of his greatest discoveries by having the good fortune to eradicate the effects of a major mathematical error by introducing a second error which nullified the effects of the first one. Kepler was one of the greatest astronomers of history; yet we could easily assemble a whole volume of his writings that would show how unscientific his thinking and his science were.

The title of Kepler's great treatise of 1609 boldly proclaims the revolutionary character of his astronomy; he says that he has produced an *Astronomia Nova*, a new astronomy. There are a number of reasons why this astronomy is new, but in the title of his book Kepler stresses only that this new astronomy is "based on causes," that it is an "Astronomia Nova ΑΙΤΙΟΛΟΓΗΤΟΣ."[1] (Kepler prints this word in Greek characters.) Or, the title continues, this book is a *Physica Coelestis* or celestial physics. By this phrase Kepler seems to have intended that he was going one step beyond Aristotle. Aristotle's metaphysics had followed his physics, and Kepler was replacing Aristotle's metaphysics by his own new celestial physics. In his forthcoming book, as Kepler wrote to Johann Georg Brengger on 4 October 1607 (1937, *16:* 54), he was setting forth his new "philosophy, or celestial physics, in place of the celestial theology, or metaphysics, of Aristotle." In a similar statement in the introduction to the *Astronomia Nova,* he explains further that he has explored or investigated "the natural causes of motions " (*3:* 20). To see how radical a program it was thus to explain planetary motions by the celestial forces causing them, we have only to note that in this Kepler had neither predecessors nor contemporaries. Even the great Galileo had no conception of a celestial dynamics, a system of forces producing motions. No wonder that Alexandre Koyré (1961, 166) was moved to write that "the very title of Kepler's work proclaims, rather than foretells, a revolution."

Kepler's astronomy was nothing less than a complete reformulation of this subject in terms of its aims, methods, and principles. Before Kepler, the goal of astronomers had been purely cinematical, that is, to produce a kind of celestial geometry (based on circles on circles) that would yield planetary positions agreeing with observations. Kepler's aim was to find the actual physical causes of the motions, that is, the reasons for the motions, and not merely to invent or to refine geometrical schemes. Since he held the sun to be the seat of the forces in question, it must be the central point of the universe. Hence it is the true sun — and not Copernicus's 'mean sun' — which has to be the common point of intersection of all the orbital planes of the planets.

As to methods, Kepler was concerned with finding by mathematics the actual orbital curve (size, shape, orientation) that the solar force produces, ultimately without any arbitrary or limiting restrictions to circles, uniform motions, and the like. After a formidable labor, he found that each of the planets moves in an ellipse, a simple convex curve. For most planets (Mercury is an exception), the ellipses are not very different in shape from true circles, but the sun is not at or even near the center; the situation is very much like a circular orbit (or quasi-circular elliptical orbit) in which the sun is notably off-center (or eccentric). Kepler also discovered that planetary motion along the ellipse is nonuniform but accords with a law of areas. This law explains at once why each planet moves more quickly at perihelion (or when in that part of the orbit near the sun) and more slowly at aphelion (far from the sun).

Kepler's astronomy is based on a set of new principles of motion: a celestial physics of forces which is directly related to his concept of body.

For him a planet or a planetary satellite (the word 'satellite' was introduced into astronomy by him) or a physical object like a stone is lifeless; it is without any internal or active force of its own. As a result of this property of inertness (which Kepler called 'inertia'), such a body cannot put itself into motion, nor can it keep itself moving. To be moved, such a body requires the action of a mover. It follows in an obvious way from this property of passivity or inertness that a body must come to rest whenever and wherever the motive force ceases to exist or to act. This may not seem a very radical conclusion to a twentieth-century reader, but it went directly against two millennia of scientific and philosophical thinking, dominated by Aristotelian ideas, according to which a body would come to rest only after it had come to its 'natural place'. The doctrine of natural places assumes a hierarchical type of space in which heavy bodies 'naturally' move downward toward a center and light bodies move upward. The spaces in which heavenly bodies move differs from the space in which 'earthly' bodies move or stay, because of the hierarchical differences in the nature and ultimate composition of these types of body. Clearly, for a convinced Copernican like Kepler, committed to the concept of a moving earth, the dogma of natural places and the associated doctrine of a hierarchical space had to be abandoned. In setting forth the new principle that space is isotropic, that space is not hierarchical, that there are no natural places, and that matter is inert, Kepler was actualizing the implications of the Copernican idea that the earth itself and its moon must share a common physics with the planets. Keplerian physical principles of inertia, force, and motion implied the end of the Aristotelian cosmos and readied the scientific stage for Newton.

If all the planets have motions governed in a direct way by the action of the sun (since all move in an ellipse with the sun at a focus and all have an orbital motion regulated by the law of areas reckoned with respect to the sun), then there must be a sun-directed force acting on the planets. This follows from Kepler's concept of the essential inertness of the planets and the consequent need of a force to keep them in their orbital motion. Kepler came to the conclusion that this force must be magnetic. He knew of William Gilbert's proof that the earth is a large spherical magnet. Since the earth is a planet, why shouldn't all the other planets also be magnets, and the sun too? The orientations of the magnetic polarities of the sun and a planet would determine the elliptical rather than a simple circular configuration for the orbit.

Kepler's concept of inertia is not the same as that developed by Galileo (and later refined by Descartes) and by Newton. But his astronomy is more like Newton's than is either Galileo's or Descartes's, because it correlates orbits and orbital motions with the forces responsible for them. What is significant is not that Kepler should have had the wrong force-function (a force varying inversely as the distance rather than inversely as the square of the distance) but rather that he should have conceived of a celestial force in the first place and should have recognized that this must be some function of the distance inversely.

In the preface to the *Rudolphine Tables,* Kepler mentions that a chief feature of his work (new and revolutionary, as we would say) is the "transfer of the whole of astronomy from fictitious circles to natural causes." Kepler said that Copernicus had produced his system *a posteriori* on the basis of observations, but he claimed that the true arrangement of the universe could be demonstrated *a priori* from the idea of creation, from the nature and properties of matter. Indeed, he held that such a demonstration would satisfy even Aristotle, if he were alive today. Kepler thus believed that in making an appeal to final causes, he had gone far beyond Copernicus. Yet, as he wrote to Fabricius on 4 July 1603 (1937, *14:* 412), his astronomical hypothesis had to be tested and confirmed by celestial observations. In this sense, as Eric Aiton wrote to me on 17 March 1979, Kepler's "a priori reasons do not imply necessary conclusions but only probable ones."

There can be no doubt that Kepler had a revolutionary program for astronomy. Because he was an introspective person, he recorded the development of his ideas and methods in some detail. We have carefully written accounts, for example, of the moment of discovery of his third law of planetary motion. In his *Astronomia Nova,* he set forth in painstaking and overwhelming detail the stages of his intellectual revolution and his revolution of commitment; he put down all the calculations that failed so that readers could follow the stages of progress of his thinking and calculation that eventually led him to abandon traditional circular astronomy and begin to explore other types of curves as possible orbits. When the reader wearies of following folio page after folio page of calculations that led to dead ends, Kepler reminds him how much more laborious it had been for him to work up these calculations by hand. As he achieved his results, he put them into print. With the publication of his major works—*Mysterium Cosmographicum* or *The Cosmographic Enigma* (1596), *Astronomia Nova* (1609), *Rudolphine Tables* (1627), *Harmonice Mundi* or *Harmony of the World* (1619), and the *Epitome of Copernican*

Astronomy (1618 – 1621) — an intellectual revolution was fully converted to a revolution on paper, in print for all to read and use.

But was there a revolution in science? Did the Keplerian revolution on paper alter the practice of astronomers and become so basic a part of astronomical thought that there was a subsequent Keplerian revolution in science? I believe the answer must be no. First of all, astronomers of the generations between Kepler and Newton did not fully accept the new Keplerian astronomy. For instance, the dominant astronomical thought shortly became centered on the system of vortices of Descartes rather than the dynamics of celestial forces proposed by Kepler. Kepler's revolution thus fails the first two tests for a revolution in science. In part this was a result of Kepler's lack of success in inventing a new mechanics sufficient for astronomical purposes — as Newton eventually did. Kepler tried to produce a celestial dynamics on a modified Aristotelian basis which just did not (and could not) work.

Additionally, there was notable opposition to the idea that there might be celestial solar forces extending out through hundreds of millions of miles. Galileo, for example, did not acknowledge and use Kepler's three laws of planetary motion in his exposition of Copernican astronomy. In his *Dialogue Concerning the Two Chief World Systems*, Galileo particularly criticized Kepler for suggesting that grasping forces can move out through space in such a manner that the moon could possibly produce tides in our seas. Although the law of elliptical orbits (Kepler's first law) came generally to be accepted by practicing astronomers, there was puzzlement as to the role of the second or 'empty' focus, and there was a very widespread and 'natural' objection, arising from hundreds of years of prejudice, to orbits not constituted of combinations of circles. The law of areas (Kepler's second law) seemed conceptually bewildering rather than helpful to many astronomers. In any event, as Kepler himself noted, this law could not serve as the basis of exact computation for planetary positions but required the use of approximations. In place of the law of areas, astronomers from Kepler's day to Newton's tended to use a direct approximation based on the uniform rotation of a radius vector centered on the empty focus (which thus served as a kind of equant). But even for those who were willing to accept and use these two laws, the laws themselves were oddities, since they were not shown to be causally or deductively associated with fundamental principles that were accepted.

Many astronomers did show an awareness of Kepler's third harmonic law (announced in the *Harmony of the World* in 1619 but not in the *Astronomia Nova* in 1609), in which Kepler was able to exhibit a constancy in the ratio between the square of a planet's sidereal period and the cube of its average distance from the sun. But this third law, however interesting, was of no practical use since it led to no predictions, had no apparent physical cause or reason or justification, and seemed to be no more than another of Kepler's many numerical curiosities. This law was of no help in computing planetary positions or in determining planetary orbits. It could, in principle, have been used to predict what the period of a planet would be at any given distance from the sun, but this would be a theoretical rather than a practical problem. As in the case of the law of elliptical orbits and the law of areas, there was no obvious principle of physics discernible in the action of this law.

In considering Kepler's astronomy, furthermore, we must remember that in his final summing up (in the *Epitome of Copernican Astronomy*), Kepler did not state only the three laws of planetary motion which we know today as Kepler's laws. Rather there was an enormous number of such laws, including relations between sizes and order of planets and orbits and rules for the eccentricities of the planetary orbits, of a sort which we would dismiss today as merely numerological. Here Kepler also included his first discovery: the law that relates the number and size of the planetary orbits to the Platonic five regular solids. Another problem in the acceptance of Keplerian astronomy was that it was a mixture of mechanical and animistic physical principles. It was not a pure dynamics of physical forces and the motions produced by them. For example, the orbital motion or revolution of the planets was accounted for by purely physical solar-planetary (magnetic) force, but the regular and continued rotation of the earth and of the sun was held to be the result of an animistic 'soul principle'. In Kepler, the "animistic principle for the explanation of the motions competes with the mechanistic" (Caspar 1959, 296).

The fact of the matter is that very few theoretical or practical works on astronomy before Newton's *Principia* (1687) even mention all three of Kepler's laws of planetary motion, to say nothing of Keplerian celestial forces producing orbital motions. Hence it seems clear that there was no Keplerian revolution in science before 1687. Looking backward, we conclude that Kepler's program constituted only a revolution on paper — not because Kepler was intellectually not fully successful in developing a system of dynamics that would adequately account for the laws of planetary motion he had discovered, but rather because he did not succeed in converting the greater part of his contemporaries and immediate successors to either his elliptical planetary astronomy or his celestial physics.

William Gilbert, Experimentalist and Spokesman

Like Kepler, William Gilbert must be reckoned among the revolutionary scientists of the early seventeenth century. In the subtitle of his book, *De Magnete* (1600), in which he declared the newness of his science, he said his work was a "Physiologia nova, plurimis & argumentis & experimentis demonstrata." That is, he had produced a "new physiology" or natural philosophy, a new science of nature, one that was "demonstrated by many arguments & experiments." The new natural philosophy was magnetism, and the title informed the reader that Gilbert was concerned with the magnet (*de magnete*) or the lodestone, and with "magnetic bodies" (such as magnetized iron), and also "the great magnet the earth." Throughout the book Gilbert stresses the notion of experimentism, a concept that implies knowledge based on experience, actual practical experience, or proof from experience. In post-classical Latin the words 'experimentum' and 'experientia' came to mean both 'experience' (even the sense of 'what everybody knows') and 'experiment', just as both meanings continue to be associated with the French 'expérience' and the Italian 'esperienza'. Gilbert was thus emphasizing actual practical experience (as of blacksmiths and navigators), the direct study of nature by experiment, and knowledge based on experience rather than intuition or speculation.

Gilbert amassed so great a quantity of new experiential information that, in addition to calling attention to this feature of his book in the subtitle, he liberally sprinkled asterisks in the margins of the book to indicate what he described as "our own discoveries and experiments," some large and some small "according to the importance and subtlety of the matter" (1900, ii). A sample of the newness of Gilbert's experiential approach to the subject can be seen when he discusses his research on the attraction in rubbed amber (ch. 2, bk. 2). He castigates the philosophers of "our own age" who, "making no investigation themselves, unsupported by any practical experience, . . . make no progress" (p. 48):

For it is not only amber and jet (as they suppose) which entice small bodies; but Diamond, Sapphire, Carbuncle, Iris gem, Opal, Amethyst, Vincentina, and Bristolla (an English gem or spar), Beryl, and Crystal do the same. Similar powers of attraction are seen also to be possessed by glass (especially when clear and lucid), as also by false gems made of glass or Crystal, by glass of antimony, and by many kinds of spars from the mines, and by Belemnites. Sulphur also attracts, and mastick, and hard sealing-wax compounded of lac tinctured of various colours. Rather hard resin entices, as does orpiment, but less strongly; with difficulty also and indistinctly under a suitable dry sky, Rock salt, muscovy stone, and rock alum.

The preface to *De Magnete*, addressed "to the candid reader," is one of the most strident declarations of the principles of the Scientific Revolution. It boasts of the superiority of "trustworthy experiments" and "demonstrated arguments" over "probable guesses and opinions of the ordinary professors of philosophy." Here Gilbert talks of "our Philosophy . . . grown . . . from things diligently observed," of "real demonstrations and . . . experiments manifestly apparent to the senses," of "the great array of experiments and discoveries (by which notably every philosophy flourisheth)." And he describes the correct method of philosophizing, whereby there is a progression from "things which are less obscure" to "others that are more remarkable" and finally to "the concealed and most secret things of the globe of the earth" so that the "causes are made known of those things which, either through the ignorance of the ancients or the neglect of moderns, have remained unrecognized and overlooked" (fol. ii).

Gilbert did not produce a mere empirical record; he was also led to devise theories and invent hypotheses. The most profound of Gilbert's own scientific insights was the discovery that the earth itself is a great magnet, having a north and a south magnetic pole. He alleged that he had shown experimentally that a perfectly spherical bipolar lodestone rotates on its axis, and he concluded therefore that the earth must rotate, just as Copernicus had taught. But Gilbert was not a Copernican in that he did not have any great interest in the revolution of the earth, which was not to him a magnetic property.

The significance of Gilbert's clear declaration that a new science is coming into being is not lessened by noting that the text of *De Magnete* does not always carry through his program in full detail. Like Kepler, Gilbert lived in an age of transition, and so we should not be too surprised to find that "beneath his rant and puffery Gilbert is a moderate peripatetic and not above plagiarizing those he criticizes" (Heilbron 1979, 169). Although Heilbron quite properly refuses to accept "Gilbert as a revolutionary hero," and will not take his "Renaissance bombast at face value," he does give Gilbert credit for having published "one of the earliest monographs devoted to a particular branch of terrestrial physics," one of the "first published reports of an extensive series of linked, reconfirmed experiments."

Yet, despite Gilbert's revolutionary fervor, he did not establish a new science. The evidence of the times and the writings on magnetism during the next half-century or more do not show that the subject had changed in a radical way. Nor did his chapter on electrical attraction, new and startling as it was, set scientists to found a new branch of physics; this occurred only in the next century. The work of Gilbert thus fails the first two tests for a revolution, and neither historians nor scientists conceive of a Gilbertian revolution in science. Thus Gilbert, decidedly a revolutionary, produced at most an imperfect revolution on paper. His *De Magnete* contains the seeds of revolution, to be sure, but it certainly did not produce a revolution in science.

But if Gilbert did not create or inaugurate a revolution, his work is both a sign and an expression of a revolution then going on, in which science was shifting from a subject that was largely philosophical and abstract to one based on experience and on that special variety of experience which is a direct interrogation of nature by experiment.

Galileo's Revolutionary Science

The scientist who, more than any other, was first and foremost in advancing the new art of experimental science was Galileo. Galileo's scientific program was certainly as revolutionary as Kepler's, and it had a greater import in that it encompassed methods and results that could potentially affect all the sciences. Unlike Kepler, Galileo wrote works that were widely read (and translated into other languages) and had a tremendous influence on the science and scientific thought of his day. This influence may even have been aggrandized by the notoriety attendant on his trial and condemnation. Galileo made a host of discoveries, but his revolutionary activities made themselves known primarily in four distinct areas: telescopic astronomy, principles and laws of motion, the mode of relating mathematics to experience, and experimental science or the science of experimentation. (One could very well make out a case for a fifth such area, philosophy of science, but a number of the revolutionary features of this aspect of Galileo's thought are subsumed under the rubrics of experimental science and the relation of mathematics to experience.)

There are many witnesses to Galileo's revolutionary work in the science of motion. Additionally, the mid-seventeenth-century writers on physics—Christiaan Huygens, John Wallis, Robert Hooke, Isaac Newton—recognized and used Galilean laws and principles. Many historians and philosophers of science during at least two centuries have hailed Galileo's revolution. Moreover, physicists and other scientists have long considered Galileo to have been a revolutionary hero, even exaggerating his role so as to make him the originator of modern science and of the scientific or experimental method and the discoverer of the first two Newtonian laws of motion. In short, Galileo seems easily to pass all tests for having produced a revolution in science.

Galileo's first public display of his revolutionary science was in 1610, when he published the initial installment of his explorations of the heavens with the telescope. In chapter 1 I referred to Galileo's transformation of individual visual experiences into intellectual conclusions about the heavens. He used principles of analogy and physical optics to show that the moon is like the earth, craggy and corrugated. He discovered that the earth shines and illuminates the moon. He found that Jupiter has a system of four moons, that Venus shows phases. His telescope not only revealed new information about previously known celestial objects—sun, earth, moon, and planets—but also brought within visible range a host of stars (and moons) that had never before been seen by human eyes.

Galileo's discoveries, and those of others, showed men and women for the first time what the heavens are like. The phases of Venus, when correlated with the apparent size of the planet, proved that Venus's orbit encircles the sun, not the earth, and that Ptolemy was wrong. All of these discoveries are consistent with Copernicus's thesis that the earth is only another planet; that is, all show that the earth is more like the planets than different from them. Galileo therefore argued at once that he had shown the truth of the Copernican system (despite the fact that all of his discoveries are also fully compatible with the system of Tycho Brahe, in which the earth stands still at the center while the other planets encircle the sun, which moves in revolution about the earth).[2]

These discoveries revolutionized observational astronomy and radically changed the level of discussion of Copernican astronomy. Before 1610, the Copernican system could appear as an intellectual exercise, a hypothetical computing scheme, something philosophically absurd to the degree that the earth does not appear to us like a planet (which we see as a very brightly shining star). But after the revelations of 1610 and their sequelae, scientists could (and did) argue that the earth and planets are really similar and so should have the same kind of motions. Copernicus was quite right in saying that the earth is only "another planet". The only defense against this new empirically oriented Copernicanism was to refuse to look through the telescope, or to declare that what one saw through a telescope must be an optical artifact or a distortion produced by the telescope lenses and not a true image of what planets are like. The fact that some very intelligent philosophers took this line shows how radical and new it was at that time to base knowledge of nature on experiential evidence.

The second area in which Galileo introduced revolutionary changes was the science of motion. This subject had always been considered central to natural philosophy; therefore Galileo boasted in his *Two New Sciences* (1638), third day, opening paragraph, that he was presenting "a brand new science concerning a very old subject" (Galileo 1974, 147). Galileo may be credited with many new laws and principles regarding motion. He discovered the isochronism of the pendulum — the fact that when a freely-swinging pendulum moves through shorter and shorter arcs, it slows down in such a way that the time for completing a whole oscillation remains (almost) constant. By dramatic experiment he showed that bodies of unequal weights fall in air at very nearly the same speed and not at speeds proportional to weight (as Aristotelians had believed and as most people untrained in physics still tend to believe today). He found that free fall is a case of uniformly accelerated motion, or that the speed increases as the time, and the distance as the square of the time. He introduced the principle of the independence of vector velocities and the method of combining (composing) vector velocities, and he applied this principle to solve the problem of the trajectory of projectiles: the path, he found, is a parabola. It follows, he showed, that the maximum range of artillery occurs when there is an angle of inclination of $45°$ between the gun barrel and the horizon.

In his analysis of the parabolic path of projectiles, Galileo adumbrated an early stage in the formulation of the principle of inertial motion. He set forth what is apparently the very first in a series of successively transformed concepts that eventually led to the Newtonian law of inertia in 1687. It must be kept in mind, however, that Galileo's analysis of motion was primarily on the level of kinematics. That is, although there were aspects or implications of the actions of forces, Galileo's treatment did not attempt either to find the forces producing (or causing) motions or to discover exact mathematical relations between forces and motions.

A third contribution by Galileo was in the area of mathematics. Modern science, especially physical science, is characterized by the expression of its highest principles and laws in mathematics. This feature of science became of major significance in the seventeenth century and reached its first high peak in the publication of Newton's *Mathematical Principles of Natural Philosophy* (the *Principia*). We may see the revolutionary aspect of Galileo's methodology in an example from the third day of his *Two New Sciences,* in the discussion of "Naturally Accelerated Motion." In introducing this topic Galileo explains that it is perfectly legitimate to invent any kind of motion and to work out its properties mathematically, as has been done frequently in the past. He, however, will follow another course — "to seek out and clarify the definition that best agrees with that [accelerated motion] which nature employs." Contemplating how a stone falls "from rest at some height," he concludes that the successive acquisitions of "new increments of speed" are "made by the simplest and most evident rule" (Galileo 1974, 153–154), which is that the addition is made constantly at the same rate. Hence, the increase in speed must be the same in either (a) each successive equal given distance of fall or (b) each successive equal interval of elapsed time. He eliminates the equal-distance rule on logical grounds and then proceeds to develop various mathematical consequences of the equal-time rule, among them the result that in uniformly accelerated motion "the spaces run through in any times whatever are to each other as the doubled ratio of their times" (that is, they are to each other as the squares of those times). Galileo then raises the question of "whether this is the acceleration employed by nature in the motion of her falling bodies."

The answer is to be found by making an experiment, a procedure that "is usual and necessary in those sciences which apply mathematical demonstrations to physical conclusions" (Galileo 1974, 169). The experiment may appear rather easy, but the design of the experiment and the interpretation of the results required an advanced level of understanding of the basic principles of modern science (see below). To appreciate how revolutionary and new Galileo's procedure was, we may compare and contrast it with the activity of medieval mathematician-philosophers who had been exploring the subject of motion actively in the twelfth, thirteenth, and fourteenth centuries (see chapter 5, above). Their mathematical developments were pitched on a plane of abstraction in which motion was a general category embracing any quantifiable change from 'potentiality' to 'actuality' (Aristotle's definition) that could include anything from love and grace to local motion (from one place to another). So it was a bold step for Galileo to develop mathematical laws of motion that would accord with (and exemplify) motions actually occurring in nature. And it was equally unprecedented to verify physical laws, so developed, by experimental test — the fourth area in which Galileo made a major contribution to science.

Galileo's mathematical development of the laws of motion, including uniform motion, uniformly accelerated motion, and the motion of projectiles, exemplifies a general feature of seventeenth-century science which cannot be overstressed, namely, the idea that the basic laws of nature must be mathematical. This stress on mathematics took various forms during the seventeenth century. For instance, on the simplest level, mathematics might merely mean quantification, the use of numerical measures. Or there might be Platonic dogma that the truths of the universe were to be found by mathematics and not by observation and experiment, that priority should be given to mathematical properties rather than agreement with the world of experience. We have seen that throughout a good bit of human history it was felt that circles embody qualities of perfection and so are particularly appropriate for the motion of heavenly bodies. Galileo argued against any such abstract notion of geometric properties, holding that there might be different geometric figures most appropriate to particular situations. Of course the idea that the highest expression of a science is mathematical was hardly new in the seventeenth century; Ptolemy had called his great astronomical masterpiece *The Mathematical Syntaxis* or *Composition*. But the difference between these traditional views of mathematics and that of the new science is that for Galileo there was to be a harmony between the world of experience and the mathematical form of knowledge, to be attained through experiment and critical observation.

When Galileo wrote about mathematics, however, he did not mean the kind of mathematics that normally comes to our minds, namely, the use of algebraic equations, mixed proportions (such as 'the distance is proportional to the square of the time'), fluxions, or the differential and integral calculus. Rather, Galileo wrote about number sequences. An example is the rule that the speeds of a freely falling body at the end of successive equal intervals of time are as the natural numbers (or integers) beginning with unity, or that the distances traversed during successive equal intervals of time are to one another as the odd numbers, or that the distances traversed in any times whatever are as the squares. In *The Assayer* (Galileo 1957, 237–238) Galileo made a famous statement about the mathematics of nature in which he showed that considerations of geometry are as important as rules concerning numbers. "Philosophy [natural philosophy, or science] is written in this grand book, the universe, which stands continually open to our gaze"; but this book "cannot be understood unless one first learns to comprehend the language and read the letters in which it is composed. It is written in the language of mathematics, and its characters are triangles, circles, and other geometric figures without which it is humanly impossible to understand a single word of it." What is important, therefore, about Galileo and mathematics is not anything innovatory on the level of mathematics itself but rather his clear and strident expression of the need for mathematical statements of natural phenomena, for mathematical laws of nature that are founded on experiment and observation.

As to Galileo and the methodology of scientific experiment, a word of caution is needed. A considerable recent scholarship (centering largely on the writings of John Herman Randall, Jr.) has sought for precursorship of Galileo's scientific methodology. Here I find that all too many historians make a fundamental error in not keeping clear the distinction between abstract statements or precepts about method and the actual doing of science. What certainly sounds like a discussion of experiment and of the manner of doing scientific research is to be found in many Italian writers of the sixteenth century, but our confidence in these as true statements concerning experiment is considerably lessened by the fact that none of these individuals ever went on to perform any scientific research. Additionally, in Latin and Romance languages, the same word is used for experiment, for experience, and in general for what everybody knows.

An experiment to answer a particular question is illustrated by Galileo's famous dropping of unequal weights from a tower. The more lurid accounts of a public confrontation of the Aristotelians by Galileo in a public exhibition at the Tower of Pisa are undoubtedly false. However, Galileo's own notebooks do record how he dropped weights "from a tower." Here, the question that Galileo was asking is whether the traditional 'commonsense' view is correct, whether heavy bodies fall freely in air at speeds proportional to their respective weights. Galileo used another kind of experiment to test his hypothesis that freely falling bodies are uniformly accelerated. As we would put it: Does the speed of a freely falling body increase in direct proportion to the elapsed time? Here we see many of the problems that arise in making an experiment in which one asks this kind of question of nature. It is impossible to make a direct test of this ratio. Galileo, therefore, tests another law, one which is a logical consequence of the one he wishes to test, namely, that the distance is proportional to the square of the time. Even this test is beyond Galileo's capabilities, since freely falling bodies move too quickly for him to make any measurements. Accordingly, he "dilutes gravity," as he says, by performing experiments on an inclined plane. Here he finds that the time-square law does indeed pass the test of experiment. Of course, Galileo was a great experimenter and he recognized fully that it is important to make a trial with a number of different angles of inclination of this plane; on all of them the law passes the test. I will not go into details concerning how Galileo worked out the mathematics of the components of gravity along an inclined plane in terms of the angular elevation of the plane. Suffice it to say that in this celebrated example Galileo shows the thought processes and the complexities of the 'science' necessary to design an experiment to test even what looks like a simple law: that distance is proportional to the square of the time.

Thus, in addition to recognizing that abstract mathematical reasoning about motion in general could be applied to real motions as observed in nature, and to understanding the technique for testing mathematical rules by experiment, Galileo also understood how to take account of the differences between ideal and experimental situations. For example, he found by experiment that a heavy body dropped from a tower hits the ground slightly before a light body; he attributed this slight difference to air friction and the relative ability of heavy and light bodies to overcome this resistance. In the ideal situation, in a vacuum or in free space, he concluded, they would fall identically.

Beyond making experiments designed to test a hypothesis, Galileo also made experimental explorations of phenomena. Stillman Drake's careful studies of Galileo's manuscripts have led him to the reconstruction of experiments of this exploratory type, which may very well have been Galileo's key to the idea of inertial motion and which seem to have led Galileo to the laws of uniform and accelerated motion in a way somewhat different from what he described in *Two New Sciences*.

Galileo was certainly not the first scientist to make experiments, but he was one of the first major scientists who made experiments an integral part of his science, along with mathematical analysis. In fact, his combination of experimental technique and mathematical analysis (as in the experiment of the inclined plane) has quite properly earned him a place as a founder of the scientific method of inquiry.

Galileo's extensive experiments and astronomical observations embody two revolutionary characteristics of his scientific philosophy (clarified for me in correspondence with Stillman Drake). One was Galileo's declared belief that "sensate experiences and necessary demonstrations" have "precedence not only over philosophical but also theological dogmas." Very likely, it was not until the nineteenth century that "most scientists adopted positions like his." A second, related aspect of Galileo's approach (which Drake says is "the main innovative aspect of his science and named by Galileo in many places") is "the worthlessness of authority in deciding any scientific question." In *Bodies in Water* Galileo went so far as to remark "that the authority of Archimedes was of no more importance than that of Aristotle; Archimedes was right because his conclusions agreed with experiment." Drake doubts that "Galileo considered anything in his science novel except his discoveries, which spoke for themselves." We may agree with Drake that Galileo simply "saw himself as applying to physics the method Ptolemy had so successfully applied to astronomy; that is, painstaking measurement applied geometrically and arithmetically to prediction capable of test, without concern over causal considerations in the old [Aristotelian] sense or [recourse] to metaphysical principles."

As Galileo's achievements became widely known, it was recognized that he had reformed or renovated the science of motion. Walter Charleton, in his *Physiologia* of 1654 — a work primarily devoted to the old and new atomistic natural philosophy and notable for its presentation of the achievements in the science of motion by Galileo, Gassendi, and Descartes — left no doubt that Galileo's studies were wholly new. It was, he said, the achievement of "the Great Galileo" to "lay the foundations of . . . the Nature of Motion," producing a "subversion" of "Aristotle's Doctrine concerning it" (p. 435). He knew of "no Enquiry at all by any one of the Ancients" into "the PROPORTION, or Rate, in which" the speed is increased in the "Downward motion of bodies," which had been discovered by Galileo. It was "the Great Galileo," furthermore, who had made an "incomparable Inquiry into the most recondite mysteries of Nature" (pp. 435, 455).

In the scientific literature of the seventeenth century, Galileo appears not only as discoverer of the laws of motion and confuter of Aristotle but also as the primary investigator of the heavens with a telescope. Joseph Glanvill, in his essay on "Modern Improvements of Useful Knowledge" (1676, 18–19), devoted a whole page to Galileo's telescopic discoveries:

The next Age after [Tycho Brahe], which is *ours*, hath made excellent *use* of his *Discoveries*, and those of his *Elder*, the famed *Copernicus;* and raised *Astronomy* to the noblest *height* and *Perfection* that ever yet it had among Men. It would take up a Volume to describe, as one ought, all the particular *Discoveries:* But my Design will permit but a short mention: Therefore briefly; I begin with *Galilaeo*, the reputed Author of the famous *Telescope;* but indeed the glory of the *first Invention* of that excellent *Tube*, belongs to *Jacobus Metius* of *Amsterdam:* but 'twas improved by the noble *Galilaeo*, and he first applied it to the *Stars;* by which *incomparable* Advantage, he discovered the *Nature* of the *Galaxy*, the 21 *New Stars* that compose the *Nebulosa* in the Head of *Orion*, the 36 that conspire to *that* other in *Cancer*, the *Ansulae Saturni*, the *Asseclae* of *Jupiter*, of whose *Motions* he composed an *Ephemeris*. By these *Lunulae* 'tis thought that *Jupiters distance* from the *Earth* may be determined, as also the distance of *Meridians*, which would be a thing of much use, since this hath always been measured by *Lunar Eclipses*, that happen but once or twice a year; whereas opportunities of Calculating by the *occultations* of these *new Planets* will be *frequent*, they recurring about 480 times in the year. Besides, (to hasten) *Galilaeo* discovered the strange *Phases of Saturn*, one while *ob-long*, and then *round;* the *increment* and *decrement* of *Venus*, like the *Moon;* the *Spots* in the *Sun*, and its *Revolution* upon its own *Axis;* the *Moons libration*, collected from the various *position* of its *Maculae;* and divers other wonderful and useful Rarities, that were strangers to all *Antiquity*.

This account, which leaves the reader breathless, may be contrasted with Glanvill's brief report on Kepler:

Kepler is next to be mentioned, who first proposed the *Elliptical Hypothesis*, made very *accurate* and *luciferous Observations* about the *Motions* of *Mars*, and writ an *Epitome* of the *Copernican Astronomy*, in the clearest and most perspicuous Method, containing the Discoveries of others, and divers considerable ones of his own; not to mention his *Ephemerides,* and Book about Comets.

Glanvill did not even mention Kepler's law of areas or his harmonic law and obviously did not value Kepler's program of a new astronomy based on the physical causes of the planetary motions.

In the *Principia* Newton said that Galileo had known not only the first two of the three laws of motion but also the first two corollaries, which deal with the composition and resolution of vector velocities. Newton thus hailed Galileo as a primary founder of his own rational dynamics, while relegating Kepler to a minor role: discoverer of the third or harmonic law of planetary motion and an observer of comets. He did not even credit Kepler with the discovery of the law of elliptical orbits and the law of areas. (For a discussion of Newton and Kepler, see Cohen 1975.)

Astronomy in the seventeenth century was certainly Galilean. Galileo's pioneering use of the telescope revolutionized the observational basis of astronomy and earned him a leading place as one of the founders of modern science. His studies of free fall and his analysis of the motion of projectiles and of motion down an inclined plane have remained classics of mathematical analysis coupled with experiment. The laws he discovered for uniform and uniformly accelerated motions are still at the foundation of this science. And the method of experiment, especially experiments in which it is possible to vary only one parameter at a time, is still known by his name. Outshining Kepler (who did not have Galileo's tremendous gift for finding knowledge by experiment) and William Gilbert (who lacked Galileo's sense for mathematics), Galileo stood for the new features of science that were characteristic of the Scientific Revolution. One of the great founders of modern science, Galileo was a heroic figure in the Scientific Revolution.

Yet Galileo's revolution was not complete. In his studies of motion, Galileo largely concentrated his attention on what we today would call the kinematical part. He began to think about the role of force in terrestrial motions, but this was not a topic in which he made most notable advances. Unlike Kepler, he did not concern himself at all with cosmic forces, with celestial or solar forces that might be responsible for planetary phenomena. He ignored Kepler's discoveries of the laws of planetary motion and scoffed at Kepler's suggestion that lunar forces, acting at a distance, might be causally responsible for the tides in the oceans. The Galilean revolution in science demanded another stage of revolution, an understanding of inertia and of the acceleration-produc-ing terrestrial and celestial forces, which were only at their beginnings in Galileo's own thoughts. Another half-century of development would be required before Newton's revolution achieved the potential of what Galileo had wrought, and a great deal more besides. It is certainly no dishonor to one who holds so high a place in the history of science to conclude that Galileo's revolution in science demanded yet another and more profound revolution, that Galileo's great discoveries of the laws and principles of motion — complete as far as they went — were only preliminary stages to the discovery of a universal dynamics that would constitute the climax of the Scientific Revolution.[3]

9

Bacon

and

Descartes

The Scientific Revolution was a period of great concern over method. In part the literature on this subject reflects the self-consciousness of the new age, in which sound principles and procedures were considered to be more important in the advance of knowledge than insight and intellect. Treatise after treatise of the seventeenth century begins with a discussion of method or concludes with a methodological statement. Thus, one of the most famous writings on this subject — Descartes's *Discourse on Method* (1637) — was written (and published) as an introduction to three scientific works: *Geometry, Meteorology,* and *Dioptrics.* One of Newton's most widely read and quoted writings was the methodological "General Scholium," written as a conclusion for the second edition of the *Principia* (1713), in which he discussed the nature of explanations in natural philosophy and the role of hypotheses.

Method was central to the Scientific Revolution because the major novel aspect of the new science or new philosophy was a combination of mathematics and experiment. Whereas older knowledge was legislated by schools, councils, learned men, and the authority of saints, revelation, and Holy Writ, the science of the seventeenth century was held to be based on an empirical foundation or good sense. Anyone who understood the art of making experiments could test the truths of science —a factor which completely differentiated the new science from traditional knowledge, whether the old science, philosophy, or theology. Furthermore, the method could be easily learned and would then permit anyone to make discoveries or find new truths. It was thus one of the greatest democratizing forces in the history of civilization. Discovery of truth no longer was vouchsafed to a chosen few—men and women of special grace or with unusual gifts of mind. In introducing his method, Descartes said that "I have never presumed that my mind is in any respect more perfect than that of the ordinary man" (Descartes 1965, 4). No aspect of seventeenth-century science was as revolutionary as the method and its consequences.

The Scientific Revolution produced two outstanding codifiers of method, Francis Bacon and René Descartes. Bacon occupies an ambiguous position in the history of science because he was not a scientist and even disparaged the great scientific discoveries of his day made by Copernicus, Gilbert, and Galileo. Descartes, on the other hand, is an honored figure in physics and mathematics, as well as being generally considered one of the foremost philosophers of the modern era. In this chapter, we will look into the question of whether there was either a Baconian or a Cartesian revolution in science in the seventeenth century, or whether Bacon and Descartes—like Copernicus, Gilbert, and Kepler—were rather primarily responsible for clarifying, emphasizing, or (to some degree only) inaugurating some basic features of the Scientific Revolution.

Francis Bacon, Herald of the New Science

Bacon's contribution to the Scientific Revolution is usually considered under four heads: as a philosopher of science, he advocated a new method of investigating nature; he focused attention on the classification of the sciences (and of human knowledge generally); he was responsible for the insight that practical applications of the new science will improve the quality of life and man's control of nature; and he envisaged an organized scientific community (stressing the importance of scientific academies and societies). As the spokesman for the inductive method, which — coupled with extensive experimentation and observation — is basic to much of science, Bacon became a spokesman for the new science.

Bacon attacked the sterility of pure deductive logic, which never can increase knowledge. He also attacked the older induction by simple enumeration, applicable only when the class of all things referred to is finite and accessible (see Quinton 1980, 56 – 57), as in the statement that the founding members of the Royal Society are all males and over the age of thirty. Bacon claimed that his new method of induction went beyond this kind of Aristotelian complete or perfect induction ("inductio . . . quae procedit per enumerationem simplicem" — *Nov. Org.* bk. 1, aph. 105) because it led to generalizations about all things, not simply to some property shared by all members of a finite enumeration. Bacon was fully aware that one cannot prove the truth of an induction in the general sense. The word 'all' must always imply the possibility that there may be found an exception to the inductive generalization, since the latter is based — as it must be — on a finite number of instances. Bacon deserves credit for his appreciation that a single negative instance is all that is needed to falsify an induction, whereas every positive confirmation only helps to strengthen our belief. Hence in his *Novum Organum* (bk. 1, aph. 46=1905, 266) he points out that the negative instance is the more powerful ("major est vis instantiae negativae"). It is no small credit to Bacon to have recognized so early the principles expounded in our own time by G. H. von Wright and Karl Popper that laws of nature or theories are not verifiable but are falsifiable.

Bacon conceived that his proposed method of experimentally based induction would provide a new tool or instrument (*novum organum*) for the sciences, to replace the older tool of Aristotelian deductive logic. He envisaged that science would develop by compiling vast tables of factual data, accumulated by experiment and observation, with a disdain for hypotheses. Bacon did appreciate, of course, that the mere accumulation of information would not necessarily yield useful inductive scientific principles; he advocated being selective, but the problem then arises how the selection rule is to be established. In varying degrees such scientists as Boyle, Hooke, and Newton expressed their adherence to the Baconian philosophy. In his *Principia* (2nd ed. 1713; 3rd ed. 1726) Newton even explored the extension of the method of induction from properties or qualities of bodies on which experiments can actually be made to "qualities of all bodies universally" (rule 3, bk. 3). And he declared emphatically, in a manner that Bacon would have fully approved, that "in experimental philosophy, propositions gathered from phenomena by induction should be considered either exactly or very nearly true notwithstanding any contrary hypotheses, until yet other phenomena make such propositions either more exact or liable to exceptions" (3rd ed., rule 4). "This rule," he said, "should be followed so that arguments based on induction may not be nullified by hypotheses."

Bacon's positive influence on seventeenth-century scientific thought can be seen in the rise of the concept of 'crucial experiment' — used so effectively by Isaac Newton in his presentation in 1672 of his experiments and theory on the analysis and composition of sunlight and the nature of color. This expression comes from Hooke's *Micrographia* (1665, 56) and is a transformation by Hooke of Bacon's 'crucial instances' (1905, 343; Bacon, *Nov. Organ.* bk. 2, aph. 36). Bacon was probably also ultimately the source of Newton's attitude against hypotheses, as expressed in the concluding general scholium to the *Principia* (2nd ed.) and epitomized in the slogan, "Hypotheses non fingo."

But if Baconian induction in general has been used by many scientists, Bacon's classification of procedures and his detailed rules have not been followed. What classical defenders say of Bacon's role as reformer and codifier of scientific method (Fowler 1881, ch. 4) turns out to apply more to philosophy than to science. Bacon's *Novum Organum* does not read like a work on modern science, and his discussion of heat (the main application of the method in book 2) seems more like the Aristotelian and Scholastic discussions Bacon was supposed to be attacking than an example of the new science. In particular, as Charles Sanders Peirce pointed out, no 'mechanical' system such as Bacon's tables of exclusion can ever generate significant new scientific knowledge. "Superior as Lord Bacon's conception [of method] is to earlier notions," Peirce (1934, 224) wrote, "a modern reader who is not in awe of his grandiloquence is chiefly struck by the inadequacy of his view of scientific procedure."

Furthermore, one of the conspicuous failures in Bacon's concept of science is in not recognizing the important role of mathematics in scientific theory. While it is all very well to stress accumulation of facts as opposed to hypothesis building, Bacon's procedure downplays the conceptual innovations that have proved to be even more important for the advancement of science than facts and restrained generalizations. The Royal Society did set forth as one of its goals to accumulate masses of factual data on minerals, professions of artisans, and so on. But the real growth modes of science have always been (and continue to be) conceptual and theoretical rather than merely factual. And what are we to make of an alleged spokesman for scientific method who rejects Galileo's discovery of the satellites of Jupiter!

In the history of science there has been one subject that traditionally developed in a true Baconian manner: meteorology. For a long time, at an enormous number of stations all over the world, meteorologists have been collecting data concerning temperature, humidity, rainfall, and wind conditions in a systematic fashion that would have delighted Francis Bacon. But it is a matter of record that this branch of science has not (inductively or in any other way) developed a useful theoretical structure as have physics, chemistry, biology, and geology. We can talk about the weather, but we can neither make very accurate predictions of it nor do anything to change it.

Perhaps Bacon revolutionized the philosophy of science, but he certainly did not produce a Baconian revolution in science. And it is much the same for Bacon's classification of the sciences, really a classification of knowledge (for which see Fowler 1881, ch. 3; Quinton 1980, ch. 6). Bacon's system was modified and then proudly displayed in both tabular and graphic form in the prospectus for, and introduction to, the great mid-eighteenth-century *Encyclopédie* of Diderot and d'Alembert. But, however great a contribution Bacon may have made in this area of philosophy, it did not constitute a revolution in *science*.

What, then, shall we conclude about Bacon and the Scientific Revolution? I believe with Quinton (1980, 83) that Bacon has a double major claim to importance: "as a prophet and a critic." He made a great contribution to the "separation of science from religion and religious metaphysics," the "transformation of the status of natural investigation from that of the forbidden, when seen as sorcery, or the despised, when seen as low drudgery" (Quinton 1980, 83–84). Even more important was Bacon's vision that science would increase human power and give a greater control of the environment. The "true and lawful goal of the sciences," he wrote in the *Novum Organum* (bk. 1, aph. 81=1905, 280), "is none other than this: that human life be endowed with new discoveries and powers." Or "the roads to human power and to human knowledge lie close together and are nearly the same" (bk. 2, aph. 4=1905, 303); "truth therefore and utility are . . . the very same things" (bk. 1, aph. 124=1905, 298). The "empire of man over things depends wholly on the arts and sciences," he wrote (bk. 1, aph. 129=1905, 300), "for we cannot command nature except by obeying her." No wonder that Bacon has been called, in the extreme, "philosopher of industrial science" (Farrington 1949). But we must remember that in these sentiments Bacon was not interested primarily in improving the conditions of life. Rather, "works themselves are of greater value as pledges of truth than as contributing to the comforts of life" (bk. 1, aph. 124=1905, 298).

Bacon was also a major prophet of the organization of scientists into societies and academies, characterized by group research. In a utopian fragment called *New Atlantis* (1627), he portrayed a central scientific research institution in which were assembled laboratories, botanical gardens, a zoo, kitchens and furnaces, and even machine shops. In this work Bacon preached that the production of knowledge is made more efficient by a division of labor in science. Those who are interested in economic history often give Bacon credit for having been the first to expound the general idea of division of labor. There can be little doubt that Bacon exerted a strong influence on the principal founders of the Royal Society, an institution that was to a considerable degree conceived under the sign of Bacon. A testimony to his influence is found in Sprat's *History of the Royal Society* (1667), where Bacon is not only mentioned by name and praised but also figures in the allegorical frontispiece. The Royal Society, we may agree, "may justly be said to constitute the greatest memorial to Francis Bacon" (Farrington 1949, 18).

Descartes's Scientific Revolution

Bacon was not the only thinker in his time to appreciate that a true science would produce advances in medicine and in the technical arts. Descartes made much the same point in his celebrated *Discourse on Method* (1637). In the concluding parts of the *Discourse* he discusses the goal to "procure to the best of our ability the general good of all men" 1965, 50). Sound science, advanced along the principles set forth by Descartes, will be the kind of "knowledge that will be of much utility in this life." Science, properly practiced, will "make ourselves masters and possessors . . . of nature." Among specific goals, he lists the invention of devices "to enable us to enjoy the fruits of agriculture and all the wealth of the earth without labor." He particularly stresses the usefulness of science for medicine, envisaging the eventual elimination of "illnesses of the body as well as the mind" and the eradication of "the debilitation of old age" (Descartes 1956, 39–40). It would thus seem that a natural consequence of the development of an experimentally or experientially based science was to conceive that advances in knowledge would yield new practical inventions and effect improvements in health.

Descartes did not, like Bacon, envisage formal societies or institutions as sponsors and providers of laboratory facilities for groups of scientists engaged in common research pursuits. But he was aware that one man could hardly do all the experiments by himself; toward the end of the *Discourse* he discussed ways in which a researcher might be helped, for example, by contributions to the "costs of the necessary experiments" and protection so that "his leisure is not interrupted by the importunities of anyone" (Descartes 1956, 47). And he even opened the question of public and private support of the sciences. In a letter of 10 May 1632 to Mersenne, he indicated that he had desired a wealthy sponsor to finance a proposed catalogue of "celestial phenomena" (Descartes 1970, 24; 1971, *1:* 249).

Bacon considered his own role to have been a herald of the new science, one whose function was to call men to the study of the new science ("Ego enim buccinator tantum": Bacon 1857, *1:* 579; *De Augmentis* 4.1). "I have only," he wrote to Dr. Playfer, "taken upon me to ring a bell to call other wits together." But Descartes was a true revolutionary, a creator of a brand new science, and he was fully aware of it. At the age of twenty-three, in March 1619, he announced (in a letter to Beckman; see 1971, *10:* 156) an imminent "completely new science"; it could, he proudly asserted, generally solve problems in mathematics. In the following November he dreamed on the occasion of the discovery of "the foundations of an amazing science" (1971, *10:* 179).

A decade later, Descartes was invited with others to hear a lecture refuting the traditional philosophy taught in the schools. There was (according to Descartes's biographer Baillet, trans. in Smith 1952, 40ff.) "almost universal applause." Among the audience only Descartes "deliberately abstained from any outward signs of approval," which brought him to the attention of Cardinal de Bérulle, the founder of the Paris Congregation of the Oratory, and others, including the Papal Nuncio, and Father Mersenne, all of whom urged him to speak out his opinion. In the ensuing dialogue he revealed his own "'universal rule', which he also called his 'natural method'" — drawn "from the treasury of the mathematical sciences." Cardinal de Bérulle was so impressed that he asked Descartes to visit him and to explain his method at greater length. Descartes revealed to him the nature of his method and "the practical advantages that might accrue, should his manner of philosophising be applied to medicine and mechanics" so as to "contribute to the restoration and conservation of health and . . . to some diminution and relief in the labours of mankind." The Cardinal exhorted him "to occupy himself with [the works of nature]," to devote his whole effort to the reformulation of science and philosophy.

This program came to fruition in 1637 with the publication of the three books on science *(Geometry, Dioptrics, Meteorology)* and the *Discourse on Method,* subtitled "for the correct use of reason and for seeking truth in the sciences." This method had been more fully expounded in an early work, completed in about 1628 (at the time he met Cardinal de Bérulle) and entitled *Rules for the Direction of the Mind;* it was published some fifty years after Descartes's death (1701). Descartes's method was a manner of thinking clearly and to some purpose, and was not in any sense a series of practical or do-it-yourself steps in making experiments and drawing conclusions from them. Yet Descartes's method, like Bacon's, was intended to serve in the making of discoveries, through the resolution of a general and complex problem into its simpler elements or component parts. His model, he said, was to be found in his new geometry, where complex curves were studied by just such a resolution into simple elements. The method was broadly conceived; it was to serve not only for science and philosophy but for "any rational inquiry . . . whatsoever" (Williams 1967, 345). In fact, Descartes believed in a firm unity of all knowledge, scientific and philosophical, which he symbolized in the metaphor of a tree whose roots are metaphysics, whose trunk is physics, and whose branches are specific topics: medicine, mechanics, morality. He said that all the sciences taken together "are identical with human wisdom, which remains one and the same, however applied to different subjects" (rule 1; 1971, *10:* 360).

Although much of Descartes's science was based on experiment and observation, his fully developed concepts of science and method are rationalist and nonempiricist. He believed that science should ultimately be based on philosophy. In Descartes's conception, elements of common experience are "composite natures" that must be reduced to "simple natures" ("naturae simplices") which he later called "principles" ("principia"), in the sense of "primary entities" such as "extension, shape, motion" (Rule 12; 1971, *10*). An example given by Descartes is the lodestone or magnet (1911, *1:* 47):

[If] the question is, "what is the nature of the magnet?" people . . . at once prognosticate difficulty and toil in the inquiry, and dismissing from mind every well-known fact, fasten on whatsoever is most difficult, vaguely hoping that by ranging over the fruitless field where multifarious causes lie, they will find something fresh. But he who reflects that there can be nothing to know in the magnet which does not consist of certain simple natures evident in themselves, will have no doubt how to proceed. He will first collect all the observations with which experience can supply him about this stone, and from these he will next try to deduce the character of that inter-mixture of simple natures which is necessary to produce all those effects which he has seen to take place in connection with the magnet. This achieved, he can boldly assert that he has discovered the real nature of the magnet in so far as human intelligence and the given experimental observations can supply him with this knowledge.

In its most extreme aspect, Descartes's philosophy would reduce all actions or phenomena in nature to principles of matter and motion.

Descartes's outstanding reform of science was the establishment of this mechanical philosophy, which sought to explain the properties and actions of bodies in terms of the parts of which they are composed. Descartes was opposed to final causes or teleological explanations, and he attacked the prevailing Aristotelian or Scholastic mode of explanation of phenomena by phrases such as 'substantial form' and 'occult property'. But he differed from others who opposed this kind of thinking in that he set up a real alternative, namely, a reduction to a small set of primary, universal, quantitative properties: "the shape, size, arrangement, and motion of material particles" (1971, *8-1:* 314; *11:* 26). There is no phenomenon in the whole universe ("in natura universa"), he declared, that cannot be explained by such "purely physical causes — i.e., ones completely independent of mind and thought."

By the time of Newton's *Principia* (1687), Descartes's mechanical philosophy had come to dominate European science (see chapter 1, above). It was the Cartesian mechanical philosophy that Boyle had in mind when he spoke of those "two grand and most catholick principles of bodies, matter and motion" (Boyle 1772, *3:* 16). Boyle's *Origins of Forms and Qualities* (1666) was devoted to an exposition of the mechanical philosophy, the action of "corporeal agents . . . by virtue of the motion, size, figure, and contrivance of their own parts." Boyle called these attributes "the mechanical affections of matter, because to them men willingly refer the various operations of mechanical engines" (Boyle 1772, *3:* 13). Huygens and Leibniz were generally adherents of the mechanical philosophy. It was on this ground that they both rejected Newton's concept of universal gravity—a force extending through space and not reducible to matter and motion.

Newton himself was intellectually reared on the mechanical philosophy. He differed from strict narrow Cartesian principles in his belief in atoms (similar to Boyle's) and the consequent acceptance of the void; Descartes did not believe in empty space and even went so far as to make an identity of extension and matter. In a day when the received philosophy demanded that all phenomena be reduced to principles of matter and motion, so that only forces of contact were to be admitted into science, it was a bold step indeed for Newton to conclude that there is a force of universal gravity extending through space. This step meant both (as Westfall suggests, 1971, 377–380) that Newton made a fundamental revision of the received mechanical philosophy and (Cohen 1980, 68–69) that the development of the 'Newtonian style' permitted him to work out the consequences of the concept of universal gravity while he was still hoping for or seeking for a way to reconcile this philosophically unacceptable new principle of force with the Cartesian concepts of matter and motion. There is testimony aplenty in Newton's *Principia* and *Opticks* to his general adherence to the Cartesian mechanical philosophy and his endeavors to seek to reduce phenomena to "the universal qualities of all bodies whatsoever" (*Principia*, 2nd ed. 1713, bk. 3, rule 3).

Descartes's *The World* (or *The Universe*), written between 1629 and 1633 but not published until after his death, contains his ideas on motion and the earliest clear expression of his principles of inertia. The bold statement that uniform rectilinear (or inertial) motion is a state that is, to a degree, dynamically equivalent to a state of rest is not yet the Newtonian principle of inertia, but the two are in part formally equivalent. Descartes, however, founded his principle on a doctrine of conservation—what motion God had created in the beginning could not be destroyed; Newton's principle arose from the nature of mass.

Descartes published his rule of inertial motion in his *Principles*, together with a set of rules of impact. But because he did not understand the vector nature of momentum, his rules were largely incorrect —as he could easily have found out by making a simple series of experiments. It was also in the *Principles* that Descartes fully expounded his system of vortices, huge swirling whirlpools of aethereal or fine matter to produce what we would call gravitational effects, including the forcing of planets into elliptical orbits. And there he developed the concept of relative space which Newton was later to oppose.

Descartes came to believe that "true physics" is a branch of mathematics, that it is only "through mathematics that knowledge of true physics can be obtained" (Descartes 1971, *11*: 315–316; Rée 1974, 31). He claimed in his *Principles of Philosophy* that his theory of science was based on his mathematics: "No other principles are required in physics than are used in Geometry or Abstract Mathematics, nor should any be desired, for all natural phenomena are explained by them." In a letter to Mersenne in December of 1637 (Descartes 1974, *1:* 478; Rée 1974, 32), he explained that the *Dioptrics* and the *Meteorology*—two of the tracts that were presented by Descartes in 1637 as "essays in this Method"—would help to persuade most people that the method was "better than usual," but he himself rather took pride in "having demonstrated this in my *Geometry*."

Descartes was one of the greatest mathematicians that ever lived. John Stuart Mill (1889, 617) hailed Cartesian mathematics as "the greatest single step ever made in the progress of the exact sciences." Descartes would have agreed. His new geometry (analytic geometry), he said in a letter to Mersenne (Descartes 1971, *1:* 479; Rée 1974, 28), was "as superior to ordinary [that is, Euclidean] geometry as Cicero's rhetoric is to the child's ABC."

Many accounts of Descartes's work in mathematics limit his contributions to coordinate geometry and the solution of 'geometric' problems by means of algebra. But perhaps his major innovation was not on any such simple level of technique but rather in his mode of thinking in general analytic terms (Rée 1974, 30). For instance, squaring a quantity traditionally meant erecting a square with a side equal to or represented by that quantity: the 'square' would be the area. Similarly for cubing. But once index notation (x^2 for xx or x-quadratum; x^3 for xxx or x-cubus) was introduced—and Descartes was the pioneer in this new mode of representing powers—then the breakthrough was Descartes's conception of such powers or exponents as abstract entities. This enabled mathematicians to write x^n where n could have values other than 2 or 3, and in fact could even have fractional values. Descartes's freeing of algebra from geometric constraints constituted a revolutionizing transformation of mathematics and produced the "general algebra" that made possible the claim (in 1628) of having achieved "all that was humanly possible" in geometry and arithmetic. Newton's earliest ideas concerning the calculus were formed during a close study of the mathematical writings of Descartes and of certain commentators on Descartes's *Geometry* (see Whiteside ed. of Newton, *Math.*, 1967, *1*). The revolutionary quality of Descartes's mathematics is seen not only by comparing mathematics before and after Descartes, but by noting that seventeenth-century mathematics (and that of the succeeding centuries) bears firmly the Cartesian imprint. Hence Cartesian mathematics passes the historical tests for a revolution.

I shall not discuss other revolutionary aspects of Cartesian science, including the explanations of animal and human physiology and human physiological psychology on a mechanical basis (see Descartes 1972). But it must be said that Descartes's goal of reducing all animal (and human) functions to machine-like actions was perhaps his boldest innovation in the sciences, lauded by physiologists of later centuries as a truly revolutionary step. Descartes accepted Harvey's general proposition about the circulation of the blood, though he disagreed with certain essentials, notably the action of the heart. He also made pioneering contributions to geology, developing a theory of stages whereby the earth was formed according to the long-term action of physico-mechanical principles.

Like Galileo and Kepler, Descartes saw himself as a revolutionary agent producing a new science. But whereas Galileo conceived that he had produced a new science of local motion and a new science of the strength of materials, and whereas Kepler declared that he had produced a new astronomy, Descartes claimed to have revolutionized *all* science and mathematics and even the methodological or philosophical underpinnings of science. His claim is of course not a sufficient ground for believing in a Cartesian revolution, but it is buttressed by the judgments of many seventeenth-century writers. Joseph Glanvill, for example, in his comparison of ancient and modern learning, not only expressed his appreciation of Descartes's formidable achievements in mathematics and in physical sciences but printed Descartes's name in a very large bold-faced type that bespoke his greatness (Glanvill 1676, *Essay* 3, 13ff.). We have seen how men of science adopted Descartes's new mathematics and his revolutionary mechanical philosophy. His radical new principle of inertia and his revolutionary concept of a state of motion became the cornerstone of Newtonian rational mechanics and celestial dynamics. In time his reductionist principles of biology came to dominate much of modern physiology. So there is no doubt that the Cartesian innovations in science pass the first two tests for a revolution in science.

Additionally, historians and philosophers have declared for a revolution associated with Descartes ever since the middle of the eighteenth century, when it became common usage to apply the concept of revolution to the development of science. This is the third test. Cartesian science also passes the fourth and final test, the opinion of active scientists. Testimony to a Cartesian revolution goes back to the eighteenth century, to d'Alembert's discussion (1751) of Descartes's revolution and Turgot's declaration that Descartes "made a revolution" (see Turgot 1973, 94). Condorcet's opinion concerning Descartes was expressed in terms of the "first principle of a revolution in the destinies of the human race." Condillac agreed that there had been a Cartesian revolution, but he expressly denied that Bacon was a revolutionary figure — instigator, or even maker, of a revolution. In the nineteenth century, William Whewell, who wrote of Descartes in relation to a counterrevolution, said that when Bacon "announced a New Method," it was not "merely a correction of special current errors" (1865, *1: 339*). Bacon's method "converted the Insurrection into a Revolution, and established a new philosophical Dynasty."

But whereas Bacon was seen by some analysts to have effected a revolution in philosophy or in the methodology of science, Descartes additionally was recognized to have had a revolutionary influence on the sciences themselves. Strong statements to this effect are to be found in the histories of science by Louis Figuier and Henri de Blainville. In an essay of 1874, "On the Hypothesis That Animals are Automata," Thomas Henry Huxley wrote that Descartes "did for the physiology of motion and sensation that which Harvey had done for the circulation of the blood, and opened up that road to the mechanical theory of these processes, which has been followed by all his successors" (Huxley 1881, 200–201). An even stronger declaration was made in our own century by Sir Charles Sherrington, nobelist in physiology. Discussing Descartes's concept of the animal body as a machine, Sherrington (1946, 187) observed that "machines have so multiplied and developed about us we may miss in part the seventeenth-century force of the word in this connection. By it Descartes said more perhaps than by any other one word he could have said, more that was revolutionary for biology in his time and fraught with change which came to stay." But L. Roth declared that "modern criticism opened with the remark of Freudenthal that the novelty of Cartesianism lay not in its psychology, its theory of knowledge, its ethics, or its metaphysics, but in its physics," and Roth concluded that "the Cartesian 'revolution' lay in the attempt to substitute a physics based on metaphysics for a metaphysics based on physics" (1937, 4).

Paul Schrecker, one of the major analysts of seventeenth-century science and philosophy of our times, wrote that although "the *Principia* of Newton . . . produced a radical change in physics," it is "nevertheless hardly a revolutionary work in the same rank as the *Principles* of Descartes" (1967, 36). Schrecker cited the opinion of the great historian Jules Michelet, who "affirmed that the Revolution of 1789 had begun with the *Discourse on Method*." John Herman Randall, Jr., referred to a Cartesian revolution again and again in his *Making of the Modern Mind* (1926, 235ff.). He had no doubt that this Cartesian revolution was the most significant revolution of the seventeenth century.

Descartes meets all of the major tests for a revolution in science. He had also been revolutionizing philosophy, but that is perhaps not wholly relevant to considerations of his effect on science.[1] Testimony by his contemporaries to the revolutionary quality of his thought is given by the fact that his *Opera Philosophica* were entered in the *Index Librorum Prohibitorum* and still remained there through the last printing of that work in the twentieth century—a century and more after Galileo's *Dialogue* had been removed.

The Cartesian revolution differs from many revolutions in science in a number of features. First of all, it did not last. The Newtonian natural philosophy was a direct frontal attack on Cartesian physics (see chapter 1, above); Newton showed in the conclusion to book 2 of the *Principia* that the system of vortices is inconsistent with Kepler's law of areas. But Descartes's influence was so great that France's leading electrical scientist of the mid-eighteenth century, the Abbé Nollet, still adhered to Cartesian vortical principles, as did his contemporary, Leonhard Euler, the greatest mathematician and mathematical physicist of his age. Descartes's denial of the possibility of a vacuum or empty space soon became a historical curiosity, but his fundamental concept of a state of motion and the attendant law of inertia became central to the subsequent development of physics. In physiology and psychology the direct influence of Descartes continued through the nineteenth century and beyond.

A second point of difference between the Cartesian revolution and other revolutions in the sciences is that there is no great scientific principle or theory that bears his name, nor is there any such principle or law or theory still taught that is associated with him. The nearest to such a particular discovery is what used to be called Descartes's law of refraction, but is now called Snel's law (or, wrongly, Snell's law) after its first discoverer, from whom Descartes has even been alleged to have plagiarized it. But the case is different for mathematics, where the Cartesian revolution was most profound and has been long lasting. We honor one of Descartes's discoveries in algebra with the name Descartes's law of signs. By calling the system of rectangular coordinates Cartesian coordinates, mathematicians continue to celebrate Descartes as the author of a great revolution at the beginning of modern science.[2]

10

The

Newtonian

Revolution

The Newtonian revolution differs from those other revolutions (actual or alleged) in science and in mathematics which we have been considering in that Newton was said in his own lifetime to have created a revolution. He was recognized by his contemporaries for the revolution of the calculus and for a revolution in the science of mechanics created by his *Philosophiae Naturalis Principia Mathematica*. From a historical vantage point, Newton was an extraordinary figure because he made so many fundamental contributions to different fields: pure and applied mathematics; optics and the theory of light and colors; design of scientific instruments; codification of dynamics and formulation of the basic concepts of this subject; invention of the primary concept of physical science (mass); invention of the concept and law of universal gravity and its elaboration into a new system of the universe gravity; invention of the gravitational theory of tides; and formulation of the new methodology of science. He also worked on heat, the chemistry and theory of matter, alchemy, chronology, interpretation of Scripture, and other topics. The range of his intellectual career never ceases to astonish.

The Newtonian revolution in mathematics had two aspects: the invention of the calculus (an honor he shares with Leibniz) and the application of mathematics to physics and astronomy. It was the latter which produced the Newtonian revolution in science (as opposed to a revolution in mathematics). Of course, Newton had great predecessors in the art of developing natural philosophy by mathematical principles: Stevin, Galileo, Kepler, Wallis, Hooke, Huygens. In this sense the Newtonian revolution in science was the culmination of a multiauthored effort, going back to the beginning of the Scientific Revolution, rather than the creation by Newton of something wholly new. Yet the simplest comparison of Newton's *Principia* with Kepler's *Astronomia Nova*, Galileo's *Two New Sciences*, Wallis's *Mechanics*, Hooke's writings on motion, or the treatment of accelerated motions in Huygens's treatise on the pendulum clock shows a difference of several orders of magnitude in depth, scope, and technique. It is because of the size of this quantum jump that Newton's *Principia* is the "epoch" (as Clairaut said in 1747) of a "revolution in physical science."

It is sometimes alleged that Newton created a synthesis, presumably putting together disparate ideas or principles of such scientists as Kepler, Galileo, or Hooke. But Newton's revolutionary science was hardly a melding or assembling of such ideas or principles, since in actual fact Newton's *Principia* declared their falsity. Surely a 'true' science cannot result from a mere amalgamation of false ideas and principles. Among such notions whose falsity is exhibited by Newton in the *Principia* are the following:

Kepler: the three planetary laws are "true" descriptions of the motion of the planets; a solar force exerted on those bodies diminishes directly as the distance and acts only in or near the plane of the ecliptic; the sun must be a huge magnet; because of its "natural inertia," a moving body will come to rest whenever the motive force ceases to act.

Descartes: the planets are carried around by a sea of aether moving in huge vortices; atoms do not (cannot) exist, and there is no vacuum or void space.

Galileo: the acceleration of bodies falling toward the earth is constant at all distances, even as far out as the moon; the moon cannot possibly have any influence on (or be the cause of) the tides in the sea.

Hooke: the centripetal inverse-square force acting on a body (with a component of inertial motion) produces orbital motion with a speed inversely proportional to the distance from the center of force: this speed law is consistent with Kepler's area law.

We may further observe that Newton also denied the existence of 'centrifugal' forces, which were basic to the development of Huygens's physics of motion. In their place Newton introduced a concept of 'centripetal' force, a name he chose because it was similar — though opposite in sense or direction — to Huygens's 'vis centrifuga'.

Comparison and contrast of Newton's *Principles of Philosophy* (the name he often used to refer to his book) and Descartes's *Principles of Philosophy* show the nature of the Newtonian revolution. For the critical reader one of the extraordinary aspects of Descartes's *Principles* is that it is devoid of mathematics, being largely devoted to philosophy and to philosophical principles of physics or natural philosophy. Only two of the four Parts deal with physics proper and the development of the cosmic system of vortices. Here Descartes does set forth the quantitative rules for impact which we have seen to be wrong in each example. Descartes included these rules as a subset of his third law of nature. But when Wallis published the true rules in the *Philosophical Transactions of the Royal Society*,[1] they bore the more restricted and more correct title of "Laws of Motion." Newton began his *Principles of Philosophy* with a set of "definitions" followed by "axioms or laws of motion," of which the first two correspond roughly to Descartes's first two laws of nature. Newton seems to have transformed the Cartesian "regulae quaedam sive leges naturae" into his own "axiomata sive leges motus." Newton's three laws of motion, the axioms to which he reduced the system of rational mechanics, were: (1) the principle of inertia, that a body will persevere in its state of rest or of uniform motion straight forward unless acted on by an external force; (2) the relation of a force to its dynamical effect, that an impulsive (or continuous) external force produces a change (change in a unit time for a continuous force) in the momentum of a body in the direction of action of the force; (3) the equality of action and reaction.

Newton also transformed Descartes's title of *Principia Philosophiae* into *Philosophiae naturalis Principia mathematica*, thus boasting that in mathematicizing the principles he had constructed a natural philosophy rather than a general philosophy. Newton's *Principia* is not only mathematical in the development of the principles and in the proofs and applications of the propositions; it also sets forth a significant new mode of using mathematics in natural philosophy.

Newton's *Principia* is a remarkable book on many levels. It contains original results in pure mathematics (theory of limits and geometry of conic sections), it develops the primary concepts of dynamics (mass, momentum, force), it codifies the principles of dynamics (three laws of motion), and it shows the dynamical significance of Kepler's three laws of planetary motion and of Galileo's experimental conclusion that bodies with unequal weights will fall freely (at the same place on earth) with identical accelerations and speeds. It develops the laws of curved motions, the analysis of pendulums, and the nature of motions constrained to surfaces, and it shows how to deal with the motion of particles in continually varying force fields. Newton also indicates the way to analyze wave motions, and he explores the manner in which bodies move in various resisting mediums. The crown of all appears in the final book 3, in which he discloses the Newtonian system of the universe — regulated by gravity, by the action of a general force, of which one particular manifestation is the familiar terrestrial weight. Here Newton treats at length of the orbits of planets and their satellites, the motions and paths of comets, and the production of tides in the sea.

As an example of the new level of thought in the *Principia,* consider the motion of the moon with its apparent irregularities. For a millennium and a half, astronomers had dealt with the moon's motion by constructing geometric schemes without reference to cause. Now, Newton showed that the chief source of the 'lunar inequalities' was the phenomenon of perturbation, chiefly the result of the gravitational action of the sun as well as of the earth on the moon. With the publication of the *Principia* in 1687, it became possible to deal with this problem by starting from first principles or causes and then studying the effects. As a reviewer of the second edition of the *Principia* observed, this was entirely a new way to deal with the problem.[2]

Perhaps the greatest triumph of all was the explanation that tides are caused by the gravitational pull of the sun and moon on the seas. "The ebb and flow of the sea," Newton declared (in bk. 3, prop. 24), "arise from the actions of the sun and moon." The magnitude of his achievement is shown by his prediction of the oblate shape of the earth on the basis of his analysis of precession and the nonsymmetrical pull of the moon on the earth's supposed equatorial bulge.

Some analysts would see the greatness of the *Principia* expressed in the commitment to an inertial physics; for Newton inertia is a property of mass. Newton is the first writer to make a clear distinction between mass and weight and to recognize, furthermore, that a body's mass has two separate and distinct aspects. Mass is a measure of the body's resistance to being accelerated or undergoing a change in its state of motion or of rest; this is its inertia. (Newton sometimes used the term 'force of inertia' or 'vis inertiae' — but this type of force differs from forces that are 'active' and that can produce accelerations.) But a body's mass is also a measure of the body's response to a given gravitational field. But why should there be a relation between a body's (inertial) resistance to acceleration and its (gravitational) response to a gravitational field? In classical physics there is no reason. Newton had the insight to recognize that this relation must rest on the foundation of experiment, and so he proceeded to prove by experiment this constancy between inertia and gravity. It is only in Einstein's relativity theory that there is a logical necessity for this equivalence of 'inertial' mass and 'gravitational' mass. Einstein greatly admired Newton for having had so deep an insight into this problem and for having recognized that the only Newtonian grounds for this equivalence were experimental.

The nature of the mathematics in Newton's *Principia* is often misunderstood. A superficial turning of the pages gives the impression that the mathematics used by Newton is geometry, particularly Greek geometry. The style seems to be that of Euclid or Apollonius. But a closer examination shows that Newton is developing the subject by the calculus, by stating relations geometrically in ratios and proportions and at once considering the 'limit' as a fundamental quantity vanishes (or is nascent). Hence, although Newton does not develop an algorithm of the calculus (or 'fluxions') which he then applies systematically, he does make extensive use of limiting procedures which are clearly equivalent to using the calculus or which can readily be translated into the symbolism of either the Newtonian or the Leibnizian algorithm. Recognizing this aspect of the *Principia*, the Marquis de l'Hôpital observed (as Newton proudly noted) that the mathematics of the book is almost entirely the calculus. This would be further evident to any careful reader from the development of the theory of limits in section 1 of book 1 and from the explicit theory of fluxions (the Newtonian version of the differential calculus) in section 2 of book 2. Additionally, the *Principia* was notable for other original uses of mathematics such as the extensive use of infinite series.

Newton's Style

The essence of Newton's revolutionary science, as I see it, is to be found in what I have called the 'Newtonian style'. This can be seen most easily in Newton's treatment of Kepler's laws in the *Principia*.[3] Newton begins with a purely mathematical construct or imagined system — not merely a case of nature simplified but a wholly invented system of the sort that does not exist in the real world at all. Here by 'real' world is meant only the external world as revealed by experiment and observation. In this system or construct, a single mass-point moves about a center of force. Newton shows by mathematics (bk. 1, prop. 1) that if in this construct or system a force is constantly directed from the orbiting mass point or particle to the immobile center of force, then Kepler's law of areas (his second law) will hold. He next proves the converse (prop. 3), that if the law of areas holds there must be such a centripetal or centrally directed force. Hence the existence of a centripetal force is proved to be both a necessary and sufficient condition for Kepler's law of areas. Then Newton shows that if the orbit is an ellipse, the central force must vary inversely as the square of the distance. Finally he proves that if under such a condition of force there are several orbiting mass points, which do not interact with each other — or (what comes to the same thing) if the motion of any given mass point is compared with what its motion would be at a somewhat different distance from the center — then Kepler's third or harmonic law will hold. Incidentally, we may observe that Newton has shown here for the first time the dynamical significance of each of Kepler's laws. Newton's procedure thus far constitutes a purely mathematical phase one.

In phase two, Newton compares his mental construct with the real world. At once, of course, he discovers that in the real world (for instance, in our solar system), orbiting bodies do not move about 'mathematical' centers of force but about other real bodies. The moon moves around the earth; the earth and the other planets move around the sun. Accordingly, in order to bring his mental construct or imagined system more into harmony with the real world, Newton modifies the system so that there are now two mass points. One is at the center and attracts the one which is moving in orbit, constantly drawing it away from its otherwise rectilinear inertial path. But according to the principle that to every action there must be an equal and opposite reaction (Newton's third law of motion), it follows that if the central body attracts the orbiting body, then the orbiting body must also attract the central body. Hence the mental construct becomes enlarged to a system of two interacting bodies. Newton proceeds to show that under these circumstances the orbiting body does not any longer move in a simple ellipse around the central body at a focus; rather, he finds that both will move in ellipses around their common center of gravity.

This two-body system constitutes a modified phase one in which Newton once again develops mathematically the properties of his (now revised) mental construct. He next compares the modified system with the external world, a modified phase two. Of course, he finds that this system also does not conform to the real world around us. For instance, in our solar system there is not just a single planet moving around the sun but several. Accordingly, to make his mental construct conform more closely to the system of the external world, Newton moves onto yet another phase one. He introduces two or more mass points orbiting about the central mass point, not just one. It follows, again as a result of the application of Newton's third law, that each of these orbiting mass points both is attracted by the central body and attracts it. In other words, a consequence is that each orbiting mass point is both a body that can be attracted and a center of attractive force. Each of these orbiting bodies will act upon and be acted upon by every other orbiting body. The system contains bodies which act by perturbations on one another, and these perturbations produce a slight departure from Kepler's laws. Newton then proceeds to find the quantitative measure of the deviation from Kepler's laws in our solar system.

In this kind of contrapuntal alternation between mathematical constructs and comparisons with the real world, between a phase one and a phase two, Newton advances from a one-body system not only to a many-body system but also to a system of orbiting bodies which have satellites, such as the moons of the earth, Saturn, and Jupiter. Thus far he has been considering mass points rather than physical bodies, because he has not yet introduced considerations of size and shape, but eventually he shifts the level of discussion from mass points to physical bodies with significant dimensions and figures.

The progression I have described is not merely a twentieth-century after-the-fact analysis of the way Newton presents his subject in the *Principia*. It also corresponds to the documented stages of development of Newton's ideas.[4] In the autumn of 1684 Newton wrote a tract (*De Motu*) in which he presented the results of his study of Kepler's laws and other aspects of the subject. There he shows that a central force is a necessary and sufficient condition for the law of areas, and that an elliptical orbit implies that the force varies as the inverse square of the distance, much as in the later *Principia*. But he has not as yet recognized that his proofs apply only to a mental construct of a one-body system and so he proudly writes: "Scholium: Therefore the major planets revolve in ellipses having a focus in the center of the sun and by radii drawn [from the planets] to the sun describe areas proportional to the times, entirely as Kepler supposed." Before long Newton realized that the planets cannot in fact move in simple Keplerian elliptical orbits. He saw that his results apply only to an artificial one-body system in which the earth is reduced to a mass point and the sun to an immobile center of force.

In December 1684 Newton completed a revised draft of *De Motu* that describes planetary motion in the context of an interactive many-body system. Unlike the earlier draft, the revised one concludes that "the planets neither move exactly in ellipses nor revolve twice in the same orbit." This conclusion led Newton to the following result:

> There are as many orbits to each planet as it has revolutions, as in the motion of the Moon, and each orbit depends on the combined motions of all the planets, not to mention the actions of all these on one another . . . To consider simultaneously the causes of so many motions and to define the motions themselves by exact laws allowing of convenient calculation exceeds, unless I am mistaken, the power of the entire human intellect.

Newton had come to perceive that the planets act gravitationally on one another. The passage cited above expresses this perception in unambiguous language: "eorum omnium actiones in se invicem" (the actions of all of them on one another). A consequence of this mutual gravitational attraction is that all three of Kepler's laws are not strictly true in the world of physics but are true only for a mathematical construct in which masses that do not interact with one another orbit either a mathematical center of force or a stationary attracting body. The distinction Newton draws between the realm of mathematics, in which Kepler's laws are truly laws, and the realm of physics, in which they are only "hypotheses" (or approximations), is one of the revolutionary features of Newtonian celestial dynamics.

In an early draft of what was to become book 3 of the *Principia,* Newton showed how considerations of the third law of motion led to the concept of a mutual force between the sun and each planet, between a planet and its satellites, and between any two planets. The same considerations lead to the revolutionary new idea that any and all bodies in the universe must "attract one another." He proudly presented this conclusion with the explanatory comment that in any pair of terrestrial bodies the magnitude of the attractive force is so small that it is unobservable. "It is possible," he wrote, "to observe these forces only in the huge bodies of the planets." Of all the planets, Jupiter and Saturn are the most massive, and so he sought orbital perturbations in their motions. With the help of John Flamsteed, Newton found that the orbital motion of Saturn is indeed perturbed when the two planets are closest together.

In book 3 of the *Principia,* which is concerned with the system of the world but is somewhat more mathematical than the earlier version, Newton treats the topic of gravitation in essentially the same way. First, in what is called the moon test, he extends the weight force, or terrestrial gravity, to the moon and demonstrates that the force varies inversely with the square of the distance. Then he identifies the same terrestrial force with the force of the sun on the planets and the force of a planet on its satellites. All these forces he now calls gravity. With the aid of the third law of motion he transforms the concept of a solar force on the planets into the concept of a mutual force between the sun and the planets. Similarly, he transforms the concept of a planetary force on the satellites into the concept of a mutual force between planets and their satellites and between satellites. The final transformation is the notion that all bodies interact gravitationally.

My analysis of the stages of Newton's thinking should not be taken as diminishing the extraordinary force of his creative genius; rather, it should make that genius plausible. The analysis shows Newton's fecund way of thinking about physics, in which mathematics is applied to the external world as it is revealed by experiment and critical observation. Because he did not assume that the construct is an exact representation of the physical universe, he was free to explore the properties and effects of a mathematical attractive force even though he found the concept of a grasping force "acting at a distance" to be abhorrent and not admissible in the realm of good physics. Next he compared the consequences of his mathematical construct with the observed principles and laws of the external world such as Kepler's law of areas and law of elliptical orbits. Where the mathematical construct fell short Newton modified it. This way of thinking, which I call the Newtonian style, is captured by the title of Newton's great work: *Mathematical Principles of Natural Philosophy.*

The law of universal gravitation explains why the planets follow Kepler's laws approximately and why they depart from the laws in the way they do. It was the law of universal gravitation which demonstrated why (in the absence of friction) all bodies fall at the same rate at any given place on the earth and why the rate varies with elevation and latitude. The law of gravitation also explains the regular and irregular motions of the moon, provides a physical basis for understanding and predicting tidal phenomena, and shows how the earth's rate of precession, which had long been observed but not explained, is the effect of the moon's pulling on the earth's equatorial bulge. Since the mathematical force of attraction works well in explaining and predicting the observed phenomena of the world, Newton decided that the force must "truly exist" even though the received philosophy to which he adhered did not and could not allow such a force to be part of a system of nature. And so he called for an inquiry into how the effects of universal gravity might arise.

Although Newton at times thought universal gravity might be caused by the impulses of a stream of aether particles bombarding an object or by variations in an all-pervading aether, he did not advance either of these notions in the *Principia* because, as he ultimately said, he would "not feign hypotheses" as physical explanations. The Newtonian style had led him to a mathematical concept of universal force, and that style led him to apply his mathematical result to the physical world even though it was not the kind of force in which he could believe.

Some of Newton's contemporaries were so troubled by the idea of an attractive force acting at a distance that they could not begin to explore its properties, and they found it difficult to accept the Newtonian physics. They could not go along with Newton when he said he had not been able to explain how gravity works but that "it is enough that gravity really exists and suffices to explain the phenomena of the heavens and the tides." Those who accepted the Newtonian style fleshed out the law of universal gravity, showed how it explains many other physical phenomena, and demanded that an explanation be sought of how such a force could be transmitted over vast distances through apparently empty space. The Newtonian style enabled Newton to study universal gravity without premature inhibitions that would have blocked his great discovery. The eighteenth-century biologist Georges Louis Leclerc de Buffon once wrote that a man's style cannot be distinguished from the man himself. In the case of Newton his greatest discovery cannot be separated from his style.

Acceptance of a Newtonian Revolution

There are numerous testimonials to the Newtonian revolution in science. The eighteenth-century historian of science Jean-Sylvain Bailly wrote that "Newton overturned or changed all ideas": his "philosophy brought about a revolution." Bailly was not content merely to state generalities concerning the Newtonian revolution in science. As he saw it, the key that in Newton's hands unlocked the celestial mysteries was mathematics: geometry. As Bailly put it: "What is supposed to make things move is what really makes things move; the demonstration was complete. Newton alone, with his mathematics [géométrie], divined the secret of nature."

With rare insight, Bailly saw that "the advantage of mathematical solutions is that they are general." The argument that if the planets move according to Kepler's laws, they must be "impelled by a force residing in the sun" depends only on mathematical or geometrical considerations and general principles of motion. No special physical properties of the sun appear in Newton's argument, which differs from Kepler's in that the latter had invoked such special qualities of the sun as its magnetic force and the orientation of its poles. Accordingly, the identical mathematical argument shows that the satellites of Jupiter and Saturn, subject to the same laws of Kepler, must be equally "impelled by forces residing in these two planets." In other words, Jupiter and Saturn are to their satellite systems what the sun is to the planetary system, the only difference being one of extent and power. And the same is true of the earth and our moon (Bailly 1785, vol. 2, bk. 12, sec. 9, pp. 486f.).

Bailly himself was willing to accept the concept and principle of a universal gravitating force, since so many phenomena were explained by its use: so many of the observed data and experiential laws could be derived by mathematics from the properties of universal gravity (sec. 41, pp. 555f.). He was aware, however, that at first many scientists (notably in France) made a distinction between the Newtonian system as mathematical and as a true natural philosophy. Thus with respect to Maupertuis, who (according to Bailly) "appears to us to have been . . . the first of our mathematicians to have used the principle of attraction," Bailly (vol. 3 ("discours premier"): 7) had to point out that "at first he considered it only in relation to its calculable effects; he accepted gravitation as a mathematician, but not as a physicist." That is, Maupertuis went along with the Newtonian mathematical system or construct (our phases one and two) but would not grant that in the system of the world (phase three) Newton was necessarily dealing with reality.

In fact, in a paper "On the Laws of Attraction" (1732), Maupertuis had been very explicit on this point. "I do not at all consider," he wrote, "whether Attraction accords with or is contrary to sound Philosophy."

Rather, "Here I deal with attraction only as a mathematician [géomètre]." Maupertuis was concerned with attraction only as "a quality, whatever it may be, of which the phenomena are calculable, considering it to be uniformly distributed through all the parts of matter, acting in proportion to the mass." Maupertuis, in other words, accepts the Newtonian style and is willing, as "géomètre," to follow out the mathematical consequences of a law of gravitational attraction. Since the results accord with the phenomena observed in nature, Maupertuis then asks himself as natural philosopher whether there is such a force as a physical entity, or whether there may be some other reason why bodies act as if there were such a force. If such a force does exist, it must have a cause; and we may observe that his thought is still so embedded in the mechanical philosophy that he restricts himself to two material causes of this gravitational action: some emanation from within the attracting body or some kind of matter outside the body.

A similar acceptance of the Newtonian style is found in the writings of Clairaut. Clairaut explained that "M. Newton . . . says expressly that he is using the term *attraction* only while waiting for its cause to be discovered; and in fact it is easy to judge by the treatise on the Mathematical Principles of Natural Philosophy that its only goal is to establish attraction as a fact" (Clairaut 1749, 330).

By the end of the eighteenth century, the concept of a universal gravity had become generally accepted. In the preface of his great *Mécanique céleste* (published 1799–1825), Laplace—the second Newton of this subject—began (1829, p. xxiii):

> Towards the end of the seventeenth century, Newton published his discovery of universal gravitation. Mathematicians have, since that epoch, succeeded in reducing to this great law of nature all the known phenomena of the system of the world, and have thus given to the theories of the heavenly bodies, and to astronomical tables, an unexpected degree of precision. My object is to present a connected view of these theories, which are now scattered in a great number of works. The whole of the results of gravitation, upon the equilibrium and motions of the fluid and solid bodies, which compose the solar system, and the similar systems, existing in the immensity of space, constitute the object of *Celestial Mechanics*, or the application of the principles of mechanics to the motions and figures of the heavenly bodies. Astronomy, considered in the most general manner, is a great problem of mechanics, in which the

elements of the motions are the arbitrary constant quantities. The solution of this problem depends, at the same time, upon the accuracy of the observations, and upon the perfection of the analysis.

Although Laplace was endowed with a philosophical turn of mind, as evidenced by his *Philosophical Essay on Probabilities* of 1814, he did not feel any need — a century after the *Principia* — to discuss whether or not it is reasonable for a force of attraction to extend itself through space. The second 'book' of the *Mécanique céleste,* "On the Law of Universal Gravitation and the Motions of the Centres of Gravity of the Heavenly Bodies," begins with a chapter "On the Law of Universal Gravitation, deduced from observation." We are "induced," he writes (1829, *1:* 249), "to consider the centre of the sun as the focus of an attractive force, which extends infinitely in every direction, decreasing in the ratio of the square of the distance." Wholly unabashed by the use of the Newtonian word 'attraction', and no longer repelled by the philosophical overtones of this word when considered at large and outside of the Newtonian context, Laplace concludes simply and straightforwardly that "the sun, and the planets which have satellites, are endowed with an attractive force, extending infinitely, decreasing inversely as the square of the distance, and including all bodies in the sphere of their activity" (p. 255). Furthermore, "analogy leads me to infer that a similar force exists generally in all the planets and comets." He has no problem with concluding "that the gravity observed upon the earth is only a particular case of a general law extending throughout the universe" and that this "attractive force" does "not appertain exclusively to its aggregated mass" but is "common to each component particle" (p. 258). He hails the Newtonian "universal gravitation" as a "great principle of nature," that "all the particles of matter attract each other in the direct ratio of their masses, and the inverse ratio of the square of their distances" (p. 259).

The success of the theory and applications of universal gravitation, or of what — since Einstein — is called 'classical' mechanics (or Newtonian mechanics), caused this subject to become the model or ideal for all the sciences. For example, much of the mid and late nineteenth-century argument about the Darwinian revolution centered on method, often focused on the question of whether or not Darwin had adhered to or abandoned the method of Newton. Scientists in as diverse fields as palaeontology and biochemistry envisioned a day when their science would have its Newton and reach the perfection of Newton's *Principia*.

Why, Georges Cuvier asked in 1812, "should not natural history also one day have its Newton?" And around 1930 Otto Warburg lamented that the Newton of chemistry (for which the need had been expressed by J. H. van't Hoff and Wilhelm Ostwald in 1887) "has not yet arrived" (see Cohen 1980, 294).

The Newtonian revolution had also a tremendous ideological component, equaled perhaps by only one other scientific revolution, the Darwinian. Isaiah Berlin (1980, 144) has summed up Newton's influence:

> The impact of Newton's ideas was immense; whether they were correctly understood or not, the entire programme of the Enlightenment, especially in France, was consciously founded on Newton's principles and methods, and derived its confidence and its vast influence from his spectacular achievements. And this, in due course, transformed — indeed, largely created — some of the central concepts and directions of modern culture in the west, moral, political, technological, historical, social — no sphere of thought or life escaped the consequences of this cultural mutation.

Newton, and his contemporary John Locke, symbolized great new ideas, comprising that "outstanding revolution in beliefs and habits of thought" (Randall 1940, 253) that marks the modern era beginning with the Enlightenment. In contemplating this effect, we today, at three centuries' remove, sometimes find it difficult to understand how unprecedented was Newton's actual achievement in producing a mathematical theory of nature. Only adjectives like 'extraordinary' or 'phenomenal' or 'amazing' can convey the awe that scientists and nonscientists felt when Halley's Newtonian prediction that a comet would appear in 1758 (long after both Halley and Newton were dead) was verified. Men and women everywhere saw a promise that all of human knowledge and the regulation of human affairs would yield to a similar rational system of deduction and mathematical inference coupled with experiment and critical observation. The eighteenth century became "preeminently the age of faith in science" (Randall 1940, 276); Newton was the symbol of successful science, the ideal for all thought — in philosophy, psychology, government, and the science of society.

The belief in a Newtonian type of "rule of nature" according to universal laws was well expressed by the eighteenth-century physiocrats. All "social facts are linked together," according to the physiocrats, "in necessary bonds eternal, by immutable, ineluctable, and inevitable laws" (Gide and Rist 1947, 2). These would be obeyed by individuals and governments "if they were once made known to them." The physiocrats not only believed that human societies are "regulated by *natural laws*," but held that there are "the same laws that govern the physical world, animal societies, and even the internal life of every organism" (p. 8). Enlightenment men and women discarded traditional concepts of human relations and the order of human society, hoping for their Newton, who — they were sure — was "just around the corner." This "Newton of social science," according to Crane Brinton (1950, 382) would produce the new "system of social science [that] men had only to follow to ensure the *real* Golden Age, the *real* Eden — the one that lies ahead, not behind." In 1748 Montesquieu published *The Spirit of the Laws*, in which he compared a well-working monarchy with "the system of the universe," in which there is "a power of gravitation" that "attracts" all bodies to "the center." As in the model of the *Principia*, Montesquieu "laid down . . . first principles" and found that the particular cases follow naturally from them.

On almost every conceivable level of thought and action in which rational principles could be applied, the Newtonian revolution had a significant impact. Even today, when Newtonian concepts of time, space, and mass, and even the Newtonian principles of gravitation, have suffered Einsteinian replacements, there are huge areas of science and of common experience in which Newtonian science still reigns supreme. These encompass all of the experience of daily life and the machines we ordinarily use (except 'nuclear' devices). The most spectacular event of our times — the exploration of space — is not an illustration of Einsteinian relativity but only a straightforward application of classical gravitational physics — the science achieved by Newton in his *Principia* and developed by two centuries of Newtonians into the great science of rational mechanics and its central core of celestial mechanics. The Newtonian revolution was not only the apex of the Scientific Revolution, it remains one of the most profound revolutions in the history of human thought.[5]

11

Vesalius, Paracelsus, and Harvey:

A Revolution

in the Life Sciences?

———◆———

Discussions of the Scientific
Revolution tend to focus on the physical or exact sciences rather than
the biological or life sciences, on the revolutions associated with Copernicus, Newton, Galileo, and Kepler, but not on the possibility of a
revolution inaugurated by either Vesalius or Harvey. For historians and
scientists alike, the major revolutions in science that occurred before
the twentieth century were — with a single exception — all in the domain of the physical sciences. Darwin was responsible for the lone revolution in biology. The present chapter examines the scientific careers of
three possible candidates for having produced a revolution in the biological or life sciences in the sixteenth and seventeenth centuries.

Andreas Vesalius, Revolt or Revolution?

Andreas Vesalius (1514–1564), founder of modern anatomical science,
published his great book, *De Humani Corporis Fabrica (On the Construction*

[or, *The Fabric*] *of the Human Body*), in the same year, 1543, that saw the appearance of Copernicus's *De Revolutionibus*. At the time of publication Vesalius was still a young man, in his prime years, while Copernicus was elderly and in fact at the point of death. Vesalius's ability was recognized at the very start of his career; he received his M.D. *magna cum laude* from the University of Padua on 5 December 1537 at the age of twenty-three, and on the following day was appointed *explicator chirurgiae* and began to give lectures to medical students on surgery and anatomy. From the start he showed his independence, for in his "annual anatomical lectures and demonstrations" — which were still "Galenic in character" — he broke with tradition and, "contrary to custom . . . himself performed the dissections rather than consigning that task to a surgeon" (O'Malley 1976, 4). A year later, in 1538, Vesalius published two works. One was the set of anatomical drawings known as *Tabulae Anatomicae Sex* or *Six Anatomical Plates*. The other was "a revised and augmented edition" of the "Galenically oriented" manual of dissection of a former teacher, an edition notable for Vesalius's own "independent anatomical judgments" (such as the "clearly anti-Galenic observation that the cardiac systole is synchronous with the arterial pulse"). In 1539 it was officially recorded that this brilliant anatomist and lecturer "has aroused very great admiration in all the students."

In that same year the judge of the Paduan criminal court turned over to Vesalius for anatomical study the corpses of executed criminals. With a sufficient supply of human bodies to dissect, Vesalius now made great progress in human anatomy and "became increasingly convinced that Galen's description of human anatomy was basically an account of the anatomy of animals in general and was often erroneous insofar as the human body was concerned" (ibid., p. 5). By the end of 1539 he was able to announce publicly in Padua and also in Bologna (where he had been invited by the medical students to make anatomical demonstrations) that the only way to learn the anatomy of the human body was by direct dissection and observation and not by reading books. He compared and contrasted an articulated human skeleton and the skeleton of an ape or a monkey to prove without any possibility of doubt that Galen's account of bones was based largely on apes and not humans. Furthermore, as Vesalius said in the preface to *De Fabrica* (trans. O'Malley 1964, 321), there are "many incorrect observations . . . in Galen, even regarding his monkeys." Since, in those days, Galen was the respected and unquestioned authority on every aspect of medical science and practice, Vesalius's bold challenge must surely be considered an act of revolt. But was it the first step of a revolution?

Vesalius's magnum opus, *De Humani Corporis Fabrica,* was a massive folio volume embellished by a large number of extraordinary plates that represent a high point in the use of art to represent scientific knowledge. They are as exciting to contemplate today as they were some four and a half centuries ago. Vesalius's subsequent role in advancing anatomical science as such may have been lessened by the fact that almost immediately after the publication of his book he gave up the academic life and abandoned his anatomical studies. With "youthful impetuosity" (O'Malley 1976, 5), he resigned his teaching post and entered medical practice as a physician to the "imperial household" of Emperor Charles V. When Charles V abdicated in 1555, Vesalius remained in Spain and became a court physician for Charles's son, Philip II. In 1564 he left Spain on a pilgrimage to the Holy Land and — apparently — died on the way home on the Greek island of Zanthos (or Zákinthos).

Vesalius's aim was to convince doctors and anatomists of the inadequacy and even falsity of the current Galenic anatomy and thus to institute a reform of the subject, which at that time — he said — was taught in such a way that there was "very little offered to the students that could not better be taught by a butcher in his shop." A true anatomy, based on dissection, was in his opinion the only sound foundation for all of medicine. C. D. O'Malley — the outstanding twentieth-century student of Vesalius's life and career — believes that even the "word 'fabrica' [in the title of Vesalius's book] could be interpreted as referring not only to the structure of the body but to the basic structure or foundation of the medical art as well." Vesalius not only sought graphically and verbally to correct the errors of Galen but also argued that every medical student and physician should personally base his knowledge of the human body on direct dissection. As O'Malley sums up Vesalius's plea, "The professor or teacher must also descend from his *cathedra,* dismiss the surgeon who had formerly performed the actual anatomy, and undertake his own dissecting" (ibid., p. 7). In one of the most striking passages in his book, Vesalius explained how and why the failure of doctors to do their own anatomies had caused the science of medicine to decline.

In older or classical Latin, the nearest expression to what we today mean by a revolution is 'novae res' (literally, 'new things'). There were of course a number of new things in Vesalius's *De Fabrica,* many of which contradicted Galen's statements or accepted views. It was also definitely new and unheard of to base anatomical knowledge on the direct experience of human dissection and on dissections of the bodies of animals for comparative purposes, and to exhort all medical students, anatomical scientists, and doctors to perform their own dissections of the human body. Not only did Vesalius show by examples that such actual dissections had produced new knowledge; he also gave explicit directions as to how the reader should proceed in making a dissection so as either to verify Vesalius's own presentation or to "arrive at an independent conclusion." This revolutionary aspect of Vesalius's book was enhanced by the beautiful and detailed artistic anatomical illustrations. And it was to emphasize the revolutionary exhortation to "do it yourself" that Vesalius even included a plate showing the tools needed to carry out the dissections he recommended that the reader perform.[1]

There can be no doubt that Vesalius successfully inaugurated a reform of the subject of anatomy and the method of teaching it. O'Malley reports that "by the beginning of the seventeenth century, with the exception of a few conservative centers such as Paris and some parts of the Empire, Vesalian anatomy had gained both academic and general support" (1976, 12). And yet O'Malley does not say that Vesalius revolutionized the subject of anatomy or that he inaugurated a revolution, even though he begins his magisterial biography with the statement that "Andreas Vesalius is now recognized by most scholars as the founder of modern anatomy" (1964, 1). Nor do I find that historians of science generally — or, for that matter, historians of biology and medicine and even of anatomy — write of a 'Vesalian revolution', even though his solid accomplishments and direct influence in changing his subject would seem more appropriately to merit such a designation than the alleged reform of astronomy epitomized in the commonly used expression 'the Copernican revolution'.[2]

A possible reason for the nonrevolutionary evaluation of Vesalius may be his natural diffidence, expressed in his actual statements about Galen, whom he refers to as the "prince of physicians." In his published writings, it is not his manner to make a frontal attack on Galen or Galenic doctrines, nor to criticize or correct Galen, except in specific cases "when he felt the facts warranted such action" (O'Malley 1964, 149). He "never went out of his way to do so" and never would have held up Galen to ridicule or "have made a public example" of him. (See, additionally, §5.2 on Vesalius's humanism.)

Vesalius did not assume a revolutionary anti-Galenic posture. He hesitated a long time before making a public expression of any disagreement with the teachings of Galen, and when he finally did so he criticized only Galen's writings on anatomy and not "the Galenic system of medicine as a whole" (Pagel and Rattansi 1964, 318). Although Vesalius stoutly criticized the followers of Galen who never departed "from him by so much as the breadth of a nail" (Vesalius 1543, pref. 4; trans. Farrington 1932, 1362), Vesalius added at once that he himself would not wish to appear "disloyal to the author of all good things and lacking in respect for his authority." Thus, after denying by descriptive statement of fact the Galenic "statement that the vena cava originated in the liver," and after pointing out that Galen "did not notice that the orifice of the vena cava has been observed to be three times the size of the orifice of the aorta," Vesalius concluded, "but I find no pleasure in pursuing these and many other matters at greater lengths" (trans. O'Malley 1964, 177). This attitude may be contrasted with that of the rebel Paracelsus (discussed below), who publicly cast the medical works of Avicenna into the burning flames as a declaration of their total lack of worth.

The nonrevolutionary attitude of Vesalius may be seen most clearly in his discussions of the pores alleged to exist in the septum (wall) separating the right ventricle of the heart from the left. These pores, or passages, were an essential part of Galenic physiology, providing a necessary pathway for the blood to ooze a drop at a time from the so-called 'arterial vein' (for us, the pulmonary artery) into the 'venal artery' (or pulmonary vein). Galen taught (and Galenists believed) that air is carried to the heart from the lungs by means of this 'venal artery', where it combines with the trickle of blood through the pores in the septum, so as to produce arterial blood. We know today that there are no such pores leading from the right ventricle into the left (or vice versa), although there are tiny pits on the septum that separate the ventricles. But these are blind pits; and "even a fine bristle cannot be made to penetrate from one ventricle to the other" (Singer 1956, 14). And yet, as Charles Singer remarks, "in spite of the evidence of their senses, men continued to believe that these channels really existed." Why? "The great Galen had believed in their existence, and that was enough!"

That there are no such passages going from one side of the heart to the other was at once made evident to Vesalius by actual dissections of the human heart. A true revolutionary would, I believe, have simply concluded that the whole Galenic physiology, and perhaps even the Galenic medicine based upon it, must be false and should be cast out at once as having no foundation in fact. But not Vesalius! Instead, he confesses in the second edition (Basel, 1555) to "a lack of that self-confidence" which would have enabled him to reform the Galenic teachings

about the heart and blood (bk. 6, ch. 15). We are told that he consciously
"'accommodated' his text to a large extent to the doctrines *(dogmata)*
of Galen." Vesalius adhered closely to the physiological doctrines of
Galen not because he sincerely believed in their truth but "because he
did not feel equal to the task of reform" (Pagel and Rattansi, 1964, 318).

Vesalius says in *De Fabrica* that the septum is "formed from the very
densest substance of the heart," and that—although the septum
"abounds on both sides with pits"—"of these pits none, so far as the
senses can perceive, penetrate from the right to the left ventricle." And
this leads him only to conclude: "We are thus forced to wonder at the art
[industria] of the Creator by which the blood passes from the right to the
left ventricle through pores which elude the sight" (trans. Singer 1956,
27). In the second edition of *De Fabrica,* this passage was somewhat
rewritten (ibid.):

> Although sometimes these pits are conspicuous, yet none, so far as
> the senses can perceive, passes from the right to the left ven-
> tricle . . . I have not come across even the most hidden channels
> by which the septum of the ventricles is pierced. Yet such channels
> are described by teachers of anatomy who have absolutely decided
> that blood is taken from the right to the left ventricle. I, however,
> am in great doubt as to the office of the heart in this part.

In yet another discussion of this subject, he indicated his gradual independence from Galen (trans. Singer 1956, 28):

> In considering the structure of the heart and the use of its parts, I
> have brought my words for the most part into agreement with the
> teachings of Galen: not because I thought that these were on every
> point in harmony with the truth, but because, in referring now and
> again to a new use and purpose for the parts, I still distrust my-
> self. Not long ago I would not have dared to turn aside even a
> nail's breadth from the opinion of Galen, the prince of physi-
> cians . . . But the septum of the heart is as thick, dense, and com-
> pact as the rest of the heart. I do not, therefore, know . . . in what
> way even the smallest particle can be transferred from the right to
> the left ventricle through the substance of that septum.

We may agree with Charles Singer's explanation of Vesalius's behavior
with respect to the heart (1956, 25): "All the physiology of his time was

based on the view of Galen, which necessitated a belief in the passage of blood from the right ventricle to the left *through the pores of the septum* and a belief that *air entered the heart through the venal artery* (our pulmonary vein)." Vesalius could not easily "cast doubts on this without explaining the action of the heart," since he would thereby "upset the whole of the current notions of the workings of the human body without putting anything in its place"; and this "Vesalius hesitated to do." Accordingly, "he hinted in his book that the passages through the septum have no real existence, yet he did not at first say so outright" (ibid.). Vesalius was not a full-fledged revolutionary. He did not simply and straightforwardly deny the possibility that the human body could function as Galen had taught and as Vesalius's contemporaries still believed.

It is true, of course, that single contradictory facts do not overthrow theories in the sense that T.H. Huxley (1894) could write of "a beautiful hypothesis killed by an ugly fact." Many historians and philosophers of science have pointed out that theories continue to exist, despite contradiction by individual facts of experiment and observation, until a better theory comes along to replace them. Or, as Max Planck put the matter (and as Joseph Lovering said some fifty years earlier), old theories never disappear until all those who have believed in them will have died (1949; see p. 467 below). But it is the accumulation of such contradictory facts that eventually may sound the death knell of a theory or a scientific system and lead to what T. S. Kuhn has called the replacement of one paradigm by another. And the fact of the matter is that in *De Fabrica* (as in the later *Epitome*) Vesalius did not adopt the bold stance of revolt that he had adopted in Padua and in Bologna in his public use of articulated skeletons of man and ape to show that Galen's anatomy of the bones was valid for the animals he had dissected and not for man.

Vesalius's nonrevolutionary attitude, even with respect to his correcting some of Galen's errors, was no doubt related to his personality. But we must also keep in mind that 1543 was a little early for a full expression of the revolutionary attitude in the sciences that we find expressed in the writings of Galileo, Descartes, and Harvey and later scientists of the seventeenth century. Furthermore, Vesalius was steeped in the humanist tradition, which was founded on a belief in the greatness of classical philosophy, letters, art, and science and which sought to restore the values of antique culture (see p. 485, §5.2). Vesalius probably saw his role as an improver of Greek anatomy and as a restorer of the Greek tradition of dissection and not therefore as the author of a revolutionary and frontal attack on the current versions of Galenic

science. We shall see that William Harvey, a true revolutionary where
Vesalius was not, was apparently willing to set aside the basis of Galenic
physiology and accept whatever consequences might be thereby implied for medical practice.

The Rebellious Paracelsus

Many historians refer to the ideas of Paracelsus, Vesalius's slightly older
contemporary, as revolutionary. And, in fact, the life and career of
Paracelsus (c. 1493–1541) show all the marks of revolt, rebellion, and
possibly revolution. Even his adopted name Paracelsus (a nickname
applied when he was about thirty-six years old) may have referred to
his having been the author of paradoxical works "that overturned tradition" (Pagel 1974, 304). 'Paradox' comes from the Greek words 'besides' and 'opinion' and so means 'contrary to opinion', that is, 'contrary
to received opinion'.[3] When Paracelsus was appointed municipal physician and professor at Basel in 1527, he declined taking the normal oath;
rather, he issued a broadside declaring his disagreement with Galenic
principles and announcing a new system of medicine. It was only a few
months later (24 June 1527) that he publicly burned a copy of the standard textbook of the day, Avicenna's *Canon of Medicine*.

In direct contravention of academic rules and traditions, Paracelsus
lectured in the vernacular German rather than Latin and even admitted
barber-surgeons to his classes. He equally rejected "organized religion
and classical scholarship" and has been described as one whose "wholesale condemnation of traditional science and medicine found its parallel
in his rough behavior and in his unwillingness to make concessions to
custom and authority" (Pagel 1974, 306). His later career was marked by
unorthodox behavior and controversy, swinging like a pendulum between good jobs with excellent facilities and periods as "a wandering lay
preacher, appearing in 'beggar's garb.'" After he died in Salzburg in
1541, his grave became "a place of pilgrimage for the sick for a long time
after" (p. 305). One of Paracelsus's given names, Bombastus, was long
thought to be the source of the word 'bombast'.

As a scientific revolutionary, Paracelsus was influential in two major
fields: medicine and chemistry. In his day, and for some two centuries to
come, almost all of medical theory and practice was dominated by the
ancient doctrine that disease is the result of an unbalance of four
humors or body fluids (blood, phlegm, choler or yellow bile, and melancholy or black bile). This imbalance, it was believed, causes a disease

which is the direct result of an excess or a deficiency of one or more of these fluids in relation to the particular 'constitution' of each individual. In principle this doctrine implies that there are as many different diseases as people, and that diseases are not caused by a specific agent and do not have specific anatomical effects or lesions. As a true revolutionary, Paracelsus took a diametrically opposite position, holding that diseases are the result of causes external to the body and that each disease has a "specific" locus. He believed that the causes of disease are to be found in the mineral world and in the air and held that a disease is "determined by a specific agent foreign to the body, which takes possession of one of its parts, imposing its own rules on form and function and thereby threatening life" — which is "the parasitic or ontological concept of diseases — and essentially the modern one" (Pagel 1974, 307). Whereas traditional medicine treated disease by causing sweating, by purging, by bloodletting, and by inducing vomiting, the aim of Paracelsian medicine was to find specific substances for curing each disease.

The search for curative chemical agents was thus closely related to Paracelsus's views on chemistry. He believed in three 'principles' — salt, related to (or responsible for) the solid state of any substance; sulfur, related to the inflammable or fatty state; and mercury, related to the smoky (vaporous) or liquid state. Although these were chemical principles, they were given spiritual connotations, related to Paracelsus's particular brand of alchemy. Paracelsus produced many new chemical compounds (chiefly in his search for curative agents) and he apparently invented the method of producing concentrated alcohol by freezing out the water content, in the manner used by Yankee farmers to convert hard cider into applejack without a still. His influence on the development of chemistry may be seen in the inclusion of Paracelsian chemicals (among them calomel) in the *London Pharmacopoeia* of 1618 and later editions. But his reputation suffered from his "uncompromisingly destructive attitude toward tradition" (Pagel 1974, 311) and his conscious revival and even development of medical lore preserved among simple and uneducated (or unorthodox) people, which caused many potential followers to be put off. Perhaps his greatest contribution to science was to deflect alchemy from its traditional goals of seeking to transmute base metals into gold and to prolong life indefinitely, and to give alchemy a new goal: finding useful substances for the treatment of disease.

The foregoing account attempts to define what we today may find best and most significant in Paracelsus's teachings and practices. But, as

Walter Pagel (1958, 344) reminds us, Paracelsus's chemistry was part of a "mythical" or "symbolistic" cosmology and philosophy, which were "decidedly unscientific," despite his sound work in the chemical laboratory, his new methods of preparing mineral compounds, and his work with heavy metals. In medicine, despite his important new theory of disease and the concomitant principles of cure, he opposed "the traditional building up of rational medicine on the basis of anatomy and physiology — subjects in which he had little interest and knowledge." His medical system, though containing the "germ cells of modern pathology," was "not scientific — taken as a whole" because it is a collection of "analogies and metaphors based on his theory of Microcosm," in which "observation and protoscientific elements" may be overlarded with "a farrago of speculations which strike us as fantastic" (ibid., p. 345).

There was a Paracelsian movement in medicine and chemistry all over Europe, beginning some thirty years after Paracelsus's death (see Debus 1965, 33–37; 1977). We can see how strong this movement was by taking note of the reaction against it. For example, in 1569 the Duke of Bavaria ordered all monasteries on his lands to "adhere to the teachings of Hippocrates and Galen instead of the new medicine." Paracelsian medicine was "a revolutionary movement" which by its sixteenth- and seventeenth-century name spread the fame of its founder, and gave emphasis to his having started the movement "single-handed" (Pagel 1958, 349). The movement, however, later dropped the name of its founder and — as developed in a more strictly scientific mode by J. B. Van Helmont and others — became 'Iatro-Chemistry' rather than Paracelsian chemistry.

In the essays of Montaigne, written in the 1570s and 1580s, the term 'revolution' does not seem to have been used in the sense of radical 'revolutionary' change. The concept of such a change appears in a striking manner (although without the actual term 'revolution') in the most celebrated of these essays, the "Apology for Raymond Sebond," composed about 1576. Here, in talking of medicine, Montaigne (1958, 429) mentions "a newcomer, whom they call Paracelsus," who — they say — "is changing and overthrowing the whole order of the ancient rules," and maintaining that up to now medicine "has been good for nothing but killing men." Montaigne finds this judgment to conform to the facts, but he wisely concludes that "as for putting my life to the test of his new experience, I think that would not be great wisdom."

In another essay, "Of the Resemblance of Children to Fathers,"

Montaigne gives an encapsulated history of ancient medicine (ibid., p. 586), which he calls "those ancient mutations in medicine," a strong word closely akin to common usage today in discussions of revolutions as they occur in science. Montaigne refers to "countless others [that is, mutations] down to our time"; these, "for the most part," he says, were "complete and universal mutations, as are those produced in our time by Paracelsus, Fioravanti, and Argenterius." Then, showing how well he understood the nature of Paracelsian medicine, Montaigne observes that the Paracelsians "change not merely one prescription, but, so they tell me, the whole contexture and order of the body of medicine." As Montaigne saw so clearly, Paracelsus and his followers had put forth a revolutionary program for medical theory and practice.

It is obvious that the Paracelsian mutation referred to by Montaigne had the makings of a revolution, but was there in truth a Paracelsian revolution? In the system of classification I have adopted in this book, Paracelsus was obviously a revolutionary. There can be no doubt that Paracelsian medicine was an intellectual revolution, a 'revolution in itself'. In a recent article, Allen Debus (1976, 307) argues for the occurrence of a post-Paracelsian Renaissance "pharmaceutical revolution" — and he designates its origin as "the Paracelsian vision of a chemical reform of medicine." Since Paracelsus published his views, which were taken up and used as guides by his followers, I believe it fair to say that there was also a Paracelsian revolution on paper. But on the question of whether or not Paracelsus produced a revolution in science, we have a negative answer given by a number of writers in the seventeenth century, a judgment in which our most acute historians concur today. Thus, Walter Pagel, the dean of Paracelsian scholars of our time, reminds us that in Paracelsus it is primarily "the desire to probe and test Nature for the validity of his cosmological and religious philosophy that forms the driving motive for his research" (1958, 350). Pagel sums up a lifetime of study of Paracelsus with the judgment that Paracelsus does not stand out "as a link in the chain of students of Nature to whom modern science owes its origin," or even as a physician with modern and revolutionising ideas" (ibid.). John Maxson Stillman (1920, 173) concluded his study of Paracelsus by observing that "his method was not that of modern science." Stillman summarized, with approval, the scholarly views of Max Neuburger (ibid., p. 129):

Neuburger appreciates the value of the accomplishments of Paracelsus, yet doubts that he is to be considered as a reformer of medicine in the sense that was Vesalius or Paré, that is, he laid no foundation-stones of importance, and the real value of much of his thought required the later developments of modern scientific thought for its interpretation. His aim was to found medicine upon physiological and biological foundations, but the method he chose was not the right method, and his analogical reasoning and fantastic philosophy of macrocosm and microcosm were not convincing and led nowhere. The disaffection and discontent with conditions in medicine produced by his campaign, can, thinks Neuburger, hardly be called a revolution. That was to come later through the constructive work of more scientific methods.

Two centuries earlier, Walter Charleton summed up a rather general disdain for Paracelsus and his works when he referred to "the stupid admirers of that Fantasticle Drunkard, *Paracelsus*" (1654, 3).

William Harvey and a Revolution in the Life Sciences

William Harvey differs from Paracelsus and resembles Vesalius in that he generally wrote of Galen with reverence and respect and seems almost to have been pained by having to correct Galenic errors. Yet his book on the circulation sets out boldly and unambiguously to establish a new basis of human and animal physiology that would completely replace the Galenic ideas that had dominated scientific and medical thought for some fifteen centuries. Harvey was fully aware of the revolutionary nature of his program and so were his admirers and detractors. Harvey not only proposed a closed mechanical system in which the heart pumps the blood through the arteries and veins; he also set forth the idea of a single circulatory system. The very idea of a single system for the blood in fact originates with Harvey. His work marks the radical transformation from imagined pathways to demonstrable circuits, from unprovable suppositions of Galen to empirically founded and quantitative biology. The contributions of William Harvey brought the life sciences into the modern era as full-fledged participants in the Scientific Revolution.

William Harvey was born in 1578, thirty-five years after the publication of Vesalius's *De Fabrica*. He was a student of Gonville and Caius College, Cambridge, from 1593 to 1599 and then went on to Padua for further education, receiving his doctorate in medicine in 1602. His teachers included the great anatomist and embryologist Girolamo Fabrici (or Fabricius), discoverer of the valves in the veins. While Harvey was at Padua, the university was a center of scientific ferment and intellectual activity; among its professors was the young Galileo, soon to discover the mountains on the moon, the phases of Venus, the satellites of Jupiter, and many other new celestial phenomena. On Harvey's return to England, he took up the practice of medicine and became a Fellow of the Royal College of Physicians (he was Lumleian lecturer in surgery from 1615 to 1656). He was appointed physician to James I and held a similar position under Charles I. A royalist in sympathy and deed, he remained in attendance on Charles I during the Civil War. Owing to Charles's interest in Harvey's work, some royal deer were made available to him for his studies on generation. Harvey died at the age of seventy-nine in 1657.

Unlike Vesalius, Harvey planned what has been described as "a vast research program" that would lead to a series of publications on a variety of subjects, to be based on his "original investigations of the motions of the heart, respiration, the functions of the brain and spleen, animal locomotion and generation, comparative and pathological anatomy," and other topics such as animal generation and embryology (Bylebyl 1972, 151). But he only completed and published two proper books, the *De Motu Cordis* (1628) — with its supplementary *De Circulatione Sanguinis* (1649) — on the circulation of the blood, and the much larger by far *De Generatione Animalium* (1651), which marked a significant advance on contemporaneous and older ideas concerning the generation and embryology of both oviparous and viviparous animals. This latter work, espousing epigenesis, was based on careful dissections to reveal all the visible stages of development. Although Harvey succeeded in formulating "the first fundamentally new theory of generation since antiquity," his views (though representing "a major advance over those of his predecessors") have been largely "undermined" by "subsequent investigation" (Bylebyl 1972, 159) and his *De Generatione Animalium* pales into relative insignificance alongside his great *De Motu Cordis*.

The full title of Harvey's book on the circulation of the blood reads *Exercitatio Anatomica de Motu Cordis et Sanguinis in Animalibus*, that is, *An Anatomical Exercise* [or *Treatise*] *Concerning the Movement of the Heart and of the Blood in Animals*. Published in Frankfurt-on-the-Main in a poorly printed volume of only 72 pages (plus 2) with two plates, Harvey's discovery of the circulation of the blood was one of the notable scientific events in the seventeenth century. *De Motu Cordis* is said to contain "a greater amount of important material in small compass than any other medical work ever published" (Dalton 1884, 163). His contemporaries were fully aware of the primary importance of his reformulation of human and animal physiology. Historians and scientists agree that he revolutionized biological and medical thought.[4] In short, Harvey's work passes all tests for a revolution in science. Furthermore, although Harvey was writing too soon to use the word 'revolution', he does make it perfectly clear in *De Motu Cordis* that he had produced a great innovation, "my new concept of the heart's movement and function and of the blood's passage round the body" (1963, pref. 5). Although many "distinguished and learned men" had illuminated some aspects of the subject, he wrote, "this book of mine was the only one to oppose tradition and to assert that the blood travelled along a previously unrecognized circular pathway of its own" (p. 6). In chapter eight (p. 57), he declared plainly and simply that his ideas "are so novel and hitherto unmentioned that, in speaking of them, I not only fear that I may suffer from the ill-will of a few, but dread lest all men turn against me." Harvey recorded that once he had understood the circulation, he began to expound his view "on this matter" both "in private to friends" and "in public, College fashion, in my anatomical lectures." Some of his colleagues "asked for fuller explanation of the novelty, asserting that it would be worth investigating and would prove of extreme practical importance."[5] (The coupling of Harvey's name with revolution has been discussed earlier, in chapter 5.)

Harvey's radical reform of biology and of the physiological basis of medicine has three significant aspects. Perhaps the greatest of the three is the firm establishment of experiment and careful direct observation as the means of advancing biology and of establishing knowledge in the life sciences. Harvey lauded Aristotle, who made experiments, and attacked Galen, who (he held) had not actually based his dogmas on experiment or even direct observation.[6] Harvey inspired a new "generation of anatomists who sought to emulate his methods in the study of animal functions" (Bylebyl 1972, 151). A second aspect of Harvey's reform of biology was the introduction of quantitative reasoning as the basis for conclusions about living processes. And, of course, there is the discovery of the circulation, which quite simply "revolutionized physiological thought" (ibid.).

I have mentioned that one of the great novelties in Harvey's book was the demonstration that the heart, arteries, and veins constitute a circulatory 'system'. In almost all discussions of this subject by historians, the

Harveyan system is contrasted with a Galenic 'system'. But in fact there was no Galenic 'system'. Galen never wrote out a single complete presentation of his physiological ideas: the system presented by historians (as Temkin has reminded us, 1973) is put together from many bits and fragments taken from different writings of Galen. Furthermore, these bits and pieces yield not one Galenic system but several. For instance, Galen viewed the liver and veins as a system wholly distinct from the system of heart and arteries. So part of Harvey's revolution was the concept of a single system.

To see how completely the Harveyan revolution altered the framework of knowledge, a brief examination of the then-prevailing ideas is in order. Galen believed that food which has been digested is transferred in the form of 'chyle' to the liver, where it is turned into blood, which then moves out through the veins in order to carry nourishment to the various organs and parts of the body. In the liver, the blood was supposed to become charged with 'natural spirit', which was believed to be necessary for the performance of the life function. This venous blood chiefly flowed out from the liver, although there was some ebb and flow (but not a circulation). Some of the venous blood, according to the system of Galen, moves through the 'arterial vein' (what we, post-Harvey, call the pulmonary artery)[7] to the lungs, where it would discharge the accumulated impurities and waste products into the ambient air. Other venous blood was thought to enter the right ventricle of the heart; Galen supposed that this blood would pass to the left side of the heart through narrow passages in the septum or muscular wall that separates the right ventricle from the left ventricle. Once the blood enters the left ventricle, it was thought to mix with air which comes from the lungs via the 'venal artery' (our pulmonary vein) and to become charged with 'animal spirit' — changing in color from dark purple to bright red. This 'new blood' would pass through the arteries into the various parts of the body. A third 'system' arose in the brain, the source of the 'animal spirit', sent out through the core of the hollow nerves.

By Harvey's day some scientists had become aware that blood goes to the lungs through the pulmonary artery and returns through the pulmonary vein. Sometimes called the lesser circulation (or, more properly, the pulmonary transit), this had been made known by Realdus Columbus, Vesalius's successor as professor of anatomy at Padua.[8] Harvey's own teacher at Padua, Fabricius ab Aquapendente, had made the significant discovery that there are valves in the veins, although he did not grasp the full significance of the valves in relation to circulation.

Harvey did more than recognize fully that Fabricius's discovery implied that blood can move in the veins only toward the heart; Harvey made a series of different kinds of experiments and tests to prove that in the veins there is only a one-way flow. Valves suggest the action of pumps and Harvey tells us that he was led to the idea of a pump (see Webster 1965; Pagel 1967, 212–213) when he examined the structure of the heart with its valve system.

The heart acts by contracting and expanding, in the actions known as systole and diastole. When a chamber of the heart contracts it expels the blood it contains; when it expands it sucks in new blood which it can then expel at the next contraction. Because of the valves in the heart, this is a one-way flow. As Harvey showed, blood is forced out of the left ventricle into the aorta, the main artery, and then is pushed further out (with each successive expulsion) into the arterial system. The blood returns to the heart through the veins into the right ventricle. Contraction and expansion force blood from the right ventricle into the right auricle and then out through the pulmonary artery into the lungs. The blood flows back into the heart through the pulmonary vein, into the left auricle. From here the blood is sent into the left ventricle and out once again into the aorta and the arterial system. Thus there is a continuous circulation — heart, arteries, and veins all part of a single system.

Harvey's new concept was confirmed by amassing a great range of evidence from vivisection, visual observation, and experiment. He could announce proudly that he had corrected "an error held for two thousand years." His findings were based not on dogma but on an empirical study of more than eighty different species, including mammals, snakes, fishes, lobsters, toads, lizards, slugs, and insects (Keele 1965, 130). The variety of his experimental and observational information is overwhelming. In chapter 5 of his book he says that Galen erred when he said that blood can pass through pores in the septum of the heart. There are no such pores; ergo "a new path must be prepared and opened."[9]

Harvey was fully aware that his quantitative studies were (as he said) "novel," and feared that all his readers might pounce upon him (ch. 8). Today such a turn to numerical argument seems a simple and natural step. But in Harvey's day this was not the case at all, although quantitative measures had entered medicine in pharmacy. We must keep in mind that at that time, numerical temperature readings and numerical measurement of blood pressure were still things of the future. But even if Harvey did not, in fact, invent the quantitative method in biology, he used numerical reasoning with telling effectiveness. Temkin (1961) has shown that Galen used a similar quantitative argument to prove that the urine is not simply "residual matter from the nutrition of the kidneys" (see Pagel 1967, 78). Van Helmont was making quantitative biological experiments at about the same time as Harvey, although they were not published until much later (Pagel 1967, 78). Santorius made a series of experiments on himself in which he recorded the quantitative measures of his solid food and liquid intake, and his liquid and solid excreta, and determined the weight loss in perspiration; his book *Statica Medicina*, describing his methods and giving the numerical data, was published in 1614, fourteen years before Harvey's *De Motu*. But the quantitative method was not then in general use and Harvey was fully aware that his numerical reasoning was as radical in its method as in its results. Harvey used quantification not merely in an empirical investigation in the life sciences, but in making "the discovery which has opened up the modern period in Biology and Medicine and on which these disciplines have remained firmly grounded" (Pagel 1967, 80). What Harvey did was to determine, by actually making measurements, the capacity of the heart in humans, dogs, and sheep. Then by multiplying this number by the pulse rate, he computed how much blood is transferred from the heart to the arteries—approximately 83 pounds of blood in each half-hour for an average man. From these quantitative measures, Harvey says, it is manifest that "the beating of the heart is continuously driving through that organ more blood than the ingested food can supply, or that all the veins together at any given time can contain." Harvey then makes the point: "Supposing even the smallest quantity of blood to be passed through the lungs and the heart, a much more plentiful amount is taken through the arteries and the whole body than could be possibly supplied by the ingestion of food—which can only be achieved by the return through a circuit" (ch. 9 Pagel trans.). In short Harvey found that he could "calculate the amount and prove the circular movement of the blood" (ch. 12). He concluded (ch. 14):

> Since calculations and visual demonstrations have confirmed all my suppositions, to wit, that the blood is passed through the lungs and the heart by the pulsation of the ventricles, is forcibly ejected to all parts of the body, therein steals into the veins and the porosities of the flesh, flows back everywhere through those very veins from the circumference to the centre, from small veins into larger ones, and thence comes at last into the vena cava and to the auricle of the

heart; all this, too, in such amount and with so large a flux and reflux — from the heart out to the periphery, and back from the periphery to the heart — that it cannot be supplied from the ingesta, and is also in much greater bulk than would suffice for nutrition.

I am obliged to conclude that in animals the blood is driven round a circuit with an unceasing, circular sort of movement, that this is an activity or function of the heart which it carries out by virtue of its pulsation, and that in sum it constitutes the sole reason for that heart's pulsatile movement.

Pagel (1967, 76ff.) found that the "high historical significance which Harvey's calculus can really claim" is confirmed by the stress placed upon the quantitative argument by both Harvey's immediate critics such as Riolan and supporters such as Andrea Argoli, Jean Martet, and Johan Micraelius. The latter gave only one reason for the new theory: "the argument from quantity" (ibid.).

There can be no doubt that Harvey's discovery "revolutionized physiological thought" (Bylebyl 1972, 151). In considering this revolution, we must be careful not to minimize it on the grounds that it did not have the cosmic implications of Newton's system of the world, nor did it alter almost all of science in the way that Newton's *Principia* was to do a generation later. But it was a revolution in biology. Not everyone accepted the new discovery, but a large number of scientists and physicians did. Harvey's arguments, after all, were compelling. The quantitative argument and the failure to find the septal pores struck at the very heart of the Galenic physiology. The valves argued for a one-way flow. The only part of the evidence missing was the existence of capillaries linking the smallest arteries to the smallest veins, and these were found eventually by Malpighi.

In evaluating the Harveyan revolution, however, we must be careful to make a distinction among the revolution in biological thought and method, the revolution in the scientific basis of medicine (that is, in physiology), and a revolution in medical practice. According to the eighteenth-century physician and historian of medicine John Freind (1750, 237), Harvey had intended to write a work on the practical applications of his discovery for medicine, but he never did so. (See §11.1 on the lack of immediate practical consequences of Harvey's discovery.) There is no want of testimony from the seventeenth century to the effect that Harvey had made a great discovery for science, that the discovery of the circulation was a great intellectual achievement, but that it had not (or had not as yet) been equally important for medical practice.[10] Hence I believe we are justified in concluding that there was a Harveyan revolution in biology (or in physiology), even though there was not a similar Harveyan revolution in the practice of medicine.

Finally, a comparison and contrast of Harvey's work and Galileo's may be revealing. Harvey created a single circulatory system with a single center (the heart) to replace the multiple systems of Galen. This was a similar achievement to the creation of a single system of the world by Copernicus, and especially by Kepler, to replace the collection of separate systems in Ptolemy's *Almagest*. Similarly, Harvey's devastating proof of the falsity of the Galenic doctrine may be likened to Galileo's proof that the Ptolemaic system for Venus does not accord with reality. But there is a fundamental difference. Although Galileo showed that Venus must move in orbit around the sun and not in an epicycle whose center moves around the earth, the conclusion was ambiguous. The new information could be accommodated not only to the Copernican system but also to the Tychonic system and even to the world-system later devised by Riccioli. Harvey's arguments, on the other hand, together with his experiments, observations, and quantitative reasoning, not only proved the falsity of the Galenic doctrines but at the same time argued conclusively for a new scientific idea — the circulation of the blood. This is the reason why we can say unequivocally that there was a Harveyan revolution in science.[11]

Changing Concepts

of Revolution

in the Eighteenth Century

IV

12

Transformations

during the

Enlightenment

The eighteenth century was notable for two major political revolutions which established the usage of 'revolution' as we understand it today—a violent social or political upheaval that leads to a wholly new and different social system or form of political organization. These were the American Revolution in 1776 and the French Revolution in 1789. But the emergence of the concept of revolution as a radical change—as a point of discontinuity or break with the past rather than a cyclical return to the better days of a bygone age—can be traced during the Enlightenment not only in the spheres of social and political thought and action but even in discussions of cultural and intellectual affairs.

We have seen that Fontenelle applied this new sense of the term 'revolution' to mathematics in the early 1700s. In 1728 the proposed Paracelsian reformulation of medicine was called a revolution in physick, and in 1747 Newton's system of dynamics was referred to as a "revolution in physics." But in the eighteenth century, as in the late Middle Ages or Renaissance, the old sense of the word 'revolution' is apt to occur side by side with the new, and even in writings of the decades just before the French Revolution it is not always transparently clear whether the word 'revolution' is being used unambiguously in the present acceptation. Careful analysis may be required to make sure that the revolution in question is really a single event, a true secular change of great magnitude, and not merely a stage in cyclical change. And there are some examples in which we shall find it impossible to decide for certain as to which sense of revolution the author had in mind.

Ambiguity in the Term 'Revolution'

The most prolific eighteenth-century author on the subject of revolutions was the Abbé de Vertot (René Aubert de Vertot d' Auboeuf). His histories went through edition after edition in French and then were translated into English, Spanish, Italian, German, and Russian. Chief among his works were a history of the Order of Malta (1st ed. 1726) and histories of the revolutions in the Roman Republic (1st ed. 1719), Portugal (1st ed. 1689), and Sweden (1st ed. 1695). The volume on Portugal, which Fontenelle encouraged Vertot to write, was apparently the most popular of all; the Bibliothèque Nationale (Paris) lists no less than 35 editions or printings of this work, and the British Library (London) records eight editions of the English translation, the first one dated 1700.

The introduction to a late edition of Vertot's book on the Roman Republic (1796) suggests a reason why his history of Portugal had at once such a prodigious circulation: the subject was perceived to be relevant to the revolution just then (1689) being completed in England. Vertot first entitled the book *History of the Conspiracy in Portugal (Histoire de la conjuration du Portugal)*, but twenty-two years later (1711), when he brought out a revised and enlarged version, Vertot changed the title to *History of the Revolutions in Portugal*. The preface to the new edition explained that the word 'revolution' was better suited to the new edition than 'conjuration' (or conspiracy), since now there had been added a number of other events ('révolutions'). Furthermore, the chief subject was "an enterprise in which the leaders had as their goal only to restore the crown to a prince whom they regarded as the legitimate heir to the throne," and in this sense 'revolution' would be more appropriate than 'conjuration'. While this particular explanation has overtones of a 'return' of the throne from the Spanish usurpers to the legitimate Portuguese rulers, Vertot tended to use 'revolution' elsewhere in this book and in his other histories for any major event producing a significant political change. Even in the first edition of the history of Portugal, with its title of 'conjuration', Vertot used the term 'revolution' for the successful revolt of Portugal in 1640, when that country became independent of Spain under John IV of the Bruganzas. This revolution, Vertot said, in the preface to the first edition, is "worthy of our attention." "There has perhaps never been seen in history," he wrote, "any other conjuration that one could call just with respect to the rights of the prince, the interest of the state, the inclination of the people, or even the motives of the greater part of the conspirators." Nor was there ever so great a participation by people "of all ages, of both sexes, and of all conditions."

More than a little ambiguity in the meaning of the word 'revolution' is found when we turn to Swift's *Tale of a Tub*, published in 1704 along with his essay on the Battle of the Books. In the opening of section 4 of the *Tale* Swift tells his readers that now they "must expect to hear of Great Revolutions." These are evidently occurrences of major significance, but there is no hint given to help the reader determine whether these are possibly stages in a cyclical process, events that punctuate the ebb and flow of life, or merely happenings of an unusual kind. We are aided a little on learning that these revolutions are associated with "the Hero of the Play," Peter, who has become subject to visions of grandeur. Peter needs funds ("a Better *Fonde* than what he was born to") to "support this Grandeur," and so Swift would have Peter "cast about at last, to turn *Projector* and *Virtuoso,* wherein he so well succeeded, that many famous Discoveries, Projects and Machines, which bear great Vogue and Practice at present in the World, are owing entirely to *Lord Peter's* Invention" (Swift 1939, *1: 65*).

The sense of a general overturning, however, rather than a purposeful violent overthrowing of an existing government or form of society, appears in a remark made by Swift a little later in section 4 (p. 75) about "all this Clutter and Revolution," a reference to the confusing and unsettling effects of the Reformation.[1] A little later on, in a metaphoric presentation of two aspects of the Reformation, Swift makes a contrast between Luther and Calvin. The latter proceeded with haste and carelessness; where "Martin" — after his first zealous acts — "resolved to proceed more moderately in the rest of the Work." The picture of

Luther's activities concludes: "And this is the nearest Account I have been able to collect, of *Martin's* Proceedings upon this great Revolution" (p. 85).

Revolution occurs in a somewhat different context in section 9 of Swift's tract "A Digression Concerning the Original, the Use and Improvement of Madness in a Commonwealth." In any "Survey of the greatest actions that have been performed in the World, under the Influence of Single Men," according to Swift, we shall find that these outstanding individuals are all "Persons, whose natural Reason hath admitted great Revolutions from their Dyet, their Education, the Prevalency of some certain Temper, together with the particular Influence of Air and Climate" (p. 102). Such "greatest actions" were of three sorts: the "Establishment of New Empires by Conquest," the "contriving, as well as the propagating of New Religions," and the "Advance and Progress of New Schemes in Philosophy." These revolutions are clearly in no sense cyclical, in no sense part of an ebb and flow. They are events producing radical change, even if not on the scale of the great political revolutions. Swift could conclude that Madness has been "the Parent of all those mighty Revolutions, that have happened in *Empire*, in *Philosophy*, and in *Religion*." Here one may sense implications of major change that are beginning to resemble the post-1789 sense of the word 'revolution'. This may be even more the case for Swift's statement that "Imagination can build nobler Scenes, and produce more wonderful Revolutions than Fortune or Nature will be at Expence to furnish" (p. 108).[2]

Swift's countrymen and successors did not consistently use this rising new concept of revolution as a single event, but continued to write also of revolutions in the older cyclical sense. Samuel Johnson, who introduced the Glorious Revolution of 1688 into his *Dictionary* in 1755, used the older concept in *The Rambler* (no. 92, 2 Feb. 1751; Bate and Strauss 1969, 2: 122), in relation to a saying of Boileau, that "the books which have stood the test of time, and [have] been admired through all the changes which the mind of man has suffered from the various revolutions of knowledge . . . have a better claim to our regard than any modern can boast." The almost identical expression occurs in Colin Maclaurin's *Account of Sir Isaac Newton's Philosophical Discoveries* (London, 1748), where it is said to be "not worth while . . . to trace the history of learning through its various revolutions in the later ages" (p. 39). Maclaurin also referred to a comparison made by Aristotle of the "revolutions of learning" and "the rising and setting of the stars" (p. 42). The sense of 'revolution' here is similar to that which occurs in

Maclaurin's discussion of the revival of learning after the cloud that had covered Europe — "the liberal arts and sciences were restored, and none of them has gained more by this happy revolution than natural philosophy" (p. 41). In context, this 'revolution' would seem to be a stage in a quasi-cyclical sequence of ebbs and flows, more of a restoration than an innovation.

The mid-eighteenth-century lack of a single clear meaning for the word 'revolution' may be seen in Jean-Jacques Rousseau's *Social Contract* (1762). In book 4, chapter 4, he writes of "the revolutions of empires" and their "causes" ; in context this proves to be a cyclic usage, an ebb and flow or a succession of empires. That Rousseau had in mind the successive appearance of nations or peoples is made apparent by a qualifying clause: "but, as new nations are no longer in process of formation, we have almost nothing beyond conjecture to go upon in explaining how they were created." But earlier in the treatise (bk. 2, ch. 8), Rousseau had written of "periods of violence" in which revolutions occur, apparently implying the noncyclical senses of the word. This interpretation of 'revolution' as a violent change in the political realm is made clear by his illustrative reference to "the State, set on fire by civil wars, . . . born again, so to speak, from its ashes." Yet even here, there are overtones of cycles of regeneration in his simile of the state, reborn from its ashes, taking "on anew, fresh from the jaws of death, the vigour of youth" (bk. 2, ch. 8). A few paragraphs later, Rousseau makes a prediction that Russia will "aspire to conquer Europe, and will itself be conquered. The Tartars, its subjects or neighbours, will become its masters and ours, by a revolution which I regard as inevitable" (p. 37). This inevitability, coupled with the succession of empires, has strong cyclic overtones, although the violent way in which the "subjects" of Russia "will become its masters" may equally suggest possibilities of a post-1789 concept of revolution. Certainly there is a cyclic context (at least a sense of succession) in Rousseau's remark that "every revolution in a royal ministry creates a revolution in the state" (bk. 3, ch. 6). But here, at least, Rousseau intended to convey a sense of radical change, for he explained the foregoing remark by reference to a "principle common to all ministers and nearly all kings": "to do in every respect the reverse of what was done by their predecessors."

In Rousseau's *Discourse on . . . Inequality among Men*, written in 1754, the word 'revolution' occurs in the description of the transition from the first or primitive stage of man to the second stage of organized society. Rousseau attributed this 'revolution' to the invention of metallurgy and agriculture (1964, 152 = pt. 2). "Metallurgy and agriculture," he wrote, "were the two arts, the invention of which produced this great revolution." And he observed that the first of the stages is the "least subject to revolutions."

Many writers of the mid-1700s invoke the cyclical concept of revolution — often an ebb and flow of cultures or a 'revolution of empires'. Typical is Jean François Marmontel, Permanent Secretary of the Académie Française, who had been given the responsibility for all the articles on poetry and literature in the *Encyclopédie* of Diderot and d'Alembert. In his *Elements of Literature* (1737), under the heading of 'poetry', he observed that historians had written about "the revolutions of empires." "How is it," he then asked, "that no one has ever thought of writing the revolutions of the arts, of seeking in nature the physical and moral causes of their birth, their growth, their splendor, and their decadence?" (1787, *9: 297*). The philosopher Condillac made a similar comparison of the stages of human thought and the 'revolutions of empires' when he said that "the revolutions of beliefs follow the revolutions of empires." (1798, *14: 17*)

But in 1755 Condillac also expressed the noncyclical view of revolutions in his shrewd observation that "Bacon proposed too perfect a method to be the initiator of a revolution; Descartes was to be more successful" (1947, *1: 776*). A somewhat similar use of the word 'revolution' is found in some early writings of the economist Turgot. In an essay of the 1750s, "On Universal History," Turgot included a brief history of scientific thought ("philosophie"). He mentioned Aristotle, Bacon, and then "Galileo and Kepler, [who] as a result of their observations, laid the true foundations of philosophy. But it was Descartes who, bolder than they, meditated and made a revolution (1973, 94)." This attribution of a revolution to Descartes is rather unusual among eighteenth-century writers, although French scientists and philosophers inevitably tend to praise him for his radical innovations. In another essay, "A Philosophical Review of the Successive Advances of the Human Mind," read at the Sorbonne in 1750, Turgot adopted a modification in his stance, when he exclaimed: "Great Descartes, if it was not always given to you to find the truth, at least you have destroyed the tyranny of error" (1917, 58). We shall see below (§13.1) that at this time there was a strong expression of belief in a two-stage revolution. Descartes had accomplished only the first stage, eradicating error, but had not advanced fully to the second stage, constructing a new theory to take the place of the old one.

Voltaire

When new concepts develop, and especially when a new concept is a transformation of an older one, there are always periods of ambiguity and confusion. The mid-eighteenth century shows this phenomenon again and again, but perhaps in no other instance as clearly as in the writings of Voltaire. One of Voltaire's earliest publications was his *Philosophical Letters* or *Letters Concerning the English Nation* (1733). Here (letter 7; 1964, *1:* 80), in a discussion of the anti-Trinitarians, Voltaire expresses the same thought that we have just encountered in Condillac: "You see what revolutions occur in beliefs, just as in empires." This example of the cyclical process of revolution is that the "party of Arius, after three centuries of triumph and twelve centuries of oblivion, is finally reborn from its ashes." Again and again in these 'letters' Voltaire indicates the greatness of seventeenth-century science and philosophy (especially Galileo, Bacon, Newton, and Locke). But he never uses the technical term 'revolution' nor does he express the greatness of the new science in terms that easily translate into the radical 'modern' concept of revolution.

Almost two decades after the *Philosophical Letters* Voltaire published his *Age of Louis XIV* (1751), a classic in the literature of history and a work notable for its integration of intellectual and political history. Voltaire introduces the notion of revolution in the second paragraph: "Every age has produced its heroes and statesmen; every nation has experienced revolutions; every history is the same to one who wishes merely to remember facts." Perhaps the sense of the word 'revolution' is to be taken here as the ebb and flow that reached its quasi-cyclical zenith in "four happy ages when the arts were brought to perfection," each of which marks "an era of the greatness of the human mind." Alternatively, Voltaire could be endorsing the new sense of revolution as an event in which something quite new was brought into being. The latter is more in keeping with what he says a few paragraphs later, while discussing what "we call the age of Louis XIV." During this period, according to Voltaire, "rational philosophy . . . came to light," that is, "from the last years of Cardinal Richelieu to those which followed the death of Louis XIV, a general revolution took place in our arts, minds and customs, as in our government." In this example, there was no real sense of a return to any antecedent state in France, although Voltaire could have had in mind that this moment of great change had features in common with the beginnings of the other three great ages (the ages of

Philip and Alexander, of Caesar and Augustus, of the Italian Renaissance). We may thus see in this sentence how the two senses of 'revolution' are associated and how the secular or noncyclical idea of innovation and alteration derives from the cyclical view or the concept of an ebb and flow, a rise and fall.

In the book on the age of Louis XIV, Voltaire uses 'revolution' to describe the Glorious Revolution in England (ch. 15, pars. 9, 20), but there is no adjective 'glorious'. As a Frenchman, Voltaire could only express the view that "in the greater part of Europe" William was known as "lawful King of England and liberator of the nation," but "in France he was regarded as . . . usurper of the kingdom of his father-in-law" (1926, 140). Voltaire introduced science, the subject of chapter 31, by referring to "this happy age, which saw the birth of a revolution in the human mind." I believe there is no ambiguity about the noncyclical sense of 'revolution' in this case, especially since Voltaire then proceeds to introduce the new creations in science made by Galileo, Torricelli, Guericke, and Descartes. Yet the discussion of Copernicus introduces a notion of revival. Voltaire does not mention him by name, but refers to "a canon of Thorn" who did "resuscitate the ancient planetary system of the Chaldeans, so long buried in oblivion" (p. 352). It is noteworthy that although Voltaire mentions a "revolution of the human mind" and "general revolution . . . in our arts, minds and customs," he never seems to have used the expressions "scientific revolution" or "revolution in the sciences," nor did he even introduce the word 'revolution' in association with a single science — such as astronomy or dynamics — or a single scientific development or individual, such as Copernicus or Newton or the introduction of the heliocentric system. This is all the more remarkable in that Voltaire recognized how significant and fundamental were the creations in science of such major founding figures as Galileo and Newton.

In Voltaire's most ambitious historical work, the *Essay on the Manners and Mind of Nations,* published in 1756, the concept of revolution occurs frequently. The introduction begins with a discussion of the changes that the earth itself has undergone, with the observation that "it may be that our world has suffered as many changes as states have experienced revolutions" (1792, *16:* 13). There seems little doubt that here 'revolution' simply means a great (or even cataclysmic) event of change. The ensuing discussion of these "great revolutions" on our globe (pp. 14, 15) makes this interpretation certain. For instance, Voltaire asserts that "the greatest of all these revolutions" would be "the disappearance of the continent Atlantis, if it were true that this part of the world had existed" (p. 15). And this clearly noncyclical use of 'revolution' occurs in chapter 197, a summing-up of the whole history, which opens with a reference to "this vast theatre of the revolutions [which the whole earth has experienced] since the times of Charlemagne" — disasters and destruction, with "millions of men slaughtered."

Revolution as Discontinuity and Change

Despite these many examples of ambiguity, by mid-century the word 'revolution' was well on its way to standing primarily for a great change, without necessary specific overtones of an ebb and flow or rise and fall or cyclical succession. The great *Encyclopédie* of Diderot and d'Alembert, although a self-styled "dictionary of the sciences, arts, and trades," gave primacy of place under the entry 'revolution' to the political sense of the word as a "considerable change in the government of a state":

> RÉVOLUTION, s.f. signifie *en terme de politique* un changement considérable arrivé dans le gouvernement d'un état.

(That is, 'revolution' is a feminine noun that in political discourse signifies "a considerable change" that has occurred "in the government of a state.")

An explanatory paragraph contains three sentences. It is said that "this *word* comes from the Latin *revolvere*, to roll," that "there have never been states which have not been subject to some *revolutions*," and that the "Abbé de Vertot has given us two or three excellent histories of the *revolutions* in different countries." There follows a paragraph on revolution and England. It is observed that "although Great Britain has, in all times, suffered many revolutions," the English particularly use this name for the revolution of 1688. This article on the Glorious Revolution is signed "D.J." (= Chevalier de Jancourt).

Following these discussions of political revolution, there occur three presentations of revolution in science. These are not devoted to revolutions that have occurred in the development of science (for which see chapter 13, below), but rather 'revolution' as a technical term in geometry (solids of revolution), astronomy (where it is indicated that there are two kinds of 'revolution': an axial rotation and an orbital revolution), and geology. Of the three, the longest, by far, is the presentation for

astronomy, written by "O" (= d'Alembert). The entry for geology is
entitled: "Earth's Revolutions." These are said to be the name given by
"naturalists" to "the natural events by which the face of our globe has
been and is still being continually altered in its different parts by [the
action of] fire, air, and water." Finally, there is a much longer article,
more than three times as great as the ones on politics and science put
together, on "*Revolution* [as used in] *Horology.*" This essay (signed by
"M. Romilly") deals with gears and combinations of gearings in clock-
work.

Of particular interest is the use of the term 'revolution' in geology.
The expression 'revolutions of the earth' or 'earth's revolutions' occurs
prominently in the writings of Buffon. For example, in his *Second Discourse* of the *Theory of the Earth,* published in 1749, he wrote (Buffon 1954:
104):

> It is not possible to doubt . . . that there have occurred an infinity
> of revolutions, of upheavals, of particular changes, and of alter-
> ations on the surface of the earth, as much by the natural move-
> ment of the waters of the sea as by the action of rains, of frosts, of
> running waters, of winds, of subterranean fires, of earthquakes, of
> floods, etc.

He thus considered the changes in the face of the earth to be the result
of a "succession of natural revolutions" (p. 105). This same usage of
'revolution' occurs in other writings of Buffon, notably his *Epochs of
Nature* (1779), which begins (1954, 117):

> Just as in Civil History, one consults titles, searches for medals, and
> deciphers ancient inscriptions in order to determine the epochs of
> human revolutions and to establish the dates of human or civil
> events [évènemens moraux], so in the same way in Natural History,
> it is necessary to dig into the archives of the world, drawing ancient
> monuments from the entrails of the earth, collecting their debris,
> and gathering together in one body of proofs all the clews of physi-
> cal changes which can enable us to regain the different ages of
> Nature.

In 1812, Buffon's comparison of the historian and the geologist was used
in a striking way by Georges Cuvier, who referred to himself as one of a
new species of antiquarian, who had "at the same time to learn to

restore the monuments of past revolutions and to decipher their meaning." Buffon refers to the changes that occurred in those earliest ages,
to events entirely forgotten, "revolutions antedating the memory of
man" (p. 118). For Buffon there is apparently a succession of revolutions,
but such revolutions—whether in the realm of politics or of natural
history—are in no sense cyclical.

Buffon's use of 'revolutions' strongly influenced the German philosopher Johann Gottfried von Herder later in the century. Chapter 3 of
book 1 of Herder's *Ideas on the Philosophy of the History of Mankind* (published in 1784; 1791) is entitled "Our Earth has experienced Many Revolutions, until it became what it is." Herder, acknowledged to be a pioneer in the study of anthropology, the science of primitive culture,
adapted an 'evolutionist' point of view with respect to the lower forms of
life, which exist for the sake of man and may exhibit the imperfections
which are absent in man; but they are not necessarily antecedent states
of existence that lead up to man. His human evolutionism is not the
biological development of man but rather his cultural development. His
book defines human history as "a pure natural history of human
powers, actions and propensities, modified by time and place." Man's
cultural development is seen as a strictly natural process, an interaction
between man and his changing physical environment. Hence Herder,
following Buffon (see Sauter 1910), discussed the history of the earth in
terms of revolutions produced by the actions of water, fire, and air (1887,
13: 21). He particularly notes that some of these helped form the earth,
and he expresses the hope that "I will live to see a theory of the first
essential revolutions of the Earth" (1887, *13:* 22), the revolutions that
"originally created the Earth." Buffon, he observed, "is only the Descartes" of this science, and his hypotheses will eventually be displaced in
the same manner that a Kepler and a Newton surpassed and refuted the
hypothesis of Descartes. Referring to "the new discoveries about heat,
air, fire, and their various effects on the components, the composition
and the decomposition of Earth-substances" and the new "simple basic
principles" of electricity and magnetism, Herder could envisage a time
when the structure of the earth would be explained as simply and surely
"as Kepler and Newton explained the structure of the solar system."

Herder quite naturally followed Buffon in using 'revolution' as a
name for the cataclysmic events that caused the development of the
Earth (*Ideen*, bk. 1, ch. 3). He concluded that "today, changes of this
terrible kind are not as frequent [as they were in the early time of the
Earth's history], because the Earth has concluded its development"; the

earth "has grown old." But such revolutions have not ceased entirely, he wrote, as the great Lisbon earthquake proved (1887, *13:* 24).[3]

Influence of the American and French Revolutions

As the eighteenth century reached its third quarter, there occurred the most notable single sociopolitical event since the Glorious Revolution. Today, after the French, Russian, and Chinese revolutions, the American Revolution — like its predecessor Glorious Revolution — may not seem very radical or even 'revolutionary'. And there has been a conservative political tendency to call the American Revolution the War of Independence or, as a compromise, the Revolutionary War. In its own time, the American Revolution had a double image. On the one hand, it was a 'revolution' in the sense of a radical change that was chiefly a return to the conditions of the Glorious Revolution and its Act or Bill of Right. Conservatives could back a revolution that implied a return or — as Bernard Bailyn likes to say — a 'revolvement' to the rights which had been guaranteed to all Englishmen a century or so earlier but which had been eroded, chiefly by the Walpole government. But certain radicals, including such diverse political figures as Thomas Jefferson and Thomas Paine, saw in the Revolution the establishment of something wholly new. This was the sense of the motto on the Great Seal of the United States, adopted soon after the Revolution, "NOVUS ORDO SECLORUM," a *new* order of the ages, or — in the reinterpretation of the late 1930s — a 'New Deal'.

The newness of the Revolution, rather than a return to an ancient condition that was better than the present, is symbolized in the ringing cadences of Jefferson's *Declaration of Independence:* "When in the course of human events it becomes necessary for one people to dissolve the political bands which have connected them with another, and to assume among the powers of the earth the separate and equal station to which the laws of nature and nature's god entitle them." Here is no backward-looking assertion of ancient rights but rather an explicit statement about the present. Furthermore, the "just and equal station" mentioned by Jefferson is one which is justified neither by the God of Revelation nor by the Holy Writ, but rather by "nature" and by "nature's god." Nor does Jefferson continue, as he once had planned to do, by invoking "truths" held to be "sacred and undeniable," but rather by declaring certain truths to be "self-evident" in the special sense in which Newton had conceived the axioms on which his *Principia* had been built to be self-evident. And the novelty of the Revolution was declared forthwith in the radical assertion that men are "endowed by their creator" with certain "inalienable rights," among them "life, liberty, and the pursuit of happiness."

The French Revolution, taking on early the name made definite by its American predecessor, went much further than either the Glorious or the American Revolution in its program of political and social reform. And, as I have mentioned earlier, after the French Revolution the word 'revolution' itself generally lost any remaining cyclical overtones, save in a purely astronomical sense. The French Revolution not only set the seal, once and for all, on the new sense of the word; its events affected thinking about revolutions in a number of ways. First of all, the extremes and violence of the Revolution caused concern about the possible evil consequences of revolutions as well as their accepted beneficial effects. Second, the French Revolution set a pattern in which profound social changes were seen to be a concomitant of political action. Third, it has been argued that this new concept of revolution had important overtones of inevitability, just as there is an inevitability about the revolutions of the planets around the sun.[4]

Although the French Revolution was forward-looking and was not, in general, conceived as a return to an antecedent state, there are important elements of the past visible in ceremonies and symbols. Thus a prominent symbol of the Revolution was a Phrygian 'liberty cap', to be seen on countless engravings of the 1790s. This cap was traditionally donned when a Greek slave received his manumission and it was a visible sign of having attained freedom (see Figures 8 and 9). Another symbol was the bundle of sticks, the Roman 'fasces', used also in the American Revolution. Here was a symbolic expression of the kinship between the new program of the French Revolution and the traditions of antiquity now being imbued with new life and perhaps also a new or extended meaning.

The late Hannah Arendt, in particular, argued that the old concept of astronomical revolution and the sense of return was a feature of the French Revolution. She took as her primary example a conversation alleged to have taken place between King Louis XVI and the Duc de la Rochefoucauld-Liancourt on the night of 14 July 1789, right after the fall of the Bastille. "It is a revolt," the King is supposed to have said. "No, Sire," was Liancourt's answer, "it is a revolution." Of course, we have

Figure 8. Allegory of the French Revolution in honor of Rousseau: an oil painting of 1794 by N. H. Jeaurat de Bertry (1728–1796). The presiding figure, in an oval frame, is Jean-Jacques Rousseau, above the all-seeing eye of Truth. On the central bundle of sticks, or fasces in the Roman style, are "force," "truth," "justice," "union." Alongside the fasces is a tree with the placard "liberty." Atop the fasces is a pole with a Phrygian liberty cap, the same cap being worn by the man in the lower right-hand part of the painting. The pages of the open book on the central broken column read "rights of Man and of the Citizen." (Musée Carnavalet, Paris; photo Bulloz.)

Figure 9. Phrygian liberty cap on a liberty pole, a watercolor made by Goethe on the return from the campaign in France, 1792. On the pole is an inscription: PASSERS-BY, THIS LAND IS FREE. A preliminary sketch, in pen and ink, appears on the back of a letter from Goethe to Herder (16 October 1792). For information on both versions, see Gerhard Femmel, ed., *Corpus der Goethezeichnungen*, vol. 6B, nos. 136 and 137 (Leipzig: Veb E. A. Seemann, 1971). (Anton und Katharina Kippenberg Stiftung, Goethe-Museum, Düsseldorf.)

no way of telling what was in Liancourt's mind; in fact, we have no good contemporaneous evidence that he made such a statement at all. Hannah Arendt was a profound student of revolutions and I for one would tend to trust her historical and analytical insight into this matter. She believed that in this reported conversation the word 'revolution' was used "politically for the last time, in the sense of the old metaphor which carries its meaning from the skies to the earth" (1977, 47). I myself found an independent confirmation of Hannah Arendt's idea in an eighteenth-century political print. Reproduced as the frontispiece to this book, this contemporaneous print displays the "Astronomical System of the French Revolution." In Liancourt's statement, furthermore, Hannah Arendt conjectured, "For the first time perhaps, the emphasis has entirely shifted from the lawfulness of a rotating, cyclical movement to its necessity." Hence she suggested that the political image of revolution continued to derive from "the movements of the stars," but that what "is stressed now is that it is beyond human power to arrest" the revolutionary movement, that it has become "a law unto itself." The alleged conversation on 14 July 1789 indicates a difference between a revolt and a revolution, which in the eighteenth century was one of scale and purpose. A revolt was held to be an insurrection or uprising, but a revolution implied a fundamental change in the political and social organization of the state. In the context of the age, Liancourt would have been saying that there was not simply an uprising against the present leaders of government but rather a movement to alter the political system. He would, in other words, have envisaged a threat to the established form of government and not just to the government in power.[5]

13

Eighteenth-Century
Conceptions of
Scientific Revolution

I n the opening years of the eighteenth century, Bernard Le Bouyer (or Bovier) de Fontenelle was in a very advantageous position to evaluate the mathematics and science of his day. As permanent secretary of the Paris Académie Royale des Sciences, he summarized the intellectual activities of the members and wrote a history of the early years of this group. Therefore Fontenelle's views on the revolution in mathematics are of particular interest to a history of the concept of revolution in science. The preface to Fontenelle's *Elements of . . . Geometry* (1727) presents a discussion of the newly invented (or discovered) infinitesimal calculus ("le calcul de l'infini") of Newton and Leibniz, and of the several ways in which "Bernoulli, the Marquis de l'Hôpital, Varignon, all the great mathematicians" carried the subject forward "with giant's steps." Then he says that the infinitesimal calculus introduced into mathematics "a facility of which one would not previously have dared to conceive a hope; and it is the epoch of an almost total revolution occurring in geometry" (1790, 6:43). The conjunction of words 'époque' and 'révolution' leaves no doubt that Fontenelle had in mind a change of such an order of magnitude as to alter completely the state of mathematics. And Fontenelle went on at once to emphasize that this revolution was "happy," that is, progressive or beneficial to mathematical science, although not unaccompanied by several problems.

Fontenelle used the term 'revolution' in the *éloge* of the mathematician Michel Rolle, which he wrote in his capacity as permanent secretary. It was first presented in 1720. 'Revolution' does not here occur in relation to the work of Rolle himself, but rather in a remark about a book by the Marquis de l'Hôpital on the *Analyse des infiniment petits,* or the infinitesimal calculus, the first textbook on the calculus (Paris, 1696; later eds., 1715, 1720, 1768). (Fontenelle was actually the author of the anonymous preface to l'Hôpital's book, though he used a style that would lead the unsuspecting reader to suppose it had been written by l'Hôpital himself.) According to Fontenelle (1792, 7: 67):

At that time the book of the Marquis de l'Hôpital had appeared, and almost all mathematicians began to turn to the new geometry of the infinite [that is, the new infinitesimal calculus], until then little known. The surprising universality of the methods, the elegant brevity of the proofs, the neatness and speed of the most difficult solutions, a singular and unexpected novelty, all attracted the mind and there was in the mathematical world a well marked revolution [une révolution bien marquée].

Fontenelle also used 'revolution' in the *éloge* of l'Hôpital (d. 1704), again in relation to his textbook and the avidity with which "l'*Analyse des infiniment petits* had been seized by all those who were in the stage of becoming mathematicians." L'Hôpital's aim had been "primarily to make mathematicians," Fontenelle wrote, and he had the satisfaction of seeing that "problems previously reserved for those who had grown old on the thorns of mathematics became the first-go's for young people." At last, "apparently the revolution will become still greater, and in time there will be found as many students of mathematics as formerly there had been mathematicians" (1790, 6: 131).

These last two uses of 'revolution' in relation to l'Hôpital's textbook differ from the former instance, in that the calculus inaugurated a conceptual revolution in mathematics, whereas l'Hôpital's *Analyse des infiniment petits* consolidated that revolution and made its methods and achievements so readily available as to revolutionize the profession of mathematician. In other words, l'Hôpital was (according to Fontenelle) primarily responsible for attracting young mathematicians ("géomètres") to the new analysis and endowing them with new powers. Fontenelle would thus seem to have made a distinction between "une révolution presque totale . . . dans la géométrie" (an almost complete revolution in geometry [mathematics]) and "une révolution bien marquée," such as l'Hôpital's book produced "dans le monde géomètre" (that is, a clearly discernible revolution in the geometrical world).

Those who study the calculus experience exactly what Fontenelle has described — the power of dealing with the most difficult problems in a simple and straightforward manner. Usually this sense of the extraordinary facility of solving complex problems is first made manifest in the study of analytical geometry and then in the calculus. The revelation of the power and profundity of mathematics thus follows the steps of the two great revolutionaries of the seventeenth century, Descartes and Newton — Newton sharing the honors with Leibniz.

As Fontenelle was well aware, Newton and Leibniz engaged in an acrimonious priority dispute in the matter of the invention of the calculus, the basis of the revolution in mathematics. In the preface to his *Elements* Fontenelle said of the calculus: "Newton was the first to find this marvelous calculus, Leibniz the first to publish it. That Leibniz was the inventor as well as Newton is a question of which we have given an account in 1716 and we shall not repeat it here."

Fontenelle's use of the word 'époque' ("it is the epoch of an almost complete revolution") indicates that 'revolution' has the sense of the creation of something wholly new (see chapter 4, above). Fontenelle has also written of a "révolution . . . totale," or complete revolution. In thinking about a change of very great magnitude, the words 'total' and 'complete' come to mind as indications that the revolution had altered everything. But this implies that users of such a phrase had become oblivious of the original cyclical sense of the word, since a complete or total revolution (as in traversing 360° or making a full circuit of an orbit) would literally mean a return to the point of starting-out, that is, no net change at all.[1]

Fontenelle wrote about revolutions in other domains of human affairs than mathematics. In a celebrated essay, "On the Usefulness of Mathematics," he said that history provides a "spectacle of continuous revolutions in human affairs." These comprise "the rises and falls of empires, of morals, of customs, and of beliefs which succeed one another ceaselessly" (1760, *6: 69*). In his *éloge* of Czar Peter the Great, Fontenelle referred specifically to revolutions occurring in Russia and to a revolution in Persia by Mahmoud.

At the beginning of the eighteenth century, Fontenelle shows us the concept of revolution (without any vestiges of the old cyclical origins of the term) as a recognized and accepted mode of scientific change — but in mathematics, not in the physical or biological sciences. I have not found any reference by Fontenelle to Descartes's having made a revolution, although Fontenelle was a strong believer in the Cartesian philosophy, nor did he invoke the concept of revolution or use the term in his biography of Newton (see Cohen and Schofield 1978, 427–474). This prominent early reference to a revolution in mathematics but not in the physical sciences is significant, I believe, indicating that neither the Cartesian nor the Newtonian natural philosophy had as yet been fully and universally accepted, as had the new mathematics of Newton and Leibniz.

As the eighteenth century advanced, Newton's revolution in natural philosophy did become recognized more and more (and, eventually, almost universally). The earliest instance I have found of an explicit statement of the revolutionary force of Newton's *Principia* occurs in the opening sentences of an essay by Alexis-Claude Clairaut, read at a meeting of the Paris Royal Academy of Sciences on 15 November 1747. Clairaut said unambiguously that Newton's "famous book of *Mathematical Principles of Natural Philosophy* has been the epoch of a great revolution in Physical Science." Here, again, we may note the use of the word 'epoch' as a strengthening factor in Clairaut's assertion of a Newtonian revolution. Clairaut's statement is all the more significant in that the essay in which it appears was devoted to the possibility that Newton's inverse-square law of gravity might not be exactly and universally true but might need to be revised.

The fact that both of these early references to revolution in science occur in relation to Newton is worthy of notice, since it was Newton's achievement in pure mathematics coupled with his analysis of the system of the world on the basis of gravitational dynamics that actually set the seal on the Scientific Revolution and caused scientists and philosophers to recognize that a revolution had in fact taken place. We might say that Newton's *Principia* of 1687 played the same role in the recognition of the occurrence of a scientific revolution that the Glorious Revolution of 1688 apparently did for political revolution.

Diderot and d'Alembert

In the great *Encyclopédie* of Diderot and d'Alembert, as we saw in the last chapter, there is a considerable discussion of political revolutions (in a secular and noncyclical sense) and of 'revolution' as a term in geometry, astronomy, geology, and horology. But there is no mention of revolutions — in the sense of a dramatic break with the past — occurring in science. For this topic, we must turn to other entries in the *Encyclopédie,* supplemented by the writings of d'Alembert and Diderot. In the "Discours préliminaire" of the *Encyclopédie* (published in 1751), d'Alembert introduced the concept of revolution in a thumbnail sketch of the rise of modern science or, rather, of a philosophy associated with modern science. But the aim of the essay was to sketch out a methodological and philosophical analysis of all knowledge, including science, which occupies a central place in his scheme, and not to portray the sciences themselves.

D'Alembert begins his historical presentation with "le Chancelier Bacon," who occupies an avuncular position, and then moves on to a brief résumé of Descartes's radical innovations. Fully appreciative of the significance of the Newtonian natural philosophy, which in fact had just overthrown and replaced the Cartesian, d'Alembert nevertheless felt the need to say some kind words for Descartes, a Frenchman and fellow mathematician. He thus called particular attention to the great "revolt" of Descartes, who had shown "intelligent minds how to throw off the yoke of scholasticism, of opinion, of authority." D'Alembert had in mind a clear image of the action of political revolutionary forces, and he portrayed Descartes (1751–1780, *1:* xxvi; d'Alembert 1963, 80–81) "as a leader of conspirators who, before anyone else, had the courage to rise against a despotic and arbitrary power and who, in preparing a resounding revolution, laid the foundations of a more just and happier government, which he himself was not able to see established." Descartes's role in thus "preparing" the "revolution," or his "revolt," was "a service to philosophy perhaps more difficult to perform than all those contributed thereafter by his illustrious successors."[2] Although d'Alembert does not say so specifically, he implies that the revolution prepared by Descartes was achieved by Newton. For d'Alembert not only proceeds at once to spell out at length the accomplishments of Newton in general physics, celestial mechanics, and optics, in the most praiseworthy terms imaginable, but he specifically says that when Newton "appeared at last," he "gave philosophy a form which apparently it is to keep." Thus, in science Newton actually achieved the revolution that Descartes had only prepared.

Furthermore, after pointing out that this "great genius [Newton] saw that it was time to banish conjectures and vague hypotheses from physical science" (1963, 81), d'Alembert observed that Newton "abstained almost totally from discussing his metaphysics in his best known writings." The significance of this remark is that it led d'Alembert to conclude his presentation of Newton by observing: "Therefore, since he has not caused any revolution here, we will abstain from considering him from the standpoint of this subject [that is metaphysics]." The implication is that from other standpoints — gravitation, celestial mechanics, the system of the world, optics, the nature and limits of scientific explanation — Newton *had* made a revolution. In fact, d'Alembert explicitly says that Newton "has doubtless deserved all the recognition that has been given him for enriching philosophy with a large quantity of real assets" (1963, 83). Then he added the appropriate comment that perhaps Newton "has done more by teaching philosophy to be judicious and to restrict within reasonable limits the sort of audacity which Descartes had been forced by circumstances to bestow upon it" (1963, 81).

The concept of revolutions in science appears explicitly and prominently in an article written by d'Alembert for the *Encyclopédie*, entitled "Expérimental." Here, as in the "Discours préliminaire," d'Alembert includes a brief history of the subject, once again emphasizing Bacon and Descartes and ending with Newton. First of all, d'Alembert observes that Bacon and Descartes had introduced "the spirit of experimental physics"; soon the Accademia del Cimento, Boyle, Mariotte, and others took up the work. Then (*Encycl.* 1751–1780, 6: 299) the science of Descartes replaced that of Aristotle, that is, the commentators on Aristotle. Newton, he held, succeeded in proving what his predecessors had only predicted — the true art of introducing mathematics into physics. Uniting mathematics with experiment and observation, Newton created a truly new science that was "exact, profound, and illuminating." At first Newton's ideas were not fully and readily accepted, according to d'Alembert, but eventually a "new generation of scientists arose" which was Newtonian. D'Alembert thus appears to have been among the first persons to recognize the generational feature of scientific revolutions, much as Max Planck was to do some two centuries later. "Once the foundations for a revolution are laid," d'Alembert wrote, "it is almost always in the following generation that the revolution is completed; rarely earlier, because the obstacles disappear of their own accord rather than give way; rarely later, because once over the barriers, the human spirit often advances faster than it itself wishes, until it reaches a new obstacle which forces it to rest for a long time." In this passage, d'Alembert not only has expressed a philosophy of historical development in science according to generations; he has also centered the great revolution in science on the work of Isaac Newton.

In another work, having no relation to the *Encyclopédie*, a brief "Tableau de l'esprit humain au milieu du dix-huitième siècle" ("Presentation of the Human Mind in the Middle of the Eighteenth Century"), d'Alembert stated a general theory of revolutions in the intellectual sphere (1853, 216–218): "It seems that for about three hundred years, nature has destined the middle of each century to be the epoch of a revolution in human thought" ("une révolution dans l'esprit humain"). He notes that the "capture of Constantinople, in the middle of the fifteenth century, caused a renewal in the world of letters in the West." Similarly, "the middle of the sixteenth century saw a rapid change in the religion and the system of a great part of Europe." Finally, "Descartes, in the middle of the seventeenth century, founded a new philosophy."

Volume 6 of the *Encyclopédie*, containing d'Alembert's article, "Expérimental," was published in Paris in 1756. The previous volume (5: Paris, 1755) contains a discussion by Diderot of a revolution in science; it occurs in his article, "Encyclopédie." Diderot was interested in the fact that changes were occurring in the sciences, so that a dictionary of the previous century would be lacking in the new words which science had either invented or brought to the fore with new meanings or new significance. Thus, under 'Aberration' the older dictionaries would not give the current astronomical meaning (associated with Bradley's discovery), and 'Electricity' would have only a line or two giving "false notions and ancient prejudices." Even so, Diderot observed, "the revolution is perhaps less strong and less perceptible in the sciences and in the liberal arts than in the mechanical arts; but there is a revolution in the sciences and in the liberal arts."

Diderot also wrote of revolutions in science in his famous essay, "On the Interpretation of Nature" (1753; enlarged ed. 1754). Here Diderot writes, "We are on the verge of a great revolution in the sciences" (1818, *1:* 420). This revolution entailed a complete rejection of geometry and of the geometric spirit in science. "To judge by the inclination that our writers have toward morals, fiction, natural history, and experimental physics," he wrote, "I feel almost certain that before 100 years are up, one will not count three great geometers in Europe."

These passages and others indicate the importance of 'revolution' (or of 'revolutionary change') in Diderot's theory of scientific development. Like d'Alembert, he conceived that the progress of science was marked by a succession of revolutions, but the concept of a "maximum interval between one revolution and another" being a "fixed quantity" was apparently original with him. Although it seems that Diderot was conceiving of revolutions primarily as radical secular changes, the foregoing passage has also some overtones of a cyclical process of revolutionary changes, in which the term maximum interval even suggests overtones of the period of revolution in the cyclic phenomena of nature. And it should be observed that although the cyclical sense of revolution in the political realm does not appear at all in the *Encyclopédie* in the entry 'Révolution', this very sense does appear in d'Alembert's "Preliminary Discourse," where he wrote (*Encycl.* 1751–1780, *1:* p. xi) of the "principal fruits of the study of empires and of their revolutions." Later on in the "Discourse" d'Alembert also wrote of revolutions as moments of radical change, but with overtones of the concept of the ebb and flow of empires, a succession of decay followed by rebirth. He begins by referring to the Middle Ages as "those dark times," when "one of those revolutions which make the world take on a new appearance was necessary to enable the human species to emerge from barbarism" (p. xx). He continues: "The Greek [Byzantine] empire was destroyed, and its ruin caused the small remainder of knowledge to flow back into Europe. The invention of printing and the patronage of the Medici and of Francis I revitalized minds and enlightenment was reborn everywhere" (1963, 62). The cyclical overtones of this passage, the sense of ebb and flow, come all the more to mind since, at that time, this would still have been a common usage of the word 'revolution'.

Two Writers on Revolutions in Astronomy

By the time the *Encyclopédie* was published, we have seen 'revolution' gaining currency — at least in French — in its new meaning of a secular, rather than a cyclical, change of great magnitude.[3] During the second half of the eighteenth century, this concept, and the word to express it, were more and more applied to realms of the mind, and in particular to writings about science. Various authors, however, dated the revolutions at different times, according to their subject. Thus in 1764, Joseph

Jérôme Le Français de Lalande (La Lande) saw a revolution in astronomy in the era after Hevelius (1764, *1:* 131):

> [This was a time when] all the nations disputed with one another the glory of making discoveries or of bringing the subject to perfection; the Academy of Sciences at Paris, the Royal Society of London, played above all the greatest part in this revolution. The number of illustrious scientists and renowned astronomers that they produced is immense.

But Lalande did not use the word 'revolution' for Copernicus's revolt against the authority of Ptolemy, nor for the radical novelties discovered or introduced by Galileo or Kepler; he apparently reserved the designation of 'revolution' for the process of discovery and improvement which he conceived to have been part and parcel of the establishment and elaboration of the subject of astronomy in more recent times. We must, of course, be wary of assuming that the distinction that appears in Lalande's text originated in a conscious and clear-cut decision concerning usage. Perhaps what is most significant is only that Lalande does introduce the notion of revolution in science.

The writings of Jean-Sylvain Bailly, published in the decade before the French Revolution, show how the concept of revolution in the sciences had achieved the form in which, with variations, it continued well established during the nineteenth century. In his *History of Modern Astronomy*, Bailly introduced revolutions of several sorts and magnitudes, ranging from the large-scale elaborations of the Copernican system of the world and the Newtonian natural philosophy to revolutionary innovations in the design and use of telescopes. As a practical astronomer, Bailly had in mind the improvement of telescopes by the addition of cross-hairs, and especially of micrometers: "This perfection added to the instruments, this exactness in practise, had an influence on all observations in a manner well enough marked to produce a revolution." Furthermore, "This revolution, the idea of this happy application, was . . . due to Picard and Auzout" (1785, 2: 272–273).

Bailly discussed revolutions of the past and of the recent present, and even made forecasts of revolutions to come, though only smaller ones, primarily the introduction of new instruments and new methods of computing (without approximations) and of integrating. He also predicted a replacement for the pendulum clock. Bailly's history also introduces a clearly worked-out concept of a two-stage revolution, applicable to revolutions in science on a grand scale, in which there is first a destruction of an accepted system of concepts, followed by the establishment of a new system (see §13.1, below). But even in Bailly's writings, the older concept of revolutionary change in cycles is present along with the new use of the term 'revolution' to indicate a radical and dramatic change in science, most often the effect of the work and thought of a single individual.

It was Bailly who put into circulation the idea that there had been a Copernican revolution, though his two-stage theory of revolutions evidently led him to conclude that neither Galileo nor Kepler had actually made a revolution of the sort he attributed to Copernicus. He did strongly believe in the Newtonian revolution, which appears in his history again and again. Bailly fully appreciated the remarkable contributions of Descartes, but he apparently did not find the Cartesian innovations revolutionary. Bailly said that astronomical observations naturally set the question as to causes: "It is a sublime idea to have dared to reduce the laws of general motion of the universe to the laws of motion of terrestrial bodies. This enterprise belongs exclusively to our modern centuries; the credit for it is due to Descartes." Furthermore, "Descartes discovered that the same mechanism must move bodies in celestial spaces and at the surface of the Earth." Even if Descartes did not quite conceive the true mechanism, Bailly went on, "We must not forget that this new and grand thought was the fruit of his genius" (1781, xi). He observed that we do not "take any glory from that great man Newton, in doing justice to Descartes." Furthermore, "if Descartes has opened the way to the most beautiful discoveries by his inventions in geometry, Kepler has foreseen and has left us more truths of physical science than he has. Descartes dared more and his audacity is the measure of the force of his genius; it was only lacking in him to have been wiser. He seems unaware of many facts known in his day" (1785, 2: 192).

Bailly also wrote in terms of a cyclical process in the development of astronomy. Thus a revolution might, on occasion, signify a return to an older idea or concept, or an older principle. But Bailly shrewdly observed that one must not assume that there had been no real change simply because an idea or concept now in current use may have occurred once before. The example he gives is a curious one: "Pagan theology supposed that the world had come out of an egg; this was not the first time that ignorance and profound knowledge, by opposite paths, have arrived at the same results" (2: 519). A more complete expression of change by cyclical revolution occurs at the beginning of the second volume of his history (2: 3–4):

> In writing this history, we perceive that on one hand men, persuaded of the simplicity of the mechanism of the universe, tend continually to this idea, even in brushing it aside: we see on the other hand that this idea is one of the most ancient that have been preserved for us. The natural conclusion is that we return to the idea from which we have set out: such is our path, we always go round in a circle. But this idea, this first beginning of known labors had itself to be the end of a revolution.

An ebb and flow of astronomical science, following the rise and fall of civilizations (or empires), appears again and again in Bailly's history (for example, 1: bk. 8, §1). Bailly believed that the astronomy of the Chaldacans, the Indians, and the Chinese was the "débris" of a science of "an earlier civilization . . . of which we do not know the greatest part. They were destroyed by a great revolution" (1781, 18). The loss of the astronomical ideas of this civilization could only have occurred "by some great revolution which destroyed the men, the towns, the knowledge, and left only debris. Everything concurs in proving that this revolution took place on the Earth" (p. 59). In the "Table générale des matières" or index, covering the three volumes of his *Modern Astronomy* and the single volume on *Ancient Astronomy*, references to these two revolutions (s.v. 'Revolution') precede the references to revolutions of stars and planets.

The fact that Bailly was aware of the possible cyclical process in revolutions, so obvious to any practicing astronomer, does not diminish the thrust of his use of 'revolution' in relation to historical events characterized by a secular rather than a cyclical change of considerable magnitude. Since Bailly not only used 'revolution' to denote a radical change in science (in the sense that d'Alembert and Diderot had done) but actually introduced both the word and the concept throughout his three-volume history of modern astronomy, we may conclude that by this time 'revolution' had become fully accepted into the discourse of the history of science and of the analysis of the growth of scientific ideas, theories, methods, and systems of thought.

Writers on Scientific Revolution at the End of the Century

By the 1780s, French authors who refer explicitly to one or another revolution in the sciences abound.[4] But the case of Condorcet may especially attract our attention since he is said (by Littré 1881–1883) to have been an originator of the term 'révolutionnaire'. The concept of a revolution in science (and the use of 'revolution' to express it) occurs frequently in the *éloges* of deceased academicians which it was Condorcet's duty to write and read, in his capacity of permanent secretary, just as Fontenelle had done earlier. Thus, of Duhamel du Monceau (1783): "He will mark an epoch in the history of sciences, because his name is found linked to that revolution in minds which more particularly directed the sciences toward public usefulness." Of Haller (1778): "The work in which von Haller published these discoveries was the epoch of a revolution in anatomy." Of d'Alembert (1783): "This principle was the epoch of a great revolution in the physico-mathematical sciences." Of Euler (1783): "He owes this honor to the revolution he produced in the mathematical sciences." (See Condorcet 1847, 2: 300, 641; 3: 58, 40, and also 7, 8, 9, 28.) And so on. In three of these examples, we see Condorcet using the term 'époque' along with 'révolution' in a century-old tradition that unambiguously defines the noncyclical sense of 'revolution'.

The major work of Condorcet's in which the term and the concept of revolution figure most prominently is his *Sketch for a Historical Picture of the Progress of the Human Mind*, first published in 1795. Condorcet wrote here of the recent American Revolution, and of the not-yet-completed revolution in France, with shrewd comments on the causes of the differences between the two. Of special interest in the present context is his discussion of Descartes, who is said to have given "men's minds that general impetus which is the first principle of a revolution in the destinies of the human race" (Condorcet 1955, 147; 1933, 173). In the account of the rise of chemistry, Condorcet introduced some of the improvements in that subject that "affecting, as they do, a given scientific system in its entirety by extending its methods rather than by increasing its truths, foretell and prepare a successful revolution." Condorcet had in mind the "discovery of new methods" of collecting and analyzing gases; "the formation of a [new] language" for chemical substances; the "introduction of a scientific notation"; "the general laws of affinities"; the use of "methods and instruments" from physics for "calculating the results of experiments with rigorous precision"; and the "application of mathematics to the phenomena of crystallization" (1955, 153–154; 1933,

180–181). Here Condorcet also spelled out his scientific version of the hotly-debated topic of our own times: the 'preconditions' of a revolution.

Condorcet's special use of 'revolution' in relation to chemistry, rather than physics or astronomy or the life sciences, was a natural result of the fact that he had actually been witness to the recent Chemical Revolution. This revolution had been invented by Lavoisier in a double sense, for he was its chief architect, and he gave the Chemical Revolution its name, referring to his own work in terms of 'revolution' in at least three manuscripts.

Lavoisier was not the only scientist of the eighteenth century to refer to his own work in science in terms of 'revolution'. Two others were Symmer and Marat. The fact that at least three scientists used the word 'revolution' to describe what they were doing is an index of the degree to which the concept of revolution in science, shorn of any overtones of cycle or return or ebb and flow, was becoming an accepted mode of conceiving how progress is made in the sciences.

Not unexpectedly, Joseph Priestley—an ardent supporter of the American and French Revolutions—was among those who transferred the concept of revolution from the political realm to science.[5] In a work on phlogiston and the decomposition of water, published in 1796, he referred to the victory of the new chemistry as one of the greatest, the most sudden, and the most general of the "revolutions in science" (see chapter 14).

Priestley differed from most of his contemporaries in that he did not believe that revolutions in science were always progressive, always causing a more rapid advance in the state of knowledge. "Nothing," he wrote, "is more common, in the history of all the branches of experimental philosophy, than the most unexpected revolutions of good or bad success." He explained his point of view as follows (1966, 300):

In general, indeed, when numbers of ingenious men apply themselves to one subject, that has been *well opened*, the investigation proceeds happily and equably. But, as in the history of *electricity*, and now in the discoveries relating to *air*, light has burst out from the most unexpected quarters, in consequence of which the greatest masters of science have been obliged to recommence their studies, from new and simpler elements; so it is also not uncommon for a branch of science to receive a check, even in the most rapid and promising state of its growth.

Others who used the concept of revolution in science were William Cullen, A.-R.-J. Turgot, and Immanuel Kant, as well as those who saw a Harveyan revolution in the life sciences. Another eighteenth-century scientist who wrote of a revolution in science was the Genevese biologist Charles Bonnet. "The book on the leaves of plants," he wrote in 1779, "had put me again in contact with another great man, who was soon to make in physiology the same kind of revolution that Montesquieu did in politics: I speak of the late M. de Haller" (Savioz 1948, 155). Bonnet thus shared the judgment of Condorcet about the revolutionary influence of Albrecht von Haller in science.

By the century's end a number of authors wrote about the revolutionary advances in science in the Enlightenment, in the century from Newton to Lavoisier and Volta. Three of them in particular developed a theoretical basis or perspective for considering such revolutions: Samuel Miller, an American; John Playfair, a Scot (for whom see §18.1, below); and Christoph Lichtenberg, a German (§14.2, below). Miller, a New Jersey clergyman, produced the first overall review of the intellectual accomplishments of the eighteenth century: *A Brief Retrospect of the Eighteenth Century. Part First; in two volumes: containing a Sketch of the Revolution and Improvements in Science, Arts, and Literature, during that period* (New York, 1803). Miller's use of 'revolution' to denote gigantic progressive steps is notable since it stresses the mode of advancement in science (as also in the arts and literature) that had come to be accepted as the norm during the century whose achievements he was reviewing. His work was more a compilation than an original essay, as he himself admitted (2: ix): "Though the greater part of this work consists of compilation; yet the writer claims to be something more than a mere compiler. He has offered, where he thought proper, opinions, reflections, and reasonings of his own." Miller would have encountered the concept of revolution in science and in the arts in the course of his readings (including many works in French, which are prominent among his footnotes and references).

In his "Recapitulation" at the end of the second volume (p. 411), Miller characterized the eighteenth century as "pre-eminently an *age of free inquiry.*" Men learned, to a greater degree than had ever been known before, to "throw off the authority of distinguished names . . . to discard all opinions, to overturn systems which were supposed to rest on everlasting foundations." Men pushed their inquiries to the utmost extent, awed by no sanctions, restrained by no prescriptions, effecting a "revolution in the human mind." The image thus conjured is one of intellectual *sans-culottes* running rampant; and Miller was at pains to point out that this "revolution . . . has been attended with many advantages, and with many evils," both of which he then spelled out.

A little later on, he returned to "the revolutions and progress of science," observing that the "last age was remarkably distinguished by REVOLUTIONS IN SCIENCE" (2: 413):

> Theories were more numerous than in any former period, their systems more diversified, and revolutions followed each other in more rapid succession. In almost every department of science, changes of fashion, or doctrine, and of authority, have trodden so closely on the heels of each other, that merely to remember and enumerate them would be an arduous task.

Miller set himself the problem of accounting for this "frequency and rapidity of scientific revolutions." His solution is a most modern one, since he saw a primary cause to be what we would call today the emergence of a "scientific community." Miller pointed in particular to the "extraordinary diffusion of knowledge" ; the "swarms of inquirers and experimenters every where" ; and—above all—"the unprecedented degree of intercourse which men of science enjoyed," the consequence of which was "the thorough and speedy investigation which every new theory was accustomed to receive," resulting in "the successive erection and demolition of more ingenious and splendid fabrics than ever previously." Thus, "the scientific world [was kept] more than ever awake and busy" by a "rapid succession of discoveries, hypotheses, theories and systems" (2: 438). With an insight that shows how far Miller surpassed the bounds of a mere compiler, he concluded his "Recapitulation" by observing: "The eighteenth century was pre-eminently THE AGE OF LITERARY AND SCIENTIFIC INTERCOURSE."[6]

Within a decade of Miller's book, there was a further recognition of the existence of revolutions in science. In the fifth edition of the *Dictionnaire de l'Académie Françoise, revu, corrigé et augmenté par l'Académie elle-même* (1811), we find the primary definition to be cyclical and astronomical:

> The return of a Planet, of a Star, to the same point from which it had set out. *The revolution of the Planets. The celestial revolutions.*

Periodic revolution. In the same sense, *The revolution of the centuries, of time, of the seasons.*

There is also mentioned a "Révolution d'humeurs." The article concludes with "memorable and violent changes which have shaken these Countries" ("changemens mémorables et violens qui ont agité ces Pays") — in reference to "Les Révolutions Romaines, les Révolutions de Suède, les Révolutions d'Angleterre." These three sets of revolutions are the ones mentioned in the article on revolution in the Diderot-d'Alembert *Encyclopédie*. In this edition of the *Dictionnaire* (1811), no mention is made of the French Revolution or of the American Revolution, although the French Revolution had been given as an example in the edition of 1793. It is pointed out by the lexicographers of the Académie that in speaking of *the* revolution, one always has in mind the establishment of a new order: "When one says simply, *The Revolution*, in speaking of the history of these countries, one designates the most memorable revolution, the one which has brought a new order. Thus, in speaking of England, *The Revolution* designates the one in 1688."[7]

In the present context, however, what is of greatest interest is the paragraph devoted to the ways in which the word 'revolution' is used figuratively: "Of the change which happens in public affairs, in wordly matters, in opinions, etc." ("Du changement qui arrive dans les affaires publiques, dans les choses du monde, dans les opinions, etc."). The examples given are:

Rapid, sudden, unexpected, marvelous, astonishing, happy revolution.

The loss of a battle often causes great revolutions in a state.

Time causes strange revolutions in affairs.

The things of this world are subject to great revolutions.

Revolution in the arts, in the sciences, in minds, in fashions.

Thus formally entered into the lexigraphic record, the expression "revolution in the sciences" attained official recognition as the name of an accepted concept to characterize scientific change.

14

Lavoisier

and the Chemical

Revolution

———◆———

The Chemical Revolution has a primary place among revolutions in science in that it is the first generally recognized major one to have been called revolution by its chief author, Antoine-Laurent Lavoisier. Scientists before Lavoisier had been aware that their program would lead to something entirely new and would run directly counter to the established norms of accepted scientific belief; but Lavoisier, unlike the others, also had in mind the concept of revolutions in science as a particular kind of change in thinking, and he made the judgment that his own work would in fact constitute just such a revolution. Others had written of revolutions in science, but these had been occurrences of the distant or recent past, not of the present. Only Robert Symmer (so far as I am aware) had preceded Lavoisier in describing his contribution to science as 'revolution'-making; but Symmer's proposed two-fluid theory of electricity did not produce a revolution, as Lavoisier's theory of chemistry did. Additionally, electricity was at most but a single branch of a science (physics), whereas chemistry encompassed the whole science of matter. A revolution in chemistry would thus shake the foundations of almost all of physical science and even the science of biology.

In writing out his plans and hopes for research, Lavoisier could not help but be conscious of their ultimate significance for science. "The importance of this subject," he wrote, in an entry to a laboratory notebook (*registre*) dated 1773, "has prompted me once more to undertake all this work, which seemed to me destined to bring about a revolution in physics and in chemistry."[1] The same concept and image of a revolution within chemistry appears in a letter that Lavoisier wrote to Chaptal in 1791, in which he concluded: "All the young scientists adopt the new theory and I thence conclude that the revolution is accomplished in chemistry."

The Chemical Revolution occurred more or less during the time of the American Revolution and reached its climax during the French Revolution. Lavoisier was not unaware of this conjunction of revolutions. On 2 February 1790, he wrote a remarkable letter to Benjamin Franklin in which he gave his American friend a succinct account of the Chemical Revolution and then wrote about the political revolution in France — thus unambiguously displaying how the two revolutions were associated in his mind. He announced to Franklin that the French scientists were divided into two camps: those who clung to the ancient doctrine and those who were on his side. The latter group included de Morveau, Berthollet, Fourcroy, Laplace, Monge, "and in general the physicists of the Academy of Sciences." After reporting on the chemical situation in England and in Germany, he concluded (Duveen and Klickstein 1955, 127; Smith 1927, 31): "Here, then: a revolution that has taken place in an important part of human knowledge since your departure from Europe," adding, "I will consider this revolution to be well advanced and even completely accomplished if you range yourself with us." Next Lavoisier turned to the political revolution: "After having brought you up to date on what is going on in chemistry, it would be well to speak to you about our political revolution. We regard it as done and without any possibility of return to the old order." By February 1790, the absolute rule of kings had been abolished and France had become a limited monarchy, with major powers vested in the National Assembly; but the new constitution was not worked out and accepted by the king until 14 July 1790.

In 1790 or in 1791, with a revolution well advanced in the political sphere, it is not surprising to find Lavoisier thinking about a revolution in chemistry. And even his earlier reference to revolution in the laboratory *registre* of 1773, before the events of either the American Revolution or the French Revolution, is not especially unexpected since the concept of political, cultural, and intellectual revolution (including revolutions in the sciences) had become fairly common in France by that time. What is remarkable about Lavoisier's note of 1773 is that (1) in it he announced an impending profound revolution in the physical sciences that then actually occurred; that is, he was able to predict a scientific revolution; and (2) the author of the note and the chief author of the revolution were one and the same.

Lavoisier's Contribution

The central feature of the Chemical Revolution was the overthrow of the reigning 'phlogiston' theory and its replacement by a theory based on the role of oxygen. This gas, Lavoisier showed, is a component of the atmosphere, which he held to be a mixture of gaseous substances rather than a single substance subject to modification. It is oxygen which is the active agent in the processes of combustion, calcination, and respiration. To see how profound a change was wrought by the Chemical Revolution, consider that at that time metallic ores were considered elementary and metals were considered compound (a compound of the ore or 'calx' and 'phlogiston'). Ever since Lavoisier, we conceive metals to be elements (if pure, that is, neither alloys nor mixtures) and calxes to be compounds of metal and oxygen. The language of chemistry reflects the new knowledge in names that include 'oxide', 'dioxide', 'peroxide', and so on. Basic to the new chemistry was the modern concept of element, compound, and mixture; the production of a table of the elements (much like our own); and a chemical analysis of known compounds.

The Chemical Revolution made use of a general principle known as 'conservation of mass', or 'conservation of matter', which explains that in a chemical reaction the total mass (or weight) of *all* the reacting substances must be identically equal to the total mass (or weight) of *all* the product substances. This principle, now basic to all sciences, was not then essential to chemical theory. If so, there would have been a paradox (assuming phlogiston to be a substance and hence — in the Newtonian sense — having mass and therefore weight). For it is a fact of experiment that the calx weighs *more* than the metal in the processes designated by the equation: *calx + phlogiston = metal*. Some phlogistonists explained this paradox by assigning a 'negative weight' to phlogiston, but others sought a way out by attempting to reduce the problem of mass or weight to one of densities (see Partington and McKie 1938, pt. 3). Priestley, much wiser than these two groups, simply said that in physical science weight (or mass) was not always a major consideration. He was right, of course; three examples of physical 'substances' that were not discussed in terms of mass or weight were the Newtonian ether, the Franklinian electrical fluid, and the fluid of heat or 'caloric' (in which Lavoisier believed). Here we may see how revolutionary were the principles of the new chemistry. We may note that Lavoisier's version of the above equation, *calx = metal + oxygen*, gives experimental verification of the basic principle of conservation of matter (since oxygen has weight).

Lavoisier's analysis of the role of oxygen (or part of the air) in combustion and calcination was recorded by him in a note of 1 November 1772 (read at the Royal Academy of Sciences on 5 May 1773), in which he observed that "sulphur, in burning, far from losing weight, on the contrary gains weight" and that "it is the same with phosphorus." This "increase of weight," he added, "arises from a prodigious quantity of air [actually, as he later found, only a part of the air: oxygen] that is fixed during the combustion."[2] This discovery, he noted, led him to believe that the same phenomenon "may well take place in the case of all substances that gain in weight by combustion and calcination" (Ihde 1964, 61; McKie 1935, 117). The note of 1773 on the "revolution in physics and chemistry" was predicated on a proposed set of experiments to be made "with new safeguards," in order to "link our knowledge" of "the air that goes into combination or that is liberated from substances" with "other acquired knowledge" so as "to form a theory" (Meldrum 1930, 9; Berthelot 1890, 48).

I have referred to the new chemical nomenclature in relation to oxides. It is characteristic of a revolution in science to change existing names in accordance with the more rigorous logic of the new theory. We have seen an example of this process in the case of veins and arteries, following Harvey's discovery of the circulation of the blood. In 1787 Louis Bernard Guyton de Morveau, Claude Berthollet, and Antoine François de Fourcroy joined forces with Lavoisier to coin a new nomenclature that would reflect the actual chemical constituents of substances, in accordance with Lavoisier's new theory of chemistry. The *Method of Chemical Nomenclature* which this quartet published in 1787 is a primary document of revolution, a key to Lavoisier's framework of ideas in action. Not only did the new names depend on Lavoisier's analysis of compounds, but the system of names could give information concerning the relative degrees of saturation of oxygen. For instance, salts containing sulfur could be sulfates (salts of sulfur*ic* acid) or sulfites (salts of sulfur*ous* acid); and in general the *-ic* acids (and the *-ates* salts) were those saturated with oxygen. But compounds that had sulfur and no oxygen were *-ides*, as in potassium sulfide. Similarly, a compound of potassium and oxygen would be potassium oxide (and so for the other metals). In his *Elementary Treatise on Chemistry* (1789; in German, 1792; in English, 1790; also in Dutch, Italian, and Spanish), Lavoisier stressed the influence of the philosopher Condillac, who had taught that "the art of reasoning depends on a well-made language." Although Lavoisier's statement may need to be taken with a grain of salt (Guerlac 1975, 112), he did say unequivocally that this final treatise grew out of considerations of language and nomenclature — which, "without my being able to prevent it," grew into a system of chemistry.

Acceptance of the Revolution

Almost at once there was public recognition in print that a revolution in chemistry had occurred. An early such reference occurs in a small book by Lavoisier's friend and collaborator Jean-Baptiste-Michel Bucquet, published in 1778 (Gough 1983) and based on an essay read a year earlier to the Paris Faculty of Medicine. The new chemical "doctrine of gases," according to Bucquet, well illustrated the principle that old ideas must be renounced when confronted by new discoveries. There is little which has "produced a greater revolution in chemistry" and has "contributed more to the progress of this beautiful science," he said, than the new discoveries concerning gases.

Gough (ibid.) has found what is very likely the first printed reference to Lavoisier's revolution in chemistry. It was made soon after Lavoisier had begun the series of experiments that would lead to a new point of view respecting combustion and the gases of the air. This reference, by Antoine Baumé, author of a three-volume treatise on chemistry, was printed in the year 1773, when Lavoisier was privately expressing his belief that his research program would "bring about a revolution in physics and in chemistry." At this time, Lavoisier was already convinced that combustion entailed chemical combination with air (or some part of the air) and that the concept of phlogiston should be abandoned, but he had not as yet published anything on the subject. Baumé mentions a revolution in chemistry in an appendix to his treatise, in a discussion of the new discoveries, notably 'fixed air' (carbon dioxide) and its properties. Some physical scientists, according to Baumé, believed that fixed air has "properties which must cause phlogiston to be thrown out and fixed air substituted in its place." Fixed air, he went on (Gough 1983), "should, according to these same physical scientists, produce a complete revolution [révolution totale] in chemistry" and even "change the order of our knowledge." Since Baumé was not particularly intimate with Lavoisier, it is not clear how he would have heard of Lavoisier's revolutionary ideas; we can only assume that the phrase 'les physiciens' (the physicists) was intended by Baumé to denote Lavoisier and his cohorts — for who else was then making such a revolution?

Henry Guerlac (1976) has traced for us the subsequent recognition of the Chemical Revolution. Bucquet's book of 1778 was not well known. Guerlac found that the author most responsible for publicizing the notion of a revolution in chemistry made by Lavoisier was Antoine François de Fourcroy. Fourcroy referred to an impending revolution even "before his conversion to Lavoisier's New Chemistry," in his *Leçons élémentaires d'histoire naturelle* (1782). Here "he writes that the proper course is to wait until further experiments convince us that all the phenomena of chemistry can be explained 'par la doctrine de gas' [by the theory of gases] without invoking phlogiston." He stresses the conviction of his fellow chemist Macquer concerning "the great revolution which the new discoveries must produce in chemistry" (Fourcroy 1782, *1*:22). In later editions, Fourcroy referred to new discoveries which daily give added force to our theories. Because of the popularity of Fourcroy's *Leçons élémentaires d'histoire naturelle* (1782), and the other writings in which Fourcroy referred to 'revolution', Guerlac concludes that it was Fourcroy who was most effective in canonizing the expression 'the revolution in chemistry' or some equivalent (see, further, Smeaton 1962). In particular, there was a long review of Lavoisier's great treatise — "signed by Fourcroy and J. de Horne" but "written and presented" by Fourcroy (Guerlac 1976, 3) — which stated as fact that "the revolution which chemistry has undergone in recent years is a result of the experiments of M. Lavoisier." This review was "printed for the first time as an appendix to the second issue of the first edition of Lavoisier's *Traité élémentaire de chimie*, and published in subsequent editions" (ibid.), so that Lavoisier's own complete statement of his theory was accompanied by a declaration of revolution.

Guerlac also found that even "before Lavoisier put the capstone on the New Chemistry with his *Traité . . . chimie* in 1789," a reference to the revolution in progress appeared in a preface to a translation into French of a book on phlogiston by the Irish chemist Richard Kirwan. The preface, attributed to Mme Lavoisier—"generally credited (on Grimaux's authority) to have been the author of the translation"— explains why a series of footnotes have been added that refute Kirwan's phlogistic arguments at every step. Without these notes, according to Mme Lavoisier, "this work would not sufficiently have advanced the revolution which is on the way in chemistry" ("la révolution qui se prépare en chimie").

This record should include one additional but significant reference in print to the revolution—by Lavoisier himself. This one (like the opinion of Macquer mentioned by Fourcroy, the preface to Kirwan, and the sentiment of Bucquet) antedates the publication of the complete theory by Lavoisier in his *Traité* (1789). The occasion was a "Memoir on the Necessity of Reforming and Perfecting the Nomenclature of Chemistry" ("Mémoire sur la nécessité de réformer & de perfectionner la nomenclature de la Chimie"), which was "read at the public meeting of the Paris Academy of Sciences on 18 April 1787 by M. *Lavoisier.* " This memoir was published as the introductory chapter to the *Méthode de nomenclature chimique* (Paris, 1787). Here Lavoisier did not say that the reform of chemical nomenclature constituted a revolution in the science of chemistry or that such a revolution was in the making. Rather, Lavoisier announced that the "new method" would "entail a necessary and even rapid revolution in the manner of teaching" chemistry ("Cette méthode . . . entraînera une révolution nécessaire & même prompte dans la manière d'enseigner"). This example reminds us that almost a century earlier, in describing the revolution in mathematics, Fontenelle had also invoked the doctrine that a truly fundamental revolution in science implies a revolution in education.

Lavoisier's prediction was rapidly verified. We may see evidence of this in a pamphlet by Joseph Priestley of 1796, nine years after the publication of the *Méthode de nomenclature chimique*, and eight years after the publication of the French translation of Kirwan's essay, with its annotations by "MM. de Morveau, Lavoisier, de la Place, Monge, Berthollet & de Fourcroy." Priestley addressed his "short defense of the doctrine of *phlogiston*" to "Messrs. Berthollet, De la Place, Monge, Morveau, Fourcroy, and Hasenfratz, the surviving Answerers of Mr. Kirwan." He began:

> There have been few, if any, revolutions in science so great, so sudden, and so general, as the prevalence of what is now usually termed *the new system of chemistry*, or that of the *Antiphlogistians*, over the doctrine of Stahl, which was at one time thought to have been the greatest discovery that had ever been made in the science.

The pace of this revolution has been so enormous, according to Priestley, that "every year of the last twenty or thirty has been of more importance to science, and especially to chemistry, than any ten in the preceding century." He then admitted that "this new theory" has been considered "so firmly established" that "a *new nomenclature*, entirely focused upon it, has been invented"—a nomenclature that "is now almost in universal use." The result is that "whether we adopt the system or not, we are under the necessity of learning the new language." For without learning this language, it would no longer be possible to "understand some of the most valuable of modern publications." Here is evidence of the close linkage between Lavoisier's revolution in the teaching and nomenclature of chemistry and the revolution in chemistry.

Finally, we may note that the publication of Lavoisier's laboratory notebooks by Marcelin Berthelot in 1890—in a book entitled *La révolution chimique: Lavoisier*—popularly and permanently fixed the name Chemical Revolution on the historical record. A century and a half earlier (as Maurice Crossland reported in 1963), there was published what is apparently the first reference to a revolution in chemistry, the prediction of such a 'révolution' by G.-F. Venel (*Encyclopédie*, 1757: "Chymie").

It is evident that Lavoisier's Chemical Revolution passes all the tests for a revolution in science. It has been recognized as a revolution by all historians and scientists, just as it was seen to be a revolution in its own time. Additionally, the whole science of chemistry and its language have followed the lines set forth in the Chemical Revolution. The Chemical Revolution is thus a paradigmatic example of a revolution in science.[3]

15

Kant's

Alleged Copernican

Revolution

———◆———

Writing at the end of the eighteenth century, Immanuel Kant would have been familiar with the idea — expounded by Montucla, Bailly, and others — that Copernicus had produced a revolution in astronomy. Furthermore, by that time the term 'revolution' was coming into fairly common use to denote radical changes in science, in taste, and in thought generally. It was a time when 'revolution' was in the air. Thus, Kant's views on revolution and revolution in science are of special interest to a study of these concepts in the eighteenth century, considering Kant's towering position in the history of philosophy. But they are even more intriguing because of the widespread belief that Kant referred to his own innovations in philosophy as a Copernican revolution.

The Myth of Kant's Copernican Revolution

In E. J. Dijksterhuis's magisterial presentation, *The Mechanization of the World Picture* (1961, 299), he declares, "Ever since Kant, 'Copernican revolution' has been a set expression for a radical change of view, and in the history of science the year 1543 is looked upon as the actual date of demarcation between the Middle Ages and the Modern Period." This notion that Immanuel Kant compared his own achievement in philosophy to a Copernican revolution occurs in a very large number of books on Kant and on the history of philosophy. A few years ago, a course on "The Age of Revolutions" was introduced in the "second level" of the program of the Open University (a mass education venture in British television, giving the equivalent of a bachelor's degree to those unable to attend a college or university in the normal way). Two of the central units were called "Kant's Copernican Revolution"; one bears the subtitle "Speculative Philosophy," the other "Moral Philosophy." In the first of these units, the author (Vesey 1972, 10) refers to "Kant's Copernican revolution in speculative philosophy," but he does not ever explicitly attribute this concept to Kant himself. In the second (Hanfling 1972, 23–25), it is said unambiguously that "Kant himself did not explicitly compare his work in moral philosophy with the 'Copernican Revolution', as he had done in the case of his speculative philosophy. Still, I think it can be fairly claimed (it often has been claimed) that the comparison applies no less to the former than to the latter."

The reader who is unacquainted with either the literature concerning Kant or histories of philosophy can have no idea as to how nearly universal (especially among British and American writers) is the belief in Kant's 'Copernican revolution'. A few examples, chosen at random:

Kant . . . in the Preface [to the *Critique of Pure Reason*] . . . speaks of the projected 'Copernican Revolution' in our modes of thought. (Bird 1973, 190–191)

Kant compares his own philosophical revolution with that initiated by Copernicus. (Paton 1936, *1:* 75)

We can now understand what Kant means when he claims to have made a revolution in philosophy like that which Copernicus made in astronomy. (Broad 1978, 12)

This new way of conceiving the possibility of a priori knowledge Kant compares to the revolution brought about in astronomy by Copernicus. (Lindsay 1934, 50)

His Copernican revolution, he insists, no more impairs the empirical reality of the world of experience than the heliocentric hypothesis alters or denies the phenomena. (Copleston 1960, *6:* 242)

In the preface to the second edition [of his *Critique of Pure Reason*] he compares himself to Copernicus, and says that he has effected a Copernican revolution in philosophy. (Russell 1945, 707)

Kant spoke of himself as having effected a 'Copernican revolution'. (Russell 1948, 9)

All that Kant means by his comparison is that in both hypotheses we find a revolution or drastic revision of a primary assumption which had long ago been allowed to pass unchallenged. In one case what is assumed is the immobility, in the other the passivity of the observer. (Weldon 1945, 77n)

It is very ironical that Kant himself signalised the revolution which he believed himself to be effecting as a Copernican revolution. But there is nothing Copernican in it except that he believed it to be a revolution . . . For his revolution, so far as it was one, was accurately anti-Copernican. (Alexander 1909, 49)

. . . an idea which Kant himself proudly calls his 'Copernican Revolution'. (Popper 1962, 180)

Kant believed that what his critique of reason had effected was a virtual 'Copernican Revolution' in philosophy. (Aiken 1957, 31)

Kant has formulated the matter succinctly in the Preface to the *Critique of Pure Reason* with his well-known allusion to the 'Copernican Revolution'. (Lukács 1923, 111)

. . . what Kant, in the Preface to the second edition of the *Critique of Pure Reason* (1787), will call his *Copernican revolution.* (Chevalier 1961, 3: 589)

I regard the teaching of Kant as a great and individual philosophical consummation of the Copernican revolution, about which Kant expressed himself several times. (Oiserman 1972, 121)

The fundamental idea of what Kant calls his 'Copernican revolution'. (Deleuze 1971, 22–23)

Kant flattered himself on accomplishing a true philosophical revolution . . . —a revolution comparable to that of Copernicus in the cosmological and mathematical order. (Devaux 1955, 434)

The revolutionary act of Kant in the history of thought, his 'Copernican revolution'. (Vuillemin 1955, 358)

This parade of quotations leaves no doubt concerning a rather general agreement among philosophers that (a) there *was* a Copernican revolution, and (b) Kant conceived that his own radical innovation in philosophy was in its turn a Copernican revolution, or was like a Copernican revolution. A half-hour's browsing on the library shelves turned up more than several score of such statements, made by eminent scholars and published in works issued by our leading scholarly and university presses. Furthermore, the "Macropaedia" (which is part of *The New Encyclopaedia Britannica,* the so-called fifteenth edition, and described as "Knowledge in Depth"; 1973, *10:* 392), states authoritatively:

> Kant proudly asserted that he had accomplished a Copernican revolution in philosophy. Just as the founder of modern astronomy, Nicolaus Copernicus, had explained the apparent movements of the stars by ascribing them partly to the movement of the observers, so Kant had accounted for the application of the mind's a priori principles to objects by showing that the objects conform to the mind: in knowing, it is not the mind that conforms to things but things that conform to the mind.

Many books on Kant or on philosophy contain chapters on "La révolution copernicienne" (Vlachos 1962, 98ff.), "Kant's Copernican Revolution" (Popper 1962, 180), "The Copernican Revolution" (Dewey 1929, 287). In the latter, the Gifford Lectures for 1929, on *The Quest for Certainty,* Dewey boldly declares, "Kant claimed that he had effected a Copernican evolution [sic: read *revolution*] in philosophy by treating the world and our knowledge of it from the standpoint of the knowing subject." Dewey concluded by rather immodestly evaluating his own contribution to philosophy as another Copernican revolution of the same magnitude as Kant's. Karl Popper, in an essay of 1954, reprinted in his *Conjectures and Refutations* (1962, 175ff.), has a section on Kant's "Copernican Revolution." Here Popper quotes Kant's statement that "our intellect does not draw its laws from nature, but imposes its laws upon nature," and remarks, "This formula sums up an idea which Kant himself proudly calls his 'Copernican Revolution'" (p. 180). A whole book (Vuillemin 1954) has been published on *L'héritage kantien et la révolution copernicienne.* In the published proceedings of The Third International Kant Congress held in 1970, at least three of the papers refer to a "Kantian Copernican revolution" (Beck 1972, 121, 147, 239), and one of the contributions (pp. 234ff.) is entitled "The Copernican Revolution in Hume and Kant."

After all this, it will surely seem as astonishing to the reader as it did (and still does) to me that Kant did *not* compare his own contribution to a Copernican revolution. And I am certain that the reader will fully understand why it was that during the final redaction of this chapter I more than once found it necessary to go back to Kant's *Critique of Pure Reason* in both the original German and the three current English translations (J. M. D. Meiklejohn 1855; Max Müller 1881; Norman Kemp Smith 1929, and many reprints) in order to keep reassuring myself that so many eminent authorities in at least three languages could have perpetuated so egregious an error. Could it be that no one in the audience of the Gifford Lectures in 1929 knew the text of Kant and so could have called Dewey's attention to his error? Was there not a single Kantian scholar present at the Third International Kant Congress who had ever read Kant in German or in English and remembered what he actually said? A paper was read in 1974 at the Copernicus symposium on "Science and Society: Past, Present and Future" (Steneck 1975), in which the Copernican revolutions of Dewey and Kant were compared (C. Cohen 1975). This paper was discussed in a learned commentary (Cropsey 1975) that takes up the question of "Professor [Carl] Cohen's . . . description of the Deweyan philosophy as the product of a true Copernican revolution" (p. 105), but the commentator does not correct the reference to Kant's Copernican revolution; again, apparently, no member of the audience did so either.

Those authors who write about Kant's Copernican revolution and who actually give a source for Kant's alleged comparison refer the reader to the preface of the second edition of the *Kritik der reinen Vernunft* (1787; 1st ed. 1781). We shall see in a moment that this new preface is extremely interesting in that it contains discussions of revolutions in the sciences (in mathematics and in experimental physics) and also revolutions in intellectual development. Here is what Kant actually says about Copernicus (quoted from Kant 1926, 20 = B xvi):

> Es ist hiermit ebenso, als mit den ersten Gedanken des *Kopernikus* bewandt, der, nachdem es mit der Erklärung der Himmelsbewegungen nicht gut fort wollte, wenn er annahm, das ganze Sternenheer drehe sich um den Zuschauer, versuchte, ob es nicht besser gelingen möchte, wenn er den Zuschauer sich drehen, und dagegen die Sterne in Ruhe liess.

One need not be a German scholar, nor even have a deep familiarity with the German language, to see that in this passage Kant says "mit den ersten Gedanken" or "with the first thoughts" of Copernicus and not "with a revolution." In what is generally considered today's reliable and authoritative translation, Norman Kemp Smith changed Kant's "first thoughts of Copernicus" so as to make it become "Copernicus's primary hypothesis." This may provide a reasonable interpretation of Kant's intentions, but it is in fact a considerable departure from Kant's own simple and straightforward expression. Accordingly, Kemp Smith also gave the original phrase in German in a footnote. His translation reads (p. 22):

> We should then be proceeding precisely on the lines of Copernicus' primary hypothesis [*mit den ersten Gedanken des Kopernikus*]. Failing of satisfactory progress in explaining the movements of the heavenly bodies on the supposition that they all revolved round the spectator, he tried whether he might not have better success if he made the spectator to revolve and the stars to remain at rest.

In Kemp Smith's volume of commentary (1923), however, the reader is given no hint whatever that Kant had written "mit den ersten Gedanken des Kopernikus" and not "mit der ersten Hypothese des Kopernikus."

This paragraph of Kant's makes his intent clear. In pre-Copernican astronomy it is assumed that all the complexities of the apparent motion of the planets are actual. But in post-Copernican astronomy it is seen that part of these complexities arise from the position of the observer on a moving earth. Earlier metaphysics had similarly assumed that all appearances of things (phenomena) had an actuality beyond the perceiving mind, just as was the case with the complexities of planetary motion for pre-Copernican astronomers. But Kant's new point of view assumed that the objects of our knowledge are not "things-in-themselves" but the result of an interaction of our minds and the objects of our sensations. Hence Kant made an important distinction between "things as they are in themselves" and "things as they appear to us" (Kemp 1968, 38).

Kant's procedure would resemble the traditional views of the Copernican revolution in that in both astronomy and metaphysics we may discern a "revolution or drastic revision of a primary assumption which had long been allowed to pass unchallenged" (Weldon 1945, 77). That is,

"In one case what is assumed is the immobility, in the other the passivity of the observer." Many philosophers have pointed out that Kant's alleged revolution was not really Copernican. As Bertrand Russell (1948, 9) said, "Kant spoke of himself as having effected a 'Copernican Revolution', but he would have been more accurate if he had spoken of a 'Ptolemaic counter-revolution', since he put Man back at the centre from which Copernicus had dethroned him."

Whatever Kant really intended, he certainly and obviously did not say that he had effected (or would effect) a Copernican revolution in metaphysics. The passage (B xvi) quoted in extenso above contains no such statement, nor does it refer either to a Copernican revolution or to any revolution (actual or impending) in metaphysics. However, although nowhere in the *Critique of Pure Reason* in either edition is there any mention of a Copernican revolution, there is an indication of a revolution in metaphysics. It is especially remarkable that Kant does not refer to a Copernican revolution since in the preface to the second edition he develops at length the concepts of revolution in science and intellectual revolution. But before presenting Kant's views on revolution, a word must be said concerning two other mentions of Copernicus — both occurring in a note to the preface to the second edition of the *Critique of Pure Reason*.¹ In this note Kant (1929, 25 = B xxii) explains how "the fundamental laws of the motions of the heavenly bodies" — presumably, Kepler's laws — "gave established certainty to what Copernicus had at first assumed only as an hypothesis, and at the same time yielded proof of the invisible force (the Newtonian attraction) which holds the universe together." If "Copernicus had not dared," Kant added, "to seek the observed movements, not in the heavenly bodies, but in the spectator," Newtonian universal gravitation "would have remained for ever undiscovered." I cannot find in these sentences any expression of the belief that there had been a Copernican revolution; they may even imply that a revolution did not occur until Kepler and Newton. These sentences do exhibit what Kant himself considered the role of the "change in point of view, analogous to this hypothesis [of Copernicus]," which Kant had "put forward in this preface as an hypothesis only, in order to draw attention to the character of these first attempts at such a change, which are always hypothetical." But this "hypothesis," he then declared, "will be proved" in "the *Critique* itself . . . apodeictically not hypothetically, from the nature of our representations of space and time and from the elementary concepts of the understanding."

The name of Copernicus appears in Kant's treatise only in the reference to the "first thoughts" and in the passages just discussed. In Kant's other writings there are additional references to Copernicus, but none in association with the idea of revolution. In short, a self-proclaimed Kantian 'Copernican revolution' would seem to have as little real existence as an alleged Copernican revolution in astronomy in the late sixteenth century. Although at least three scholarly articles in prominent journals have attempted to inform the community of philosophers that Kant did not compare his contribution to a Copernican revolution (Cross 1937; Hanson 1959; Engel 1963), books and articles by eminent philosophers continue to appear with a prominent place given to 'Kant's Copernican revolution'.

Kant's Views on Revolution in Science

The preface to the second edition of the *Critique of Pure Reason* is notable for its discussion of revolutions in science. Kant is a member of that company of eighteenth-century scholars who believed that the sciences advance by revolutions, by sudden and dramatic leaps forward that create something entirely new in science, something that had never existed until that time. The first of the revolutions to which Kant refers is a dramatic alteration of our knowledge, a revolution in the new sense of the word that was coming into general use. With respect to his use of the term 'revolution', Kant is strictly a modern and not a traditionalist; by 'revolution' he does not mean a cyclical change or an ebb and flow, or a return to some better antecedent state, but rather a radical forward step that makes a clean and thorough break with the past.

The first revolution, according to Kant, occurred in mathematics and consisted in transforming an empirical knowledge of earth measure into a deductive system. There was discovered the "true method" which was like a "new light" which "flashed upon the mind of the first man (be he Thales or some other) who demonstrated the properties of the isosceles triangle." Concerning this event, Kant says (1929, 19 = B xi – xii):

> The true method, so he found, was not to inspect what he discerned either in the figure, or in the bare concept of it, and from this, as it were, to read off its properties; but to bring out what was necessarily implied in the concepts that he had himself formed *a priori*, and had put into the figure in the construction by which he presented it

to himself. If he is to know anything with *a priori* certainty he must not ascribe to the figure anything save what necessarily follows from what he has himself set into it in accordance with his concept.

Kant is here contrasting "the sure path of science" and "groping." This contrast is not always easy to grasp, but basically Kant seems to be saying that in logic, reason is concerned with itself as subject and nothing else, but in scientific geometry reason is applied to something outside itself — for example, geometric figures — in particular isosceles triangles. The revolution in thinking (*Revolution der Denkart*) consisted in the recognition that "neither empirical observation nor analysis of concepts will help us to demonstrate any mathematical truth" (Paton 1937, 366). It is not enough to use one's eyes and to determine the properties of the isosceles triangle by inspection, nor to examine the concept of such a triangle. Rather, "we must employ . . . what Kant calls the 'construction' of concepts; that is, we must exhibit *a priori* the intuition corresponding to our concept." Hence "the discovery attributed by Kant to the earliest mathematician seems to be this" (ibid.).

> that it was necessary to produce the figure by means of what he himself *thought into it* and exhibited *a priori* in accordance with concepts; and that to have certain *a priori* knowledge he must attribute nothing to the figure except what followed necessarily from what he himself had *put into* it in accordance with his concept.

According to Kant (1929, 19 = B xi), this radical transformation of geometry "must have been due to a *revolution* brought about by the happy thought of a single man" ("diese Umänderung einer *Revolution* zuzuschreiben sei, die der glückliche Einfall eines einzigen Mannes . . . zustande brachte"). This man thereby marked out "the path upon which the science must enter, and by following which, secure progress throughout all time and in endless expansion is infallibly secured."

Kant held that "this intellectual revolution [Revolution der Denkart]" was "far more important than the discovery of the passage round the celebrated Cape of Good Hope." Then he referred to "the memory of the revolution [das Andenken der Veränderung]." There are thus three distinct references to revolution (twice as 'Revolution' and once as 'Veränderung') within a few lines of one another on a single page (p. 19 = B xi).

In the following short paragraph (1929, 19–20 = B xii), Kant turns from mathematics to "natural science in so far as it is founded on *empirical* principles." It took much longer for natural science to enter "upon the highway of science" than it did for mathematics. It was "only about a century and a half" before, Kant declares, that Bacon "partly initiated" this transformation and "partly inspired fresh vigour in those who were already on the way" to making an empirically based science — something which can "be explained as being the sudden outcome of an intellectual revolution [Revolution der Denkart]."

In the next paragraph, Kant made no pretense at "tracing the exact course of the history of the experimental method," mentioning as examples only the experiments of Galileo, Torricelli, and Stahl. He concludes that physics underwent a "beneficent revolution in its point of view [vorteilhafte Revolution ihrer Denkart]." For Kant, "the happy thought" on which "the beneficent revolution" in physics hinges is that "while reason must seek in nature, not fictitiously ascribe to it, whatever as not being knowable through reason's own resources has to be learnt, if learnt at all, only from nature, it must adopt as its guide, in so seeking, that which it has itself put into nature." It is in this way "that the study of nature has entered on the secure path of a science, after having for so many centuries been nothing but a process of merely random groping" (pp. 20–21 = B xiv).

Origins of the Myth

Having discussed mathematics and experimental or empirically based physics, Kant turns to metaphysics, "a completely isolated speculative science of reason" (p. 21 = B xiv). He compares this subject to mathematics and natural science, which, he notes, "by a single and sudden revolution [Revolution] have become what they now are" (pp. 21–22 = B xv–xvi). On the third page of this discussion occurs the phrase "mit den ersten Gedanken des Kopernikus," an expression which, we have seen, may be translated word for word as "with the first thoughts of Copernicus." In context, Kant's point is clearly that Copernicus had made a shift from the perspective of a stationary observer to that of a revolving observer; he had shown that a change occurs when one disengages the observer's own motion from the observed or apparent motion of the sun, planets, and stars. Thus Kant appears to have in mind Copernicus's "first thoughts" in the sense of logical priority and not of historical sequence. Furthermore, if Kant had wanted to say that Copernicus had instituted or inaugurated a revolution — in astronomy, in science, or in the realm of intellect — would he not have done so? Since only a few pages earlier he *had* been discussing revolution in science, which is referred to again at the beginning of this page, the concept of such revolution was evidently in the forefront of his mind. Whether or not Kant believed that there had been a Copernican revolution, he certainly did not say so in the preface to the second edition of his *Critique of Pure Reason*, a fact which seems especially significant in the context of the discussions of scientific and intellectual revolutions in which Kant's remarks on Copernicus are imbedded. It is possible, of course, that the reference to revolution at the beginning of the paragraph mentioning Copernicus may have led commentators to think that Kant mentioned a Copernican revolution.

Kant says that he believes his book gives to metaphysics the certainty of the method of the sciences. Philosophers, he holds, should try to imitate the procedure of mathematics and the physical sciences, at least "so far as the analogy which, as species of rational knowledge, they bear to metaphysics may permit" (1929, 22 = B xvi). Later on, Kant talks of leaving "to posterity the bequest of a systematic metaphysic," which he says is "a gift not to be valued lightly" because "reason [will] be enabled to follow the secure path of a science, instead of, as hitherto, groping at random, without circumspection or self-criticism" (p. 30 = B xxx).

Would such a change in metaphysics constitute a revolution? Kant replies in the affirmative. The aim of his treatise, he declares, is "to change the prevailing procedure of metaphysics and thereby effect a complete revolution [eine gänzliche Revolution] in it [metaphysics] after the example of the geometers and natural philosophers" (B xxii). Kant thus joins the ranks of the eighteenth-century scientists — Symmer, Lavoisier, Marat — who said their own work was revolution-making. But Kant neither says that this revolution was Copernican, nor refers to Copernicus or to astronomy as examples. Since Kant did not (in any known letter, printed work, or manuscript) ever refer to a Copernican revolution, he could not have said that his momentous contribution to philosophy was (or would inaugurate) a Copernican revolution.[2]

How then can the literature be so nearly unanimously wrong? One possible explanation is this: The statement about a revolution in metaphysics opens a paragraph; the preceding paragraph has a long footnote appended to it which mentions both Copernicus and Newton. Perhaps the commentators' error arose from conflating the sentence about a revolution in metaphysics with the previous footnote. But since Kant uses "the example of the geometers and natural philosophers" and not astronomers, it would seem that any possible (though improbable) connection would be with a Newtonian revolution rather than with a Copernican revolution. Whatever the source of the original error, one author has obviously picked it up from another, without carefully checking the sources. Despite three warnings that Kant never invoked the image of a Copernican revolution, much less said that he was creating a Copernican revolution in metaphysics, this error is perpetuated in the philosophical literature year after year.[3]

Just as I was completing my final revision of this chapter, I came upon four new books perpetuating this long-standing error. One, by Roger Scruton, is published by Oxford University Press in its series entitled "Past Masters." Here (1982, 28) stress is given to "what Kant called his 'Copernican Revolution' in philosophy." Another, a great classic by the late Ernst Cassirer (first published in 1918), has just been made available in English. A new "Introduction to the English Edition" (1981, vii) opens with a discussion of "Kant's Copernican revolution in philosophy." "The Copernican revolution," we are told, "is based on an entirely new conception of philosophy and philosophical method which Kant describes as critical and transcendental" (p. viii).

In a major study of Goethe, Kant, and Hegel, Walter Kaufmann (1980, 87–88) writes that "Kant claimed to have accomplished a Copernican revolution." It is Kaufmann's opinion, however, that in the *Critique of Pure Reason* Kant "brought off an *anti-Copernican revolution*. He reversed Copernicus' stunning blow to human self-esteem" in that he "restored man to the center of the world." *A Dictionary of the History of Science* (1981) contains a very perceptive article on the Copernican revolution which stresses that this expression can have two meanings: either Copernicus's "introduction of a heliocentric system in astronomy" or the "firm establishment of such a system during the 17th century in the revised form with elliptical orbits due to Johannes Kepler." The article concludes with the observation that "'Copernican revolution' has also been used, as by Immanuel Kant (1724–1804), to describe generally any fundamental recasting of ideas which makes intellectual progress possible." But there is no reference to an alleged Copernican revolution in the later article on Kant in the same dictionary.

The attribution to Kant of a self-proclaimed Copernican revolution in philosophy or in metaphysics is not a recent invention. Between 1799 and 1825 at least four writers on Kantian philosophy stated publicly — in print or in lectures — that Kant himself either had desired or had undertaken a Copernican revolution in philosophy. Charles de Villers (1765–1815), a Frenchman who lived for many years in Germany, devoted a number of publications to an explanation of Kant's ideas to his countrymen. In an article on "Critique of Pure Reason," in *Le Spectateur du Nord* in 1799, Villers said of Kant that his reflection on human knowledge and reasoning "made him think that there was needed in metaphysics a revolution nearly the same as that which Copernicus accomplished in astronomy" (p. 7). Villers then explained the nature of the Kantian revolution in terms similar to those used by Kant himself in the preface to the second edition of the *Critique of Pure Reason* in describing the "first thoughts of Copernicus" (B xvi). In another work, *Philosophy of Kant* (1801, viii–x), Villers implied that an intellectual revolution had been produced by Descartes and Lavoisier as well as Copernicus and Kant.

Sixteen years later, Victor Cousin (1792–1867) repeated the theme of Kant's Copernican revolution. Cousin was one of the most widely read popularizers of philosophy of his age, and his books went through many editions and reprints. In 1817 he made a connection between Kant and the Copernican revolution during a course of lectures for the Faculty of Letters of the University of Paris. These lectures were not published until 1841, when the editor's "avertissement" stated that these were the first university presentations of the Kantian system in France (1841, iv–v). In the second edition (1846, *1:* 105–113) it is made clear that Cousin had also lectured on Kant in 1816, but at that time his knowledge of German had been so poor that he had had to rely on the Latin translations of Kant's works and on French secondary works. In 1817, when he could read Kant in the original German (*1:* 255 n.2), Cousin explained that "Kant undertook to bring about in metaphysics the same revolution which Copernicus had brought about in astronomy." In his lectures of 1820 (editions in 1842; 1846, *5;* 1857; Engl. trans. 1854), Cousin says that "Kant was conscious of the revolution which he was undertaking; he had judged his epoch and had understood its needs." Then he repeated the summary of Kant's preface to the second edition, using almost the same words as in his lectures of 1817.

In 1818, an important article on Kant, by Philipp Albert Stapfer (1766–1840), appeared in vol. 22 of the reference work, *Biographie Universelle.* A footnote explains that Charles de Villers had been writing this article but had asked Stapfer to do it instead, since approaching death would prevent him from giving the article the form he desired. Stapfer discusses the second edition of Kant's *Critique* (which, p. 239, n.1, he calls the third edition) and explicitly attributes to Kant the idea of accomplishing a Copernican revolution in philosophy. Kant, he says, "saw himself called to bring about in the speculative sciences the revolution which his illustrious compatriot, the Prussian Copernicus, had effected in the natural sciences — a parallel which was the idea of Kant himself" (p. 239). Stapfer develops the idea at some length on pp. 239–240, finally returning (p. 240) to explicit mention of Copernicus: "We will no longer turn around things: by constituting ourselves their center we will make them turn around us. This is the revolution of Copernicus." This presentation, similar to Cousin's, would make Kant's position more Ptolemaic than Copernican. But neither Cousin nor Stapfer seems to feel, as others will later, that this revolution could be designated a Ptolemaic counter-revolution.

In 1825, in an article on philosophy in the *Encyclopaedia Londinensis* (vol. 20), Thomas Wirgman quotes in English translation the passage from the *Biographie Universelle* connecting Kant with a Copernican revolution. Although this is a passage from Stapfer, Wirgman attributes the article to Villers. Kant, according to Wirgman's translation (p. 151), "saw that he was destined to accomplish in the speculative sciences a similar revolution to that which his illustrious countryman, Copernicus of Prussia, had produced in natural philosophy; a parallel, the idea of which was first conceived by Kant himself." Wirgman continues with Stapfer's development, which leads to the conclusion (ibid.): "We shall thus no longer revolve around the things, but, making ourselves their centre, we cause them to revolve around us. This is the Copernican revolution."

Wirgman connects Kant with revolution and with Copernicus elsewhere in the *Encyclopaedia Londinensis,* both in the article on philosophy and in other articles. In the article on philosophy (1825, 129), Wirgman develops at length a comparison of Kant and Copernicus, saying in particular that "I cannot refrain from indulging myself in running the prospective parallel of these two great men. Kant is the author of a theory equally bold with that of Copernicus; and if, like that, it shall stand the test of ages, the revolution it will effect will be equally glorious."

Villers (1799), Cousin (1817; 1820), Stapfer (1818), and Wirgman (1825) are not the only writers of this early period to link Kant to a Copernican revolution in philosophy. Another is Mme de Staël (1766–1817). In the

1813 London edition of *De l'Allemagne*, which may be considered the first edition because the 1810 Paris edition was suppressed when the printing was not yet complete, she declares (*3:* 13–14):

> "The human spirit," says Luther, "is like a drunk peasant on horseback; when he is picked up on one side he falls down again on the other." So man has fluctuated ceaselessly between his two natures: sometimes his thoughts have disengaged him from his sensations, sometimes his sensations have absorbed his thoughts, and he has wanted in alternation to ascribe everything to thoughts or to sensations. It seems to me, nevertheless, that the moment for a stable doctrine has arrived. Metaphysics must undergo a revolution like that which Copernicus effected in the system of the world. It must put our soul back in the center and make it just like the sun around which exterior objects trace their circle and whose light they borrow.

This passage occurs in a discussion of Bacon, but the presentation of German philosophy further on in the same part of *De l'Allemagne* suggests that Mme de Staël here has in mind German idealist philosophy in general and Kant in particular.

Mme de Staël did not assert that Kant himself wanted a Copernican revolution, but Villers said that Kant thought that such a revolution was needed in metaphysics. Cousin wrote that Kant undertook to bring about a Copernican revolution in metaphysics, and Stapfer that Kant saw himself called to bring it about in the speculative sciences. Stapfer dramatized his discussion by summarizing: "C'est la révolution de Copernic," which Wirgman translates as, "This is the Copernican revolution." It may be noted that all of these writers except Villers develop Kant's analogy "with the first thoughts of Copernicus" into a metaphor that by far exceeds in vividness and implication anything that Kant actually says in the immediate context of his Copernican references.

The case of Karl Leonhard Reinhold is especially interesting, since he was a noted promoter and explicator of Kantian philosophy in the 1780s and since he referred to both revolution and Copernicus in Kantian contexts. Reinhold does not appear to have written expressly about a Copernican revolution, but there is at least one passage which may have led others to connect Kant with a Copernican revolution. Reinhold's discussion of Kant's *Critique of Pure Reason* is one of the early secondary presentations of this work. This appeared in 1794, in the second volume of his *Beyträge zur Berichtigung bisheriger Missverständnisse der Philosophen.* In the seventh section, "Veber das Fundament der Kritik der reinen Vernunft," Reinhold discusses the preface to the second edition of Kant's *Critique of Pure Reason,* saying that Kant here indicates in an interesting manner the "Umänderung der Denkart" which is imminent for metaphysics through the *Critique* (p. 411). He continues by excerpting at some length and commenting on Kant's remarks about revolution (pp. 411–415). Then on p. 415 he juxtaposes two statements which in the preface to Kant's second edition are separated by several pages (occurring on B xvi and xxii):

> "It is the situation with this just as with the first thoughts of Copernicus, who, after it was not going well with the explanation of the heavenly movements if he assumed that the whole starhost revolved around the spectator, tried whether it might not succeed better if he had the spectator revolve and, on the other hand, the stars be at rest." — "In this attempt to change the up-to-now procedure of metaphysics and thereby that, in accordance with the example of the geometers and physicists, we undertake a complete revolution in it [metaphysics] consists now the business of this critique of pure speculative reason."

These passages are given here in an awkwardly literal translation, so that it may be seen how their juxtaposition possibly influenced later readers to call the Kantian revolution in metaphysics a Copernican revolution. Reinhold himself, though he does attribute to Kant "the foundation and introduction of the now wholly inevitable revolution" (pp. 415–416), does not speak explicitly of a Copernican revolution.

Yet as early as 1784 Reinhold (p. 6) referred to the Enlightenment as a revolution. Moreover, in the first of his famous "Letters on the Philosophy of Kant" — a piece which appeared in August 1786 and hence earlier than the preface to the second edition of the *Critique of Pure Reason* — Reinhold already had connected Kant with revolution (pp. 124–125) and with Copernicus (p. 126), but he did not combine the two to make Kant the author of a Copernican Revolution.

In the middle of the nineteenth century, William Whewell was very careful to summarize Kant's own statement without distortion. In his *Philosophy of the Inductive Sciences* (1847, 479), he wrote: "The revolution in the customary mode of contemplating human knowledge which

Kant's opinions involved, was most complete. He himself, with no small justice, compares it with the change produced by Copernicus's theory of the solar system." Referring the reader to Kant's "*Kritik der Reinen Vernunft,* Pref., p. xv," Whewell clearly kept separate Kant's statement about a revolution in metaphysics and the new standpoint introduced by Copernicus.

16

The Changing Language

of Revolution

in Germany

———◆———

In earlier chapters the development of the concept and name of revolution and the application of this concept to scientific change has been presented primarily through writings in French and in English — the two major vernacular languages for intellectual discourse in the seventeenth and eighteenth centuries. At this time German was not yet the language of the international community of scientists, political and social thinkers, philosophers, historians, and theologians that it was to become in the nineteenth century. But discussions of revolution and a serious debate on the language to be used in that discussion began in the late eighteenth century in a manner that has affected German thought and writing on this subject ever since.

Looking at the German literature on revolution, one is struck by the contest between two usages: the French word 'révolution' ('Revolution') and proposed German substitutes, of which the chief one was 'Umwälzung', and others were 'Umdrehung', 'Umsturz', 'Umschwung', 'Umlauf' (and 'Kreislauf'), and the verb 'umkehren'. Other words include 'Veränderung' (change or alteration), 'grosse Veränderung' (great change), 'Staatsveränderung' (change or alteration of the state).

The word 'Umwälzung' derives from 'um-' (around) and 'walzen' (to roll, to turn, as in the English word 'waltz') and hence is a Germanic equivalent of 'revolution'. 'Umdrehung' is commonly used in the direct sense of turning around (revolution); for instance, the German equivalent of 'RPM' (revolutions per minute) is 'U/min' (Umdrehungen in der Minute); 'Drehung' means turning (as of a wheel), hence rotation or revolution. 'Umsturz' has the sense of 'throw down', 'tip over', 'downfall', 'overthrow', 'turn upside down', 'cataclysm', and hence social and political 'subversion', 'overturn', 'upheaval', and 'revolution'. 'Umschwung' means 'swing around', 'sudden change', 'change-over', and 'revolution'. 'Umlauf' has the sense of 'revolution' in astronomy and also means circulation; the latter sense is also expressed by 'Kreislauf' (as in 'Blutkreislauf' for 'circulation of the blood'). The verb 'umkehren' also means to 'turn around' or 'to turn upside down' (overturn). It is perhaps the nearest word to 'Umwälzung' and the closest in meaning to 'revolution' of German words current in the sixteenth century. Thus, when Luther attacked the Copernican system (which he had heard about but had never studied), he is reported (in his *Tischreden* or *Table-talk*) to have said, "Der Narr will die ganze Kunst Astronomiae umkehren" (The fool wants to overturn the whole art of astronomy). In a Latin report of this same discussion, Copernicus is referred to as a man "who wishes to overturn all of astronomy" ("qui totam astrologiam invertere vult"), where the sense of 'invertere' is 'to invert' or 'to turn upside down' and 'to overturn'.

Because 'Revolution' is a word of French origin (ultimately Latin), it is excluded from the Grimm brothers' celebrated *Lexikon* (the volume for *R* was published in 1893), which enables scholars to trace the ways in which almost all major words have been used by German writers from the Middle Ages to recent times. But in the mid-eighteenth century, the lexicographer Johann Heinrich Zedler was not so nationalistically narrow-minded, and in his gigantic *Grosses Universal-Lexicon*, published in fifty-four volumes from 1732 to 1750, he included 'die Revolution' in vol. 31 (1742). Two definitions were given, without any quoted examples or citations. The first had reference to political change: Revolution "is said of a country which has suffered a considerable change in its administration and policy." This definition is so broadly couched that it embraces equally the notion of a revolution as the act of establishing something wholly new and unprecedented and the idea that a political revolution is a cyclical phenomenon, a return to some antecedent state. The second definition is obviously cyclical: the period of a planet's motion in orbit around the sun.

Although Zedler had introduced 'die Revolution' as if it had become completely assimilated as a German expression, later lexicographers did not so simply accept this foreign word. There is no entry for 'Revolution' in Johann Christoph Adelung's four-volume *Grammatical and Critical Dictionary of the High-German Dialect* of 1774–1780, although 'Revolution' does appear in the six-volume expanded version of 1793–1801. Between the two editions, there had occurred two notable events: the French Revolution and an accelerated movement of replacing foreign words by their German equivalents. For instance, the French word 'édition' was abandoned at this time in favor of a translation into German of the two roots of the word: 'Ausgabe'. This nationalistic outburst of the late eighteenth century was paralleled during the Hitler times when, for example, words of foreign origin (such as 'Telephon') were ordered replaced by their nearest pure German equivalents (in this case 'Fernsprechapparat').

Adelung reacted strongly to the two aforementioned events. He made direct reference to the French Revolution and he discussed the attempted Germanization of 'Revolution'. Like Zedler, Adelung identified the types of revolutionary change (defined as "a total change [gänzliche Veränderung] in the course or the relationship of things"): (1) Natural revolutions, or revolutions in nature, are said to be the great occurrences that have changed the face of the earth. (2) Civil revolutions are said to be complete changes in the constitution of a state, usually accompanied by violence, as in the replacement of a monarchy by a republic; examples are the Glorious Revolution in England and the French Revolution. Adelung expressed strong disapproval of attempts "in recent times" to replace 'Revolution' by a German word. The "most unfortunate" of the proposed substitutes, he noted, were 'Umwälzung' and 'Staatsumwälzung' ('overturning' and 'overturning of the state'), since they were only "literal translations of a foreign word." Other such words were 'Veränderung' or 'Umänderung' (change, alteration), 'Umschaffung' (transformation), 'Hauptveränderung' (chief or primary change), and 'Staatsveränderung' (change or alteration of the state). If one had to find a German equivalent to 'Revolution', he concluded, 'Umwandlung' (change, transformation) might be preferred.

A reply to Adelung appeared in Joachim Heinrich Campe's two-volume dictionary in 1801. Campe attacked Adelung for being a "hair-splitting linguist" and demanded that Germans use their own word 'Umwälzung' in place of the foreign word 'Revolution' as the term for political change. Campe also boasted that it was he himself who had introduced 'Umwälzung' in a political context[1] in 1792, in order to denote the French Revolution, and he immodestly concluded that thousands of German writers of the past decade had "honored" him by using 'Umwälzung' in the new sense that he had introduced. A critic had argued that, 'Umwälzung' implies a corporeal and regular motion, as in the turning of the earth on its axis, and hence is used improperly in relation to political changes. While some of the latter occur in a peaceful or regular pattern, many do not — especially those which are influenced by votes or by mob actions. Campe implied that the only permissible meaning of revolution is the original etymological sense of a cyclical event, and hence this word could not signify the types of political changes that occur in political revolutions. Additionally, he argued, the syllables in 'Um-wäl-zung' follow one another more euphoniously than do those in 'Re-vo-lu-tion'.[2]

Campe did not discuss the possibility that either 'Revolution' or 'Umwälzung' might be legitimately applied to any radical changes other than in the realm of politics. This is all the more curious in that he used quotations from Kant (along with some from Herder) in order to illustrate the use of 'Umwälzung' by leading German thinkers. Yet, as we have seen, Kant wrote about 'revolutions' in science and in philosophy. A few years ago, a computerized index verborum was made of the first nine volumes of Kant's *Gesammelte Schriften* (Martin 1967–1969), which includes all of Kant's major writings. For words which convey the sense of revolution, the number of occurrences was found to be as follows:

die Revolution	57
der Umlauf	33
die Umdrehung	25
die Drehung	15
die Umwälzung	12
der Umschwung	10
der Umsturz	7
der Kreislauf	6

Evidently, Kant used 'Revolution' nearly five times as often as 'Umwälzung' and far more often than any other synonym.

Kant invoked the notion of revolution in a number of works. In his early scientific writings — notably those on astronomy, of which the most celebrated is his *Universal Natural History and Theory of the Heavens*

(1900; 1970; orig. German ed. 1754)—there are references aplenty to orbital revolutions of astronomical bodies. He discussed political revolutions in his *Der Streit der Fakultäten* (1798; see Kant 1902, 7: 59, 85, 87, 88, 93). In a tract on religion, he twice specifically invoked "a revolution in man's disposition" (1793; Kant 1960, 41–43). But, except for the discussion of two great revolutions in the sciences in the second edition of his *Critique*, Kant does not seem to have given much thought to revolutions in science.

Kant's contemporary, Johann Wolfgang von Goethe, made many references to the French Revolution and to revolutions in Italy, Spain, and Portugal; these occur in letters, poems, essays, and accounts of travel. In the introduction to his translation of an essay by Diderot (1798–1799), Goethe argued (1902, *33:* 206–207) that Diderot had been responsible for a "revolution in art." Every "revolution in arts" ("Revolution der Künste"), according to Goethe, tends to promote a "thorough knowledge of nature" ("gründliche Kenntnis der Natur"). In literature, too, Goethe said in his autobiography, *Dichtung und Wahrheit* (1811–1831), there are revolutions; he himself had participated in more than one "German literary revolution" ("deutsche literarische Revolution"; ibid. 1902, *24:* 52). In 1820 he said that certain developments in perspective are "revolutionary" (*37:* 119–120). Those who give up their adherence to the past "always bring forth a revolutionary change" ("revolutionären Übergang"). In one of his aphorisms (*4:* 221), he referred to the foolish "revolutionary views" ("revolutionäre Gesinnungen") of weak men who believe that they can govern themselves.

Goethe's beliefs concerning revolutions in science are chiefly expressed in his challenging *Farbenlehre (Science of Colors)* in 1810 and the ancillary work, *Materialien zur Geschichte der Farbenlehre (Materials for the History of the Science of Colors)*, Goethe's only extended foray into the history of science. In this work Goethe praised Bacon for his advocacy of a science based on experiment. That Bacon could do so in spite of his having been educated in the traditionally Aristotelian philosophy was an example for Goethe of the way in which "revolutionäre Gesinnungen," that is, "revolutionary ways of thinking," may develop as a contribution of single individuals rather than a slow imbibition from the general milieu (1947–1970, *6:* 147). Goethe also characterized Bacon's energy and activity as "striving against authority" ("gegen die Autorität anstrebende") and characterized his thought and influence by the phrase "revolutionärer Sinn" ("revolutionary sense"). He had had a revolutionary mentality that showed itself in its highest form in his writings about natural science (p. 152). This is the clearest statement Goethe ever made on the subject of revolution in science in the sense of a radical change.

But Goethe did endorse a cyclical theory of scientific change, although in his discussion of this topic in *Science of Colors* he does not use the actual word 'Revolution' in this context. He argues that history, like living organisms, never stands still: "Nothing stands still" ("Nichts ist stillstehend"). There is progress, but it is never linear, so that the forward movement is circular, actually helical or spiral—just as in plants, to whose "Spiraltendenz" Goethe devoted much original study. In this point of view toward history, and the historical development of science, Goethe was influenced by the *Scienza nuova (New Science)* of Giambattista Vico, the major expression of the cyclical view of history during the Enlightenment. Goethe had read this work and approved its doctrines (Vietor 1950, 131). In his notebooks, Goethe developed in detail a set of cycles that describe the growth of science (see Groth 1972, 14–18).

In the history of German writings about revolution, the position of Alexander von Humboldt is particularly important, because his writings were so widely read. He published a great five-volume work entitled *Kosmos* between the years 1845 and 1862, in which he attempted to give an accurate general picture of the physical structure of the entire universe in terms that could be understood by the educated public without special scientific training. This work includes a presentation of the history of science for each major subject discussed. It has been estimated that in the 1850s more than 80,000 copies of this work had been sold; several different English translations were made and published.

Humboldt divided his history of attempts to understand the universe into seven epochs. The first occurred in Greek antiquity and the most recent was the series of discoveries beginning with the invention of the telescope in the seventeenth century. He wrote of the possibility of a revolution ("eine Revolution") which might have arisen in the mathematical knowledge of the universe if the Indian (Hindu-Arabic) number system had become known to the Greeks (1845–1862, 2: 198). Humboldt declared that Copernicus had produced a revolution ('Umwandlung', p. 198) in the astronomical world view, and had also originated a "scientific revolution" ('wissenschaftliche Revolution', pp. 350–351). "The scientific revolution originated by Copernicus," wrote Humboldt, "had the rare fortune (setting aside a brief retrogressive movement by the Tychonic hypothesis) of proceeding uninterruptedly towards its goal of discovering the true structure of the universe." When Humboldt turned to the seventeenth century and post-telescopic astronomy, he did not use the term 'revolution' in relation to Galileo and Kepler as he had for Copernicus. In this, perhaps he was following the lead of the historian J.-S. Bailly. Nor did Humboldt refer to Newton's achievement as a 'revolution', perhaps because in this work Newton and the law of universal gravitation are not given much prominence and only mentioned in passing. This could be an influence of the writings of Goethe, whose *Farbenlehre* had been devoted in large measure to a confutation of Newton's celebrated theory of light and a replacement of it by Goethe's own theory. Goethe and Humboldt were very good friends. Much of Humboldt's philosophy was influenced by or parallels that of Goethe.

Humboldt discussed the way in which "the rate of progress rapidly increases," and he invoked the idea "of the periodical, endless transformations ['Umwandlungen'] which await all the physical sciences" (ibid., *3:* 24). But if there are only a few references to revolutions in science in the German edition of *Kosmos,* readers of the English version, in the translation made by E. C. Otté, would have found more references to revolutions than had been written by Humboldt in the original. For instance, Otté (1848–1865, *1:* 48) wrote of "the happy revolution," where Humboldt had merely said "die glückliche Ausbildung."

In the early writing of another major thinker of the nineteenth century, Georg Wilhelm Friedrich Hegel (1817), one finds discussions of revolutions in the earth in the style of Buffon, Herder, and Schlözer. In his *History of Philosophy,* Hegel called the period of German philosophy centering around Immanuel Kant a "revolution in the form of thought" (1927–1940, *19:* 534). He also held that "metaphysical empiricism," deriving from the work of Newton and Locke, could be considered the "complete Revolution" in the operations of the mind. Although Hegel lauded Newton's and Locke's "metaphysical empiricism" as revolutionary, he was extremely critical of Newton — a stance which was later carried over into the writings of Engels. He ridiculed the "barbarous" Newtonian theory of light (1970: 2, 139) and criticized Newton severely for his "ineptitude and incorrectness" in experimenting (ibid.; cf.1927–1940, *19:* 447). In particular, he castigated Newton for his alleged mathematical proof of Kepler's law of areas in the beginning of the *Principia.* He regarded Newton's assumption that sines and cosines may be considered equal in infinitesimally small triangles as a violation of the fundamental principles of mathematics (1969, 273). Furthermore, and more seriously, "mathematics is altogether incapable of proving qualitative determinations of the physical world in so far as they are laws based on the qualitative nature of the subject matter." Hegel, however, did refer to revolutions in science in relation to revolutions in history. In his *Encyclopedia*, a general work on the philosophy of nature (1970, *3:* 18–21), Hegel said: "All revolutions, whether in the sciences or world history, occur merely because spirit [Geist] has changed its categories in order to understand and examine what belongs to it, in order to possess and grasp itself in a truer, deeper, more intimate and unified manner."

Friedrich Engels' famous book on science, *Anti-Dühring*, also bore the title *Herrn Eugen Dührings Umwälzung der Wissenschaft*, which has been translated into English as *Herr Eugen Dühring's Revolution in Science*. Yet there was some uncertainty about Engels' intent, since in the text he uses both words, 'Umwälzung' and 'Revolution'. This problem apparently troubled one of the French translators. In the first French version (1911), the translator hesitated to use the "mysterious title" *La révolution de la science par Eugène Dühring* and adopted the more descriptive *Philosophie, économie politique, socialisme*, with "Contre Eugène Dühring" as subtitle. But the author of the revised version changed this so as to make it read *M. E. Dühring bouleverse la science* (1932). Nowhere in this work does Engels refer to either a 'Revolution' or an 'Umwalzung' of the sciences (see chapter 23, below).

In the nineteenth century, as in the twentieth, 'Umwälzung' gained currency as a rival to 'Revolution'. The major difference between the two seems to have been (and remains) that 'Umwälzung' is rarely used —to my knowledge—for cyclical events, such as the revolution of a planet in its orbit, and is not usually used to denote 'great' political revolutions, such as the French Revolution or the Russian Revolution.[3] Scientists (for example, Albert Einstein) have used both 'Umwälzung' and 'Revolution' in reference to revolutions in science. But I have not encountered any references to *the* Scientific Revolution or the Industrial Revolution that used 'Umwälzung'.

17

The

Industrial

Revolution

———◆———

The Industrial Revolution was not a revolution in science, nor even a revolution based directly and primarily on the applications of science—in the sense that is true of the revolution in the manufacture of dyes that occurred in the second half of the nineteenth century.[1] But it was a revolution whose time span embraces the American and French Revolutions as well as the Chemical Revolution, and like the latter, it was recognized at that time as constituting a revolution in human affairs. Therefore it must be considered in any history whose central themes include the growing consciousness of the occurrence of revolutions outside of the political domain.

I shall not attempt to explore here the nature and significance of the Industrial Revolution, a subject that has generated a tremendous scholarly literature. Such an inquiry would take us far afield, into topics beyond the major focus of this book. In our present context, the Industrial Revolution is of interest primarily for the similarities (and dissimilarities) in its historiography and that of the Scientific Revolution and the concept of revolutions in science.

The major historiographic problem that the Industrial Revolution has in common with both the Scientific Revolution and other revolutions in science is to define exactly what is meant by the name; next comes the double question of when and in fact whether such a revolution occurred. The change in point of view on these questions is apparent on making a comparison of the treatment of Industrial Revolution in the *Encyclopaedia of the Social Sciences* of 1932 and the successor *International Encyclopedia of the Social Sciences* in 1968. Whereas the former devotes thirteen double-column pages to 'Industrial Revolution', the latter merely tells the reader to "*See* Economic Growth; Economy and Society; Industrialization; Modernization." In the new style of thinking about the social sciences, Industrial Revolution is no longer a primary category. In fact, of these four articles, only one ('Industrialization') even mentions the Industrial Revolution. Two paragraphs are devoted to this episode. The first declares that this phrase "has long been used to identify the period roughly from 1750 to 1825," during which "mechanical principles, including steam power" were applied "to manufacturing in Great Britain" so as to produce "an identifiable change in economic structure and growth." The second paragraph underlines the lack of "agreement among scholars about the origins of the industrial revolution in Britain," and points to recent scholarship (Deane and Cole 1962) that "questions the uniqueness of the classical period of the industrial revolution in the long-term evolution of the industrial structure of the British economy" (p. 253).

Even the older article of 1932 began with a notice that "as a label," the name "is admittedly unsatisfactory," even having been called "an unhappily chosen epithet." The "chief objection," it is noted, "is to the word revolution." Economic historians "use the phrase," but they do so "with increasing hesitation and many mental reservations" (p. 4):

> They dislike the suggestion that revolutions in any generally acceptable sense of that term happen in economic affairs. "Sudden catastrophic change is inconsistent with the slow gradual process of economic evolution," says Birnie; "On the vast stage of economic history no sudden shift of scene takes place," says Sée; while Lipson emerges from a study of the seventeenth and eighteenth centuries with the conclusion that there is "no hiatus in economic development, but always a constant tide of progress and change, in which the old is blended almost imperceptibly with the new."

Nevertheless, it was conceded in 1932 that "despite all hesitation the term stands and no better one has been devised." In a lecture on "The Idea of the Industrial Revolution," G. N. Clark (1953, 6) wryly observed that when Sir John Clapham "wrote the great standard book on British economic history from 1820," he "avoided the pedantry of offering a substitute [for Industrial Revolution], but (I believe unintentionally) he never used the term at all."

There is a wide range of opinions of historians of science concerning the first use of the word 'revolution' in relation to science. With regard to the Industrial Revolution, Anna Bezanson (1921–22) found that Paul Mantoux in 1905 credited Arnold Toynbee (uncle of the renowned historian) with the concept and the name: "The expression is, we believe, from Arnold Toynbee"; and that about a decade later, W. E. Rappard (1914, 4) gave 1845 as the "date of the first printed form of the term" and concluded that Friedrich Engels was the first "accredited user." More recently (in 1962), E. J. Hobsbawm wrote in his *The Age of Revolution 1789–1848*, "The very name of the Industrial Revolution reflects its relatively tardy impact on Europe. The thing existed in Britain before the word. Not until the 1820s did English and French socialists — themselves an unprecedented group — invent it, probably by analogy with the political revolution of France" (p. 45).

Perhaps the first unambiguous reference to the Industrial Revolution appears in 1788, when Arthur Young observed that "a revolution is in the making." He had in mind the specific example of the introduction into the woolen industry of the recently invented machines for weaving cotton. Others had apparently used such phrases as "great and extraordinary," "most wonderful," or "beyond the power of calculation" to describe the new technologies (including steam power, coke-smelted iron, new ceramics, and textile machinery), though only Young actually used the term 'revolution'. But a year later, in 1789, the events in France gave currency to the concept and name of revolution in its present most common usage, and before long there were many references in France to 'revolutions' in technology and the Industrial Revolution.

We know more about the history of the name and concept in France because the only detailed study of this topic (by Anna Bezanson 1921–1922) deals exclusively with French sources. It appears that the term 'Industrial Revolution' was rather common in France by the 1820s. For instance, in an article in *Le Moniteur Universel* for 17 August 1827, the words "Grande Révolution Industrielle" stand out in italic type in the middle of the page. The replacement of flax culture by sugar beets was described by Prosper de Launay in 1829 as an example of an "another victim of this industrial revolution." Even earlier, there are references to a revolution in the industrial sphere, although the term 'révolution industrielle' was not used explicitly. The earliest instance found by Anna Bezanson was a proposed regulation of the Chamber of Elboef (on 27 December 1806), recognizing that "this revolution has been useful to industry." In 1819 the French chemist Jean Antoine Chaptal referred to changes in yarn manufacture as "a great revolution in the arts." In a notable discussion of tariffs in 1836, Lamartine says, "It is a complete revolution, the 1789 of commerce and industry," thus linking the "complete revolution" in the economic realm to the changes effected by the French Revolution in the political realm.

Of particular interest to historians of science is the opinion expressed in France in the early nineteenth century that applied science and technical expertise had had a decisive role in the Industrial Revolution in France. This is to be contrasted with the general opinion that the earlier Industrial Revolution in England did not depend on applications of science so much as on technical and mechanical ingenuity.* Anna Bezanson cites a number of examples, of which one comes from a work on dye technology of 1804:

A favorable revolution occurred among us in this respect; our manufactories are no longer entrusted to ignorant workers; rather in most of them one finds very enlightened men, well-educated physicists, and it is to them that one must turn in order to promote progress in the useful arts.[3]

The Industrial Revolution apparently entered the literature of economic history in 1837, with Jérôme Adolphe Blanqué's *Histoire de l'économie politique*. Within a decade, Friedrich Engels had linked the rise of a proletariat with the Industrial Revolution. The opening paragraph of his *The Condition of the Working Class in England* (1845; Eng. trans. 1958, 9) begins:

The history of the English working classes begins in the second half of the eighteenth century with the invention of the steam engine and of machines for spinning and weaving cotton. It is well known that these inventions gave the impetus to the genesis of an industrial revolution. This revolution had a social as well as an economic aspect since it changed the entire structure of middle-class society. The true significance of this revolution in the history of the world is only now beginning to be understood.

Karl Marx used "industrielle Revolution" at least twice in *Das Kapital*, but did not either call special attention to this word or elaborate its meaning; he merely introduced it "incidentally as something that the reader would understand"[4] (Clark 1953, 14).

Arnold Toynbee's posthumously published *Lectures on the Industrial Revolution in England* (1884) set the pattern for those later historians who saw in the British experience of industrialization a pattern of revolution similar to the great political revolutions of history. Toynbee chose 1760 as the date for the beginning of the revolution,[5] whereas other writers have preferred 1750 to 1760 as a starting date, and still others (for example, John U. Nef) have traced the revolution back to the mid-sixteenth century.[6] The extent of Toynbee's influence was symbolized at the opening of the present century when Paul Mantoux's important general work, *La révolution industrielle au XVIIIᵉ siècle* (1905; Eng. tr. 1928, 12th ed., 1964), set forth on its first page that Toynbee had invented the name. Since then there has been a veritable flood of books and articles in which the Industrial Revolution is the central topic and may even appear prominently in the title, though considerable disagreement exists over the meaning of the term.[7] For example, in an oft-reprinted general work first published in 1948, *The Industrial Revolution 1760–1830*, T. S. Ashton (who took his date of 1760 from Toynbee) questions the appropriateness of the word 'revolution' (since "'revolution' implies a suddenness of change that is not, in fact, characteristic of economic processes") and also insists that the "changes were not merely 'industrial,' but social and intellectual" as well. A problem about this revolution is that, unlike political revolutions but like the Scientific Revolution, the Industrial Revolution was spread over a long period of time, covering some seven or eight decades in two centuries. Additionally, the 'revolution' was not wholly industrial since some of the most 'revolutionary' aspects of the Industrial Revolution are demographic (the change in the size of the population and the shift in the traditional ratio of rural and urban populations), agricultural, and economic (growth of commerce, trade, and the modern system of competition).

At least for those writers who believe that 'Industrial Revolution' is a meaningful concept, it seems to give rise to outbursts of eloquent rhetoric. We will see in chapter 26 some examples (Butterfield, Smith, Ornstein) of rhetorical eloquence in relation to the Scientific Revolution similar to these sentiments expressed by Cipolla (1973, 7):

> Between 1780 and 1850, in less than three generations, a far-reaching revolution, without precedent in the history of Mankind, changed the face of England. From then on, the world was no longer the same. Historians have often used and abused the word Revolution to mean a radical change, but no revolution has been as dramatically revolutionary as the Industrial Revolution — except perhaps the Neolithic Revolution. Both of these changed the course of history, so to speak, each one bringing about a discontinuity in the historic process. The Neolithic Revolution transformed Mankind from a scattered collection of savage bands of hunters, whose life in Hobbes's famous phrase was 'solitary, poor, nasty, brutish and short,' into a collection of more or less interdependent agricultural societies. The Industrial Revolution transformed Man from a farmer-shepherd into a manipulator of machines worked by inanimate energy.

Other examples are: the Industrial Revolution "transformed in the span of scarce two lifetimes the life of Western man, the nature of his society, and his relationship to the other peoples of the world" (Landes 1969, 1); "for the first time in human history, the shackles were taken off the productive power of human societies, which henceforth became capable of the constant, rapid, and up to the present limitless multiplication of men, goods, and services" (Hobsbawm 1962, 45). Hobsbawm (1968, 13) flatly declares that "the Industrial Revolution marks the most fundamental transformation of human life in the history of the world recorded in written documents." Even Toynbee's book contains examples of high prose. A chapter devoted to "The Chief Features of the Revolution" begins with the following two sentences: "The essence of the Industrial Revolution is the substitution of competition for the mediaeval regulations which had previously controlled the production and distribution of wealth. On this account it is not only one of the most important facts of English history, but Europe owes to it the growth of two great systems of thought — Economic Science, and its antithesis, Socialism" (p. 58). In this chapter Toynbee stressed the rapid growth of population and an "agrarian revolution," which "plays as large part in the great industrial change of the end of the eighteenth century as does the revolution in manufacturing industries, to which attention is more usually directed" (p. 61).

As the twentieth century unfolded, some writers began to conceive of other industrial revolutions. The *Encyclopaedia of the Social Sciences* (1932) called attention to two works in which post-World War I "efforts toward rationalization and the changes resulting from the coming of electric power and new chemical processes" were described "as the New Industrial Revolution" (Walter Meakin) and "the Second Industrial Revolution" (H. Stanley Jevons). It was also suggested that there had been other industrial revolutions even before the Industrial Revolution. As early as 1894 Mrs. J. R. Green wrote of an English industrial revolution of the fifteenth century, and in our own times H. Van Werveke has conceived of a "sort of industrial revolution" in the eleventh century, while V. Gordon Childe has written of an "industrial revolution of the late bronze age." This process of conceiving that there have been other industrial revolutions is not dissimilar to the discussions by historians of science that there have been more than one Scientific Revolution (see chapter 6, above).

The Industrial Revolution resembles the Scientific Revolution also in the way in which some historians have tended to see both revolutions as continuing processes, lasting up to the twentieth century or even to our own days. Thus Eric Hobsbawm (1962, 46) notes that the Industrial Revolution "was not indeed an episode with a beginning and an end. To ask when it was 'complete' is senseless, for its essence was that henceforth revolutionary change became the norm. It is still going on."

In light of these shifting concepts, almost all writers on *the* Industrial Revolution have sensed a need to be precise about their terms. A most careful and extensive discussion of this problem has been given by David Landes (1969, 1). He observes that: "The words 'industrial revolution' — in small letters — usually refer to that complex of technological innovations which, by substituting machines for human skill and inanimate power for human and animal force, brings about a shift from handicraft to manufacture and, so doing, gives birth to a modern economy." This type of "industrial revolution" has "already transformed a number of countries, though in unequal degrees." Landes observes further:

The words sometimes have another meaning. They are used to denote any rapid significant technological change, and historians have spoken of an 'industrial revolution of the thirteenth century,' an 'early industrial revolution,' the 'second industrial revolution,' an 'industrial revolution in the cotton south.' In this sense, we shall eventually have as many 'revolutions' as there are historically demarcated sequences of industrial innovation, plus all such sequences as will occur in the future; there are those who say, for example, that we are already in the midst of the third industrial revolution, that of automation, air transport, and atomic power.

Finally, Landes notes a third meaning that occurs when this pair of words is capitalized. The term Industrial Revolution is used to "denote the first historical instance of the breakthrough from an agrarian, handicraft economy to one dominated by industry and machine manufacture." This Industrial Revolution (or *the* Industrial Revolution) "began in England in the eighteenth century, spread therefrom in unequal fashion to the countries of Continental Europe and a few areas overseas."

When G. N. Clark (1953, 29) gave his Glasgow lecture in 1952 on the concept of the Industrial Revolution, he could not help concluding, "From the chronological point of view the idea of the Industrial Revolution has collapsed":

> It must indeed be conceded that the short period to which that name used to be given was a period of rapid change . . . There *were* wonderful new machines; there *was* a great growth and displacement of population; there *was* a new kind of social discontent. But these were not aspects of one unique mutation in economic life which can be summed up in a formula of a single sentence.

In particular, Clark found that there were antecedents of the Industrial Revolution, so that it had no sharp beginning. Furthermore, the actual timetable was not the same in different — though neighboring — locales. Finally, what kind of revolution, he asked, could begin in the seventeenth century and be still unfinished in the twentieth? We will see that almost these same considerations arose when twentieth-century historians grappled with the Scientific Revolution (Chapter 26, below).[8]

Scientific Progress
in the
Nineteenth Century

V

18

By Revolution

or

Evolution?

———◆———

The nineteenth century — an age in science that extends from Dalton's atomic theory to Planck's quantum theory, and that includes Darwinian evolution — was replete with revolutionary ideas and revolutionary political and social movements. The roster of radical theories and systems includes the social and political concepts of Marx and Engels, Darwin's theory of evolution, Comte's positive philosophy, and Freudian psychoanalysis. Ushered into being by the French Revolution, this century saw additional revolutions in 1820–1824, 1830, 1848, and 1871, and the rise of revolutionary movements all over Europe on a national and international scale. The year 1848 was particularly volatile. Culminating in the abortive Russian revolution of 1905, the nineteenth century was unquestionably an "age of revolutions" (Hobsbawm 1962). And yet the nineteenth century was also the age of evolution. Darwinian evolution, the major new scientific concept of the century, not only changed the course of biology and current notions of how science progresses, but also influenced theory in fields ranging from sociology, political science, and anthropology to literary criticism. It is a paradox that this dominant idea of evolution was put forth in the context of one of the greatest revolutions in science's history.

From the very start of the century, 'revolution' seems to have been generally understood in the sense of the French Revolution — the establishment of a new order, the creation of a new system, the putting forth of a new set of ideas. For almost all thinkers and doers, the word 'revolution' completely lost its original etymological connotation of return, cycle, or ebb and flow.[1] But a new twist to the concept arose around the middle of the nineteenth century: "permanent revolution". This expression was introduced by Karl Marx and Friedrich Engels in 1850 in a discussion of what the stance of 'proletarian' organizations should be under a 'petty bourgeois' democratic government (Marx and Engels 1962, *1:* 106–117). The answer given by Marx and Engels is that "their battle-cry must be: The Revolution in Permanence" ("Die Revolution in Permanenz"). But even earlier, in October 1848, P.-J. Proudhon (1923, *3:* 17) had publicly declared: "Who speaks revolution necessarily speaks progress." It follows, he continued, "that the revolution is *en permanence* and that to speak properly there have not been several revolutions but only one and the same and [it is] perpetual." The concept of permanent revolution later came to have more than mere ideological significance when, in Russia after the death of Lenin, it became a point of major intellectual division between Trotsky and Stalin and their respective followers (see Tetsch 1973, 84–92, 97–105). Permanent revolution was certainly a radical transformation of the eighteenth-century concept of revolution as a single event or sequence of related events that could overturn an existing political, social, or economic system and set up a new one.[2]

Those who wrote in the nineteenth century of revolution in science did not specifically use the Marxian phrase of 'revolution in permanence' or 'permanent revolution', and it was not science that created the image of this long-term revolution in the minds of Proudhon, Marx, and Engels. Yet in the nineteenth century, many scientists and analysts of science began to conceive of science as a permanent or never-ending quest. This aspect of the scientific quest has been expressed by a mathematical metaphor, that truth lies on an asymptote, implying that there is no simple finite endpoint to science, that truth is an ever-distant goal toward which we may approach closer and closer but never fully attain.

Thus as the nineteenth century progressed, there was an acceptance of the idea that revolutions occur in science and that science advances by revolutions (perhaps an unending series of revolutions), but it came to be recognized that such revolutions may be of long duration — not the brief span of years of a political revolution. And it was at this time that the concept arose of *the* Scientific Revolution: a series of events spread out over perhaps a century or more, from Copernicus to Newton, in which modern science was created. This concept appears clearly in the writings of Auguste Comte (see Guerlac 1977, 33). But, as is the case with many of Comte's ideas, the germs of this one can be found in the writings of Henri de Saint-Simon. (Comte had been Saint-Simon's secretary; see chapter 22 below.) We have also seen that in the opening decades of the nineteenth century, there had been a general awareness of a long-term industrial revolution. In the twentieth century, the view of science as a continuing process or as a long-term, even permanent, revolution figures in Herbert Butterfield's widely-read lectures (1949) and in Rupert Hall's *The Scientific Revolution 1500–1800* (1954).

Not all nineteenth-century thinkers on the progress of science accepted the view that revolution was desirable or inevitable, however. In the final quarter of the century a hope began to be expressed that revolutions in science could be avoided, and in some circles it was believed that revolutions in science do not occur at all. Scientists of note — Mach, Boltzmann, Newcomb, Einstein — argued that major breakthroughs were part of a process of evolution rather than revolution. At the Congress of Arts and Sciences at the St. Louis world's fair (Universal Exposition) in 1904, Simon Newcomb delivered an introductory address on "The Evolution of the Scientific Investigator." This evolution, he held, is a "worthy theme" (1905, 137): "From this viewpoint it is clear that the primary agent in the movement which has elevated man to the masterful position he now occupies, is the scientific investigator . . . As the first agent which has made possible this meeting of his representatives, let his evolution be this day our worthy theme. As we follow the evolution of an organism by studying the stages of its growth, so we have to show how the work of the scientific investigator is related to the ineffectual efforts of his predecessors." Newcomb saw revolutions as climaxes to long periods of evolutionary development; they may not be obvious and they may require deep study to be revealed.

This shift in perspective from revolution to evolution toward the end of the century was in part a response to political and social developments which made thinkers more and more aware of the negative aspects of political revolutions. However one might feel about the aims and ideals of the French Revolution, there was the inescapable fact that the Republic had given way to a directorate and eventually to rule by an emperor. The persistence of the nobility, with the additions created by Napoleon, made a mockery of 'equality', and it would be a long time indeed before the excesses of the Terror would be forgotten. It must be remembered that the revolutions of nineteenth-century Europe were accompanied by violence: in 1848 there was fighting at the barricades, recalling the extremes of the French Revolution.

In 1830 the historian B. G. Niebuhr wrote in the preface to the second volume of his history of Rome (Niebuhr 1828–1832, 2: 2; cf. Schieder 1950, 237) that the world was on the verge of a new destruction "if God does not intervene," much like "the one that occurred in the Roman world in the middle of the third century: an annihilation of prosperity, freedom, education, and science." Four decades later, in November 1871, Jacob Burckhardt gave a series of lectures on the era of the French Revolution, beginning with these observations: "It can be remarked about this course that actually everything to our days is a time of revolution, and perhaps we are relatively close to the beginning or in the second act; the three seemingly peaceful decades from 1815 to 1848 have turned out to be a mere interlude in the great drama. But it seems that this is going to become one movement that is in contrast to all the known past of our Globe" (Burckhardt 1942, 200).

In light of these and similar remarks about the destructiveness of revolution, we are not surprised to see the concept of revolution, which was a dominant image in the first three-quarters of the century, give way somewhat to the concept of evolution in the final quarter. In the sciences we may sample this movement from revolution to evolution in the theory of geological change. This example is particularly noteworthy because it illustrates the effect upon actual developments in scientific thought (not on views concerning scientific progress or the history of science) of the changing perception and experience of revolution in the political domain. This change may be seen by contrasting three uses of the term 'revolution' by geologists: in the eighteenth century, in the early nineteenth century, at the century's end.

In the eighteenth century, considerations of the earth's history tended to follow Buffon's notion of revolutions that changed the nature of the earth, that altered the structure and surface. In keeping with Enlightenment tradition, such revolutions tended to be conceived more as punctuated stages of orderly development that were particularly significant than as catastrophes marked by violence. In the early nineteenth century, the image of revolution changed, in response to the French Revolution. Hence the sense in which Cuvier used this term was wholly different from that of his predecessors. Cuvier was very much aware of the effects of the French Revolution, especially on science, and wrote an insightful study of this topic in 1827. We are, accordingly, not surprised to discover (following up a suggestion by Martin Rudwick 1972, 109) that Cuvier transformed Buffon's concept of revolutions on the earth to a post-1789 sense. No longer were such revolutions merely a succession of alterations of the earth's crust, of which the last was being caused (according to Buffon) by man. Now they became violent catastrophic events, with the destruction of life itself. In this way, Cuvier's revolutions encompassed not only geological changes but the extinction of ancient species of fauna and flora, whose existence in past ages has become known to us from the study of the fossil records.

By the nineteenth century's end, there was a widespread feeling of revulsion against 'revolution', and it was hoped that geologists could now dispense with revolution altogether in accounting for the earth's history. The aim was to supplant the older view of such revolutions by a geological analogy of the Darwinian explanation of the "evolution of species," which had replaced Cuvier's succession of catastrophes or revolutions for explaining the succession of plants and animals found in fossils. This point of view is clearly expressed in an address given in 1904 at the St. Louis Congress of Arts and Science by William Morris Davis, who used the concepts of both evolution and revolution in assessing the progress of the earth sciences during the nineteenth century. He was, of course, fully aware of "the revolution that replaces the teleological philosophy of the first half of the nineteenth century by the evolutionary philosophy of the last half" (1906, 494). He asserted that "our conception of the earth as well as of its inhabitants has been profoundly modified by this revolution." Of more significance in the present context is his insistence that geologists should use the term evolution in a "broader meaning" than Darwinian natural selection. In discussions of geological changes, he concluded, "we are glad to replace the violent revolutions of our predecessors with the quiet processes that evolution suggests" (p. 496).

It might seem rather extraordinary that at a time of bitter revolutionary activity and reaction to it, essentially conservative men like Charles Darwin and the astronomer and philosopher Sir John Herschel could be so radical in their view of science as to consider 'revolution' to be a praiseworthy achievement. Both Darwin and Herschel referred to

Charles Lyell's impact on geology as a revolution, and Darwin went so far as to predict, rightly, that when his own ideas became generally accepted, there would be a "considerable revolution" in the biological sciences. This notion of 'revolutionizing' a science was in fact quite common during the nineteenth century, despite a movement away from the concept toward the end of the century. An address on the microscope and histology in 1845 exclaimed about the way the discovery of the electric current had "revolutionized all chemical and a great part of physical science" (Bennett 1845, 520). In the year in which Darwin published his first paper on evolution (1858), the president of the Linnean Society of London predicted that biology was ripe for a revolution. In a discussion of the germ theory of disease in 1888 (Conn 1888, 5), it was explained that when physicians of that day had been students, this theory had been treated with ridicule: "It is not surprising, therefore, that they should still refuse to accept a theory which so revolutionizes the conceptions of disease." In a biography of Laplace (Arago 1855, 462; 1859, 309), François Arago referred to the achievements of Kepler and of Newton as "admirable revolutions in astronomical science." In an article in *Harper's New Monthly Magazine,* an American journalist (Rideing 1878) said that Lister's "antiseptic method of treating wounds" has "almost revolutionized surgery."

The tension between the idea that science grows by slow cumulation and the more radical notion of revolution is illustrated in writings of Justus von Liebig, one of the most prominent scientists in the middle of the nineteenth century. In an essay of 1866 called "The Development of Ideas in Science," Liebig argued the rather novel proposition that there had been a steady advance of science over centuries as a result of the cumulative contributions of large numbers of researchers (in Liebig 1874). An example was the set of current ideas on the nature of the gases of the atmosphere, the result of the work of hundreds of men over several thousand years. This is, perhaps, one of the first formal statements of the 'cumulative' or 'incremental' view of the development of science.

Of course, as Liebig recognized in another essay, the contributions of great scientists are crucial to scientific progress. In order to indicate the exact nature of such contributions, he used the analogy of a movement in a circle, which he said was a circle of varying radius. "Progress is a circular movement," he wrote (ibid., p. 273), "in which the radius lengthens and a new fruitful thought must necessarily be added to the existing ones, if the scope of our knowledge is to grow." He explained the process as follows: "Take away from the most influential scientific achievements of the greatest men the thoughts received from others and something will always remain that the others did not have — usually only a small part of a new thought, but it is this that makes the man great." This particular view of science precludes the concept of development by revolutions. But in an "autobiographical sketch" (1891, 36; 1891a, 277), Liebig wrote that when he returned to Germany from Paris in 1824, he found that "through the school of Berzelius, H. Rose, Mitscherlich, Magnus, and Wöhler, a great revolution ["Umschwung"] in organic chemistry had already commenced."

In an oft-reprinted and influential *History of European Thought in the Nineteenth Century,* published in 1903, the historian J. T. Merz aligned himself with those students of the nineteenth century who refuse to see the period primarily in terms of revolution. Merz declined "to consider nineteenth century thought as essentially revolutionary" because "the work of destruction belongs in its earlier and more drastic episodes to the preceding age," to a "period [that has] . . . not incorrectly been termed a century of revolution" (1896, *1:* 77–78). In succeeding pages, Merz explores the destructive character of "the revolutionary spirit." Thus, he said, "The work of destruction is indeed still going on; in the midst of this constructive or reconstructive work we still witness the workings of the revolutionary spirit." As an example of "these destructive influences," he pointed to the "new thought, which grew up in Kant's philosophy and the idealistic school" which "degenerated in its further development into a shallow materialism and a hopeless scepticism."

So obsessed was Merz with the synonymity of revolution and destruction that he even declared his aim to be "to look upon thought as a constructive, not a destructive agency." Hence, although he recognized that "no age has been so rich in rival theories, so subversive of old ideas, so destructive of principles which stood firm for many ages, as ours" (p. 80), he would nevertheless stress ("gather my observations and my narrative on") the "prominent and constructive ideas which have sprung up in the course of the century" (p. 81): "Such constructive ideas are those of energy, its conservation and dissipation; the doctrine of averages, statistics, and probabilities; Darwin's and Spencer's ideas of evolution in science and philosophy; the doctrines of individualism and personality, and Lotze's peculiar view of the world of 'values' or 'worths.'" Thus it is not in any way surprising that Merz's development of his subject only rarely uses the concept of revolution in science (or in philosophy), even as a metaphor. We may accordingly give a special importance to his use of the adjective 'revolutionary' in relation to Maxwell's electromagnetic theory. It is curious that by the time Merz has advanced in his narrative to Maxwell, he has forgotten his earlier equation of revolution and destruction and seems to use 'revolutionary' in the more ordinary sense of his age to indicate a radical innovation of exceptional fruitfulness.

Merz illustrates a phenomenon to which I have already referred: the fact that many of the utterances of historians and scientists in relation to revolutions in science are expressions of opinions that may not represent a carefully and fully worked-out and consistently adopted philosophical position. Thus, having equated revolution and destruction in volume 1 of his history, Merz—in volume 2, the second of the two volumes devoted to science—introduces the concept of revolution in a wholly different and more common sense. Maxwell's electro-magnetic theory is not the only example of Merz's coupling of science with 'revolution'. Arguing that "the ideas of energy" have reacted more powerfully "on general thought" than any others from the sciences except "the conceptions introduced by Darwin," Merz (1903, 2: 136–137) points out that a "new vocabulary had to be created," textbooks "had to be rewritten," established "theories had to be revised and restated in correcter terms," and "problems which had lain dormant for ages [had] to be attacked by newly invented methods." These results of "regarding nature as the playground of the transformations of energy," he declared, are to be considered "revolutions in the domain of scientific thought." But in his subsequent presentation of these developments, the word and concept of 'revolution' is conspicuously absent.[3]

Despite the view of Leon Errera and others that science was a permanent or never-ending quest (be it revolutionary or evolutionary), by the century's end there was a growing movement of thought that held science to be finite and nearly complete in some areas. This opinion seems to have been held for the most part by physicists, although there were also expressions of it by chemists and astronomers (see Badash 1972). One of the images of this sense of completeness was the opinion reported by Maxwell in his inaugural address, as Cambridge University's first Cavendish professor (1890, 2: 244), "that in a few years . . . the only occupation which will then be left to men of science will be to carry on these measurements [of the great physical constants] to another place of decimals." Maxwell himself may have been putting forth this opinion only to counteract it, but L. Badash (1972) has shown that the point of view may have been more widespread than is customarily supposed, especially among physical scientists in the English-speaking world.

An often-cited example of this 'next-decimal-place' syndrome is A. A. Michelson, famous for his own measurements of the speed of light and his participation in the Michelson–Morley experiment. In the University of Chicago's catalogue for 1898–99, he published an excerpt from his address at the dedication of University of Chicago's Ryerson Physical Laboratory, reading in part (quoted in Badash 1972, 52): "While it is never safe to affirm that the future of Physical Science has no marvels in store even more astonishing than those of the past, it seems probable that most of the grand underlying principles have been firmly established . . . An eminent physicist has remarked that the future truths of Physical Science are to be looked for in the sixth place of decimals." According to Michelson's colleague Robert A. Millikan (1950, 23–24), this eminent physicist was Lord Kelvin. Millikan says that Michelson later would "upbraid himself roundly for this remark"; yet Michelson repeated it on more than one occasion. In 1903, in his book on *Light Waves and Their Uses*, he said (pp. 23–24):

> What would be the use of . . . extreme refinement in the science of measurement? Very briefly and in general terms the answer would be that in this direction the greater part of all future discovery must lie. The more important fundamental laws and facts of physical science have all been discovered, and these are now so firmly established that the possibility of their ever being supplanted in consequence of new discoveries is exceedingly remote. Nevertheless, it has been found that there are apparent exceptions to most of these laws, and this is particularly true when the observations are pushed to a limit, *i.e.*, whenever the circumstances of experiment are such that extreme cases can be examined. Such examination almost surely leads, not to the overthrow of the law, but to the discovery of other facts and laws whose action produces the apparent exceptions.

In 1897, there appeared in London a *Theory of Electricity and Magnetism* by Charles Emerson Curry. I do not know who Curry was (his name does not appear in the *Dictionary of National Biography*, the *Dictionary of Scientific Biography*, nor the *World Who's Who in Science*). But his book was published by Macmillan and Company and the author was evidently eminent enough to merit a preface by Ludwig Boltzmann. The opening sentence reads: "All branches of theoretical physics, with the exception of electricity and magnetism, can be regarded at the present state of science as concluded, that is, only immaterial changes occur in them from year to year." Pessimism about the future of physics was also reported by two physicists who later became quite famous for their research—Planck and Millikan. Planck, in 1875, had difficulty choosing a career among the subjects of classical philology, music, and physics. His choice of physics overrode the advice of Professor Philipp J. G. von Jolly, who told him there were no new discoveries to be made in that science (Meissner 1951, 75). Millikan (1950, 269–270) said that in 1894 at Columbia, he was "razzed continuously" by his fellow graduate students "for sticking to a 'finished,' yes, a 'dead subject' like physics, when the new 'live' field of the social sciences was just being opened up."

A complete history of nineteenth-century ideas concerning revolution in science could easily fill a whole book. Three prominent French thinkers—Saint-Simon, Comte, and Cournot—will be considered in a later chapter, as will the influence of Marx and Engels. But first we will turn to the most significant scientific revolution in the nineteenth century, the Darwinian Revolution—which, ironically, made popular the very concept (evolution) that eventually helped to undermine some scientists' confidence in the existence of scientific revolutions.[4]

19

The

Darwinian

Revolution

———◆———

Т he Darwinian revolution was the major revolution in the sciences in the nineteenth century. It destroyed the anthropocentric concept of the universe and "caused a greater upheaval in man's thinking than any other scientific advance since the rebirth of science in the Renaissance" (Mayr 1972, 987). The Darwinian revolution is the only biological revolution to be mentioned in the usual list of great revolutions in science, which are traditionally associated with the names of physical scientists: Copernicus, Descartes, Newton, Lavoisier, Maxwell, Einstein, Bohr, and Heisenberg. The Darwinian revolution, as Sigmund Freud (1953, *16:* 285) percipiently observed, was one of the three that gave significant blows to man's narcissistic self-image — the other two being the Copernican and the one Freud himself had initiated. The Darwinian revolution, furthermore, differs from all other revolutions in science in that it is the only one, to my knowledge, in which the first full presentation of the theory contained a formal announcement that it would produce a revolution.

The tremendous revolutionary impact of Darwinian evolution arose in some measure from the extra-scientific component, from what has been called the concomitant ideological revolution. This is even true of the reaction of scientists, since scientists—like other human beings—tend to be strongly influenced in their judgments by philosophical, religious, and other preconceptions. Thus, one of Darwin's critics held that the *Origin of Species* "greatly shocked" his "moral taste." Darwin, he said, had departed from the view that "causation [is] the will of God." The critic said he could "prove" that God "acts for the good of His creatures," and he feared that the alternative view proposed by Darwin would end up in causing humanity to "suffer a damage that might brutalize it." He was concerned that Darwin would cause "the human race [to sink] into a lower grade of degradation than any into which it has fallen since its written records tell us of its history." These fears occur in a letter written to Darwin (Darwin 1887, 2: 247–250) by the Woodwardian Professor of Geology at Cambridge University, who signed his letter, "your true-hearted old friend," Adam Sedgwick. This sentiment underlines the prophetic truth of Huxley's warning to Darwin (ibid., p. 231) of "the considerable abuse . . . which, unless I greatly mistake, is in store for you."

Darwin's Views on Revolution

Charles Darwin published the *Origin of Species* in 1859, a year and a decade after the revolutions of 1848 had swept over Europe. He wrote the final draft of the *Origin* just ten years after *The Communist Manifesto*, which not only announced an imminent revolution but also institutionalized action toward political and social revolution. The periodicals Darwin was reading during the 1840s and 1850s were filled with references to political revolution, revolutionary activity, and even revolutions in science. But despite some signs of industrial unrest in England, Englishmen did not feel threatened by revolution; their only experience of revolution went back to the days of 1688, and by the standards of 1789 or 1848, the Glorious Revolution had been a rather peaceful change. So British scientists and philosophers could contemplate revolution, at least in the sciences, with an aloof equanimity. In the decades before the publication of the *Origin*, Darwin would have become familiar with the image of revolutionary change (for details see §19.1), and he introduced several striking references to revolution in science into his book.

One of them occurs in chapter 10, where Darwin praises Lyell's "revolution in natural history." Again, in discussing "the imperfections of the geological record" in chapter 9 (1859, 306), Darwin wrote that there had been a "revolution in our palaeontological ideas." In the concluding chapter of the *Origin*, containing the full and formal announcement of his own theory, Darwin said simply and straightforwardly that "when the ideas advanced by me in this volume, or when analogous views on the origin of species are generally admitted, we can dimly foresee that there will be a considerable revolution in natural history." This statement has a special Darwinian ring to it. It is couched in a kind of modesty for which Darwin is well known, in the words "we can dimly foresee," but it then moves on to the bold and forceful declaration of "a considerable revolution."[1]

This event, a declaration of revolution in a formal scientific publication, appears to be without parallel in the history of science. A number of scientists have written in correspondence or in manuscripts, notebooks, or private journals of research that their own work was either revolutionary or revolution-making. But only Lavoisier and Darwin evaluated their own contributions in print as revolutionary or revolution-making.[2] Lavoisier read a paper at the Paris Académie des Sciences, which he later printed, that referred to the new chemistry and the consequent production of a new language of chemical nomenclature in terms of revolution (a revolution in the basis of chemistry, thus affecting education), but he did not use 'revolution' in the full presentation of the new theory, as Darwin did.

We have no direct evidence concerning the development of Darwin's thoughts on revolutions, or on revolutions in science. He was certainly familiar with the concept of revolutions in the geological sense as used by Cuvier. Lyell's writings continued this tradition. Lyell's *Geological Evidence of the Antiquity of Man* (1914) contained a chapter on "Vast Geographical Revolutions" in past times. We also know from Darwin's autobiography that he associated the French Revolution with violence. Describing a frightful event that he had witnessed in Henslow's company in Cambridge, Darwin wrote that it was "almost as horrid a scene as could have been witnessed during the French Revolution" (1958, 65). What happened was that two body-snatchers, who were being brought to prison, "had been torn from the constable by a crowd of the roughest men, who dragged them by their legs along the muddy and stony road." The victims "were covered from head to foot with mud" and "their faces were bleeding," either "from having been kicked or from the stones," so that "they looked like corpses." This experience of violence, so long remembered, strengthens our conviction that for Darwin the concept of revolution in science was not an idle metaphor for change but implied a complete alteration of basic beliefs that did violence to the established order of scientific knowledge.

As early as 11 January 1844, a decade and a half before the *Origin*, Darwin wrote to the British naturalist Sir Joseph Hooker (1887, 2: 23), "At last gleams of light have come." "I am almost convinced," he said, "(quite contrary to the opinion I started with) that species are not (it is like confessing a murder) immutable." We may agree with the late Walter Faye Cannon (1961) that it was indeed murder that Darwin was contemplating, "murder of everything that Lyell had stood for with his uniformitarian principle of eternal stability."

During the next decade and a half, Darwin advanced from this pre-1848 conception of the violence of revolt in science as "murder " of established ideas to the proud announcement of "a considerable revolution" in 1859. The dozen years between these two expressions, murder and revolution, include the revolutionary activities of 1848 and their sequelae. These events figured prominently in the journals Darwin was reading during those years (see §19.1).

We have direct evidence that by 1859, just at the time when Darwin was completing the writing of the *Origin*, the idea of revolution in science was in the air. The president of the Linnean Society (London), Thomas Bell, discussed revolutions in science in his Presidential Address in May 1859, as part of a review of the activities of the society during the previous twelve months. It is "only at remote intervals," he said (Gage 1938, 56), "that we can reasonably expect any sudden and brilliant innovation which shall produce a marked and permanent impress on the character of any branch of knowledge." The appearance of a "Bacon or a Newton, an Oersted or a Wheatstone, a Davy or a Daguerre, is an occasional phenomenon," he continued, "whose existence and career seem to be specially appointed by Providence, for the purpose of effecting some great important change in the conditions or pursuits of man." These remarks about scientific revolutions and revolutionaries, four of the six being living contemporaries, were a kind of gloss on his main point: that the "year which has passed . . . has not, indeed, been marked by any of those striking discoveries which at once revolutionize, so to speak, the department of science on which they bear." These comments are all the more significant in that during the year in question there had been read at the Linnean Society both Darwin's preliminary report on evolution and Alfred Russel Wallace's paper, "On the Tendency of Varieties to depart indefinitely from the Original Type."

Bell had been in the Chair during the meeting when these papers had been read. The historian of the Linnean Society has noted, "Bell had apparently little or no idea that he was presiding over the start of a revolution in ideas of life in general, and of human life in particular" (Gage 1938, 56). True enough! But in the present context it is more significant that Bell was aware that revolutions occur in science and that the life sciences were ready for revolution. Darwin's statement in the *Origin* about an impending revolution in natural history may be read as a direct reply to Bell's presidential summary.

The Early Stages of the Darwinian Revolution

Darwinian evolution plainly exhibits the stages of growth of a revolution from the early intellectual roots to a revolution on paper. Darwin's experience on the voyage of the *Beagle* (1831–1836) was of crucial importance, especially his study of fossils and the "confirmation of the law that existing animals have a close relation in form with extinct species"; but as Ernst Mayr (1982, 395) has insisted, "The Darwin who joined the Beagle in 1831 was already an experienced naturalist." We have good evidence (ibid., 408–409; Sulloway 1983) that Darwin did not become an evolutionist on the voyage. His conversion occurred in 1837, at a time when he opened his first notebook on "Transmutation of Species."

Darwin worked out the consequences of his ideas slowly. By 1844 he wrote an essay of 230 handwritten pages (Darwin 1958), which contains the essence of what eventually became the *Origin*. We have the extraordinary paradox, then, of Darwin becoming an evolutionist in 1837, conceiving of the theory of natural selection in September of the following year, and not publishing his ideas in any form for some two decades. In brief, the intellectual revolution was achieved in 1836–37; the second stage of the commitment to the revolution, the private revolution, took form in 1844; but the public stage of revolution on paper had to wait for another decade and a half until Darwin received Wallace's paper, with its independent conception of natural selection, in 1858.

One aspect of the transition from the private revolution to the public revolution on paper that should be noted is Darwin's commitment to this transition at the time of writing the 1844 essay. On 5 July 1844 he wrote a letter to his wife, stating that he had "just finished" his "sketch" of his "species theory." He requested that, in case of his "sudden death," she "devote £400 to its publication," specifying that Lyell would be the best editor to prepare the work for the press ("if he would undertake it"), and that Forbes, Henslow, Hooker, and Strickland would be next on the list, in that order. Darwin even advised his wife on the steps to be taken should "none of these" be willing to take on this assignment, and told her how to proceed should there "be any difficulty in getting an editor."

The first announcement of Darwinian evolution, as is well known, took the form of a joint communication by Darwin and Wallace, after Wallace had sent Darwin a short paper to be forwarded to the geologist Charles Lyell, should Darwin find it "sufficiently novel and interesting." The paper, in fact, to Darwin's shock and amazement, contained what Sir Gavin de Beer (1965, 148) has called "a succinct but perfect statement of Darwin's own theory of evolution by natural selection." Darwin's first and honorable instinct was to suppress his own work and publish Wallace's short paper. He was, however, finally convinced by Lyell and by the botanist Joseph Hooker — both friends of Darwin and, even more important, friends of science and truth — to publish jointly the paper by Wallace and a portion of Darwin's unpublished "Essay" of 1844 together with an extract from a letter Darwin had written in 1857 to Professor Asa Gray at Harvard, containing a "short sketch" of the book on which Darwin had been working. These communications, together with Wallace's paper, were read at the meeting of the Linnean Society in London on 1 July 1858 and were published on the following 20 August in the Society's *Journal of Proceedings* under the title, "On the Tendency of Species to Form Varieties; and on the Perpetuation of Varieties and Species by Natural Means of Selection."

As to the reception of the new ideas, Darwin later wrote that "our joint publications excited very little attention, and the only published notice of them which I can remember was by Professor Haughton of Dublin, whose verdict was that all that was new in them was false, and what was true was old" (1887, *1:* 85). (Darwin himself did not attend this famous meeting of the Linnean Society.) Hooker later reported to Francis Darwin (in 1886) that he and Lyell both "said something impressing the necessity of profound attention (on the part of Naturalists) to the papers and their bearing on the future of Nat. Hist. etc. etc. etc." (1887, *2:* 125–126). The "interest excited was intense," he said, but there was "no semblance of discussion." The new doctrine was talked over after the meeting "with bated breath: Lyell's approval, and perhaps in a small way mine, . . . rather overawed the Fellows, who would otherwise have flown out against the doctrine." George Bentham, who later became President of the Linnean Society, was, however, so "perturbed" by the reading of the Darwin – Wallace papers that he withdrew his own communication, scheduled for later in the meeting, in which he had drawn on his study of British flora to "support the idea of fixity of species" (Darwin 1887, 2: 294).[3]

This episode illuminates a question that has often been discussed, namely: how much credit should be given to Alfred Russel Wallace for Darwinian evolution? Is it fair to give Darwin such exclusive credit for the 'Darwinian revolution'? Wallace's paper certainly was of primary importance as the immediate cause that galvanized Darwin into rapidly completing for publication a readable version of the *Origin*.[4] And let me say at once that this alone is a major contribution to evolutionary science! But it is evident from the modest reaction to the 1858 paper published by the Linnean Society that mere publication of the idea of evolution by natural selection by Darwin and by Wallace did not launch the revolution. The revolution awaited the form of the argument as presented in Darwin's book, the *Origin*, buttressed by an overwhelming mass of evidence.[5] For here was set forth a new way of thinking in biology and a wholly new kind of science (see Scriven 1959).[6] The date of publication was 24 November 1859, and the whole edition was bought up on publication day. A second edition was needed and appeared about a month and a half later on 7 January 1860, followed rapidly by a third edition. Within a couple of years more than 25,000 copies were sold.

One scientist did make use of the papers in the Linnean Society in a scientific communication. This was Canon Henry Baker Tristram, an Anglican priest and ornithologist who had been studying the larks and chats of the Sahara. He had been particularly struck by the "gradual" variations he had observed in their colorations and in the size and shape of their beaks. In 1858 he showed his results to a friend, Alfred Newton, later the first professor of zoology at Cambridge, who was then returning from an ornithological expedition to Iceland. When Newton returned home, he found awaiting him the issue of the *Journal of the Linnean Society* for August, containing the Darwin – Wallace papers. He was converted at once and saw immediately that the new doctrine of evolution by natural selection could account for Tristram's findings and for certain other variations he had encountered. He sent the news on to Tristram. Tristram's report in *Ibis* for October 1859 refers to the communications of Darwin and Wallace to the Linnean Society, and explains how natural selection accounts for the birds having a coloration matching either the sand or soil of their environment, a factor that would give them protection from predators and so favor them in the process of natural selection; and it was the same with regard to the differing size and shape of the birds' bills, which could accordingly be more favorable to food-gathering in relation to the type of soil in which they would dig for worms.

Tristram's later history provides a very interesting commentary on the famous Huxley–Wilberforce debate at the meeting of the British Association for the Advancement of Science at Oxford in 1860. This debate is usually presented as if Bishop Samuel Wilberforce ("soapy Sam") was humiliated and defeated by Huxley and retreated from the scene in intellectual disgrace. The fact of the matter is, however, that Wilberforce made a deep impression on a number of those scientists who were present. Among them was Tristram, the first public convert in print to the new theory of evolution by natural selection. Tristram was so convinced by Wilberforce's argument that he then and there became an anti-Darwinian and remained so for the rest of his life, despite repeated attempts by his friend Newton to reconvert him. It may be added, furthermore, that far from being ashamed of his performance, Wilberforce published an enlarged and corrected version of his talk in the *Quarterly Review*. This paper was later proudly reprinted in the two-volume collection of Wilberforce's papers. (For information on Tristram and on Wilberforce, see Cohen 1984.)

Recently having occasion to reread Wilberforce's essay, I discovered that although Wilberforce did attack Darwin with strength and vehemence, he also praised Darwin for important contributions to science in the *Origin*. In Wilberforce's view, the chief innovation in biological thinking for which Darwin should be given credit was — believe it or not — the idea of natural selection. Wilberforce did not believe in evolution, of course, and so he interpreted natural selection as God's process of weeding out the unfit. This seems to me all the more remarkable in that Thomas Henry Huxley, one of the chief defenders of Darwinian evolution, sometimes called "Darwin's bulldog," never fully accepted this particular part of the theory (see Poulton 1896, ch. 18).

There is no want of evidence that scientists and other thinkers of Darwin's own time considered his theory of evolution and natural selection to be revolutionary. After receiving a pre-publication copy of the *Origin,* the British botanist Hewett C. Watson wrote to Darwin that natural selection "has the characteristics of all great natural truths, clarifying what was obscure, simplifying what was intricate, adding greatly to previous knowledge." And although he pointed out to Darwin the "need, in some degree, to limit or modify, possibly in some degree also to extend, your present applications of the principle of natural selection," he concluded by telling Darwin, "You are the greatest revolutionist in natural history of this century, if not of all centuries." Twentieth-century scientists, philosophers, and historians (for example, Ernst Mayr, Michael Ruse, D. R. Oldroyd, and Gertrude Himmelfarb) now also agree on the existence of the Darwinian revolution in science and on the deep and long-term influence of Darwin's theory on the history of biology and paleontology since 1859. The history of biology since Darwin's day, and especially in the last two decades, shows how deeply Darwinian evolution has affected the subject. Here then is a great revolution in science which easily passes all the tests for such revolutions.

The Nature of the Darwinian Revolution

But what precisely were the revolutionary features of the Darwinian doctrine? Everyone is aware that Darwin was not the first person to believe in evolution. Historians, in fact, seem to take a kind of perverse pleasure in seeking out predecessors of Darwin who believed in a kind of evolution in general, and even those who may have anticipated the idea of natural selection. It must be pointed out, however, that the expression of these ideas prior to 1859 did not radically alter the nature of science in the way that Darwin's *Origin* did. One of the main reasons for this difference, it seems to me, lies in the fact that Darwin presented not merely another essay, another statement of a hypothesis, however plausible, but rather showed by careful reasoning and a mountain of observational evidence that the doctrine of the evolution of species by natural selection was a sound and plausible one. Among other things, he brought together the tremendous experience of breeders, who practiced (as he said) a sort of artificial selection — from which one could get the idea of nature producing a "natural selection." He also adduced a great variety of evidence from the geographical distribution of plants and animals, from geological history, and other fields related to natural history. Furthermore, Darwin presented in a striking and convincing fashion the fact of nature's almost limitless variation among the individuals of any single species. This fact was coupled with the rule of the natural increase in populations and with the lack of a similar increase in available food supply. The result seemed inescapable to him as it does to us: a struggle for life, leading to a process of "natural selection," which he later also called the "survival of the fittest," adopting — at the suggestion of A. R. Wallace — an unfortunate expression originating with Herbert Spencer.

In other words, Darwin did not merely restate some old general ideas of evolutionary development but set forth new and challenging specific arguments for further discussion and for the advancement of science. An example may be seen in the problem of the sequence of different species found in the fossil records of successive geological eras. A number of explanations had been put forth to account for this phenomenon. Cuvier had proposed a series of 'revolutions', life-destroying catastrophes that were followed by new life. Charles Lyell proposed what seems an obvious and logical explanation, namely, that there was a contest among species for survival, that some species disappeared during this struggle and have become known to us only through the fossil or geological record. Lyell proposed what Ernst Mayr (1972, 984) has called "a kind of microcatastrophism," a "concept of a steady extermination of species and their replacement by newly created ones." The chief difference between Lyell's ideas on this topic and Cuvier's is that Lyell pulverized "the catastrophes into events relating to single species, rather than to entire faunas." Darwin transformed this concept of Lyell's from a contest among species into that of a contest among individuals.

The individual members of a species differ from one another in various characteristics according to well-established facts of variation. But some variations are more suited for survival in relation to the nature of the environment. In the ensuing fight for life some variations are more favorable than others; for example, a coloration that blends with the background may help to save an individual from the searching gaze of a predator and may thus favor survival, whereas a coloration that contrasts with the background makes it easy to be spotted and eaten. Darwin saw in such phenomena that the chances of an individual's survival depend on the particular variations that the individual possesses. He gave the process of differential survival the name of natural selection: a process in which an eventual success in reproduction occurs among those individuals whose variations are best suited to the environment and who therefore have the greatest probability of reproducing their own kind. This concentration on the single individual, a "stress [on] the uniqueness of everything in the organic world," is, according to Ernst Mayr (1982, 46), the key to a revolutionary new way of considering the world of nature: "population thinking." Population thinkers "emphasize that every individual in sexually reproducing species is uniquely different from all others." In this new mode of doing biology or natural history there are no "ideal types," no "classes" of essentially identical individuals. Darwin's theory of evolution by natural selection was based squarely on the "realization of the uniqueness of every individual," which Ernst Mayr has described as "revolutionary" in relation to the development of Darwin's thinking.

The transition from Lyell's interspecies competition to Darwin's intraspecies competition is a primary illustration of the creative process that I have called the transformation of ideas (1980, ch. 4, esp. §4.3). The occasion of this momentous and revolutionary step was Darwin's chance reading of Malthus. We are indebted to Sandra Herbert (1971; and, especially, see Ghiselin 1969) for pointing out Malthus's special role in calling Darwin's attention to the "terrible pruning . . . exercised on the individuals of one species," which "impelled Darwin to apply what he knew about the struggle at the species level to the individual level." Darwin then saw "that survival at the species level was the record of evolution, and survival at the individual level its propulsion." In short, Lyell's "concentration on competition at the species level" apparently numbed Darwin "to the evolutionary potential of the 'struggle for existence' at the individual level." Thus Herbert concludes that Malthus should be considered "as contributor rather than catalyst" to the "new understanding" achieved by Darwin, after 28 September 1838, "of the explanatory possibilities of the idea of struggle in nature." Since Darwinian natural selection is based on three elements — "individual variability, the tendency toward overpopulation, and the selective factors at work in nature" (ibid., p.214) — we can see how crucial this transformation was as a stage in Darwin's creative thought. And, furthermore, we can now make precise the exact role of Malthus, not in adding yet another factor to a supposed Darwinian synthesis, nor in supplying Darwin with a mathematical law of population increases, but rather in directing Darwin to transform Lyell's concept into a struggle among individuals by bringing him "to concentrate on the competitive edges to nature — predation, famine, natural disaster — as they played upon the individual differences of members of the same group." This was the crucial moment of the "conceptual shift" to a recognition of a struggle for existence (as Mayr insists, 1977, 324) "among individuals of the single population," of the crucial shift to what is known today as 'population thinking'.

Of course, there are additional factors that must be taken into account for a full understanding of the receptivity of Darwin's mind to Malthus and the recognition of the significance of competition that led to population thinking, among them the principles of individualism and competition in Adam Smith's economic thinking (as Schweber 1977 and Gruber 1974 have shown). In this context we must also take cognizance of Darwin's own statement that the concept of natural selection arose from what we would call a transformation of artificial selection — the long-term practice of breeders who would select for reproduction those individuals exhibiting desirable characteristics. And there was in the air a feeling that a divinely ordained process weeded out misfits in a manner that is somewhat like 'selection'.

Reactions to Darwin's Theory

The revolutionary quality of Darwin's thinking is made manifest in the attacks upon Darwin for not having followed the simple prescribed model that was supposed to be the accepted way of doing science. In order to see the extent to which Darwinian evolution by natural selection represented a departure from traditional norms of scientific thought, for instance as found in the Newtonian natural philosophy, one has only to take account of the fact that Darwinian evolution is nonpredictive, but nevertheless causal. That is, although by natural selection and various other subdoctrines Darwinian evolution assigns a cause to the process whereby present species result from natural selection, this science is unable to predict what the future course of evolution will be, even given the conditions of the environment, with some degree of precision. In other words, Darwin showed that a science can give a "satisfactory explanation of the past," even when "prediction of the future is impossible" (Scriven 1959, 477).

In his public attacks on Darwin, Adam Sedgwick said that "Darwin's theory is not inductive — is not based on a series of acknowledged facts" (Darwin 1903, *1:* 149n) and that Darwin's method "is not the true Baconian method" (Darwin 1887, *2:* 299). He wrote to Darwin that "you have *deserted* . . . the true method of induction." But Darwin asserted in his *Autobiography* (1887, *1:* 83) that he "worked on true Baconian principles, and without any theory collected facts on a wholesale scale." Darwin was particularly pleased to learn that the "method of investigation pursued was in every respect philosophically correct" (1903, *1:* 189). Henry Fawcett reported to him that in the opinion of John Stuart Mill Darwin's "reasoning throughout is in the most exact accordance with the strict principles of logic." Furthermore, Mill said that "the method of investigation" that Darwin had followed "is the only one proper to such a subject."[7] We may understand why Huxley (Darwin 1887, 2: 183) especially took umbrage at the criticism of Darwin in the *Quarterly Review* for July 1860, in which "a shallow pretender to a Master of Science" had the audacity to scorn Darwin "as a 'flighty' person, who endeavours 'to prop up his utterly rotten fabric of guess and speculation,' and whose 'mode of dealing with nature' is [to be] reprobated as 'utterly dishonourable to Natural Science.'" Huxley showed the incompetence of this critic by exposing his ignorance of paleontology, his utter lack of knowledge of comparative anatomy; only after Huxley had written these passages did he discover that the author was his old Oxford adversary, Bishop Wilberforce (Darwin 1887, 2: 183)

Darwin's admirers, on the other hand, compared him to Newton and to Copernicus — authors of acknowledged great revolutions of the past. The German physiologist Emil DuBois-Reymond said that Darwin was most fortunate in living to see his ideas generally accepted (1912, 2, ch. 29), in contrast with Harvey, who died before the scientists of his day were willing to acknowledge the circulation of the blood. T. H. Huxley had no doubt that "the name of Charles Darwin stands alongside of those of Isaac Newton and Michael Faraday" and, like them, "calls up the grand ideal of a searcher after truth and interpreter of Nature" (Darwin 1887, 2: 179). Darwin's name, he added, is bound up as closely "with respect to that theory of the origin of the forms of life peopling our globe" as Newton's is "with the theory of gravitation." The *Origin*, furthermore, is "the most potent instrument for the extension of the realm of natural knowledge" that has come into being since "the publication of Newton's *Principia*" (p. 557). A. R. Wallace (1898, 142) held that the *Origin* "not only places the name of Darwin on a level with that of Newton, but his work will always be considered as one of the greatest, if not the very greatest, of the scientific achievements of the nineteenth century."

Even Darwin, on a number of occasions, compared himself to Newton with respect to the acceptance or rejection of the "Newtonian theory of gravitation" (1903, 2: 305). He was careful and modest enough to insist that he did not wish to imply that natural selection in any way was the equivalent of universal gravitation. Yet he did invoke in his own defense the fact that "Newton could not show what gravity . . . is." Darwin (1887, 2: 290) added that Newton argued against Leibniz "that it is philosophy to make out the movements of a clock, though you do not know why the weight descends to the ground."

Later Stages of the Darwinian Revolution

During the two decades following the publication of the *Origin*, most biologists in England and many elsewhere (but with notable stand-outs and a general lack of adherents in France) became converted to the evolution of species. In 1878 Darwin wrote that "now there is almost complete unanimity amongst Biologists about Evolution" (1887, *3*: 236). But there was less enthusiasm for natural selection and for Darwin's ideas of sexual selection and common descent (see Mayr 1982, 501ff.; Ruse 1979, ch. 8; and especially Bowler 1983). In the letter just quoted, Darwin admitted, "There is still considerable difference as to the means, such as how far natural selection has acted, and how far external conditions, or whether there exists some mysterious innate tendency to perfectibility." As R. W. Burkhardt has remarked (*Science*, 1983, *222*: 156), "The most ardent champions of Darwin in his own day — T. H. Huxley in England and Ernst Haeckel in Germany — differed from Darwin (and each other) in their understanding of how evolution works."

A major question at issue was whether evolution proceeds by the cumulative effect of small variations in generation after generation, or whether large variations were crucial. Another major problem arose over the issue of heredity, which complicated selection in two ways: what mechanisms cause the variation on which natural selection acts, and how are variations passed on to offspring? By the twentieth century, Mendelian genetics shifted attention away from natural selection and small variations to large variations, mutations, and saltations (see Allen 1978; Provine 1971; Ruse 1979). Then began the decline of natural selection and Darwinism, the period Julian Huxley (1974, 22ff.) called the "eclipse of Darwinism." The historical judgment was unambiguous in the 1930s, when I was a beginning graduate student. A standard work which all of us read, Erik Nordenskiöld's *History of Biology* (2nd Eng. ed., 1935), stated that it is "quite irrational" to raise the theory of natural selection, "as has often been done, to the rank of a 'natural law' comparable in value with the law of gravity established by Newton, . . . as time has already shown" (p. 476). In fact, Nordenskiöld warned his readers, "Darwin's theory of the origin of species was long ago abandoned. Other facts established by Darwin are all of second-rate value." On what grounds, then, could "the proximity of his grave to Newton's" in Westminster Abbey be "fully justified"? Nordenskiöld's answer was that such an honor might be deserved if we do not consider his place in science but "measure him by his influence on the general cultural development of humanity" — that is, his influence on philology, philosophy, the concept of history, and man's general conception of life.

In recent decades, however, there has been a resurgent acceptance of natural selection and the emergence of an "evolutionary synthesis" (for which see Mayr and Provine 1980, esp. Mayr's prologue). In other words, the original Darwinian revolution lost ground to the extent that there was an anti-Darwinian counterrevolution, which was not against evolution in general but only Darwinian evolution with its primary concept of natural selection. Ernst Mayr has discussed this rift between Darwinists or neo-Darwinists and their opponents in terms of "conceptual differences between geneticists and naturalists" and has argued that these two groups "belonged to the two different biologies that I characterize as the biology of proximate causes and that of ultimate causes" (Mayr and Provine 1980, 9; Mayr 1961). It would seem to an outsider that the 'evolutionary synthesis' that has characterized recent evolutionary biology — a result of joint activity by geneticists and naturalists — may well constitute a second Darwinian revolution or a second stage of the Darwinian revolution, or perhaps a transformed Darwinian revolution. But it should not be thought that the revolution is over. A major revision has been suggested that once again challenges simple natural selection and that suggests an explanation in terms of 'punctuated equilibria' (see Eldredge and Gould 1972; Gould and Eldredge 1977).

The Impact of the Darwinian Revolution Outside of Science

Darwin's ideas have had a revolutionary impact outside the field of science, far beyond their importance for biology or natural history. Who is not familiar with the proliferation of 'evolution' into every aspect of human thought or endeavor, from studies of the evolution of the novel to the evolution of society? Woodrow Wilson, in a famous study of the Constitution of the United States, wrote that it had been a mistake to apply to this subject the scientific principles of the Newtonian natural philosophy. Rather, he said, the way to understand the Constitution is through evolution: "Government is not a machine, but a living thing. It is accountable to Darwin, not to Newton" (1917, 56). It is well known that in the late nineteenth century a particular form of social thought arose that was called 'social Darwinism', and that attempts were made to link socialism with evolution, a connection which Darwin described in a famous letter (1887, *3:* 237) as "foolish."

But of course in Darwin's day what really shook people about evolution was the challenge which the theory gave to the literal interpretation of Scripture. I do not believe that there would have been quite the outcry against Darwin had the issue been merely one of plants and animals and even the age of the earth. That is, if it had not been necessary to include man himself in the evolutionary scale and the evolutionary process, or if it had not been necessary to conclude that human beings are the result of natural selection, then probably religious believers would not have had quite so strong a reaction. Of course, there were certain fundamentalists (as there still are) who so believed in the literal account of Holy Writ that they would have risen up in arms to challenge even an assumption that the age of the earth is greater than the age computed from the Bible. And we must not forget that the same kind of fundamentalist believers are battling right now in American state legislatures and the courts to establish a doctrine of "equal time" in the classroom for "creationism" alongside of evolution.

Darwin had tried to skirt the issue of man in the *Origin* by only hinting in a single sentence that "much light will be thrown on the origin of man and his history" (1859, antepenultimate paragraph). But Darwin's critics from then till now have stressed the obvious implication of the theory of evolution for ourselves, the apparently inescapable conclusion that man is merely a temporary end-product of an everlasting evolutionary process. Indeed, it is a matter of record that even Alfred Russel Wallace could not bring himself to believe that natural selection could account for man's development in history, and thought it was necessary to invoke the active hand of a Creator (see Kottler 1974). This occurred for the first time in an article on "Man" in the *Anthropological Review* in 1864 and again in a book review in the *Quarterly Review* (1869) in which Wallace discussed the tenth edition of Lyell's *Principles* (1867–1868) and the sixth edition of his *Elements of Geology* (1865). He argued that natural selection alone could never have produced the human brain, the human organs of speech, the hand, and so on. Darwin, in anguish, wrote to Wallace in March 1869 that "I hope you have not murdered too completely your own and my child." In his own copy of the *Quarterly* (see Darwin 1903, 2: 39–40), Darwin marked this passage with "a triply underlined 'No,' and with a shower of notes of exclamation."

The Darwinian revolution was probably the most significant revolution that has ever occurred in the sciences, because its effects and influences were significant in many different areas of thought and belief. The consequence of this revolution was a systematic rethinking of the nature of the world, of man, and of human institutions. The Darwinian revolution entailed new views of the world as a dynamic and evolving, rather than a static, system, and of human society as developing in an evolutionary pattern. We shall see that Karl Marx even foresaw an evolutionary history of technology or invention, in which Darwinian concepts introduced for animal organs would be used in analyzing the development of human tools.

The new Darwinian outlook denied any cosmic teleology and held that evolution is not a process leading to a "better" or "more perfect" type but rather a series of stages in which reproductive success occurs in individuals with characters best suited to the particular conditions of their environment — and so also for societies. No longer were there to be grounds for special creation. An end was heralded to any "absolute anthropocentrism," since a principle of "common descent" was proposed for all living creatures, including man. To these implications we must add that the Darwinian revolution sounded the death knell of any argument about design in the universe or in nature, since variation is a random and nondirected process. In the life sciences, there was a dramatic shift from the older biological concepts to the new population thinking. And in addition to these many new directions, Darwin initiated innovations with respect to method, introducing a new kind of scientific theory in which the role of prediction differed from the classical Newtonian model.

All of these implications were not evident at once, but enough of them were so inescapably obvious that there was an immediate explosion of opinions. Never before in history did the announcement of a scientific theory give rise to such immediate heated debates in countries all over the world — an index of the truly revolutionary character of Darwinian evolution by natural selection. The translations, reviews, commentaries, and attacks began almost at once and continue to our own day. There is only one other scientific author of modern times who can be compared to Darwin in this regard, and that is Sigmund Freud — a fact that shows the tremendous insight that Freud had early on when he likened the prospective impact of his own ideas to the effect of Darwin's (see chapter 24, below). That historical, philosophical, and even scientific debates on evolution and its implications continue to exercise the minds of serious thinkers a century after Darwin's death gives us continuing evidence of the extraordinary vitality of Darwin's science and of the profundity of the Darwinian revolution.[8]

20

Faraday,

Maxwell,

and Hertz

T he nineteenth century was witness to many revolutionary advances in physics, though none had the global impact — in either its scientific or its ideological component — of the Darwinian revolution. The triumphs of nineteenth-century physics include the new doctrine of energy and the laws of its conservation, the undulatory theory of light, the kinetic theory of gases and statistical mechanics, the laws of electric currents, the theories of magnetism and electromagnetism, the principles of electric motors and generators, the new science of spectroscopy, discoveries about the radiation and absorption of heat, the extension of radiation into the infrared and ultraviolet, and much else. Most physicists and the new generation of historians of modern physics agree, however, that one of the most profound revolutions, if not *the* most profound, was the one named for the theories of James Clerk Maxwell — sometimes attributed to Maxwell and Michael Faraday and also with some justice to Faraday, Maxwell, and Heinrich Hertz. The significance of the Maxwellian revolution is that it not only produced a fundamental revision of the theories of electricity, magnetism, and light but was the first large-scale revision of the Newtonian framework for physical science.

Although some features of this revolution can be understood by all readers, the central core of Maxwell's thought is difficult even for many historians trained in physics. A major problem is to ascertain the relationship between the ideas of Michael Faraday and the theory that Maxwell developed. Certainly Faraday's contribution was of tremendous importance, including the significant conception that a magnetic field is composed of lines of force, as well as the extraordinary insight that electrical and magnetic actions are not transmitted instantaneously but require time. Faraday's essentially nonquantitative and nonmathematical formulation did not lead to a numerical value of that supposed time of transmission, however. In his *Treatise* Maxwell paid Faraday the highest tribute possible for his seminal ideas, and went so far as to say that it was "perhaps for the advantage of science that Faraday, though thoroughly conscious of the fundamental forms of space, time, and force, was not a professed mathematician." Faraday expressed his ideas "in natural untechnical language," and, Maxwell concluded, "it is mainly in the hope of making these ideas the basis of a mathematical method that I have undertaken this treatise." Yet, we are warned by all who have studied the history of this subject that to regard Maxwell's "contribution as solely one of translation would grossly undervalue it" (Tricker 1966, 102). As Max Planck has eloquently stated, "Maxwell, with bold phantasy and mathematical insight, went far beyond Faraday, whose standpoint he both generalized and made more precise." Maxwell "thus created a theory which not only could compete with the well established theories of electricity and magnetism but surpassed them entirely in success" (1931, 57).

Historians and historically minded scientists agree that Faraday's papers would never have produced a revolution without Maxwell's profound transformation of Faraday's ideas in the process of constructing a mathematical theory—one that can accordingly be called a Faraday–Maxwellian theory. Maxwell not only transformed Faraday's thoughts into mathematical form but developed a quantitative expression that linked the fundamentals of electrostatics and electromagnetism with the speed of light—an achievement which made the electromagnetic theory plausible and opened up the possibility of an experimental test by means of the actual production of electromagnetic waves. Recognition of Faraday's role in Maxwell's thought gives emphasis to the creative process of transformation that produced Maxwell's theory but in no way lessens the central contribution of Maxwell to the Maxwellian revolution. This may be even truer of William Thomson's contribution to the revolution (see below), since "it was Thomson's peculiar genius to generate powerful disconnected insights rather than complete theories" (Everitt 1974, 205). By applying Thomson's method of electric images and the results of Thomson's "applications of energy principles to electricity" Maxwell was able to realize their importance.

Maxwell's ideas were developed in a series of papers dated 1855–1856, 1861–1862, 1863, 1864, and 1865 and were given a more or less final form in his *Treatise on Electricity and Magnetism* in 1873.[1] But for some years thereafter, this revolutionary new doctrine remained no more than a revolution on paper and did not become a revolution in science until the work of Heinrich Hertz. For this reason, the revolution is sometimes known as the Faraday–Maxwell–Hertz revolution; even those who discuss the revolutionary work of Maxwell point out that the revolution was not his alone. For example, Albert Einstein discussed "the great change [Umschwung], which will be associated for all time with the names of Faraday, Maxwell, and Hertz" (1953, 161; 1954, 268). But he was quick to add that the "lion's share in this revolution [Umwälzung] fell to Maxwell." On another occasion he omitted Hertz and merely wrote about "the revolution [Umwälzung] in electrodynamics and optics brought about by Faraday and Maxwell," which he said was "the first great fundamental advance in theoretical physics since Newton" (1953, 154–155; 1954, 257). In his autobiography, however, Einstein spoke simply about "Maxwell's theory" ("die Maxwell'sche Theorie"), which he said appeared to be "revolutionary" ("revolutionär") when he was a student (Schilpp 1949).

Maxwell's Transformation of Faraday's Ideas

The process of transformation can be seen in Maxwell's famous paper "On Physical Lines of Force" (1861–62). In considering Faraday's ideas that there must be some kind of stresses in a space in which there are magnetic lines of force, Maxwell started out by asking in effect: what kind of medium would there have to be for space to exhibit the actual stress distributions required by Faraday's hypothesis? C. W. F. Everitt has traced the way in which Maxwell drew on ideas of the Scottish engineer W. J. M. Rankine along with the conclusions of William

Thomson (Lord Kelvin) to produce his own theory of physical lines of force.[2] Here we may see the ingredients of a classical transformation of scientific ideas, resulting in a wholly new idea, that electricity may be "disseminated through space" and need not necessarily be only "a fluid confined to conductors." Maxwell concluded his paper with what has been called a "stunning discovery" — that vibrations of this newly proposed medium would not only account for the lines of magnetic force but also would have "properties identical with light." Maxwell conveyed the extraordinary quality of his result by using italic type. We "can scarcely avoid the inference," he wrote (1890, *1*: 500), "that *light consists in the transverse undulations of the same medium which is the cause of electric and magnetic phenomena.*"

Even here, however, seeds of Maxwell's ideas may be found in a remarkable paper of Faraday: one that was printed in the *Philosophical Magazine* for May 1846 as "Thoughts on Ray-vibrations." In this communication, Faraday put forth as "the shadow of a speculation" the bold view of "radiation as a high species of vibration in the lines of force." What may interest us most about the ideas in this paper is that, as Silvanus P. Thompson observed in 1901 (p. 193), it did not attract much notice, even on the part of Faraday's earlier biographers. Writing before the general acceptance of Maxwell's electromagnetic theory of light, they did not recognize in it the importance later attributed to it. John Tyndall (in 1868) merely dismissed Faraday's speculation as "one of the most singular speculations that ever emanated from a scientific man." In 1870 Henry Bence Jones alluded to it in passing, in half a line of type. John Hall Gladstone, in 1872, did not even mention it. But Maxwell later said the "conception of the propagation of transverse magnetic disturbances to the exclusion of normal ones is distinctly set forth by Professor Faraday in his 'Thoughts on Ray Vibrations (1890, *1:* 535).'" According to Maxwell, "The electromagnetic theory of light, as proposed by him [Faraday], is the same in substance as that which I have begun to develop in this paper, except that in 1846 there were no data to calculate the velocity of propagation." I agree with C. W. F. Everitt that Maxwell's remarks about Faraday's "Thoughts on Ray Vibrations" may be taken *cum grano salis* insofar as "any direct influence" of that paper can be discerned in the development of Maxwell's thought. "These remarks were made several years after the event and are an example of Maxwell's quixotic generosity. His comments at the time in letters to Faraday and Thomson disclose no such influence."

In a review of Maxwell's contribution to physics (1896, 204–205), R. T. Glazebrook called attention to five fundamental features of Maxwell's theory and "admitted that in Maxwell's day there was direct proof of very few" of them. One of the boldest assumptions made by Maxwell was that the same medium that sustains the waves of light must be capable of being the medium for electric and magnetic fields. He concluded that there must be a possibility of electromagnetic waves in space. Furthermore, Maxwell, a pioneer in dimensional analysis, showed that the factor linking two systems of electrical units, electrostatic and electromagnetic, was a velocity and in fact had a numerical value very close to the velocity of light. This implied that light is itself an electromagnetic phenomenon, a series of electromagnetic waves. In 1864 Maxwell could report that numerical results seemed "to show that light and magnetism are affections of the same substance, and that light is an electro-magnetic disturbance propagated through the field according to electro-magnetic laws."

Max Planck (1931, 57) saw in this insight the best possible illustration of the "criterion of the value of a theory, that it explains quite other phenomena besides those on which it was based." Planck assumed that neither Faraday nor Maxwell "originally considered optics in connection with their consideration of the fundamental laws of electromagnetism," and yet "the whole field of optics, which had defied attack from the side of mechanics for more than a hundred years, was at one stroke conquered by Maxwell's Electrodynamic Theory," so that "since then every optical phenomenon can be directly treated as an electromagnetic problem." For Planck, this must "remain for all time" one of "the greatest triumphs of human intellectual endeavour."

The Contribution of Heinrich Hertz

Here then was a test — not only to see if electromagnetic waves could be produced, but to find out if they had the speed of light. So we may understand the importance of the experiments of Heinrich Hertz, conducted in the years up to 1888, which ultimately confirmed the predictions of Maxwell's theory. Hertz not only produced electromagnetic waves and found their speed (by measuring the wavelength of standing waves of known frequency); he showed that these waves were like light in properties of reflection, refraction, and polarization, and that they could be focused. Hertz himself referred to this theory as "a theory which Maxwell constructed upon Faraday's views, and which we call the Faraday–Maxwell theory" (1893, 19).

Hertz's contribution was not merely to devise and to execute a clever experiment, great as that achievement was. He showed also how his experiments were significant as the "first demonstration of the finite propagation of a supposed action at a distance" (McCormmach 1972, 345). His experiments thus had the effect of causing physicists to make a radical shift in their views about electromagnetism from "instantaneous action at a distance" to "Maxwell's view that electromagnetic processes take place in dielectrics and that an electromagnetic ether subsumes the functions of the older luminiferous ether" (ibid.). But to complete the revolution, Hertz had to clarify precisely what theory physicists were subscribing to "when they declared themselves followers of Maxwell." (On this point, see the excellent summary by McCormmach on p. 346, especially the discussion of Hertz's treatment of Maxwell's 'vector potential'.) In the end he eliminated certain physical features of the theory that "unnecessarily complicated the formalism" (1893, 21) and concluded (in the introduction to his *Electric Waves*) that "Maxwell's theory" is merely "Maxwell's system of equations." Because the adoption of the Maxwellian theory, especially on the Continent, followed along lines suggested by Hertz,[3] we can understand why Einstein and others have included Hertz's name in discussing the revolution.

Maxwell's theory was difficult to accept for a number of reasons. First, it was conceptually new, with radical notions such as the 'displacement current'.[4] Second, Maxwell presented the theory not merely as a mathematical elaboration of new principles but in terms of physical models. At first these were expressed in such mechanisms as cogwheels and pulleys in what his sincere admirer Glazebrook could not help but call (1896, 166) a "somewhat gross conception," although he does stress that these were to their author nothing more than "a model." Rotating tubes and aethereal vortices were never completely abandoned by Maxwell. In his *Electricity and Magnetism* (2: §831; 1881, 2: 428), he wrote of "magnetic force" as "the effect of the centrifugal force of the vortices" and "electromotive force" as a product of "the stress on the connecting mechanism." The French mathematician Henri Poincaré, who wrote in glowing terms about Maxwell's theory (see below), could not help introducing a book of *Lectures on Maxwell's Theories and the Electromagnetic Theory of Light* (1890, p. v) by expressing how, when "a French reader opens Maxwell's book for the first time," a feeling of uneasiness, often even of distrust, is mingled with his admiration. In another work (1899; Eng. trans., 1904, 2), Poincaré admitted that the "complicated structure" which Maxwell attributed to the aether "rendered his system strange and unattractive." In fact, according to Poincaré, one "seemed to be reading the description of a workshop with gearing, with rods transmitting motion and bending under the effort, with wheels, belts, and governors." Furthermore, according to Poincaré, it is "the taste of the English for conceptions of this kind, whose concrete appearance appeals to them." But he also observed that Maxwell himself "was the first to abandon his own extraordinary theory," and that "it does not appear in his complete works," by which he may have meant Maxwell's *Papers.* Poincaré is quick to add that we must not regret that the mind of Maxwell "followed this by-path, since it was thus led to the most important discoveries," and Poincaré insisted (p. 12) that "the great element of permanency in Maxwell's work" is the fact that "it is independent of all particular explanations."

The experiments that led to the verification of Maxwell's prediction were undertaken by Hertz at the suggestion of the great German physicist Hermann von Helmholtz. On the Continent, especially in Germany, the tendency (of Gauss, Weber, et al.) was to seek the "completion of Electrodynamics" exclusively—as Planck explained (Planck 1931, 58–59)—"in terms of Potential Theory, which had been derived by Gauss from Newton's law of action at a distance for statical electric and magnetic fields" and which "had been brought to a high degree of mathematical completion." The Faraday–Maxwell conception that there can be no such "immediate action at a distance," and that the field of a force has "an independent physical existence" was so strange and difficult to comprehend that the new theory, according to Planck, "found no foothold in Germany and was scarcely even noticed." Helmholtz had developed a theory of his own, in which he tried to keep the mode of instantaneous action at a distance and yet encompass Maxwell's equations. He encouraged Hertz to make his experiments not only to find out whether electromagnetic waves exist or can be produced (as both his theory and Maxwell's demanded) but to decide between the two versions, which led to very different predictions about the physical properties of the waves. (For a succinct account of the differences between Helmholtz's theory and Maxwell's, see Turner 1972, 251–252.)

In a popular—that is, nonmathematical—work on *Maxwell's Theory and Hertzian Oscillations* (1899; Eng. trans. 1904, ch. 7), Poincaré explained how Hertz's experiments provided the 'experimentum crucis' between Maxwell's theory and its rival. Both theories agreed on many predictions that were verified (for example, that the velocity of propagation of electrical disturbances along a wire is the velocity of light, that there are electromagnetic disturbances transmitted through space), but they differed on the time of propagation of these effects in space. If there do not exist Maxwellian "displacement currents," then the propagation should be instantaneous; but according to Maxwell's theory, the velocity in air or in empty space should be the same as along the wire — that is, it should be the speed of light. Poincaré put the question thus: "Here, then, is the *experimentum crucis:* we must determine with what velocity electromagnetic disturbances are propagated by induction through the air. If this velocity be infinite, we must adhere to the old theory; if it be equal to the velocity of light, we must accept the theory of Maxwell." Hertz's initial experiments did not give an easy answer. The results "seemed undeniably to condemn the old electrodynamic theory," but "seemed nonetheless to condemn the theory of Maxwell." Writing in 1899, Poincaré reported that "this failure is still unsatisfactorily explained." He guessed that Hertz had used a mirror that was "too small with respect to the wave-length," so that "diffraction entered to disturb the phenomena." In any event, later experiments (first by Sarasin and de la Rive) proved conclusively that Maxwell's theory was correct. This marked the end of theories based on instantaneous action at a distance, and the beginning of a general acceptance of field theories in the Maxwellian mode, with finite speed of propagation equal to that of light. Thus was the Faraday–Maxwell revolution on paper converted into the Faraday–Maxwell–Hertz revolution in science.

Testimony to the Revolution

In a lecture given in 1888, the year in which Hertz sent to Helmholtz his final results of his experiments on electric waves, Helmholtz (1907, 3) spoke of a "complete revolution" ("eine vollständige Umwälzung") that the "Faraday-Maxwellian ideas" had brought about in theoretical physics ("the theoretical physics of the aether"). Helmholtz (p. 4) then went on, in Kuhn-like terms, to discuss the "crisis" that the theory of electricity must first go through ("eine Krisis, die erst durchgemacht werden muss").[5] The difference between Helmholtz's 'crisis' and 'revolution' and Kuhn's is, however, that Helmholtz appears to have seen the 'crisis' arising from the 'revolution' and not being the antecedent condition.

A somewhat more restrained expression of 'revolution' appeared in 1894 in August Föppl's textbook, *Einführung in die Maxwell'sche Theorie der Elektricität*, the textbook from which Einstein, in his student days in Zürich, learned Maxwell's theory. (The significance of Föppl in the development of Einstein's thought is discussed in Holton 1973, 205–212.) In the preface Föppl stressed how Hertz had not only demonstrated the existence (and speed) of electromagnetic waves but had established a "turning point" in theory which effectively shifted physicists away from the old theory (of Weber and others) based on action-at-a-distance. The Hertzian discoveries produced an "Umschwung der Meinungen," a "reversal [that is, an overturn; *possibly*, a revolutionary change] in opinions" (pp. iii, iv).

Similar views were soon after expressed by the French philosopher-scientist Pierre Duhem. Duhem's discussion is all the more interesting in that he was a distinguished historian of science in addition to being a scientist of note and a philosopher of distinction. He called his book a "historical and critical study" of the electrical theories of Maxwell. In describing the effects of Maxwell's work, Duhem (1902, 5) used two words in rapid succession, 'bouleverser' and 'révolution'—the very same words that appeared in successive French versions of Engels' *Anti-Dühring* as translations of 'Umwälzung'. Duhem said simply and straightforwardly, "This revolution [cette révolution] was the work of a Scottish physicist, James Clerk Maxwell" (1902, 5). In a historical aside Duhem noted that "Maxwell had reversed the natural order according to which theoretical physics evolves; he did not live long enough to see the discoveries of Hertz transform his audacious temerity into a prophetic prediction" (p. 8). In discussing Maxwell's first paper, in which he had introduced the analogy of electrical phenomena and the motion of a fluid in a resisting medium, Duhem observed, however, that Maxwell's language would seem to indicate that it had not at all been "his intention to revolutionize this branch of physics" (p. 55). Duhem also highly praised the work of Ludwig Boltzmann, published in 1891 and 1893, in which he attempted, "by means of wholly new concepts, to construct a system in which the equations of Maxwell are logically connected" and which Duhem saw as a means of eliminating a major problem in Maxwell's own development of his different equations, which Duhem found replete with "contradictions and paralogisms" (pp. 223–224).

A year after Duhem's discussion of Maxwell and revolution, John Theodore Merz published the second volume of his *History of European Thought in the Nineteenth Century* (1903). Here he referred to Maxwell's papers on electromagnetic theory as a "revolutionary series of works," and recorded that the "influence of Maxwell's ideas on scientific — nay, even on popular — thought has been very considerable" (pp. 77–78, 88).

I have already referred to Einstein's continued references to Maxwell in terms of revolution. In a conversation recorded in 1920 (Moszkowski 1921, 60), Einstein summarized the Maxwellian revolution as follows:

> Classical mechanics referred all phenomena, electrical as well as mechanical, to the direct action of particles on one another, irrespective of their distances from one another. The simplest law of this kind is Newton's expression: 'Attraction equals Mass times Mass divided by the square of the distance.' In contradistinction to this, Faraday and Maxwell have introduced an entirely new kind of physical realities, namely, *fields of force*. The introduction of these new realities gives us the enormous advantage that, in the first place, the conception of action at a distance, which is contrary to our everyday experience is made unnecessary, inasmuch as the fields are superimposed in space from point to point without a break; in the second place, the laws for the field, especially in the case of electricity, assume a much simpler form than if no field be assumed, and only masses and motions be regarded as realities.

In his "autobiographical notes" (Schilpp 1949, 32–33), Einstein elaborated this theme:

> The most fascinating subject at the time I was a student was Maxwell's theory. What made it seem revolutionary was the transition from forces acting at a distance to fields as fundamental quantities. The incorporation of optics into electromagnetic theory, with its connection between the speed of light and the absolute electrostatic and electromagnetic systems of units, as well as the relation of the index of refraction to the dielectric constant, and the qualita-

tive connection between the reflection coefficient and the metallic conductivity of a body—it was like a revelation.

Einstein's perceptive evaluation of a Maxwellian revolution, made a half-century or so after Hertz's experimental verification of the prediction of electromagnetic waves, has been stated anew by Karl Popper in his trenchant survey of revolutions in science (1975, 89). "The revolution of Faraday and Maxwell," he said, "was, from a scientific view, just as great as that of Copernicus." The reason is that "it dethroned Newton's central dogma—the dogma of central forces."

Many commentators have pointed out that Maxwell's theory gained more general adherence in England than on the Continent. Yet there were dissenting voices. Lord Kelvin was one. In his *Baltimore Lectures,* given at Johns Hopkins in 1884, he said bluntly: "If I knew what the electro-magnetic theory of light is, I might be able to think of it in relation to the fundamental principles of the wave theory of light." Additionally, "I may say that the one thing about it that seems intelligible to me, I do not think is admissible." In analyzing the situation in England from 1875 to 1908, Sir Arthur Schuster said that in England no experimental test of Maxwell's predictions was made because "we were perhaps over confident in the inherent truth and simplicity of Maxwell's conception." Why should one undertake an "extended experimental investigation," one which "would certainly have absorbed much time and labour," since we did not "consider [it] worth undertaking, considering the indirect evidence in favour of the electro-magnetic theory"? In fact, there seemed little point in making an experiment of which it seemed obvious that "the result" would be "a foregone conclusion." Schuster, however, says that the young men at the Cavendish "were wrong," because they "forgot that the great body of scientific thought abroad, and to some extent in this country, was apathetic and even reluctant to abandon an elastically solid aether which had done good service, and to accept in its place a medium, the properties of which were unlike those of any known body."

The Maxwellian revolution differs somewhat from those we have been considering, which can be associated easily with the scientific thought of a single individual such as Lavoisier or Darwin. This revolution—which developed over a long period of time, more than half a century—required three notable contributions: by Faraday, Maxwell, and Hertz. Opinions may differ concerning the importance of these three great physicists. Perhaps the name Maxwellian revolution arose from the fact that the theory is epitomized in Maxwell's equations, which could have been the reason why Einstein attributed to Maxwell the "lion's share" in the revolution. But Einstein also revered Faraday, and on the walls of his study he had portraits of them both. This revolution may seem to be like the revolution attributed to Copernicus, in which the concepts of Copernicus were transformed by Kepler and then developed by Newton. But there is a fundamental difference in that Kepler essentially abandoned Copernicus's principles whereas Maxwell enshrined Faraday's in his own theory, gave Faraday's concepts new precision and significance, and developed Faraday's ideas in the sense that Newton built on Kepler's foundation.

Maxwell's contributions to a new physics were not limited to his theory of electromagnetism. They encompassed many other topics including molecular physics, thermodynamics, and kinetic theory of gases. He made scientists aware of the significance of dimensional analysis, and he gave currency to the concept of a model in physical theory, which has become so prominent a feature of the physics of our own day. We have seen evidence that Maxwell's electromagnetic theory easily passes three tests for revolution: the testimony of witnesses, the judgment of historians, the opinion of scientists. The fourth test — the record of physical thought — shows that the Maxwellian revolution (or the revolution of Faraday, Maxwell, and Hertz) has been a central factor in the transition from the classical physics of the eighteenth and nineteenth centuries to the new physics of relativity and quantum theory of the twentieth century. It constitutes a great revolution in human thought, on a par with the Newtonian revolution and with other revolutions in science that have introduced new ways of understanding the phenomena of the external world.[6]

21

Some Other

Scientific

Developments

---

 he Darwinian and Maxwellian revolutions were not the only upheavals in the biological and physical sciences that were thought to be revolutionary in their day, and which we today would still generally consider to have been revolutions. A number of candidates for scientific revolution in the nineteenth century have been proposed by historians and scientists, in fields ranging from mathematics and statistics to geology and medicine.[1] In this chapter we will examine briefly some of these developments, concluding with a few words about the great revolution in the field of applied science.

Lyell's Revolution in Geology

Looking at progress in the earth sciences during the nineteenth century, Leonard Wilson has made a case that a "revolution in geology" occurred in the "years to 1841," when Charles Lyell produced his 'uniformitarian' theory, which he elaborated in his three-volume *Principles of Geology* (1830–1833). Lyell's aim was grandiose, as he explained in a letter of 1829 (Wilson 1972, 256). He said that, while his book "will not pretend to give an abstract of all that is known in Geology," it "will endeavour to establish the *principles of reasoning* in the science & all my Geology will come in as illustration of my views of those principles, & as evidences strengthening the system necessarily arising out of the admission of such principles." Basically, he believed that *"no causes whatever have from the earliest time to which we can look back to the present ever acted but those now acting"* and that these causes "never acted with different degrees of energy from that which they now exert." Seventeen chapters of his book, according to Wilson, "fulfilled the promise in the title 'to explain the former changes of the earth's surface by reference to causes now in operation'" (p. 280). Additionally, there were four chapters in which Lyell presented "ideas that were distinctly new and original." Wilson concludes that the book was "revolutionary" (pp. 280, 281, 293) and then goes a step further, emphasizing that the book was read and that copies sold rapidly. We may add that further editions appeared in rapid succession (2nd ed., 3 vols., 1832–1833; 3rd ed., 4 vols., 1834), a sign of the interest in, and importance of, this treatise. It would thus appear that if this was indeed a revolution, it was not merely a revolution on paper.

But not all historians of geology agree with Wilson's conclusion that "Lyell had started a revolution in men's thinking about the history of the earth" (p. 293). In a review of Wilson's biography (in *Science*, 5 Jan. 1973, *179:* 57–58), Cecil Schneer discusses evidence with which one may "challenge the biographer," and he argues that "Lyell's uniformitarian ideas were neither so new nor so relevant to the emerging secular history of the world as to justify the designation revolutionary." It is true that Wilson does not support his own judgment by citing any reviewer or contemporaneous commentator who declared Lyell's *Principles* to be revolutionary or revolution-making.[2] But as we have already seen, only a couple of decades after the first volume of Lyell's treatise had been published, Charles Darwin, near the beginning of chapter nine of the *Origin of Species* (1859, 282), evaluated "Sir Charles Lyell's grand work on the Principles of Geology" by declaring that "the future historian will recognise [it] as having produced a revolution in natural science." A gloss on Darwin's statement is provided in an earlier letter (1844) to Leonard Horner (Darwin 1903, 2: 117; quoted in chapter 29, below), in which Darwin said that after reading Lyell, even new phenomena would be seen as if "through his eyes." Additional contemporaneous testimony concerning a Lyellian revolution is found in a letter to Lyell from the astronomer and philosopher John Herschel on 20 February 1836, in which he said that "your Geology [is] a work which appears to me one of those productions which work a complete revolution in their Subject" (see Babbage 1838, app. 1, 226).

Since Lyell's geology was recognized as revolutionary by contemporaries, a crucial historical test is whether or not the subsequent history of geology and the allied science of paleontology show an effect of Lyell's work commensurate with a revolution. I do not believe that this is in question. The argument between historians rather centers on the degree of newness. In science, absolute newness tends not to be a defining feature of revolutions. Most (if not all) revolutions exhibit features of continuity so that even the most radical ideas in science prove again and again to be mere transformations of existing traditional ideas. (This theme is developed at large in my *Newtonian Revolution*, 1980.) This is so characteristic a feature of science that some scientists, such as Albert Einstein, have ended up by conceiving their own work to exhibit evolution rather than revolution: a radical transformation and restructuring of what is known or believed rather than the invention of something new. The only argument against a Lyellian revolution is that all thinking in earth science was not equally conditioned by the ideas he put forth, but strictly speaking this would limit the extent and the effect of the revolution, not deny its existence altogether.

Advances in the Life Sciences

In a study entitled *Biology in the Nineteenth Century* (1977), William Coleman discusses a number of major revolutions in the life sciences. He compares the actions of pathological anatomists "to revolutionize the concerns of traditional topographic and organ anatomy" to the later transformation of pathological anatomy by the cell theory (p. 20). In particular, he directs our attention to the physicians in the Paris hospitals who, around 1800, "effected a revolution in medicine" by combining "postmortem physical examination of the cadaver with a clinical description of the patient's affliction." In a chapter on "Man," Coleman begins by asserting that "a revolution in man's awareness of his past" occurred between Lamarck and Haeckel (p. 92). In this context, Coleman finds Durkheim's conclusions to have been "indeed revolutionary" (p. 114). In a chapter on "Function: The Animal Machine," he describes how four German 'reductionists' met in Berlin in 1847, "a year before the outbreak of revolution, and there, it is related, cast a plan for a revolution in physiological aspiration and methodology" (p. 151). The book ends with an account of the situation at the century's close, with "new recruits to biology . . . inclined to be vocal partisans of a physiological view of biology's concerns." Experimental physiology "had established a model approach" in "experiment" to an understanding of the "vital process, the day-to-day, second-to-second events whose sum total was life." In the name of experiment, Coleman concludes, there "was set in motion a campaign to revolutionize the goals and methods of biology."

In 1858 Rudolf Carl Virchow published his great treatise, *Cellular Pathology,* which is thought by many today to have heralded a revolution in biology. While there is not universal agreement on this point, there is little or no doubt that Virchow's theory constituted a revolution in the biological basis of medicine — as Virchow himself declared. Virchow is of special interest to us because he combined an active political career as a radical reformer with his scientific career in medical pathology. Sent by the government in early 1848 to Upper Silesia, which had suffered from a typhus epidemic, he was greatly affected (as he himself has told us) by the precarious conditions of life of the Polish minority. This experience turned him from a man of liberal social and political beliefs into a radical who recommended an extensive program of social and economic reform. Not surprisingly, he participated in the Berlin uprisings that were part of the general revolution of 1848 and fought on the barricades. He became a member of the Berlin Democratic Congress and edited a weekly entitled *Die medicinische Reform.*

Suspended from his academic position in Berlin as a result of his revolutionary political activities, he moved to Würzburg, where he was appointed in 1849 to the first chair in Germany in the new subject of pathological anatomy. It was here that he achieved his great status as a scientist, developing the concept of what he called "cellular pathology." He returned to Berlin in 1856 as professor and director for a new Institute of Pathology. He became famous for his teaching and for his doctrine that the cell is the fundamental unit equally under ordinary conditions of health and extraordinary conditions of disease, and that diseases are disturbances of the living cells. His later career embraced development of his biomedical concepts, activity in politics and in public health, and production of a sociological theory of disease. He even became a founder of the new science of anthropology.

In 1861 Virchow was elected a member of the Prussian Parliament, representing the German Progressive Party, of which he was one of the founders. He was a vigorous opponent of Bismarck, who in anger challenged him to a duel; Virchow did not accept the challenge. Hence he was an uncommon example of a great scientist who was a political activist and social reformer and who achieved professional reform to the extent of changing the rules of the medical profession and improving conditions of public health and medical care. Other scientists have also been political activists, but few have achieved the important or high political position that Virchow did as the leader of the opposition to Bismarck in Parliament (see Fleming 1964, p. x).

In the first issue of the weekly which he founded, *Die medicinische Reform* (10 July 1848), Virchow combined ideas of political revolution and medical reform. He wrote (p. 1) that the "revolution [Umwälzung] in the conditions of the state" and the "building of new institutions" were part of the "political storms" affecting all thinking men and women in Europe, hence marking "radical changes in the general life-view [Lebensanschauung]." He insisted that medicine could not be unaffected by these storms, that "a radical reform can no longer be put off." Erwin Ackerknecht (1953, 44) has observed that for Virchow "freedom and science are natural allies," that "the revolution in 1848 was obviously as much a scientific as a political event." In his weekly, Virchow wrote: "Eventually the days of March arrived. The great fight of criticism against authority, *of natural science against dogma,* of the eternal rights against rules of human arbitrariness, this fight which had already twice shaken the European world, broke out for the third time and victory was ours." Ackerknecht sees this unity of politics and medicine as a characteristic of Virchow's thought (p. 45):

His theory of cellular pathology was important to Virchow, as it seemed to show objectively in the human body a situation he strove for and regarded as 'natural' in society . . . thus cellular pathology was to Virchow far more than a biological theory. His political and biological opinions reinforced each other mutually at this point. Cellular pathology showed the body to be a free state of equal individuals, a federation of cells, a democratic cell state. It showed it as a social unit composed of equals, while an undemocratic oligarchy of tissues was assumed in humoral, or solidistic (neuro) pathology. Just as Virchow fought in politics for the rights of the 'third estate,' thus in cellular pathology he fought for a 'third estate' of cells (connective tissue) not duly recognized in their value and function.

Hence we are not surprised to find Virchow saying such things as: "The last task of medicine is the constitution of society on a physiological basis" (quoted ibid., p. 46). Believing that social science is a subdivision of medicine, he stated explicitly that "medicine is a social science, and politics nothing else but medicine on a large scale," and "the physicians are the natural attorneys of the poor, and the social problems should largely be solved by them."

In his writings about the practice of medicine, according to Acker-knecht (1953, 47), Virchow "preferred the expression 'reformer' to 'revolutionary,' as it seemed to him the better characterization for the combination of destruction and construction, of criticism and respect for past accomplishments which he stood for." But, as in 1848, he did engage in revolutionary political action.

In the preface to his great treatise, *Cellular Pathology* (1858; Engl. trans. 1860), Virchow spoke of the obligation of the medical scientist to make known to his "professional brethren" the new knowledge that is rapidly accumulating. Then he said dogmatically: "We would have reform, and not revolution" ("Wir wollen die Reform, und nicht die Revolution"). He held, furthermore (1858, p. ix; 1860, p. x), that his work might seem to "savour more of revolution than reformation" ("mehr revolutionäre, als reformatorische Einwirkung"), but that was primarily because of "the necessity of combating rather the false or exclusive doctrines of the more recent [modern], than those of the older writers." And yet when, in the text proper, he came to describe the radical new ideas that he had been developing—and just prior to his declaration (1860, 27) that "where a cell arises, there a cell must have previously existed" ("omnis cellula e cellula")—he used the more dramatic image of revolution. He referred explicitly to 'der Umschwung' (translated in 1860 as 'the revolution') in pathology which had occurred in the "last few years." Here he chose 'Umschwung', although in relation to political/social events he used 'Umwälzung' or even 'Revolution'. But what is significant about Virchow is that he was one of the very few scientists who both made a revolution in science and were active in a political revolution. Additionally, he stands out for his expressed opinion that revolutionary politics and revolutionary science might influence and even reinforce each other.

Mathematics, Probability, and Statistics

Tremendous advances were made during the nineteenth century in mathematics. New fields opened up (for example, non-Euclidean geometry, mathematical statistics, vector analysis, and quaternions), and new standards of rigor completely transformed classical analysis or function theory (functions of complex variables). At the century's end Georg Cantor produced a new branch of mathematics, transfinite number theory. His great contribution has been described as a "bold advance into the realm of the infinite" which "ignited twentieth-century research on the fundamentals" (Meschkowski: 1971, 56). Obviously, this was a revolution in mathematical thinking. Cantor himself was fully aware of the revolutionary significance of his work. In a letter to Cantor in 1885, the Swedish mathematician Mittag-Leffler wrote that Cantor's "work was no less revolutionary" than Gauss's research on non-Euclidean geometry (Dauben 1979, 138). Joseph Dauben found that in a letter to the French historian of science Paul Tannery (1934, *13*: 304), furthermore, Cantor stated simply and clearly that his work was revolutionary.

Cantor was not the only nineteenth-century mathematician to express the belief that he had made (or would make) a revolution. Another was the Irish mathematician William Rowan Hamilton. Thomas L. Hankins discovered that in 1834 Hamilton wrote a remarkable letter about what he called (in a previous letter to his uncle) "his hope and purpose to remodel the whole of Dynamics, in the most extensive sense of the word." The letter in question was written in 1834 to William Whewell. The new dynamics, Hamilton wrote (Hankins 1980, 177–178), "will make, perhaps, a revolution." Hamilton's work is not generally familiar to nonmathematicians. The paper which was the occasion of the remarks just quoted was entitled "On a General Method in Dynamics" (1834). In it Hamilton developed properties of what he called the "characteristic function" and unfolded "methods of approximating the characteristic function in order to apply it to the perturbations of planets and comets" (Hankins 1972, 89). The characteristic function was one of Hamilton's two great 'inventions', the second being quaternions, a system of three-dimensional complex numbers which could be used in a way similar to vector analysis, the invention of J. Willard Gibbs which eventually displaced quaternions as the language of dynamics and mathematical physics. (Hamilton's quaternions were so popular in their day and so well-suited to physics that J. C. Maxwell used them in his famous treatise on electricity and magnetism for the mathematical expression of the subject.) Hamilton's essay contained the "first general statement of the characteristic function applied to dynamics" (p. 88) and developed the principle of what we call today the "Hamiltonian". This paper was indeed revolutionary, since in it he derives the 'canonical equations' of motion, 'Hamilton's principal function', and Hamilton's own version of what has come to be known as the Hamilton–Jacobi equations. Hamilton's essay, "General Method in Dynamics" (1834; supplement, 1835), gave classical mechanics the formulation which has become standard for today's quantum theory and statistical mechanics.

The Hamilton method, particularly as developed by Jacobi, has proved to be especially useful for celestial mechanics, for example, in the classical problem of determining the motion of three bodies, each of which attracts the other two according to Newton's inverse square law of universal gravitation. Hamilton's quaternions have been eliminated from physical science by the general acceptance of vector and then tensor analysis. According to J. D. North (1969), in the long run perhaps the "overwhelming importance" of Hamilton's theory of quaternions "was its introduction of a law of non-commutative multiplication" which "inspired other algebraists" to reject "from their axiom sets" the commutative law. (The commutative law of multiplication states that the order in which two numerals are multiplied does not affect the product — eight multiplied by two gives the same result as two multiplied by eight.)

Three major areas saw notable developments in relation to probability and statistics during the nineteenth century. The first is the mathematical theory (beginning with Laplace); the second is the application of statistics to the analysis of society, starting from so-called 'moral statistics'; the third is the introduction of a statistical basis for science. The second of these is associated generally with the name of the Belgian statistician Adolphe Quetelet, who startled readers all over the world with his revelations concerning certain numerical constancies or regularities: in marriages, deaths, births, crimes, and so on.

We have an eloquent testimony, by a most competent witness, to the revolutionary force of the new statistical findings about society. "Men began to hear with surprise, not unmingled with some vague hope of benefit," as Sir John Herschel wrote in 1850 (pp. 384–385),

> that not only births, deaths, and marriages, but the decisions of tribunals, the results of popular elections, the influence of punish-

ments in checking crime — the comparative value of medical reme-
dies, and different modes of treatment of diseases — the probable
limits of error in numerical results in every department of physical
inquiry — the detection of causes physical, social, and moral, —
nay, even the weight of evidence, and the validity of logical
argument — might come to be surveyed with that lynx-eyed scrutiny of a dispassionate analysis, which, if not at once leading to the
discovery of positive truth, would at least secure the detection and
proscription of many mischievous and besetting fallacies.

This extract comes from a widely read and widely discussed article in
the *Edinburgh Review* (July 1850) on the translation (1849) just published
of Quetelet's Letters to Prince Albert on the *Theory of Probabilities* (see
Herschel 1857, 365ff.).

But was there a revolution? One way of gauging whether the new
statistical analysis of society was sufficiently profound to be reckoned a
statistical revolution is to take cognizance of the intensity of reactions
against the new statistical way of thinking. Two opponents of statistically based science or knowledge were Auguste Comte and John Stuart
Mill. Comte, in his *Course of Positive Philosophy* (bk. 6, ch. 4), scorned "the
pretension of some geometers to render social investigations positive by
subjecting them to a fanciful mathematical theory of chances" (1855,
492). Comte castigated James Bernoulli and particularly Condorcet for
seeking to apply probability and statistics to social theory (or to sociology). There was, he wrote (p. 493),

no excuse for Laplace's repetition of such a philosophical mistake,
at a time when the general human mind had begun to discern the
true spirit of political philosophy, prepared as it was for the disclo-
sure by the labors of Montesquieu, and Condorcet himself, and
powerfully stimulated besides by a new convulsion of society. From
that time a succession of imitators has gone on repeating the fancy,
in heavy algebraic language, without adding anything new, abusing
the credit which justly belongs to the true mathematical spirit; so
that, instead of being, as it was a century ago, a token of a premature instinct of scientific investigation, this error is now only an
involuntary testimony to the absolute impotence of the political
philosophy that would employ it. It is impossible to conceive of a
more irrational conception than that which takes for its basis or for
its operative model a supposed mathematical theory, in which,

signs being taken for ideas, we subject numerical probability to calculation, which amounts to the same thing as offering our own ignorance as the natural measure of the degree of probability of our various opinions.

Comte's opposition to statistics and probability was very likely based on his conviction that "all sciences aim at prevision" (that is, precise prediction), as he wrote in his essay of 1822 on "reorganizing society" (Fletcher 1974, 167). To this end, "the laws established by the observation of phenomena" should enable the scientist to foretell the succession of phenomena. It follows that "observation of the past should reveal the future in politics as it has done in astronomy, physics, chemistry, and physiology." Comte expanded this theme in book 6 ("Social Physics") of his *Course of Positive Philosophy*, where in chapter 3 he argued that "social phenomena are subject to natural laws, admitting of rational prevision." What Comte had in mind was the simple causal predictions of classical rational mechanics — which he saw as the antithesis of the 'inexact' predictions of statistics and probability.

In John Stuart Mill's *System of Logic,* his most important or "principal philosophical work," a stand is taken against statistical arguments or improper use of probability in science or social science. According to Mill (1973–1974, 1142), "It would indeed require strong evidence to persuade any rational person that by a system of operations upon numbers, our ignorance can be coined into science." Mill added that it was "doubtless this strange pretension" which "has driven a profound thinker, M. Comte, into the contrary extreme of rejecting [this doctrine] altogether," despite the fact that it "receives daily verification from the practice of insurance and from a great mass of other positive experience." This statement, like others in the first edition (1843) of the *Logic,* was eliminated in the second and later editions; but no reader can escape the obvious conclusion that Mill took a rather unfavorable view of the foundations of probability and the usefulness of applying it (see Mill 1973–1974, *8–9:* bk. 3, ch. 17–18, app. F, app. G, pp. 1140–1153). Mill left no doubt concerning his position when he said in his *Logic* (1973–1974, bk. 3, ch. 18, §3) that "misapplications of the calculus of probabilities" have made it "the real opprobrium of mathematics."

Many scientists, as well as philosophers, either came out directly against the use of probability and statistics in science or voiced strong doubts concerning their use. As late as 1890, in the second edition of his *Properties of Matter,* Peter Guthrie Tait could still take an antistatistical posture and write of the "remaining difficulties" in the kinetic theory of gases as "greatly enhanced by an apparently unwarranted application of the *Theory of Probabilities,* on which the statistical method is based" (p. 291).

A more continual and outspoken critic of the use of statistics and probability in science was Claude Bernard, often called the founder of modern experimental physiology. In his *Introduction to the Study of Experimental Medicine* (1927, 131–139), Bernard said simply that he could not possibly understand "how we can teach practical and exact science on the basis of statistics." The use of statistics, he believed, must necessarily "bring to birth only conjectural sciences," and "can never produce active experimental sciences, i.e., sciences which regulate phenomena according to definite laws." Furthermore, he argued, "By statistics, we get a conjecture of greater or less probability about a given case, but never any certainty, never any absolute determinism." Since "facts are never identical," statistics can serve only as "an empirical enumeration of observations" (pp. 138–139). Hence if medicine were based on statistics, it could "never be anything but a conjectural science; only by basing itself on experimental determinism can it become a true science, i.e., a sure science." Here Bernard was expressing the difference between what he denominated the point of view of "so-called observing physicians" and that of "experimental physicians." Bernard saw experimental science leading to a strict determinism which he and other physiologists believed to be incompatible with probabilistic or statistical considerations.

In an address read to the Congress of Arts and Science at the Universal Exposition in St. Louis in 1904, the philosophically minded theoretical physicist Ludwig Boltzmann discussed briefly the applications of statistics to science and to social science. Defending the "theorems of statistical mechanics," which he claimed to be as valid "as all well-founded mathematical theorems," he noted that there was a difficulty in other applications of statistics, for instance, in assuming "the equal probability of elementary errors." Alluding to the broadening application of statistics to "animated beings, . . . human society, . . . sociology, etc., and not merely . . . mechanical particles," he called attention to "difficulties of principle" that arise from basing such studies on the theory of probability. This subject, he said, "is as exact as any other branch of mathematics if the concept of equal probabilities, which cannot be deduced from the other fundamental notions, is assumed" (1905, 602).

During the academic year 1983–84, an international interdisciplinary seminar and workshop was conducted at the University of Bielefeld. The topic was "The Probabilistic Revolution 1800–1930." The various studies undertaken there provide convincing evidence that the changes going on in social and scientific thought during the nineteenth century exhibit a revolutionary force. But I have seen no evidence that the revolution, if any, was ever more than a revolution on paper until the century's end, with the development of statistical mechanics. On the other hand, physics and biology both underwent a very radical change in the twentieth century, with the introduction of a probabilistic or statistical basis for genetics and evolution and for quantum theory. The quantum revolution is often acknowledged to have been one of the greatest ever to have occurred in science, and the transition from simple causality to statistical considerations is generally considered to be one of its most revolutionary features. Hence I would conclude that there was no 'probabilistic revolution' (or, better, 'probabilizing revolution') in the nineteenth century, in the sense of a full-scale revolution in science. This was at most a revolution on paper that did not achieve the potentialities of a revolution in science until the early years of the twentieth century.[3] By 1914, in a book entitled *Chance*, a nontechnical exposition of probability and statistics "in different branches of scientific knowledge," the French mathematician Emile Borel indicated that "we have been present, almost without being aware of it, at a genuine scientific revolution" (p. ii).

The Revolution in Applied Science

Historians agree that one of the great revolutions in the nineteenth century was the rise of science as a major force fueling technological and social change. This revolution was succinctly characterized by Alfred North Whitehead, when he observed that the greatest invention of the nineteenth century was the invention of the method of invention. The fecundity of this kind of technological innovation may be seen in such a simple fact as that nearly half of the gross sales of the Du Pont Company for 1942 consisted of products that either did not exist in 1928 or were not then manufactured in large commercial quantities. Such was the influence of one company's research program.

Although we hold it to be a commonplace today that advances in fundamental scientific knowledge have a major thrust in altering the materials of our lives, our food, our health, our communication and transportation, the way we earn our living and conduct our national defense, this was not generally the case a hundred years ago. Scientists and philosophers ever since Bacon and Descartes had predicted that advances in knowledge would make man the master of his environment, but the number of convincing instances of such a process were few. There was one major example prior to about 1800 in which the research of a scientist undertaken purely for the advancement of knowledge had led, as an unanticipated by-product, to a practical invention useful for mankind. This was the basic research of Benjamin Franklin on the nature of conductors and insulators, the phenomena of electrostatic induction, the effect of the shapes of bodies on their electrical properties, the role of grounding in electrical effects, and the nature of glow, spark, and brush discharges. This research led Franklin to the identification of the lightning discharge as an electrical phenomenon, then to an experiment to test this conclusion, and finally to a device — the lightning rod — to disarm charged clouds so as to prevent a stroke or even to conduct a stroke safely into the ground. As late as the early nineteenth century, in a public discussion in France, this case history of the lightning rod could still be cited as a primary example of the way in which basic scientific research leads to unexpected practical inventions. But this example was not as convincing as it would have been if the resulting practical invention had been directly related to food or health, communication or transportation, national defense, or modes of earning a living.

Revolutionary change in the effects of science on technology occurred in the nineteenth century, first in the dye industry. Prior to the mid-nineteenth century, dyes were obtained from natural sources: vegetables, insects, shellfish, and some minerals. By the end of the century, synthetically produced dyes had all but replaced these natural products. The first stage of the revolution was the discovery in 1856 by William Henry Perkin of a new dyestuff which dyed silk a mauve color. He was then only a student, and the coloring matter he found was the end product of a failed experiment to produce synthetic quinine. The raw material was coal tar, a by-product of the process for producing illuminating gas by distillation from coal. Perkin began to manufacture the new mauve coloring material, and the succeeding years saw the growth of a new industry based on the researches of chemists who were able to synthesize existing dyes that had been obtained traditionally from natural products or to create wholly new synthetic dyes. The new dyes were cheaper and more color-fast. We may see the revolutionary effect of this new technology in the history of a single dye, alizarin or 'Turkey red'. In the 1860s alizarin was obtained from the madder plant, which composed the principal agricultural crop in Provence and was grown extensively in southern Spain, Italy, Greece, and North Africa. Within a few decades, synthetic alizarin had all but wiped out madder agriculture, and today the madder plant is grown in botanical gardens as a curiosity.

In contrast to many of the earlier synthetic dyes, alizarin was — in the mind of the dye chemist Witt (Haber 1958, 83) — "the first fruit of a new trend in chemical research, that of purposive chemistry" ("Das Prinzip der zielbewussten Synthese"; see O. N. Witt 1913, 520). Chemists were now organized to direct their research toward specified technological goals. One of the last of the natural dyes to surrender to a synthetic product was indigo, whose production was almost wholly in British hands. Indigo was actually synthesized as early as 1880, but the process was slow and costly. It took seventeen years of directed research, a combination of scientific effort on the part of industrial research chemists and their academic counterparts, before synthetic indigo appeared on the market in 1897. The cost to the Badische Anilin- und Soda-Fabrik has been reckoned at about $5,000,000, the largest sum ever spent until that time on a single research project. Within three years, German production was equivalent to the yield of indigo that would have been obtained from a quarter of a million acres (Brunck 1901).

It was in the dye industry that science first showed its technological power on a grand scale. Almost overnight the whole economy of large regions was radically altered, as when land formerly devoted to madder culture was either turned over to growing grapes or other products or allowed to lie fallow. The fate of nations and of the world was affected by the fruits of applied chemical research. Germany barely had any dyestuffs industry in the early 1860s but by 1881 had become the producer of about one-half of the world's dyestuffs; by 1896 this figure had risen to 72 percent, by 1900 to 80–90 percent. The success of the German manufacturers in capturing the world market was in good measure the result of their being "able to draw on a large reservoir of extremely capable chemists whose enthusiasm for research, often of a painstaking nature, was unmatched in other countries, except Switzerland" (Haber 1958, 129). It should be noted, finally, that since unstable dyes are explosives, the government-sponsored dyestuff industry in Germany was creating a potential arsenal for world war.

Another way of seeing the profound effect of the revolution in applied chemistry is to note that British India exported naturally produced indigo in 1896 at a value of more than £3,500,000, a figure which dropped to a mere £60,000 by 1913. Furthermore, the value of the indigo exported in 1913 by Germany (the chief producer of synthetic indigo) was about £2,000,000. But the full scope of the revolution is revealed by the additional datum that during these seventeen years the price of indigo dye had dropped from about 8 sh. to only some 3.5 sh. per pound (see Findlay 1916, 237).

22

Three French Views:

Saint-Simon,

Comte, and Cournot

———◆———

The concept of revolution in science occurs in a significant way in the writings of three nineteenth-century French philosophers and social thinkers: Saint-Simon, Comte, and Cournot. All three developed a philosophy of historical change, in which science was of special importance, and all three envisaged a time not too distant when social science would reach the advanced and certain state already achieved by astronomy and mathematics and in process of being attained by "physiology" (biology).

Henri Saint-Simon: Revolution and the Religion of Science

Henri Saint-Simon (Claude Henri de Rouvroy Comte de Saint-Simon, 1760–1825) is an interesting figure in the history of thought because, although he was virtually ignorant of science, he wrote eloquently about the importance of science and he envisaged that scientists would play a crucial role in reordering society. Although later in life he became somewhat disenchanted with science, and — more particularly —

disappointed with the reception of his own ideas by his contemporary scientists, his blueprints for a better society always stressed the importance of scientific ideas and ideals. He even dreamed of a religion of science with scientist-priests and a physicist as analogue of a pope. More importantly, he looked forward to a not-too-distant time when science would be reorganized, along with the system and method of education, so as to ensure that science would "perfect industrial arts" to the advantage of all who labor (see Manuel 1956; 1962, 113ff.).

Today Saint-Simon is remembered, if at all, as a pre-Socialist 'socialist' thinker, as an early preacher of the cult of scientism, and as a precursor of the positive philosophy of Auguste Comte. Saint-Simon's political and social ideas were lauded by Friedrich Engels in his booklet *Socialism: Utopian and Scientific*, where he says that "in Saint-Simon we find a comprehensive breadth of view, by virtue of which almost all the ideas of later socialists, that are not strictly economic, are found in him in embryo" (1935, 38). Emile Durkheim called Saint-Simon "the founder of positivism and sociology." We can see the beginnings of modern positivist philosophy in such a statement as the following one, taken from Saint-Simon's *Memoir on the Science of Man* (1865–1878, *40: 25–26*; quoted in Manuel 1956, 133):

> All sciences began by being conjectural. The great order of things has ordained that they shall all become positive. Astronomy began by being astrology; chemistry was in its origins nothing but alchemy; physiology, which for a long time floundered about in charlatanry, today rests on observed and verified facts; psychology is beginning to base itself on physiology and to rid itself of the religious prejudices on which it was founded.

He also predicted, in his *Letters to an Inhabitant of Geneva* (written in 1813), that the social sciences would become sciences on the same level as astronomy, physics, chemistry, and physiology. (He did not, in this work, use the term 'positive' to describe the exact sciences, a term that had been first used by him in 1807; see Manuel 1956, 132). He set forth a hierarchy of the sciences, according to their successive "emancipation from superstition and metaphysics" (ibid.) that anticipated the later formulation of Auguste Comte. Like Comte, he saw that physiology was only just entering or about to enter the 'positive' state. "Physiology," he wrote, in the *Letters to an Inhabitant of Geneva* (1865–1878, *15: 39–40*; trans. in Manuel 1956, 133), "is still in the unfortunate position through which the astrological [sic] and chemical sciences have passed." He added, "The physiologists have now to expel from their midst the *philosophers,* the *moralists,* and the *metaphysicians,* just as the astronomers drove out the astrologers and the chemists drove out the alchemists." Comte was to invoke the same image of the astrologers and alchemists.

Saint-Simon wrote three major works directly relating to the theme of science: *Introduction to the Scientific Works of the XIXth Century* (1808), *Work on Universal Gravitation* (December 1813), and *Memoir on the Science of Man* (written in January 1813 but not published until 1858). It is in the *Memoir* that he develops most fully his theory of revolutions in science. The discussion of revolutions, which occurs in a supplement to the first of the two parts of this work, takes the form of a "Letter to Physiologists" (1858, 382–386). If they will "boldly back me up," there "will occur in a few years a great and useful scientific revolution." Saint-Simon then explains that history shows an alternation of scientific revolutions and political revolutions; successively, each is the cause of the next revolution and the effect of the previous one. The recapitulation, says Saint-Simon (1858, 382–386), "will prove that the next revolution must be a scientific revolution, just as my work will prove to you, with more and more evidence, that it is principally you [the physiologists] who must produce this revolution and that it is particularly to you that it must be useful."

Saint-Simon begins his historical succession with the scientific revolution associated with Copernicus, followed by the political revolution of Luther. The next scientific revolution encompassed the work of Bacon and Galileo's demonstration of "the diurnal rotation of the Earth about its axis, which completed the system of Copernicus." The subsequent political revolution occurred in England with Charles I being "judged by his subjects" and a "new order of social organization, unknown to the peoples of antiquity" being established; at the same time Louis XIV "undertook to submit all of Europe to his lay jurisdiction." In the scientific revolution which followed, Newton and Locke appeared and "gave birth to important new ideas which made a great step forward in science"; their ideas were developed and used in France in the *Encyclopédie.* The political revolution that came next was the French Revolution which "began a few years after the publication of the *Encyclopédie.*"

Now Saint-Simon had to make a prediction concerning the next scientific revolution. There would be a revolution in "the science of man," based on "physiological knowledge." Saint-Simon envisaged that this new science would become part of public education and that those who would be brought up on this new scientific fare would be able to treat political matters by the method used in the other sciences (astronomy, physics, chemistry). Whereas the writings of the eighteenth century had tended to disorganize or disrupt society, those of the coming nineteenth century would "tend to reorganize Society." I have reproduced two pages of the original publication of Saint-Simon's *Memoir on the Science of Man,* so that the reader may appreciate the typographical counterpoint between the two types of revolution (see Figure 10).

Saint-Simon omitted the Chemical Revolution from his list. The sequence of past revolutions begins with Copernicus, reaches a fruition with Bacon and Galileo, leading up to Newton (and Locke and the encyclopédistes). This triad of achieved scientific revolutions, the only actual such revolutions mentioned by Saint-Simon, comprise what we today know as *the* Scientific Revolution, a unitary concept that seems to

— 382 —

LETTRE AUX PHYSIOLOGISTES.

MESSIEURS,

Agréez, je vous prie, la dédicace de cette première partie de mon travail; vous êtes directement intéressés au succès de mon entreprise, vous êtes parfaitement en mesure de l'accréditer, vous êtes de tous les savants ceux dont je puis recevoir les avis les plus utiles, et avec lesquels je puis combiner et coaliser mes forces de la manière la plus avantageuse pour l'amélioration du sort de l'Espèce humaine, qui est le but commun de nos travaux. Sans votre secours, il me serait impossible de réussir; si vous m'appuyez franchement, il s'opérera en peu d'années une grande et utile révolution scientifique.

L'histoire constate que les révolutions scientifiques et politiques ont alterné, qu'elles ont successivement été, à l'égard des unes et des autres, causes et effets. Récapitulons celles qui ont été les plus marquantes depuis le xvᵉ siècle. Cette récapitulation vous prouvera que la plus prochaine révolution doit être une révolution scientifique, de même que mon travail vous démontrera, avec de plus en plus d'évidence, que c'est principalement vous qui devez opérer cette révolution, et que c'est particulièrement à vous qu'elle doit être utile.

RÉVOLUTION SCIENTIFIQUE.

Copernic renverse l'ancien système du monde, il en établit un nouveau; il prouve que la terre n'est point

— 383 —

au centre du système solaire, et qu'elle est par conséquent encore moins au centre du monde; il démontre que c'est le soleil qui est au centre du système dont nous faisons partie, et dans lequel nous ne jouons qu'un bien petit rôle. La démonstration de Copernic sape dans sa base tout l'échafaudage scientifique de la religion chrétienne.

RÉVOLUTION POLITIQUE.

Luther change le système politique de l'Europe, en soustrayant la population du Nord à la juridiction religieuse papale; il affaiblit le lien politique qui unissait les peuples européens. Le succès de la révolution de Luther fait concevoir à Charles-Quint le projet de soumettre toute la population européenne à sa juridiction laïque; il en tente l'exécution, il échoue dans son entreprise.

RÉVOLUTION SCIENTIFIQUE.

Bacon, par son ouvrage sur la dignité des sciences, et encore plus par la conception à laquelle il donne le titre de *Novum Organum,* culbute l'ancien système scientifique. Il prouve qu'on ne pourra organiser solidement le système de nos connaissances, qu'après lui avoir construit une base uniquement composée de faits observés.

Galilée démontre le mouvement journalier de rotation de la terre autour de son axe, ce qui complète le système de Copernic.

Figure 10. A counterpoint of political and scientific revolution. A pair of facing pages from Claude-Henri de Saint-Simon's *Memoir on the Science of Man* (written in 1813; Paris, 1858). (Harvard College Library.)

have been given its first clear expression in Auguste Comte's transformation of Saint-Simon's ideas.

Auguste Comte and the Positivist Philosophy

Auguste Comte (1798–1857) was one of the most original and important thinkers of the nineteenth century. His profound influence on science, philosophy, and the social sciences was widespread. He inaugurated the movement of thought called 'positivism' and he invented the name 'sociology' for a discipline that had not yet come into being. His philosophical ideas were set forth at length in his *Course of Positive Philosophy,* published in French in 1830–1842 and translated into English by Harriet Martineau. Comte's influence in the Anglo-American world was not as penetrating as it has been, and still is, in France and on the Continent and in Latin America. In the twentieth century, some parts of Comte's philosophy were given a wholly new lease on life in the doctrines of 'logical positivism' which were influenced strongly by the ideas of Ernst Mach and were disseminated through the Vienna Circle. In this new incarnation, references to Comte as founder of positivism are conspicuously absent.

Comte introduced two major novel concepts in the historical development of the sciences. The first was his law of the three stages. According to Comte, the human mind has advanced through three stages which constitute an inevitable progression in the understanding of, and way of accounting for, the phenomena of the external world. The first is 'theological', in which events are ascribed to the action of gods; the second is 'metaphysical', in which the will of gods or divine forces are replaced by abstractions; finally, the third or 'positive' stage is reached when scientific explanations take the place of metaphysics. Comte explored the succession of these three stages through an extended historical presentation of the development of culture or civilization, of thought, and especially of science. He was "convinced that a knowledge of the history of the sciences is of the highest importance," and he even went so far as to say "that a science is not completely known if we are ignorant of its history" (1970, 49). Comte thus became one of the first advocates of the serious and systematic study of the history of science. George Sarton always lauded Comte for being the founder of the discipline of history of science.

Comte's second historical concept was part of a new and highly original classification of the sciences. He set forth a scheme in which the sciences are arranged in a historical and analytical hierarchy of "decreasing generality and increasing interdependence and complexity." Thus the system of classification not only was determined by a logical analysis but was confirmed by history. Mathematics is the base of all, the most general of all the sciences and historically the first to become 'positive'. Next in order was astronomy, in which physical bodies were considered in freedom of motion or motion without constraints of resistance to motion by surrounding fluids, collisions, friction, and all the other complexities that arise in the terrestrial physics of gross bodies. Following astronomy were physics, chemistry, physiology (in the process of becoming 'positive' in Comte's time), and eventually the final science of 'sociology'.[1] There was no place in Comte's hierarchy for psychology, which he assumed would be part of human biology ("physiology"). Such a scheme was in harmony with Comte's training as a mathematician and his study of the exact physical sciences at the École Polytechnique. Because of his background in mathematics and in physics, Comte saw physics (terrestrial physics) as the model of all the sciences, in which observation and experiment are combined with mathematics to produce a genuinely 'positive' system of knowledge. In his early writings, accordingly, Comte saw the future science of society as a 'social physics'—a term later used in a quite different sense by Quetelet.

Comte's law of the three stages, like all original thoughts, was to some degree a transformation of concepts of his predecessors, notably Condorcet, Cabanis, and Saint-Simon, whom Comte had served as secretary. The degree of this transformation, and accordingly a measure of Comte's true originality, may be seen by juxtaposing Saint-Simon's views and those of Comte. Saint-Simon had seen the final stage of development when philosophy had become 'scientific' in the sense of rejecting whatever is not verifiable. But for Comte, the establishing of the last of the sciences—sociology—as a 'positive' discipline was not yet the ultimate stage; the separateness of the several sciences would not yet have been overcome so as to produce a total positivist system, a "conception of the world and man" that would at last be a synthesis worthy of the name of "philosophy." In this final stage all of knowledge would be 'positive' and unified, held together in the science of man and society, that is, the new science of sociology. There would then arise not only an understanding of the problems and needs of man and society but also a clear conception of the steps to be taken to reform and improve the condition of man and his society. Inevitably this line of thought led positivism into a sort of religion, even to the extent of having churches, a priesthood, and a calendar of 'positivist' saints, including Moses, Homer, Aristotle, Archimedes, Julius Caesar, Saint Paul, Charlemagne, Dante, Gutenberg, Shakespeare, Descartes, Frederick the Great, and Bichat.

The influence of Saint-Simon and Comte upon each other is difficult to determine because there was a bitter falling-out between them. Both Comte and Saint-Simon expressed their indebtedness to Condorcet for the doctrine of the successive maturation of the sciences, but Comte carefully refused to acknowledge any indebtedness to Saint-Simon and referred to him only in terms of contempt. It is often assumed that the influence of Saint-Simon on Comte must have been more significant than any possible influence of Comte on Saint-Simon. As near as I can determine, the only reason for this position (undocumented by any real evidence) is that Comte was younger during the period of their intellectual closeness, when Comte was Saint-Simon's secretary. But considering that Comte was one of the most brilliant and influential thinkers of his day, and taking into account that young people tend to have a greater fund of original ideas than older people, is it not likely that the influence of Comte upon Saint-Simon would have been more significant than the reverse? In any event, the near coincidence of many of their ideas (including the law of the three stages, the successive development of the sciences, the concept of 'positivism' or 'positive' science) does not minimize the estimate of Comte's creative genius. What is of significance is not that Comte was in part transforming some ideas of his older associate but rather that he made creative use of these ideas. (A good review of this topic may be found in Manuel 1962, 251–260.) In the end, Comte had no doubt that others (Jean-Baptiste Say and Charles Dunoyer) had been more important in his intellectual development than the man he called a "foolish old philosopher" and a "depraved juggler." His "spiritual antecedents" (Manuel 1962, 257) were Hume, Kant, Condorcet, de Maistre, Gall, and Bichat.

In Comte's discussions of the development of science, he frequently introduced the notion of revolution in science and the concept of a general revolution in the sciences in the sixteenth and seventeenth centuries. He invoked the idea of *the* Scientific Revolution, for example, in an essay of 1820, "A Brief Appraisal of Modern History" (Fletcher 1974, 99), in which he stated that

down to a recent period they [the natural sciences] suffered from an admixture of superstition and metaphysics. It was only towards the end of the sixteenth and the early years of the seventeenth centuries that they succeeded in entirely disengaging themselves from theological beliefs and metaphysical hypotheses. The epoch at which they began to be truly positive must be referred to Bacon who gave the first signal of this great revolution; to Galileo his contemporary who furnished its earliest exemplification; and lastly to Descartes who irrevocably emancipated the intellect from the yoke of authority in matters of science. Then it was that natural philosophy arose and the scientific capacity acquired its true character, that of contributing the spiritual element of a new social system.

Furthermore,

From this epoch, the sciences successively became positive in the natural order of sequence, that is to say according as they were more or less closely related to man. Thus, astronomy first, then physics, later chemistry and finally, in our own day, physiology, have been constituted as positive sciences. This revolution, then, has been completely accomplished for all special branches of knowledge, and evidently approaches its consummation for philosophy, morals, and politics.

In an essay published in May 1822, entitled "Plan of the Scientific Operations Necessary for Reorganizing Society," Comte set forth the doctrine that "scientific men ought in our day to elevate politics to the rank of a science of observation" (Fletcher 1974, 135). His analysis was based on his law of the three stages. Declaring that "four fundamental sciences" — astronomy, physics, chemistry, and physiology — "as well as their dependent sciences" had already become positive, he could not help but observe that some aspects of physiology still existed in all three states. For example, "the phenomena specially called *moral*" were "conceived by some people as the result of a continuous supernatural action; by others as connected with organic conditions susceptible of demonstration and beyond which it is impossible to go." This line of thought is developed at greater length in an essay of November 1825 on "Philosophical Considerations on the Sciences and Savants" (Fletcher 1974, 182ff.). In this presentation Comte, evaluating "the progress of the human mind during the last two centuries," notes that "the Moral Phenomena were the latest of all to pass out of the domain of theology and metaphysics and enter that of physics." He actually believed that "physiologists [or biologists], in our day, study moral phenomena exactly in the same spirit as the other phenomena of animal life." And, although he would not come out in favor of one or the other of the conflicting theories in the domain of "moral physiology," he did unequivocally declare that "the very existence of this diversity of theory, evincing an uncertainty unavoidable in every young science, clearly proves that the great philosophic revolution has been accomplished for this branch of our knowledge, as for all others."[2]

There can thus be no doubt that Comte saw the development of the sciences — their transformation into a positive state — as a revolutionary sequence; he believed the establishment of modern science to have been a "great" revolution. But I have not found any discussion by Comte of the revolutionary process itself whereby science develops, nor have I been able to determine whether or not Comte ever worked out a well-grounded comparison or contrast between scientific or philosophical revolutions and social or political revolutions.[3] Nevertheless, for Comte there are simple reasons why the "passage from one social system to another can never be continuous and direct" and why there "is always a transitional state of anarchy" (1975, 24; trans. in Lenzer 1975, 201). First, the new system is apt to be stimulated to a greater degree by "the experience of the evils of anarchy" than by considering the failings of the old system. Second, before the destruction of the old system, "no adequate conception could be formed of what must be done," since

> short as is our life, and feeble as is our reason, we cannot emancipate ourselves from the influence of our environment. Even the wildest dreamers reflect in their dreams the contemporary social state: and much more impossible is it to form a conception of a true political system, radically different from that amidst which we live. The highest order of minds cannot discern the characteristics of the coming period till they are close upon it; and before that, the incrustations of the old system will have been pretty much broken away, and the popular mind will have been used to the spectacle of its demolition.

As an example, Comte invoked the case of Aristotle, who "could not conceive of a state of society that was not founded on slavery, the irrevocable abolition of which took place some centuries after him." Of his own times, Comte said that "the destined renovation [is] so extensive and so thorough" that "never before . . . was the critical preparatory period so protracted and so perilous." For the "first time in the history of the world," he said, "the revolutionary action is attached to a complete doctrine of methodical negation of all regular government." For the student of the history of revolutions in science, Comte's three-stage analysis of revolutionary political reform is interesting because two of the three stages had been introduced into discussions of science by J. S. Bailly a century earlier. Comte's trinity comprised the destruction of the old, a resultant state of anarchy, and the establishment of the new; Bailly had conceived of a two-stage process, whereby every revolution in science would consist first of the destruction of an existing system of knowledge and then the creation and adoption of a new system.

Cournot

Antoine-Augustin Cournot (1801–1877), a contemporary of Auguste Comte, was a mathematician and administrator who is remembered today chiefly for his contributions to the theory of probability, but also for his general or philosophical analyses of scientific knowledge and his studies of the nature of scientific explanation. He differed from Comte in that his epistemology was characterized by probabilism, whereas Comte took a strong stand against probability and statistics as the key to social science or science.

Cournot was like Comte in proposing a classification of the sciences that was linked to history, to the stages by which the sciences have actually developed. But Cournot rejected Comte's formulation of the "alleged fatal order" of three stages of "successive appearance of religious, philosophic, and scientific doctrines" (1973, *4:* 27). And whereas Comte saw a one-dimensional or linear progression, Cournot proposed a two-dimensional matrix, which he called a "double-entry" table (see Cournot 1851, §237, 289; Granger 1971, 452–453). Here the vertical categories are somewhat similar to Comte's historical classification: mathematical sciences; physical and cosmological sciences (equivalent to Comte's astronomy, physics, chemistry, plus geology and engineering); biological and natural sciences (Comte's physiology); noological and symbolic sciences (absent as such from Comte's scheme); political and historical sciences (including Comte's sociology).

In his *Essay on the Foundations of Our Knowledge* (1851), Cournot does not explicitly state that the vertical column represents a historical sequence, although this may be implied by logical dependencies which require that some parts of the sciences be antecedent in time to others.

The *Essay* is replete with historical examples, but there is not much if any discussion of the process of scientific change. Great changes, such as those occurring in the mathematics of calculation, are referred to simply as "great innovations," "invention,""important discovery" (§200, 201; pp. 246–249). What was obviously an event of revolutionary significance to Cournot was Galileo's rejection of the age-long search in vain by "philosophers from Pythagoras to Kepler" who had hoped to find "the explanation of the great cosmic phenomena" in "ideas of harmony," which they "mysteriously connected with certain properties of numbers considered in themselves and independently of the application that could be made of them to the measurement of continuous magnitudes" (p. 246):

> True physics was founded on that day when Galileo rejected these long-sterile speculations and thought not only of the idea of examining nature by means of experiments — which Bacon, for his part, also proposed — but of stating precisely the general form to be given to experiments, by setting up as their immediate object the measurement of everything in natural phenomena which is capable of being measured.

Cournot then likened Galileo's bold innovation to Lavoisier's, which he called "a similar revolution." This "revolution was made in chemistry a century and a half later," according to Cournot, "when Lavoisier ventured to submit to balances, that is to say, to measurement or quantitative analysis, materials to which chemists before him had applied only the type of analysis which they call qualitative." Thus, Cournot held that both Galileo and Lavoisier had been responsible for a revolution in science. But at this point in his *Essay,* in a discussion of "Continuity and Discontinuity," Cournot was more interested in "number and quantity" than in revolutions in science.

One of Cournot's books has the attractive title *Treatise on the Sequence of Fundamental Ideas in the Sciences and in History* (1861). This work, though drawing on many historical examples, is not so much a historical enquiry as a study in the logic or philosophy of the sciences and of history; for Cournot the "sequence of fundamental ideas" is a logical and not a chronological sequence. Although this book contains some discussions of political and social revolutions (notably the English and French revolutions), the concept of revolution is not introduced in relation to such commanding scientific personalities as Copernicus,

Descartes, Galileo, Leibniz, and Newton. But there is a passing reference to revolutions in the opening paragraph of chapter 5, in which Cournot compares and contrasts physical mathematics with chemistry and physics. Here he observes that "chemistry and physics achieve their progress, undergo their revolutions, without there being progress or corresponding revolutions in geometry and in mechanics" (1861, 120). In the subsequent discussion, however, Cournot neither specifies what these revolutions are nor the degree to which such revolutions are characteristic of scientific progress.

Cournot's most historical work is his *Considerations on the Progress of Ideas and Events in Modern Times,* first published in 1872. Here a major theme is the role of revolutions in the development of science and technology, social science, and human societies. Three chapter headings indicate the importance of the concept of revolution: "the revolution in mathematics" in chapter 1 of book 3 (on the seventeenth century), "the revolution in chemistry" in chapter 1 of book 4 (on the eighteenth century), and "the economic revolution" in chapter 6 of book 5 (on the nineteenth century). The whole of the concluding book 6 is devoted to the French Revolution and its consequences.

Following a general introduction and a discussion of the Middle Ages (bk. 1), book 2 opens with an analytical narrative of "scientific progress" in the sixteenth century. A preliminary account of mathematics yields to the Copernican revolution (1872, 99): "In the history of the sciences of the XVIth century, everything pales into insignificance before the name of Copernicus and the importance of the revolution he brought about in astronomy." Indeed, "the revolution brought about in astronomy by Copernicus will remain forever the most perfect example of a great victory achieved by reason over the senses, over the imagination, and over all sorts of prejudices, the decisive proof that such a victory is possible, and the best example to which one can compare all critical discussions of the same genre" (p. 101). It was "absolutely fitting [bien dans l'ordre] that this model was furnished by the oldest in time and the most perfect of all the sciences."

Book 3, on the seventeenth century, opens with a reference to "future revolutions of opinions, beliefs, institutions, languages, and taste" (p. 172). It is said that "the progress and revolutions of science during the seventeenth century give that age that singular and exceptional quality of greatness that neither religion, nor politics, nor philosophy, nor literature, nor the arts would confer upon it to the same degree of eminence." This century was marked by "great scientific discoveries" and a "revolution in mathematics" (ibid.). Cournot sums up his views of this century and its revolutions in science as follows (pp. 173–174):

> The history of the sciences in the seventeenth century exactly marks off an epoch when the abstract sciences, long cultivated for their own sake and for the charm which certain minds found in them, or by a secret and vague presentiment of their future role, give all of a sudden the key to that which is most fundamental, most simple, most great, and consequently most impressive in the order of the universe. The general laws of motion, the action of weight, in the end the theory of the shape and of the motions of the heavenly bodies, or . . . "the system of the world" — these are the results established and explained (in so far as it is given to mankind to explain anything) by the marvelous union of abstract speculations and critical observations [observations judicieusement discutées]. From then on, discoveries crowd one another in the domain of the abstract sciences as in the field of observation and experiment; discoveries become revolutions [les découvertes deviennent des révolutions], in geometry as in astronomy and in physics; and these revolutions, at least for geometry and astronomy, are of the order of those which, each in its field, have not had and cannot again have any equals. As a result, the names of the great scientists whom these revolutions recall are names without equal, that no more recent glory will lessen that will always hold in the memory of mankind the same rank that the capital truths and higher laws they have had the great fortune to discover and to make known hold in the economy of the divine plan.

Cournot, while aware of the great significance of the calculus (bk. 3, ch. 1, p. 177), does not refer to its invention by Leibniz and Newton as a 'revolution', as Fontenelle had done, even though he cites Fontenelle's ideas concerning some of the novelties in seventeenth-century mathematics (p. 180). Nor does Cournot label as revolutions the discoveries in "the physical and natural sciences in the seventeenth century" (bk. 3, ch. 2), although he does mention a "revolutionary crisis" that then arose and was concentrated in the domains of pure mathematics and physical mechanics (p. 192). But he lauds Galileo for having set science on new paths: Galileo, according to Cournot, showed how to draw significant scientific conclusions from "the most common phenomena" such as the falling of a stone or the swinging of a hanging lamp (pp. 186–187). He showed the way "to force Nature to give up her secret, to unveil the simple and fundamental mathematical law." Galileo was "the creator of experimental physics and of mathematical physics" and, in particular, "the creator of physical mechanics." But apparently he did not produce a 'revolution'. And it is the same with respect to Newton (pp. 189–190).

The only discovery in the whole range of the physical and natural sciences of the seventeenth century to give rise to the use of the term 'revolution' by Cournot is the Harveyan circulation of the blood. He observed that after Harvey's discovery, "there was some grounds for expecting a revolution in medicine of the kind which modern chemistry much later brought about in industry." But this discovery did not for a long time exercise any "decisive influence on the vicissitudes of medical theory and practice." Hence Cournot concluded that the effective import of a scientific discovery is related less to the intrinsic importance of what is actually discovered than to the stage of maturity of the science of which it becomes a part and to the property which it can have of giving rise to one of those new ideas which contain the germ of scientific reforms or revolutions" (pp. 194–195).

In presenting the mathematics and science of the eighteenth century, Cournot signaled the work of Lavoisier as "the revolution of chemistry" (p. 271). Lavoisier's researches caused "chemistry truly to change its face"; this science "underwent a revolution" ("elle subissait une *révolution*") (p. 278). He then asked: "Why has chemistry — which has made such progress since Lavoisier, and in which theories have changed so often — not had more revolutions?"

In the nineteenth century (that is, up to 1870), Cournot did not find any scientific advances that merited the use of the term 'revolution'. One must use caution in giving too much importance to this simple statement of fact. Cournot most probably did not carefully evaluate each of the discoveries or innovations which he entered into his discussion in order to find out whether or not it constituted a revolution. Yet the fact remains that his book contains a considerable discussion of the English Revolution (pp. 90, 94, 242–251, 543, 549), the French Revolution (pp. 461–550), parallels between the English and French revolutions (pp. 540–550), political revolutions (pp. 91, 93, 111), the economic revolution of the nineteenth century (pp. 418–427), and a number of revolutions in mathematics and science as well as the general nature of such revolutions. Accordingly, any failure on Cournot's part to characterize scientific events as 'revolutions' must, in context, be significant.[4]

23

The Influence

of Marx

and Engels

———◆———

In any study of the revolutions of the nineteenth century, or of the development of the concept of revolution, the ideas and actions of Karl Marx assume a primary place. Even those revolutions which occurred too early to be influenced by Marx often tend to be interpreted nowadays from a 'Marxian' point of view. I have referred to the Marxian concept of 'the revolution in permanence' and to the fact that Marx was the spearhead in setting up national and international organized groups with the avowed explicit purpose of revolution. In this chapter my purpose is not so much to explore Marxian ideas on revolution or Marxian revolutionary activity as it is to examine the specific topic of Karl Marx's expressed views on scientific change and on revolution in science, and to compare and contrast Marx's ideas on these topics with those of Friedrich Engels. This subject differs fundamentally from a study of Marx's influence in the twentieth century on the interpretation of the history of science.

Whoever looks into this question quickly recognizes that Marx was neither particularly well educated in (nor deeply concerned with) the traditional physical sciences, with the technical content of astronomy, physics, chemistry, geology. His humanistic education embraced some mathematics, but he never had any formal training in any of the sciences listed above — say at the 'Gymnasium' or university level. In his mature life he became interested in some aspects of the life sciences and did a certain amount of reading in the works of German popularizers or vulgarizers, including Georg Büchner, Jakob Moleschott, and Karl Vogt. Although Marx criticized the "crudely mechanical form of materialism" endorsed by this group (see Schmidt 1971, 86ff.), he was apparently influenced by Moleschott's "conception of nature as a process of circulation," which he found to have much in common with the ideas of Pietro Verri, which are quoted with approval in *Das Kapital*.

In the light of the importance given to the adjective 'scientific' (used by Engels and all Marxists, and especially by the orthodox writers in the Soviet Union, to describe what is alleged to be 'scientific' socialism or 'scientific' communism), it is not without interest to see what Marx himself meant by this word. A clue is given in part 2 of *Theories of Surplus-Value* (1968; see Marx 1963–1971), the draft of an unfinished fourth volume of *Das Kapital*. In chapter 9 (§2), Marx contrasts Ricardo's economics and Malthus's. Ricardo, he says, "puts the proletariat on the same level as machinery or beasts of burden or commodities," because from his point of view, "their being purely machinery or beasts of burden is conducive to 'production,'" or because "they really are commodities in bourgeois production." This is not, according to Marx, "a base action." It "is stoic, objective, scientific." Furthermore, "In so far as it does not involve *sinning* against his science, Ricardo is always a philanthropist, just as he was in *practice* too."

Ricardo is contrasted with "the parson Malthus," who "reduces the worker to a beast of burden for the sake of production and even condemns him to death from starvation and to celibacy." Furthermore, "whenever it is a question of the interests of the aristocracy against the bourgeoisie or of the conservative and stagnant bourgeoisie against the progressive," Marx says that "in all these instances 'parson' Malthus does not sacrifice the particular interest to production but *seeks*, as far as he can, to sacrifice the demands of production to the particular interests of existing ruling classes or sections of classes." It is "to this end," according to Marx, that Malthus "*falsifies* his scientific conclusions." And then Marx concludes, "This is his *scientific* baseness, his sin against science, quite apart from his shameless and mechanical plagiarism." Marx added that Malthus's "scientific conclusions" are "'considerate' toward the ruling classes in general and towards the reactionary elements of the ruling classes in particular" and, in short, "he *falsifies* science for these interests."

Hence it would appear that 'scientific' was intended by Marx to mean 'impartial' and 'true' and thus has no direct connotation of a particular kind of method of inquiry or of proof. Nor does 'scientific' seem to imply any particular limitation in subject matter. This is made clear in the next section (pt. 2, ch. 9, §3), where Marx gives three examples to "exemplify Ricardo's scientific impartiality."

I have been unable to find in Marx's published and edited writings any discussions of either *the* Scientific Revolution or revolution in science in general or any particular revolution in any of the sciences.[1] (There are, however, references aplenty to the Industrial Revolution and to revolutionary mechanical or industrial inventions.) Nor have I been successful in locating any analysis by Marx of the ways in which the progress of science takes place or even a list of major events in a sequence of scientific discoveries.[2] But there is an interesting discussion by Marx of the application of the Darwinian theory of evolution to the historical development of technology, which appears to have been the earliest proposal for an evolutionary history in this area.

For a number of years the literature of historical scholarship propagated the legend that Karl Marx wished to dedicate *Das Kapital* to Darwin and wrote to Darwin to ask permission to do so, but that Darwin refused this honor. It is now definite that Darwin's draft of a letter of refusal was addressed to Marx's son-in-law, Edward Aveling, and not to Marx himself. Marx did present a paper-bound copy of volume 1 of *Das Kapital* to Darwin. This volume, still preserved with other books from Darwin's library, tells a curious story. It is inscribed on the upper right-hand corner of the title-page:

Mr. Charles Darwin
On the part of his sincere admirer
Karl Marx
London 16 June 1873
Modena Villas
Maitland Park

It would appear that Marx's decision to send an inscribed copy of *Das Kapital* to Darwin was not made until some time after publication, since the book presented to Darwin was not the first edition of 1867 but rather the second of 1872. Darwin did not read the whole book. When I examined it in Down House, in Down (Kent), the leaves of this book had been cut only up to page 105 (out of 822 pages). There is, additionally, no evidence concerning what Darwin thought (if anything) of Marx's work.

Marx does not mention Darwin in the first edition of *Das Kapital*, published in 1867, eight years after the appearance of the *Origin of Species* (1859); Darwin and evolution appear for the first time in two footnotes to the second edition (which may explain why Marx sent Darwin a copy of his book after the second edition had appeared). These footnotes contain the only explicit references to Darwin in *Das Kapital*. In one (vol. 1, ch. 14, §2), Marx quotes Darwin on the comparison of organs of plants and animals and tools. In the other (vol. 1, ch. 15, §1), he refers again to the organs of plants and "Nature's Technology." But in another footnote, as we have seen, Marx suggested that the history of technology be written from an evolutionary point of view. In other writings, Marx is unstinting in his praise of Darwin. A letter to Engels on 19 December 1860, only a few months after the publication of the *Origin of Species* (Padover 1978, 359) refers to his reading in "Darwin's book on *Natural Selection*," which he praised as "the book that contains the natural-history foundation for our viewpoint [historical materialism]."[3] He repeated the very same sentiment in a letter to Lassalle on 16 January 1862, about a year later (McLellan 1977, 525): "Darwin's book is very important and serves me as a natural-scientific basis for the class struggle in history." In this letter, Marx stressed the importance of Darwin's having dealt "the death-blow" ("for the first time") "to 'teleology' in the natural sciences." Thanks to Darwin, the "rational meaning [of 'teleology'] is empirically explained." In a letter to Engels on 7 December 1867, Marx referred to a "transformation process in society," which was similar to what "Darwin has demonstrated from the point of view of natural history."

Before long, Marx said in a letter to Engels (18 June 1862; Padover 1978, 360), he was "amused at Darwin, into whom [he] looked again." He had found that Darwin applied "the 'Malthusian' theory *also* to plants and animals, as if the joke in Herr Malthus did not consist of the fact that he did *not* apply it to plants and animals but only to human beings — in geometrical progression — in contrast to plants and animals." This idea was further developed by Marx in *Theories of Surplus-Value* (1963–1971, 2: 121). Marx quoted an extract from the *Origin* (ed. of 1860, pp. 4–5) in which Darwin referred to his treatment of the "high geometrical ratio of . . . increase" among "all organic beings throughout the world," which inevitably leads to a struggle for existence. "This," according to Darwin, Marx wrote, "is the doctrine of Malthus applied to the whole animal and vegetable kingdoms." Marx commented that Darwin evidently "did not realize that by discovering the 'geometrical' progression in the animal and plant kingdom, he overthrew Malthus's theory." The reason is that "Malthus's theory is based on the fact that he set Wallace's geometrical progression of man against the chimerical *'arithmetical'* progression of animals and plants." Thus "Darwin's work" contains a "detailed refutation, based on natural history, of the Malthusian theory."

Marx, however, should not be overly credited for special percipience in evaluating the worth and significance of Darwin's theory of evolution. In a letter to Engels on 7 August 1866, one year before the publication of *Das Kapital*, Marx was singing the praises of another "very important work" (Padover 1978, 360–361). The new book, he wrote, constitutes "a *very important* advance over Darwin." He is sending the book to Engels, so that he too may learn its message. "In its historical and political application," he declared, this book "is much more important and copious than Darwin." The book so highly praised by Marx was P. Trémaux's *Origine et transformations de l'homme et des autres êtres* (Paris, 1865). The judgment of history does not accord with Marx's laudation. For example, Trémaux does not rate an entry in the recently completed 16-volume *Dictionary of Scientific Biography,* nor is his name even mentioned in the standard histories of biology and of evolution (as by Bodenheimer, Carter, Eiseley, Fothergill, Mayr, Nordenskiöld, Rádl, Singer). Furthermore, in the international *Critical Bibliography of the History of Science,* compiled and published by George Sarton, by me, and by our successor editors from 1913 to 1975, there is no entry to record a single scholarly article or book on Trémaux's life or contribution to science. As the lawyers say, "res ipsa loquitur." Why did Marx become so beguiled by Trémaux that he considered his book superior to Darwin's? One reason is that, like Herbert Spencer and unlike Darwin, Trémaux evidently believed in progress. As Marx explained to Engels (ibid.), "Progress, which in Darwin is purely accidental, is here a necessity, on the basis of the periods of developments of the earth."

Yet it was to Darwin, and to Darwin alone, that Engels compared Marx in the speech he delivered at Marx's graveside at Highgate Cemetery in London on 17 March 1883. "Just as Darwin discovered the law of development of organic nature," he said, "so Marx discovered the law of development of human history."[4] Engels repeated this comparison in the preface to the fourth edition (1891) of his *Origin of the Family, Private Property, and the State*. Praising Lewis Henry Morgan's *Ancient Society* (1877), the "book upon which the present work is based," Engels called attention in particular to Morgan's discovery "of the original mother-right gens as the stage preliminary to the father-right gens of the civilised peoples" as having the "same significance for the history of primitive society as Darwin's theory of evolution has for biology and Marx's theory of surplus value for political economy" (Marx and Engels 1962, 2: 181–182). In a book review of the first volume of *Das Kapital* (27 Dec. 1867, quoted in Schmidt 1971, 45), Engels stressed that Marx was "simply striving to establish the same gradual process of transformation demonstrated by Darwin in natural history as a law in the social field." Engels also said that "science was for Marx a historically dynamic, revolutionary force." But in the preface to the second volume of Marx's *Das Kapital*, Engels compared Marx to Lavoisier rather than Darwin.

This same comparison of Marx and Darwin appears in a book by Marx's son-in-law, Edward Aveling. Aveling published two companion books: *The Students' Marx* (1892) and *The People's Darwin* (1881). In the introduction to *The Students' Marx* Aveling wrote, "That which Darwin did for Biology Marx has done for Economics" (p. viii). Each of these great men produced "a generalisation the like of which their particular branch of science had never seen" (p. ix). Each generalization, furthermore, "not only revolutionised that branch, but is actually revolutionising the whole of human thought, the whole of human life." Writing in 1892, Aveling could not help observing that Darwin's generalization had become "much more universally accepted than that of Marx." The reason, he believed, was that Darwin's work "affects our intellectual rather than our economic life," and so can "be accepted in a measure alike by the believers in the capitalistic system and by its opponents."[5]

Friedrich Engels

Whereas Marx wrote very little about the sciences as usually understood (that is, the physical and biological sciences), Engels has much to say about these sciences, their development, and their revolutions. One of his most widely read works is his *Anti-Dühring*, also entitled *Herr Eugen Dühring's Revolution in Science*. Published in German in 1878 (2nd ed., 1885; 3rd ed., 1894), this book was not, the author stated (1959, 9), "the fruit of any 'inner urge' " to discuss the sciences, but rather the product of his wrath against "the laws of economics, world schematism, etc.," which Dühring claimed to have discovered and which Engels found to be characterized by "erroneousness or platitudinousness," as was the case also for the "laws of physics and chemistry put forward" by Dühring (1959, 12).[6] Before analyzing what Engels has to say about revolutions in science, we must take cognizance of the fact that the title in German does not use the word 'Revolution' but rather 'Umwälzung': *Herrn Eugen Dührings Umwälzung der Wissenschaft.* Whether or not 'Umwälzung' is the equivalent of 'Revolution' — a question explored immediately below — Engels was being ironic. He certainly did not believe that Dühring had actually generated a revolution in science. In fact, the full title is quite obviously a parody of a polemical work in which Dühring had attacked the ideas of the American economist Henry C. Carey: *Careys Umwälzung der Volkswirthschaftslehre und Socialwissenschaft* (1865), although this is not one of the three books toward which Engels's polemic is primarily directed.[7] Engels mocked Dühring's claims, put forth in the latter's *Course of Philosophy* (1875), and wrote that "we are still waiting for the 'earths and heavens of outer and inner nature' which this philosophy promised to reveal to us in its mighty revolutionizing sweep [in ihrer mächtig umwälzenden Bewegung]" (1980, 134; 1959, 198).

We have seen earlier that in the late eighteenth and early nineteenth centuries there had been a movement in Germany to replace the Latinate word 'Revolution' by a Germanic equivalent 'Umwälzung'. Engels used both terms almost as if they were interchangeable. His writings do not show a real preference for 'Umwälzung' over 'Revolution'. His usage is disclosed by an examination of his *Dialectics of Nature,* most of which was apparently written in the decade between 1872 and 1882 and which was intended to contain his most mature thoughts about science. This work was never completed and was not published until 1927 (Engels 1940, xiv). The opening paragraphs describe the great changes that occurred in the fifteenth and sixteenth centuries, "the greatest progressive revolution [die grösste progressive Umwälzung] that mankind has so far experienced" (1975, 10–11; 1940, 2–3). At that time, "natural science developed in the midst of the general revolution [in der allgemeinen Revolution]," and it was "thoroughly revolutionary [durch und durch revolutionär]." This work thus not only begins with a statement of revolution, but in successive paragraphs uses the new German word 'Umwälzung' and the older French word 'Revolution' apparently interchangeably. Shortly afterwards, Engels (1975, 13; 1940, 6) makes a contrast between "revolutionary science [revolutionäre Naturwissenschaft]" and "conservative nature." Despite such a beginning, the remainder of Engels's brief history of the sciences (the "Introduction") does not refer to the great innovations as revolutions. Thus Kant made the "first breach in this petrified outlook on nature" (1975, 16–17; 1940, 8), Lyell "first brought sense into geology" (1940, 10), physics "made mighty advances . . . [in] 1842, an epoch-making year for this branch of natural investigation" (1940, 10–11), in chemistry there was a "wonderfully rapid development" (1940, 11), and so on. The only possible exception to this rule is Cuvier whose "theory of the revolutions of the earth" is said to have been "revolutionary in phrase and reactionary in substance" (1940, 10); but Engels is here most likely referring to Cuvier's use of the actual word 'revolution' in phrases concerning the geological 'revolutions of the earth' rather than characterizing Cuvier as having used phrases that were revolutionary in either their denotation or connotation.

Some light may be thrown on Engels's use of 'Umwälzung' and 'Revolution' by comparing these opening paragraphs of the finished introduction to *Dialectics of Nature* with some preliminary historical notes (1940, 184–186). Here Engels uses the word 'Revolution' and not 'Umwälzung' both when he writes of "the greatest revolution [die grösste Revolution] that the world had so far experienced" and when he says that "natural science . . . moved and had its being in this revolution [in dieser Revolution]," and "was revolutionary [revolutionär] through and through" (1975, 187; 1940, 184). It is anyone's guess whether Engels, in his final draft, changed the first 'Revolution' to 'Umwälzung' because he did not want to use the word four times in such close proximity. But it is notable that in the draft Engels merely writes "die grösste Revolution" without any further qualifying adjective, whereas in the final version he not only substitutes 'Umwälzung' for 'Revolution' but also changes "die grösste Revolution" to "die grösste progressive Umwälzung" (1975, 10). It is as if an 'Umwälzung' was any radical overturning or complete change, and so would have required a qualifying adjective to specify whether it was a necessarily progressive change. For Engels, a 'Revolution' would never need an adjective to express its progressive character.

A sentence in *Anti-Dühring* shows the difficulty of making a meaningful distinction between Engels's use of 'Revolution' and 'Umwälzung'. In it he writes that while "the hurricane of the Revolution [der Orkan der Revolution]" swept over France, in England "a quieter, but not on that account less tremendous, revolution [eine stillere, aber darum nicht minder gewaltige Umwälzung] was going on" (1959, 358). This 'Umwälzung' was the transformation of old-fashioned manufacturing into "modern industry," which "revolutionized [revolutionierten] the whole foundation of bourgeois society." 'Revolution' here is used, as always, for the French Revolution, but 'Umwälzung' is introduced for what Engels more often calls 'die industrielle Revolution' — though its effect is described by the verb 'revolutionieren'. And, indeed, about one page later, Engels (referring to Robert Owen) talks about 'die industrielle Revolution', the term commonly used by Marx. Engels also writes of both a bourgeois 'Revolution' and a bourgeois 'Umwälzung', a 'Revolution' in production and an 'Umwälzung' in production (tending to favor the latter by a ratio of 6 to 1).

Despite its title, *Herr Eugen Dühring's Revolution in Science* contains very few references to revolutions in science and does not disclose a developed or coherent theory as to how science progresses. In the whole book there are only two occurrences of 'Revolution' in relation to science. The first of these appears in the preface to the second edition (1885), in which Engels refers to a coming "revolution which is being forced on theoretical natural science by the mere need to set in order the purely empirical discoveries" now being accumulated (1959, 19); the second is in the discussion of production in part three, where Engels introduces the "revolutionary""technical basis of modern industry" (1959, 407). The latter is illustrated by a quotation from Marx's *Das Kapital*, in which Marx discusses "machinery, chemical processes and other methods." According to Engels's summary, science contributes to the "technical basis of modern industry," which "also revolutionizes the division of labour within the society, and incessantly launches masses of capital and of workpeople from one branch of production to another." It is to be observed that in this second instance, reference is made to the revolutionizing effect of science, rather than to a revolution in science.

In the preface to the second edition, in the paragraph preceding the one referred to above, Engels says of "natural science itself" that it is in a "mighty process of being revolutionized," using the word 'Umwälzungsprozess' — which seems to give yet further evidence of the interchangeability of 'Revolution' and 'Umwälzung' for processes of change other than traditional political revolutions (1980, 12; 1959, 19). Later on, in poking fun at Dühring, Engels (1980, 205–206; 1959, 305) uses the verb 'umwälzen' to describe how allegedly "it becomes possible for anybody, even the editors of the Berlin *Volkszeitung,* to 'lay deeper foundations' and to revolutionize science" and to believe that "we have only to say that eating is the fundamental law of all animal life, and we have revolutionized the whole of zoology."

It can thus hardly be concluded that the concept of revolution of science, even though expressed as such, was of fundamental significance for Engels. Even in a fragment relating to the "great discoveries" and "advances" made in natural science, which was omitted from Engels's published tract on Ludwig Feuerbach, there is no mention of 'revolution' (neither 'Revolution' nor 'Umwälzung'), nor does Engels use the term or concept of revolution in his many references to Darwin's great restructuring of thought in biology. In the discussion of Lavoisier (in Engels's introduction to vol. 2 of *Das Kapital*) there is no mention of the phrase 'Chemical Revolution'. Engels, however, did become remarkably well educated in many aspects of science (see R. S. Cohen 1978, 134), and was seriously concerned with questions relating to the history of science (ibid. 135).

The extracts just quoted, and others, demonstrate that Engels recognized the revolutionary force of science. In a number of examples he showed an awareness that revolutions actually occur in science, and he had some important insights into the revolutions of science.[8] For example, he recognized that one of the results of a scientific revolution is a revolution in the technical terms (though he never developed this thesis). But there is no evidence that he ever gave serious thought to, or wrote as much as a couple of consecutive paragraphs on, the theory or the process of revolution in relation to the advance of science.[9]

24

The

Freudian

Revolution

Threee greatest intellectual
revolutions of the past century are associated with the names of Karl
Marx, Charles Darwin, and Sigmund Freud. The Darwinian revolution
constituted a radical restructuring of natural science that has had major
repercussions outside the narrow confines of evolutionary biology,
especially in the social sciences. Marxism, with its intellectual and political sequelae, has been a revolutionary force in the social sciences (and in
social and political action); its advocates claim it to be 'scientific'. The
Freudian revolution is, for many, ambiguous in that there is not agreement on its status: Is Freudian psychoanalysis science, or is it social
science, or is it not even science at all?

The literature concerning Freud, psychoanalysis, and the Freudian
revolution is vast, confusing, and extremely contradictory. This situation is in good measure a result of the constant splintering off of groups
from the orthodox central core as established by Freud. Psychoanalysis
has aroused a continued stream of hostile criticism, from philosophers
or scientists concerned with method[1] to prudish men and women who

cannot abide Freud's open discussions of sexual matters. These continuing extremes of hostility may be taken as an index of the profound impact of the Freudian revolution.

In addition to the factors already mentioned, yet other problems arise in analyzing and evaluating this revolution. Not the least of these is the present unavailability of many of the original crucial documents (among them Freud's complete letters to Wilhelm Fliess) which would shed considerable historical light on the developmental stages of Freud's theories, particularly his controversial seduction theory (see below)— an episode in the psychoanalytic revolution that some interpret as undermining the very validity of psychoanalytic science and therapy.[2] Not until well into the twenty-first century, when the Freud archives become unsealed and available for full scholarly inspection, will we be able to assess critically this and other crucial episodes in the development of Freud's thinking and the use of these ideas by other members of the psychoanalytic movement.

The Freudian revolution differs from all other revolutions in science presented in this book in that the central core of the science was created almost entirely by a single individual, Sigmund Freud (but see Whyte 1960; Ellenberger 1970). Additionally, it is the only such revolution in which the original documents (Freud's own books and articles) are still highly regarded and carefully studied today by practitioners for their scientific content rather than their historical value. Not only are Freud's own writings still read by orthodox Freudians — psychoanalysts, psychiatrists, psychologists, social workers, sociologists, anthropologists, and others — but many of his works are also fundamental texts to scientists, practitioners, and social scientists who do not necessarily agree with Freud's concepts and theories and who differ in varying degrees from the central orthodoxy. The methods of therapy (centering around the process of psychoanalysis) remain essentially those same ones developed and used by Freud. Largely for this reason, critics often charge Freudian psychoanalysis with being a closed system, more nearly akin to philosophy or even religion than to true science.

Freud was a dramatic and forceful writer, a truly gifted master of German prose — an aspect of his scientific style that is lost in English translations. Although Freud's goal, as he said on many occasions, was to create a scientific psychology divorced from its historical philosophical burdens, he purposely chose "simple pronouns" to describe the three mental agencies (1953, 20: 195) — *das Ich* (the I), *das Es* (the It), and *das Über-Ich* (the Over-I). The reason, he said, was that in psychoanalysis "we like to keep in contact with the popular mode of thinking and prefer to make its concepts scientifically serviceable rather than to reject them." There was "no special merit" in using such ordinary rather than recondite expressions; the reason was practical: psychoanalysts want their theories "to be understood" by their patients, "who are often very intelligent, but not always learned." He explained: "The impersonal 'it' is immediately connected with certain forms of expression used by normal people. 'It shot through me,' people say; 'there was something in me at that moment that was stronger than me.' *'C'était plus fort que moi.'*" But in English, these ordinary names are lost. They become the recondite Latin pronouns 'ego', 'super-ego', and 'id' — known today perhaps to more people for their Freudian than their Latin context. Freud was here following the tradition of physicists who had used terms from ordinary language — work, force, energy — in new specific and restricted scientific contexts. Freud also introduced classical expressions, such as Oedipus complex and libido.[3]

Robert Holt (1968, 3) has suggested that Freud's work can best be understood by considering three rubrics. One is the "general theory of psychoanalysis" (Rapaport 1959), sometimes referred to as metapsychology. The subject is a possible set of "theoretical assumptions on which a psycho-analytic system could be founded," expounded by Freud in his "Project for a Scientific Psychology" of 1895 (1954, 347–445), the "Papers on Metapsychology" of 1915 (1953, *14:* 105–235), and *The Interpretation of Dreams* (1900; 1953, *4–5*). Another is what Holt calls "Freud's phylogenetic theory," comprising Freud's "grand speculations, largely evolutionary and teleological in character." The works in this category abound in literary allusions and metaphors, rather than rigid or "explicit models of a psychic apparatus." To this category belong such general works of Freud as *Totem and Taboo* (1913), *Beyond the Pleasure Principle* (1920), *The Future of an Illusion* (1927), *Civilization and Its Discontents* (1930), and *Moses and Monotheism* (1934–1938).

Finally, the most scientifically important of all Freud's contributions is "the clinical theory of psychoanalysis, with its psychopathology, its accounts of psychosexual development and character formation," based on a subject matter consisting of "major events (both real and fantasized) in the life histories of persons." For practicing psychoanalysts, it is this theory that guides clinical diagnosis and therapy. Even those who may not be strict Freudians — psychiatrists, psychiatric social workers, clinical psychologists — have been strongly influenced by the theory, which — "loosely referred to as 'psychodynamics'" — has "even penetrated into general academic psychology via textbooks on personality."[4]

In a valuable study of Freud's influence, David Shakow and David Rapaport (1964) have shown how deeply Freud's revolutionary ideas have permeated psychological thought; not necessarily "the specific concepts and the explanatory theory in which they are embedded" (Holt 1968, 4) but rather the "general conceptions and the observations." Foremost among Freud's radical innovations was his recognition of the unconscious and the influence that psychological forces beyond our rational control exert over behavior, wishes, fantasies, and motivations. He called attention to the significance of all psychological phenomena, from dreams and fantasies to mere slips of the tongue, giving primary attention to the role of sexuality in the psychological development of the individual from infancy.

The Stages of the Psychoanalytical Revolution

Like all revolutions in science, the initial stage of the Freudian revolution involved an intellectual revolution, or revolution-in-itself. This occurred in the early 1890s, when Freud began his studies on hysteria with Joseph Breuer using hypnosis, a technique he had begun to study clinically during a short but fruitful period spent with Jean-Martin Charcot in Paris. Freud's ideas on the power of the unconscious underwent a rapid and radical development during a series of intellectual exchanges with Wilhelm Fliess, a Berlin nose-and-throat specialist. Not only did Fliess have considerable influence on Freud's physiological and psychological ideas, but he also turned Freud into a nonrational bionumerologist, an aspect of Freud's development that his biographers have played down (Sulloway 1979, 144). The documents that Freud wrote while in communication with Fliess constitute the revolution of commitment, including the composition of the "Project for a Scientific Psychology" (Freud 1954, 355–445).

In May of 1896, in an address to the Vienna Society of Psychiatry and Neurology, Freud discussed the etiology of hysteria. As described in his autobiography (1952, 62–64), Freud had at first believed the stories female patients told him about their having been seduced in early childhood by a father (most often), an uncle, or an older brother. Later he found that his patients' "neurotic symptoms were not related directly to actual events but to phantasies embodying wishes" and that "as far as the neurosis was concerned psychical reality was of more importance than material reality." This was the first glimpse Freud had of "the *Oedipus complex,* which was later to assume such an overwhelming importance."

At about the same time that Freud was abandoning his seduction theory of hysteria, he began his famous self-analysis.[5] The process extended over many years, but the most intensive part came during the summer and fall of 1897, shortly following the death of Freud's father in October of 1896 (Jones 1953, *1:* 324). Freud's analysis of his repressed childhood feelings for his parents led to the conclusion that young males have Oedipal feelings for their mothers and hostility toward their fathers as a normal stage in their development.

Freud introduced the example of Oedipus in a letter to Wilhelm Fliess (Freud 1954, 223) of 15 October 1897 and developed this theme at large in *The Interpretation of Dreams* in 1900. Freud had not as yet introduced the term 'complex'; he used the story of Oedipus merely to confirm his discovery, to show that strong evidence for it goes back to classical antiquity, to a legend of "profound and universal power." In this presentation Freud wrote of "children . . . in love with the one parent and hating the other" as "among the essential constituents of the stack of psychical impulses." While Freud was stressing the experience of his psychoneurotic patients, he expressed his belief that "psychoneurotics [do not] differ in this respect from other human beings who remain normal." Psychoneurotics, he concluded, "are only distinguished by exhibiting, on a magnified scale, feelings of love and hatred to their parents which occur less obviously and less intensely in the minds of most children" (Freud 1953, *4:* 260–261). In his autobiography, Freud wrote that during the early years of life, when "the relation known as the Oedipus complex becomes established," boys "concentrate their sexual wishes upon their mother and develop hostile impulses against their father as being a rival, while girls adopt an analogous attitude." Hence, from the start, the Oedipus complex was not conceived as being wholly restricted to the male sex (see §24.1).

In a paper of 1898 on "Sexuality in the Etiology of the Neuroses," Freud discussed his ideas on infantile sexuality publicly for the first time. But it was not until 1900 that he formally announced the psychoanalytic revolution in his first great book, *The Interpretation of Dreams.* I believe this was the last revolution in science to be made public in a printed book rather than a paper in a scientific journal or in a series monograph. Published in German in Vienna in 1900, the book was enlarged and revised again and again (1901, 1911, 1914, 1919) and was first published in English translation in 1913.

The next years saw other important works on *The Psychopathology of Everyday Life* (1901), *Jokes and Their Relation to the Unconscious* (1905), and *Three Essays on the Theory of Sexuality* (1905). By this time a complete body of theory and practice was available for scientific evaluation and acceptance or rejection. At first the medical community of psychiatrists and neurologists and the academic psychologists were extremely hostile to Freud's ideas. Those who "expressed vehement opposition . . . up to about 1910" form (according to Shakow and Rapaport 1964) a veritable "'Who's Who' of psychiatry and neurology," and "the reactions" of those in "other branches of science and medicine" were "equally negative" (see Freud 1913, 182, also 166). Shakow and Rapaport suggest that the lack of interest (or, when there was interest, the opposition) of educated laymen reflects the strong rejection by professionals; they also find that in these early days Freudian ideas did not especially occupy the attention of the clergy.

Freud's discovery of infant sexuality especially was greeted with tremendous antagonism. In his autobiography Freud said that "few of the findings of psychoanalysis have met with such universal contradiction or have aroused such an outburst of indignation as the assertion that the sexual function starts at the beginning of life and reveals its presence by important signs even in childhood" (1952, 62). And yet, "no other findings of analysis can be demonstrated so easily and so completely." We may see how new and revolutionary Freud's discovery was by taking account of the then-current views of childhood. Freud explains clearly: "Childhood was looked upon as 'innocent' and free from the lusts of sex, and the fight with the demon of 'sensuality' was not thought to begin until the troubled age of puberty. Such occasional sexual activities as it had been impossible to overlook in children were put down as signs of degeneracy and premature depravity or as a curious freak of nature."

The Freudian revolution in science did not, therefore, proceed by converting established professional men and women but rather by attracting and convincing young, less conventionally-minded practitioners at or near the start of their careers, who later became psychoanalysts. A gathering of adherents to the new idea assembled in 1909 at Clark University in Worcester, Massachusetts, invited by the president of the university, G. Stanley Hall. Included were Freud himself, A. A.

Brill (American translator of some of Freud's works), Sandor Ferenczi (a Hungarian psychoanalyst who for many years was one of Freud's close associates), Ernest Jones (later to become Freud's biographer), and Carl G. Jung. A little over a year earlier, in April 1908, a group of psychoanalysts met in Salzburg for their first International Congress. Attending were one American (Brill), twenty-six Austrians (among them Freud, Alfred Adler, Otto Rank, Wilhelm Stekel, and Fritz Wittels), two Britons (Jones and the surgeon-psychologist Wilfred Trotter), two Germans (including Karl Abraham), two Hungarians (Ferenczi and F. Stein), and six Swiss (including Jung). After the sessions, the first journal to be dedicated to the new science was founded, the *Jahrbuch für psychoanalytische und psychologische Forschung (Yearbook for Psychoanalytic and Psychological Research)*. The Second Psycho-Analytical Congress met in Nuremberg in March 1910 and there has been a regular succession of international psychoanalytic meetings ever since. Local groups enrolled themselves as Branch Societies of the International Psycho-Analytical Association, and by 1911, a year after its founding, this professional group contained 106 members. The revolution in science was launched. Because this group was composed of strong divergent personalities, there was soon a constant stream of deserters from the Freudian group, each of whom formed a dissident movement of his own, among them Adler (in 1911), Stekel (in 1912), Jung (in 1913), and Rank (in 1926). But even they continued to be influenced by Freudian concepts, however revised, giving added testimony to the complete change in thinking about the mind and in treating its disorders that has been characteristic of the Freudian revolution. Critics of orthodox Freudian psychoanalysis have argued that there have not been enough significant departures from Freud's original ideas. Others, including Alfred Kazin (1957, 16), have held that "for much of this 'Freudian' revolution, Freud himself is not responsible."

A Nineteenth- or Twentieth-Century Revolution?

I have assigned the Freudian revolution to the nineteenth century because its first three stages — the revolution-in-itself, the revolution of commitment, and the revolution on paper — were achieved by 1900. Since Freudian science and its implications have been so significant in our own times, we might rather have concentrated on the revolution in science, which occurred during the twentieth century.

In an essay written in 1923 and published in the next year (1953, *19:* 191), Freud himself considered the question of whether the movement should be considered a nineteenth- or a twentieth-century phenomenon. He said that "psycho-analysis may be said to have been born with the twentieth century; for the publication in which it emerged before the world as something new — my *Interpretation of Dreams* — bears the date '1900.'" Freud then explained that psychoanalysis "did not drop from the skies ready-made" — "It had its starting-point in older ideas, which it developed further; it sprang from earlier suggestions, which it elaborated. Any history of it must therefore begin with an account of the influences which determined its origin and should not overlook the times and circumstances that preceded its creation." Freud began with mid-nineteenth-century treatments of "what were known as the 'functional' nervous diseases"; then discussed the work of Bernheim, Charcot, and Janet; and then the advances made by Breuer, leading up to the joint publication in 1895 of *Studies on Hysteria* by Breuer and himself. He then recounted his own contributions, which reached a climax in 1900.

But the question of nineteenth or twentieth century is not so clearly determined as Freud would indicate. In his essay Freud stressed the twentieth century because he was writing it in 1923 as a chapter in a volume entitled *These Eventful Years: The Twentieth Century in the Making, as Told by Many of Its Makers* (London and New York, 1924). As the editor of the standard edition of Freud's psychological works points out (1953, *19:* 191; *4:* p. xii), the *Interpretation of Dreams* does (as Freud says) bear the date of 1900, but it was actually published early in November of 1899. In an essay of 1932, Freud (1953, *4:* p. xii) said, "It was in the winter of 1899 that my book on the interpretation of dreams (though its title-page was post-dated into the new century) at length lay before me." And, in a letter to Wilhelm Fliess of 5 November 1899 Freud announced that "the book at last came out yesterday" (Freud 1954, 302).

This example may serve only to illustrate how difficult it is to fit intellectual history and the history of science into arbitrary chronological divisions such as centuries. In any event Freud shared a common error in believing that the year 1900 began the twentieth century. Since the first year of our era is the year 1, the hundredth year (completing a century) is the year 100 and not the year 99. Thus the final or hundredth year of the nineteenth group of hundred years (the nineteenth century) is 1900 and not 1899, and the first year of the twentieth century must properly be 1901.

Freud's Views on Scientific Revolution and Originality: Comparisons with Copernicus and Darwin

The hostility to Freud's ideas, particularly those related to sexuality, naturally caused Freudians to see their master's travails as analogous to those of any daring pioneer. Freud's biographer, Ernest Jones, wrote that "Copernicus and Darwin dared much in facing the unwelcome truth of outer reality" (1940, 5), but "to face those of inner reality," as Freud did, "cost something that only the rarest of mortals would unaided be able to give." Freud himself was acutely aware of his revolutionary position in the history of the science and therapy of the mind. On a number of occasions, he compared his own scientific theory to the theories of Copernicus and Darwin, which interested him not for their scientific impact so much as for what we might call today their 'ideological' component. Although Freud did not ever (in recorded conversations, published correspondence, or published writings) invoke the expressions 'Copernican revolution' or 'Darwinian revolution', he did imply that what Copernicus and Darwin had done was radical and of enormous significance for man's concept of himself. Nor, apparently, did Freud ever say explicitly that he was a revolutionary or that psychoanalysis constituted a revolution. In *The Future of an Illusion* (1953, 21 : 55) Freud wrote, "The transformations of scientific opinion are developments, *advances, not revolutions.*"

Freud declared in 1907 that if asked to name "the 'ten most significant books,'" he would place "among them scientific achievements like those of Copernicus, of the old physician Johann Weier on the belief in witches, Darwin's *Descent of Man,* and others" (1953, 9: 245). This collocation of Copernicus, Johannes Weier, and Darwin is of more than passing interest, since these men represent the three areas in which Freud believed man had received staggering blows to his narcissistic self-esteem: cosmology, psychology, and evolutionary biology. Copernicus, according to Freud, had dethroned man from his fixed and central place in the universe, whereas Darwin had shown man's close kinship to other animals. Weier, a sixteenth-century physician of rare insight and remarkable courage, fought valiantly against the excesses of the witch-hunt mania, in particular by explaining that pseudocyesis ('false pregnancy') is not an indication of a woman's having had intercourse with the devil but is a medico-physiological condition that arises from what we would call today psychological or psychosomatic causes. It is surprising that Freud should have cited a rather obscure sixteenth-century physician who is not even mentioned in standard histories of medicine (Singer and Underwood; or Shryock; or Zilboorg and Henry 1941), however we may esteem his modernity, reasonableness, and courage (see Zilboorg 1935). But then again, there were not many students of the mind worthy of so exalted a position alongside Copernicus and Darwin. He might have chosen Charcot, for whom he again and again expressed admiration (1953, *1:* 135; *3:* 5, 9–10; *6:* 149; *12:* 335; *19:* 290; *24:* 411), describing him as neurology's "greatest leader" and the "master teacher" of "neurologists of every country." Another curiosity about this list is Freud's choice of Darwin's *Descent of Man* over his *Origin of Species.* It is not at all clear whether Freud chose *The Descent of Man* deliberately or whether he merely jotted down the Darwinian title that first came to mind. But Freud may have consciously intended *The Descent of Man,* since in that work Darwin most pronouncedly set forth the doctrine of the closeness of man and animal species. Given Freud's special interest in the blows to man's self-image, *The Descent of Man* would have obviously been a more important book than *The Origin of Species,* however much greater the latter might have been for evolutionary biology or for science as a whole.

For Freud, the Copernican shift from an earth-centered to a sun-centered universe — like the Darwinian theory of descent, which "tore down the barrier that had been arrogantly set up between men and beasts" — was of interest primarily in analogy to the reception being given to psychoanalysis. Freud showed how "the relation of the conscious ego to an overpowering unconscious was a severe blow to human self-love," just as "the *biological* blow delivered by the theory of descent and the earlier *cosmological* blow aimed at it by the discovery of Copernicus" had previously injured our narcissistic self-image (1953, *19:* 221). Freud believed that the barriers to acceptance of these three theories had arisen from emotional rather than intellectual sources, which explained "their passionate character." He suggested that people "in the mass" reacted to the theory of psychoanalysis just as did "individual neurotics under treatment for their disorders." Resistances to Freud's theories — like earlier resistances to the theories of Copernicus and Darwin — were not "of the kind which habitually arise against most scientific innovations," but rather stem from "the fact that powerful human feelings are hurt by the subject-matter of the theory."[6]

The most famous instance of Freud's coupling of the effects of Copernicus and Darwin with the hostility to psychoanalysis occurs in part 3 of Freud's *Introductory Lectures in Psycho-analysis* (1916–1917), dealing with the "General Theory of the Neuroses." Here Freud discusses how "the *naïve* self-love of men has had to submit to . . . major blows at the hands of science." From Copernicus men "learnt that our earth was not the centre of the universe but only a tiny fragment of a cosmic system of scarcely imaginable vastness" (1953, *16:* 285). Darwin's research "destroyed man's supposedly privileged place in creation and proved his descent from the animal kingdom and his ineradicable animal nature." But the "third and most wounding blow" to "human megalomania" comes, according to Freud, "from the psychological research of the present time which seeks to prove to the ego that it is not even master in its own house, but must content itself with scanty information of what is going on unconsciously in its mind."[7]

Curiously enough, Freud appears not to have ever referred to his own dramatic overthrow of classical psychology and traditional psychotherapy in terms of revolution. But (in 1916–1917) he did introduce the notion of a "general revolt against our science," which he said was characterized by "the disregard of all considerations of academic civility and the releasing of the opposition from every restraint of impartial logic" (1953, *16:* 285). This expression is especially interesting to the historian of revolutions in that 'revolt' suggests an uprising against established authority, whereas Freud has just been complaining of the resistance to the acceptance and establishment of his own radical and new ideas.

Freud was fully aware that Copernicus had not been the first to assert the mobility of the earth. In his *Introductory Lectures* he noted that "something similar [to the Copernican system] had already been asserted by Alexandrian science" (1953, *16:* 285) and he also made it clear that long before Copernicus, "the Pythagoreans had already cast doubts on the privileged position of the earth, and in the third century B.C. Aristarchus of Samos had declared that the earth was much smaller than the sun and moved round that celestial body." Thus "even the great discovery of Copernicus . . . had already been made before him." The cosmological blow to "the self-love of mankind," accordingly, occurred not when "the discovery" was made but only when it "achieved general recognition." It was the same with Darwin in relation to man's being not "different from animals or superior to them," but being "himself . . . of animal descent . . . more closely related to some species and more distantly to others" (1953, *17:* 141). These results came not from Darwin alone but derived from "the researches of Charles Darwin and his collaborators and forerunners."

In thus setting forth the predecessors of Copernicus and Darwin, Freud was not in any sense denigrating the originality of these two men. Rather he was giving expression to a general theory of creativity. Freud believed that many (if not all) of our most 'original' ideas could be traced back to some earlier thinker, often someone we might have forgotten about in our conscious minds. A striking example given by Freud was that of Ludwig Börne, whose essay of 1823 on "The Art of Becoming an Original Writer in Three Days" contains a striking presentation of the method of free association, so important in psychoanalysis. This essay was brought to Freud's attention after he had been made aware of Havelock Ellis's claim that the Swedenborgian mystic, poet, and physician Garth Wilkinson had been the 'true' inventor of free association (Freud 1955, *18:* 264). Although Freud had completely forgotten Börne's essay, he later recalled "that when he was fourteen he had been given Börne's works as a present, that he still possessed the book now, fifty years later, and that it was the only one that had survived from his boyhood." Börne, furthermore, "had been the first author into whose writings he had penetrated deeply." Freud was particularly astonished to find in Börne's essay a discussion of "the censorship exercised by public opinion over our intellectual productions," which was declared to be more oppressive than the "censorship of governments," a notion developed in a way that reminded Freud of the "'censorship' which reappears in psycho-analysis as the dream-censorship." Freud concluded, "Thus it seems not impossible that this hint may have brought to light the fragment of cryptomnesia which in so many cases may be suspected to lie behind apparent originality."

In another work, Freud invoked the notion of 'cryptomnesia' in relation to "the dualistic theory [1937] according to which an instinct of death or of destruction or aggression claims equal right as a partner with Eros as manifested in the libido" (1953, *23:* 244)—a theory that, he noted, had not been generally accepted. How pleased he was, he recorded, to have come upon this theory of his in the writings of Empedocles of Acragas. "I am very ready," Freud said (pp. 245–247), "to give up the prestige of originality for the sake of such a confirmation." And this was especially the case, he added (p. 245), "as I can never be certain, in view of the wide extent of my reading in early years, whether what I took for a new creation might not be an effect of cryptomnesia."

In 1923 Freud asserted that "the originality of many of the new ideas employed by me in the interpretation of dreams and in psycho-analysis" had proved to have been conceived and expressed by others. "I am ignorant of the source of only one of these ideas," he said, the concept to which "I gave the name of 'dream censorship'" (1953, *19:* 261–263). He could now report that "precisely this essential part of my theory of dreams was . . . discovered by [Josef] Popper-Lynkeus independently" (1953, *19:* 262; also *4:* 94–95, 102–103, 308–309n; *14:* 13–20). Freud did not, however, go on from this statement of independent discovery to conjecture a common source, nor did he investigate (or speculate about) the differences rather than the similarities among successive appearances of a scientific idea and the ways in which the genius of a scientist so transforms an idea as to make it an essentially original creation. (On this general topic, see Cohen 1980.)

In 1956 Nigel Walker published in *The Listener* an article, based on a BBC radio talk, entitled "Freud and Copernicus," a title which he changed (in the reprint of the article, 1957 and 1977) to "A New Copernicus?" This article overemphasizes Freud's debt to the psychological ideas of Johann Friedrich Herbart and others, concluding that Freud's comparison of himself with Copernicus and Darwin was not valid, since what Freud held to be "a scientific revolution in our conception of the mind" was rather "a technical advance" which in a spectacular fashion "popularized" a conception already suggested by nineteenth-century German thinkers. Freud's role in history therefore seemed to Walker to be like that of the "circumnavigators of the globe," since they "did more to convince people of the earth's roundness than all the geographers' arguments." Walker thus compared Freud to the eighteenth-century British navigator and explorer Captain Cook rather than Copernicus or Darwin. In 1957 he upgraded this comparison from Captain Cook to Magellan, with the declaration, "In comparing Freud to Magellan rather than Copernicus I am not devaluating his achievement." In defense of his position, he asserted that technicians (such as Watt and Marconi) "probably had a greater effect upon the next generation's way of life than Newton or Dalton."

Walker's article, reprinted again and again, is replete with errors of history (for instance, that it was John Dalton who overthrew "the notorious phlogiston theory"). One such error may serve to direct our attention to a widespread misunderstanding concerning Freud's statements about Copernicus and Darwin: that Freud *compared himself* to these two great scientists.[8] Rather, on each of the three occasions where Freud discussed Copernicus and Darwin, he very carefully did not make a personal comparison but rather stressed the similarities among the Copernican, Darwinian, and psychoanalytic theories and their effects.

His biographer Ernest Jones reported (1953, *2:* 415), "I doubt very much if Freud ever thought of himself as a great man, or that it ever occurred to him to measure himself with the men he considered great: Goethe, Kant, Voltaire, Darwin, Schopenhauer, Nietsche." When Marie Bonaparte once remarked that Freud was "a mixture of Pasteur and Kant," Freud replied: "That is very complimentary, but I can't share your opinion. Not because I am modest, not at all. I have a high opinion of what I have discovered, but not of myself. Great discoverers are not necessarily great men. Who changed the world more than Columbus? What was he? An adventurer. He had character, it is true, but he was not a great man. So you see that one may find great things without its meaning that one is really great."

It was Jones (1953, *3:* 304) who boldly and simply "bestowed on Freud the title of the Darwin of the Mind." Jones actually gave Freud this "highly appropriate title" in 1913 (see Sulloway 1979, 4) and discussed this theme again in 1930, when he argued that "Freud's work, that is, the creation of psycho-analysis, signifies a contribution to biology comparable in importance only with that of Darwin's." Sulloway comments (p. 5) on the irony that Jones "was later to play a key role, along with other Freudians, in establishing Freud's subsequent identity as a 'pure psychologist.'"

Soon after Freud published in *Imago* (in 1917) the essay on "A Difficulty in the Path of Psycho-analysis," in which he discussed the three blows to man's self-image (1953, *17:* 139–143), with the bold notion "that the ego is not master in its own house," his friend and associate Karl Abraham "mildly remarked" that this paper had "the appearance of a personal document" (Jones 1953, *2:* 226). Freud replied, in a letter of 25 March 1917, that Abraham was "right" in saying that he had given "the impression of claiming a place beside Copernicus and Darwin." But he commented that he had not wanted "to give up the interesting train of thought on that account," and so he had "at least put Schopenhauer in the foreground." Freud was referring here to the fact that he had not directly referred to himself, but in the concluding paragraph had rather introduced his predecessors. Following a statement concerning the revolutionary (Freud used the word 'momentous') "significance for science and life of the recognition of unconscious mental processes," he had declared that "it was not psycho-analysis" which "first took this step"(1953, *17:* 143). There had been those "among philosophers" who should "be cited as forerunners — above all the great thinker Schopenhauer." Freud held that Schopenhauer's "unconscious 'will' is equivalent to the mental instincts of psycho-analysis." Furthermore, it was also Schopenhauer who had "admonished mankind of the importance, still so greatly under-estimated by it, of its sexual craving." Psychoanalysis, Freud concluded, had merely "demonstrated" the "psychical importance of sexuality and the unconsciousness of mental life" on "an *abstract* [that is, scientific rather than philosophical] basis" and had "demonstrated them in matters that touch every individual personally."

One might argue that Freud, and — following Freud — his biographer Ernest Jones, have been overly sensitive in denying that Freud compared himself to Copernicus and Darwin.[9] Shakow and Rapaport (1964) find this "sensitivity difficult to understand, considering Freud's repeated juxtaposition, if not complete identification, of psychoanalysis with the other two historical developments." They surmise that "modesty of both author and biographer perhaps forbade their pointing to the amount of objective justification for the comparison."[10] Yet a careful analysis of what Freud actually wrote shows that he was not concerned with his own image as a creator or revolutionary so much as with the blows (cosmological, biological, psychological) to the human "narcissistic self-image." Freud was concentrating on the revolutionary implications of the jolts to geocentrism, anthropocentrism, and egocentrism and only perhaps by indirect implication — if at all — suggesting that his own place in the history of science might be on a par with that accorded to Copernicus and Darwin.[11]

The Twentieth Century,

Age of

Revolutions

VI

25

The

Scientists

Speak

———◆———

The nineteenth century was an age of political, social, scientific, industrial, intellectual, and artistic revolutions — successful and unsuccessful — that for the first time in history made men and women aware that change can be dramatic and revolutionary rather than gradual. The twentieth century is an age of revolution in a different sense, because revolutions have become more frequent and their effects more profound, not only disturbing man and his society and institutions but even threatening nature. Hardly an area of human activity has escaped tremendous changes by revolution. There have been revolutions in communications (radio, television), in manufacturing (synthetic fabrics and plastics), in electronics (solid-state transistors, printed circuits, chips), in warfare (atomic and nuclear bombs, guided missiles), in painting (Picasso, Matisse, Miró), in music (Stravinsky, Schoenberg, Stockhausen), in literature (Joyce, Virginia Woolf), in navigation (radar, loran), in every branch of science (Einstein, Bohr, Crick and Watson), in medicine (Salk vaccine, psychoanalysis, pacemakers and heart surgery), and in data and information processing, where we are currently observing the stages of the computer revolution. We have also witnessed a seemingly endless sequence of political and social revolutions. More people have been affected (and to a greater degree) by the Russian Revolution of 1917 and the Chinese Revolution than by all previous revolutions recorded. The news from Latin America and Africa is regularly punctuated by announcements of great and small uprisings that may range from the assumption of power by a military group to a genuine social and political revolution.

The nineteenth century had been born in the post-1789 turmoil and had witnessed the political unrest of 1848 and the rise of the Marxist-directed revolutionary movement. In science, we have seen that Darwinian evolution itself had been formally announced in the *Origin of Species* in a context of impending revolution. Even so, the main intellectual currents of the latter half of the nineteenth century (and early part of the twentieth) tended to be evolutionary rather than revolutionary, exemplified in such works as Benjamin Kidd's *The Evolution of Society* (1894) and L. Houllevique's *L'évolution des sciences* (1908). By and large the political and social revolutions of the nineteenth century were unsuccessful, culminating in a major 'failed' revolution in Russia in 1905 (a year that coincidentally marks the beginning of twentieth-century science). In the nineteenth century, political and social change was dramatic and sometimes even violent, but progress was still conceived as being generally an orderly process, and this attitude spilled over into discussions of scientific advance.

The twentieth century, by contrast, has been shaken by tremendous upheavals, true breaks in the continuity of history. The social, political, and economic effects of the Russian and Chinese revolutions alone have by far transcended those of the French Revolution, engendering international revolutionary movements on a worldwide scale. The early years of this century also witnessed great revolutions in science, chiefly in physics: x-rays, quantum theory, radioactivity, relativity, the electron, the nuclear atom. In 1905, the same year as the abortive Russian revolution, Einstein published his epoch-making paper on relativity and another paper which revolutionized the physics of matter and radiation and established quantum theory (inaugurated in 1900 by Max Planck) in a direction that has dominated physical thought ever since. In the arts, the pre-1914 era produced the savage rhythms of Stravinsky's *Le sacre du printemps* and the shocking brutal paintings of Picasso and Braque, works which began an age of constructivism, modernism, and abstract art, of dissonance and atonality.

It is not surprising that theories and opinions about revolution would abound in a century which saw such dramatic political, social, artistic, and scientific upheavals in its early decades — and that revolution, not evolution, would emerge as the dominant concept of scientific change in our own time. But the idea that revolutions are a desirable or necessary characteristic of scientific progress was not as easily accepted during the first half of the century as it is by most people today. Many observers — including historians and scientists themselves — were troubled by the revolutionary changes taking place in the very foundations of science, particularly physics, just as they were troubled by the upheavals shaking political and social structures throughout the world. Some, such as Einstein, responded by rejecting the concept of revolutions in science (see Chapter 28); others, such as R. A. Millikan, by rejecting both the concept of revolutions in science and the revolutionary advances themselves.

In this chapter we will sample some of the opinions about scientific revolution that have been expressed during this century, concentrating particularly on the views of scientists. The following chapter will trace the growing acceptance of the concept of revolution in science among twentieth-century historians of science, culminating in the publication of Thomas S. Kuhn's highly influential *Structure of Scientific Revolutions* in 1962. In chapter 27 I shall indicate the stages of revolution in relativity theory, in most people's minds the paradigmatic revolution in science of our time, and in quantum theory, held by those in the know to be one of history's greatest revolutions in science. In chapter 29, I shall discuss the revolution in earth science, which is of special importance because those in the midst of it have recognized it as a revolution, have written its history in the language and structure of revolution, and have even used Kuhn's theoretical analysis of revolutions in order to understand its structure. Moreover, this revolution exhibits in a clear and dramatic fashion some of the major characteristics of all great revolutions in science.

Political and Scientific Radicalism before World War II

In 1908 the political revolutionary V. I. Lenin published a philosophical treatise in which a major concern was the nature and effect of the revolutions going on in physics. Entitled *Materialism and Empirio-Criticism*, this book's professed goal was primarily to defend "the philosophy of Marxism" against some recent "attacks on dialectical materialism" (p. 9), but in the present context Lenin's discussion (embodied in a chapter entitled "The Recent Revolution in Natural Science and Philosophical Idealism") is primarily noteworthy as an instance of the early widespread sense that a revolution had occurred in physics.[1]

Lenin's central example of radium appears prominently in many other writings of this period. What seemed shocking to Lenin and his contemporaries was that a sample of radium apparently could maintain itself constantly at a temperature higher than its surrounding environment, whereas according to classical thermodynamics and the law of conservation of energy, a warm body must give off heat to its cooler environment until equilibrium is reached, that is, until the body and environment have the same temperature. Hence, the properties of radium not only presented scientists with a new phenomenon — radioactivity — which had to be integrated into the conceptual scheme of science; there were aspects of this new substance that destroyed the very foundations of science. Perhaps most notable was the fact that in radioactivity, atoms of one element were found to decay and to be 'transmuted' by nature into atoms of a quite different element.

Among the many authors Lenin drew upon for information was the French mathematician and philosopher Henri Poincaré. In the latter's philosophical book on *The Value of Science* (1907) Lenin found a discussion of the "serious crisis" in physics. The main culprit, according to Poincaré, was "radium, the great revolutionary." Poincaré's opinions commanded tremendous respect, since he was one of the most distinguished scientists in France, if not the whole world. Heed was paid to his gloomy announcement that the new discoveries were not only undermining the principle of the conservation of energy; equally endangered by the revolution were "Lavoisier's principle, or the principle of the conservation of mass," the foundations of mechanics, including "Newton's principle, the equality of action and reaction," and other foundations of accepted physical science.

The awareness of a revolutionary destructive force which had been sensed with respect to radium and radioactivity in general was characteristic of many other new discoveries as well. This theme found expression in a most dramatic way in the *Education of Henry Adams* (1907), in the discussion of Adams's reactions to the Great Exposition of 1900. Using the image of "The Dynamo and the Virgin," Adams mused about the gap between the older force of steam and the new force of electricity. He found a "break of continuity" that "amounted to abysmal fracture for a historian's objects" (p. 381): "No more relation could he discover between the steam and the electric current than between the Cross and the cathedral. The forces were interchangeable if not reversible, but he could see only an absolute *fiat* in electricity as in faith." In his bewilderment, Adams went for help to Samuel Pierpont Langley, the astrophysicist who was secretary of the Smithsonian Institution in Washington.

> Langley could not help him. Indeed, Langley seemed to be worried by the same trouble, for he constantly repeated that the new forces were anarchical, and especially that he was not responsible for the new rays, that were little short of parricidal in their wicked spirit towards science. His own rays, with which he had doubled the solar spectrum, were altogether harmless and beneficent; but Radium denied its God — or, what was to Langley the same thing, denied the truths of his Science. The force was wholly new.

Lenin, Adams, Poincaré, and Langley were not alone in seeing a revolution in the advances being made in physics from about 1890 to 1905, though not everyone was troubled by the implications of these new developments. For example, in an essay on "Space and Time" (1963, 23), Poincaré wrote of the new theory of relativity as the major issue of "the revolution" that had occurred during "the recent progress of physics"; and in another essay (ibid., ch. 6) he implied that the quantum theory was potentially "the most profound revolution that natural philosophy has experienced since Newton."

In the twenties, the word 'revolution' acquired a new radical connotation from the Russian Revolution of 1917 — the second or Bolshevik revolution, which introduced a new noun, 'bolshevism', into the language of general discourse. This revolution not only put a complete end to the old tsarist form of government but dramatically altered the property system and economic life of the Russian people. These revolutionary changes must have been intensified by the fact that, as Crane Brinton observed (1952), in the Russian Revolution "events were telescoped together in a shorter period" than in any other revolution of modern times.

In the minds of most Americans and Europeans, the two archetypal revolutions are the French Revolution and the Russian Revolution. But the latter had perhaps a wider significance because it raised the specter of an exportable bolshevism, promoted by an international movement of revolution and subversion. Additionally, the French Revolution did not produce a stable revolutionary republic; within a decade and a half France accepted the rule of an emperor, whereas the Soviet system has lasted for over half a century and is more potent today than in its first years. It is not surprising, then, that some scientists, witnessing the destruction of the old value systems in Russia and fearing a threat to established ways of life in their own countries, would become equally uneasy about the situation in science. When quantum physics and new concepts of the atom added further crises to a physics already beset by x-rays, radioactivity, and relativity, some scientists saw a parallel between the new science and bolshevism. The fear of bolshevism, and even the admonitions of a possible 'taint' of bolshevism, appeared in discussions of science and of revolution in science during the 1920s.

The revolution in psychology which arose during the 1920s exemplifies the way in which revolutionary science was associated in the minds of some people with political radicalism. John B. Watson's *Behaviorism* (1924) was hailed in the American press as "perhaps . . . the most important book ever written," a work that "marks an epoch in the intellectual history of man" (Watson and McDougall 1928, 102). In Britain it was noted that Watson's system claimed "to revolutionize ethics, religion, psycho-analysis — in fact, all the mental and moral sciences." These extracts were quoted by Watson's adversary William McDougall, who added that Watson's book "claims, not merely to revolutionize, but to abolish, all these august things."

McDougall's statement has an element of truth in it. Watson's *Behaviorism* did conclude with a declaration that behaviorist psychology would lead to a replacement of the known principles and practices of psychotherapy. The final section of the concluding chapter boasted (in a subhead) "Behaviorism a Foundation for All Future Experimental Ethics" (1924, 247), a statement that was reinforced by the opening two sentences, in which Watson envisioned behaviorism as "a science that prepares men and women for understanding . . . their own behavior" and thus would help "men and women . . . to rearrange their own lives" and "prepare themselves to bring up their own children in a healthy way" (p. 248). He held out the utopian ideal of a changed universe, if children — brought up according to behavioristic principles, "in behavioristic freedom" — would "in turn bring up their own children in a still more scientific way" so that the world would finally become "a place fit for human habitation."

Unlike B. F. Skinner—the primary behaviorist psychologist of our era, who proposed a 'Walden II'—Watson made fun of people who "go out to some God-forsaken place, form a colony, go naked and live a communal life" with "a diet of roots and herbs." His Utopia was to be the whole world; his plan, he said, "if acted upon, will gradually change this universe." Yet, perhaps to ward off possible criticism (remember that the year was 1924), Watson insisted that "I am not asking here for revolution." In a preface to the 1930 edition of his book, Watson acknowledged that "we have been accused of . . . being bolshevists" and that the "literature of criticism" has been "personal, even vituperative" (p. x). He supposed that the vehemence of the opposition to his ideas arose from an abhorrence to his basic belief "that man is an animal different from other animals only in the types of behavior he displays" (p. ix). He suggested that he was encountering the same kind of resistance that Darwin faced, because "human beings do not want to class themselves with other animals." Timid souls, he concluded, are driven "away from behaviorism" because a psychologist who is "to remain scientific" must "describe the behavior of man" in exactly the same terms used to describe "the behavior of the ox you slaughter."[2]

Behaviorism still connotes revolution. Writing on this topic in 1983, Peter and Jean Medawar say that Watson "and those persuaded by him carried out a virtually Baconian revolution in psychology," because they substituted "the empirical for that which, because it was not presented to the senses, could only be known by inference." The Medawars had in mind the traditional "states of mind such as joy, misery, or malevolence, or indeed (for where do we draw the line?) consciousness itself." They define the extent of this revolution and its influence by observing that behaviorism "has supplanted the presumption of privilege in introspective psychology by empirical narrative and reportage."

Another example of a tendency during the first half of the century to associate scientific developments with political radicalism occurred in regard to Einstein's theory of relativity. To many of its scientific critics and opponents it seemed to be nothing more than a manifestation in science of the same kind of destructive or anarchic bolshevism that was rampant in Russia, that had appeared in Germany and in Hungary, and that seemed to be threatening all the accepted values of Western civilization and society. The staid *New York Times* (16 Nov. 1919, p. 8) printed an article entitled "Jazz in Scientific World," beginning with four questions: "When is space curved? When do parallel lines meet? When is a circle not a circle? When are the three angles of a triangle not equal to two right angles?" The answer: "Why, when Bolshevism enters the world of science of course." The article then proceeded to a series of quotations from an interview with Charles Poor, Professor of Celestial Mechanics at Columbia University. Here is a sample:

"For some years past," Professor Poor said the other day, after reading the cable dispatches about the Einstein theory, "the entire world has been in a state of unrest, mental as well as physical. It may well be that the physical aspects of the unrest, the war, the strikes, the Bolshevist uprisings, are in reality the visible objects of some underlying, deep mental disturbance, world-wide in character. This mental unrest is evidenced by the widespread intent in social problems, by the desire, on the part of many, to throw aside the well-tested authors of Governments in favor of radical and untried experiments.

"This same spirit of unrest has invaded science, and today there is just as great a conflict in the realm of scientific thoughts as there is in the realm of political and social life. There are many who would have us throw aside the well-tested theories upon which have been built the entire structure of modern scientific and mechanical development in favor of psychological speculations and fantastic dreams about the universe."

Then, the Columbia professor gave an extended discussion of the history of gravitation theories from Newton to Einstein and concluded:

"The fact that such a bending effect [of light rays by the sun] has now been measured is of great scientific importance, and the results may change some of the hitherto accepted ideas as to the density and distribution of matter near the sun, but I fail to see how such an observation can prove the existence of a fourth dimension, or can overthrow the fundamental concepts of geometry.

"I have read various articles on the fourth dimension, the relativity theory of Einstein and other psychological speculation on the constitution of the universe: and after reading them I feel as Senator Brandegee felt after a celebrated dinner in Washington. 'I feel,' he said, 'as if I had been wandering with Alice in Wonderland and had tea with the Mad Hatter.'"

Einstein was asked by a reporter what he thought of the opinions of "Professor Charles Lane Poor," who had gone on record that Einstein's theories "cannot be proved," and that—despite Einstein—we can "explain all physical phenomena, even the irregularities of Mercury, by Newtonian law." Einstein wisely replied (*New York Times*, 4 April 1921), "I did not see Professor Poor's statement."

Arthur S. Eddington, a British astronomer, had the honor of introducing general relativity to the English-speaking world. In 1916, in the midst of the war, Eddington had received a copy of Einstein's 1915 paper from the Dutch astronomer de Sitter and, recognizing the importance of the subject, had studied the 'absolute differential calculus' used by Einstein so that he could understand general relativity. His celebrated *Report on the Relativity Theory of Gravitation* (1918), prepared for The Physical Society of London, called general relativity "a revolution of thought, profoundly affecting astronomy, physics, and philosophy, setting them in a new path from which there could be no turning back" (280). Later, Eddington published a less technical account of relativity called *Space, Time and Gravitation* (1920) and also a work for scientists called *The Mathematical Theory of Relativity* (1923), said by Einstein in 1954 to be "the finest presentation of the subject in any language" (p. 281). And so it is of more than ordinary interest that in Eddington's expositions of the new ideas in physics, notice is taken of the charge that physics has been invaded by a kind of scientific bolshevism.

The opening paragraph of Eddington's Gifford Lectures, published under the title *The Nature of the Physical World* (1928, 1), discusses the "protests against the Bolshevism of modern science and regrets for the old-established order." Eddington had just contrasted the "fundamental changes in our ideas of time and space" (introduced by Einstein and Minkowski between 1905 and 1908) with Rutherford's introduction in 1911 of "the greatest change in our idea of matter since the time of Democritus." The general public, he said, experienced "no great shock" from Rutherford's work, whereas the "new ideas of space and time were regarded on all sides as revolutionary." And yet in considering any such alleged bolshevism, he was "inclined to think that Rutherford, not Einstein, is the real villain of the piece."

Like others in the twenties, Eddington was deeply conscious of revolution as a feature of science. He explained at length how it was that the astonishing new discoveries relating to atomic structure were not generally included in the rubric of revolutionary developments. "The epithet 'revolutionary,'" he said, "is usually reserved for two great modern developments—the Relativity Theory and the Quantum Theory." And, in explanation, he added that the latter pair were not only "new discoveries as to the content of the world," they involved fundamental "changes in our mode of thought about the world" (ibid., p. 2).

Eddington, who led the eclipse expedition in 1919 which verified one of the predictions of general relativity (see ch. 27, below), saw the revolution of relativity, like the revolutions in atomic structure and quantum theory, to be merely instances of the way in which scientific knowledge moves forward in revolutionary stages. He raised the question in his conclusion as to whether there is any "guarantee that the next thirty years will not see another revolution, perhaps even a complete reaction," again having in mind the political image of a reaction in the form of a counterrevolution. He then introduced the concept of a succession of revolutions, and ended his *Nature of the Physical World* (1928, 352–353) with the observation (in language and image similar to Kuhn's usage some four decades later) that "scientific discovery is like the fitting together of the pieces of a great jig-saw puzzle." In this process, "a revolution of science does not mean that the pieces already arranged and interlocked have to be dispersed; it means that in fitting on fresh pieces we have had to revise our impression of what the puzzle-picture is going to be like." His final word was that the systems of Euclid, Ptolemy, and of Newton "have served their turn," and that in the future the systems of Einstein, Bohr, Rutherford, and Heisenberg "may give way to some fuller realisation of the world": "But in each revolution of scientific thought new words are set to the old music, and that which has gone before is not destroyed but refocussed. Amid all our faulty attempts at expression the kernel of scientific truth steadily grows; and of this truth it may be said — The more it changes, the more it remains the same thing."[3]

Many other scientists of this era between two World Wars were also writing about revolutions. In a book about her late husband Pierre Curie, Marie Curie (1923, 133–134) wrote that when he had been "named professor at the Sorbonne," he lectured on symmetry, vectors and tensors, and crystals, and also "set forth the discoveries made in this new domain [of radioactivity] and the revolution they had caused in science."

A prolific writer on the new science during the 1920s and 30s was James Jeans—like Eddington, a British astronomer. He opened one of his last books, *Physics and Philosophy* (1943, ch. 1), with a discussion of revolutions in science. The first sentence, reminiscent of George Sarton and Lord Rutherford (see ch. 1 above), announced that "science usually advances by a succession of small steps." There is a "fog" of the unknown, through which "even the most keen-sighted explorer" can rarely see "more than a few paces ahead." But "occasionally the fog lifts" and a "wider stretch of territory" can be seen "with startling results." Then a "whole science" may "seem to undergo a kaleidoscopic rearrangement," that is, there may be a revolution causing a "shock of readjustment [that] may spread to other sciences" and that may even "divert the whole current of human thought." For Jeans, such classical "rearrangements" or revolutions "are rare"; he mentioned only three that "come readily to mind." These are the Copernican, the Darwinian, and the Newtonian revolutions. "A fourth such revolution," he continued, "has occurred in physics in recent years." The consequences of this revolution "extend far beyond physics," since they "affect our general view of the world in which our lives are cast—in a word, they affect philosophy." To use Karl Popper's categories, each of these four revolutions has had a significant ideological component. For Jeans (1943, 14), the revolutionary "new physics" was centered on two theories: relativity and quanta.

Opposing Views among Physicists on Revolution in Science

The examples just discussed show us that, in the first half of our century of political and social revolutions, of scientific and intellectual revolutions, of revolutions in art, music, and architecture, the image and metaphor of revolution were everywhere. But throughout the century many people have denied that scientific revolutions, whether constructive or destructive, ever occur. An early opponent of the idea that science progresses by revolutions was the physicist R. A. Millikan, long considered the 'dean' of American scientists. His first discussion of revolutions in science appeared in *The Popular Science Monthly* for May 1912. He began this article, on "the kinetic theory of matter" and "the atomic theory of electricity," with the open declaration of his "wish vigorously to combat the point of view" that there are "revolutionary discoveries," as is believed by "too many of those who are not engaged at first hand in scientific inquiry." Concerning the "revolutionary discoveries which are continually being announced," Millikan said that "nine tenths of them are just as revolutionary as was the discovery of the seven-year-old boy" who had only recently learned "that 3 and 4 made seven" when his teacher "told him that 5 and 2 make seven" (p. 418).

Millikan's attack was primarily directed at the idea that fundamentally new discoveries were continually being made that completely overturned the existing structure of knowledge. But in a lecture given in February 1917, he was a little more specific. "The progress of science," he said, "is almost never by the process of revolution" (1917, 175). The "newspaper headings" are often "about revolutions," but they "almost never happen." No, he reiterated, "the growth of science is in general by a process of accretion, almost never by that of revolution.""Once in a while we have something revolutionary," he concluded, "but not often." It is typical, however, of the inconsistency often noted when scientists discuss revolution that just a couple of pages earlier than his dismissal of revolutions, Millikan said that within the last hundred years ("or 130 years at most"), "all the external conditions" of life have been "more completely revolutionized" than during all the preceding "ages of recorded history" (p. 172).

Similar views to Millikan's were expressed by Karl K. Darrow, for many years the secretary of the American Physical Society. In one of his books, *The Renaissance of Physics* (1937, 15), Darrow, who was essentially a conservative, stressed the point that the "ways of thinking" of Newton, Laplace, and Fourier still serve us well, thus "extolling the conservatism of physics" rather than presenting "its radical new ideas, its violent breaches with its classical past, and its many astounding discoveries." Of course, he was aware that the changes physics had been undergoing were "so great as to justify strong language to describe them," but the appropriate language would not be "quite so strong as one frequently hears used." He argued that it was wrong to "speak of the classical physics as having been upset, or overthrown, or repudiated, or revolutionized." He then explained that "no one ought to talk of a revolution in physics, unless he adds at once that never has there been a revolution anywhere more gradual, more cautious, more tenacious of the virtues of the old regime." No, he concluded, " 'revolution' is not the proper word!" In modern physics there is no revolution, but rather "a tremendously rapid *evolution*" (p. 16).

Like others who adopted a similar viewpoint, Darrow assumed that a revolution implies a complete break with the past and "physics had never broken with its past." The fact is, he said, that "physicists simply hate to give up any theory which ever served them well; and actually, we seldom do." And he concluded this discussion by asserting that, as a rule, innovators in theoretical physics are "supremely eager to have themselves accepted as legitimate heirs of the classical royal line." I do not know who it was that Darrow particularly had in mind, but by 1937 the literature about science was replete with references to revolution. In fact, 'revolution' and Darrow's "tremendously rapid *evolution*" may be taken as synonyms, except for the radical political connotations of the word 'revolution' and the implication that revolutions in science destroy or wipe out the past and substitute something wholly new in its place. In fact, this did occur in the transition from Aristotelian to Newtonian physics and also in the eventual transition from Ptolemaic to Keplerian astronomy. But in many revolutions the break with the past is not so complete as implied in Darrow's statements.

A somewhat similar, but less extreme, attitude toward revolution pervades the writings of the Polish-French chemist and philosopher Emile Meyerson, who exerted a powerful influence on the history and philosophy of science during the 1930s. Meyerson introduces the concept of revolution only rarely. He does write of revolutions in a passing reference to the "revolution" of the physics of quanta, which "overturns the image of reality" (1931, 69). Most often he wrote of the "evolution of science" (for example, 1931, 116) or "the evolution of mathematics" (p. 326). He referred to others who have had similar ideas, such as John Dewey, who produced a schema to "explain the evolution of science" (p. 416). He quotes, with evident approval, a statement made in 1927 by Madame Curie (p. 758), in an éloge of H. A. Lorentz, about "the disturbing evolution of the theory of quanta and the new mechanics" (1928, p. vi). Meyerson's purpose was not so much to write a history of science as to trace and describe exactly the thought processes of philosophers and scientists. He stresses that his aim was conditioned by his realization that "the evolution of the sciences" is a history of constantly changing views of the world, punctuated by "scientific revolutions," events in which scientists have altered basic concepts, such as the abandonment of phlogiston by chemists and of the caloric fluid by physicists (p. xii). He was concerned to understand how scientists could so readily give up basic premises so as to adopt a new theory which is often the direct opposite of what has been believed. An example, occurring in his own day, was the "evolution" (not revolution!) being wrought by the theory of relativity. He concluded that "the crucial advances [progrès décisifs]" or "the revolutions" in science exhibit a process that goes counter to the "fundamental evolution of the sciences." These "revolutions" frequently occur because a "great innovator (such as Lavoisier)" breaks the shackles which the methods of research and of thought had seemed to impose." But Meyerson is more concerned with the process of evolution than with revolution, even to the extent of conceiving certain revolutions to be evolutions.

Very recent physics has led to many discussions of revolution. In the spring of 1963, Eugene Rabinowitch, editor of the *Bulletin of the Atomic Scientists,* gave a series of four public lectures at the University of Chicago on scientific revolution. Our generation, he began (1963 [Sept.], 15) has had "the unique experience of living through three simultaneous revolutions." The first two were "a social revolution" ("replacing old ruling groups . . . by new ones") and "a national revolution" ("overthrowing colonial empires"), but the "third revolution" was "engendered by science and its child, technology." He claimed that "the scientific revolution" had altered the character of traditional social and political revolutions of the past, by replacing "localized" or "transient" national upheavals by "a worldwide and largely irrevocable change," and by converting "local national upheavals into a worldwide 'revolution of rising expectations.'" The "scientific revolution suggests to all peoples that poverty need not be permanent." This led him to a central but somewhat different theme: that the scientific revolution had changed our view of our "habitat" and our "position in the universe." In a kind of unconscious parody of Freud, he introduced three revolutions. For him the first were the Copernican and Darwinian but the third one was not the revelations of psychoanalysis but "the widening expanse of the universe." (He mistakenly believed that the "homocentric view of the world" has to do with man being central.) Then he discussed the subject he really knew well, the "revolution in nuclear physics" and the capacity now given to "the human race . . . to destroy itself." One "effect of the scientific revolution" to which he called attention is worth noting (pp. 16–17):

> Modern science has dispelled the apprehensive expectation that in foreseeable time all technological progress will have to stop because of exhaustion of coal and oil. Man may still contemplate with dismay the certainty of his extinction when the solar system itself will end, but the present change in our outlook can be compared to the difference between the life expectation of an old man whom death stares in the face and a youth just entering life.

Rabinowitch called the duality of modern physics and its "renunciation of strict causal interpretation" a "major revolution in man's concept of the world" (p. 18). Relativity (1963 [Oct.], 11) was another "revolutionary thought."

In one of his lectures, Rabinowitch invoked the concept of "the successful permanent revolution in science" and prophesied that it would have to "influence . . . the thinking of men [and women] in other fields." He held that wars have been made "irrational" and that diplomacy has been deprived of "its most important tool — plausible war threats" (1963 [Nov.], 9; 1963 [Dec.], 14) — as a consequence of the "scientific revolution," which has culminated in "the discovery of nuclear bombs and ocean-spanning missiles." And he concluded on a chiliastic note by examining what he held to be a major consequence of "the scientific revolution of our time": "the beginning of world community" led by the "international community of scientists" (1963 [Dec.], 14).

In the preface to his *From Being to Becoming* (1980, p. xii), Nobel Prize winner Ilya Prigogine states his aim "to convey to the reader my conviction that we are in a period of scientific revolution." In this revolution, there is a reappraisal of the "very position and meaning" of the "scientific approach." Prigogine compares this period to two other dramatic moments in the history of science: the birth "of the scientific approach" in ancient Greece and "its renaissance" in the age of Galileo. Prigogine wants his readers to be certain that "when I speak of a scientific revolution," more is meant than the sequence of radical innovations in science, such as quarks, pulsars, and molecular biology. Rather, for him the revolution consists in abandoning the long-standing belief "in the 'simplicity' of the microscopic — molecules, atoms, elementary particles." This leads him to three main theses: (1) that "irreversible processes are as *real* as reversible ones," (2) that they "play a fundamental *constructive* role in the physical world," and (3) that "irreversibility is deeply rooted in dynamics." Such a 'revolution' is obviously different from the normal "evolution of science" (p. xvi). Like many other scientists, Prigogine introduces the idea of revolution but does not develop its consequences. The term is thereafter conspicuous by its absence. It occurs in one prominent position, however, at the start of chapter 2, on "Classical Dynamics," a subject which Prigogine sees as "the starting point of the scientific revolutions of the twentieth century, such as relativity and quantum theory" (p. 19).

In a survey of physics published in 1979, Arthur Fisher reported Murray Gell-Mann's opinions about the problem of 'unification' in physics. Looking forward to "understanding in a deeper and deeper way the nature of the universe in which we live," Gell-Mann said he expects "that there will be an intellectual revolution" in physics (Fisher 1979, 12). This revolution, he said, will be "comparable to those that have taken place in the past with the heliocentric idea, evolution, special relativity, quantum mechanics." Steven Weinberg (1977, 17f.) has written of "the development of special relativity and of quantum mechanics" as "great revolutions." But he warns us against applying the concept of revolution inappropriately to every aspect of physics in the twentieth century. In "the development of quantum field theory since 1930," for example, he finds that "the essential element of progress has been the realization, again and again, that a revolution is unnecessary."

An interview with the physicist Wolfgang Pauli, shortly before he died in 1958, gives us some insight into the currency of revolution in the context of science, and the goal of young physicists to make such revolutions. This concept of revolution carries no destructive taint of bolshevism but rather expresses a common feeling that revolutions are a creative force in scientific progress. The interview was conducted by Jagdish Mehra, to whom Pauli said: "When I was young I thought I was the best formalist of the day. I thought I was a revolutionary. When the great problems would come I shall be the one to solve them and to write about them. The great problems came and went. Others solved them and wrote upon them. I was of course a classicist rather than a revolutionary." Then, apparently as an afterthought, he remarked, "I was so stupid when I was young" (Mehra and Rechenberg 1982, xxiv).

Views on Revolution from Outside the Physical Sciences

Similar opinions on the pros and cons of revolution have been expressed regarding the biological sciences.[4] Molecular biology and the allied technology of genetic engineering are second only to the computer in evoking the name 'revolution' in the daily press. For instance, an article in *The Boston Globe* on 4 March 1981 was headlined "Coming Revolution in Science of Biology." The story centered on a "new California laboratory" which "is about to revolutionize the science of biology" by methods whereby "proteins can be analyzed accurately" and "genes can be built from scratch to make them." An article in the "Science Times" (*New York Times,* 12 April 1983) was headed "DNA's Code: 30 Years of Revolution." The occasion was a forthcoming meeting to celebrate the thirtieth anniversary of the letter published on 25 April 1953 in *Nature,* in which James D. Watson and Francis H. C. Crick announced their discovery of the structure of "the master chemical of heredity of all living things," an event that is believed by many scientists to be "clearly the century's most important discovery in medical science." Few observers of the current scene would disagree with this judgment, or with Peter Medawar's opinion (*New York Review of Books*, 27 Oct. 1977) that "the greatest scientific discovery of the twentieth century" is, "without qualification, . . . that the chemical makeup of the compound deoxyribonucleic acid (DNA) — and in particular the order in which the four different nucleotides out of which it is assembled lie along the backbone of the molecule — encodes genetic information and is the material vehicle of the instructions by which one generation of organisms governs the development of the next" (p. 15). This is "the great revolution of molecular genetics" (p. 19). The revolutionary quality of this advance was apparent to at least one of the two co-discoverers even before the discovery had been made. James D. Watson, in his chronicle of those days (1980, 116), has recorded his recognition that the "double helix . . . would revolutionize biology."

In psychology, as we have seen, revolution appeared early in the twentieth century. Wilhelm Wundt, who founded the first laboratory for experimental psychology, discussed revolution in the fifth and sixth editions of his influential *Grundzüge der Physiologischen Psychologie* (*Fundamentals of Physiological Psychology*, 1902; 1908, *1*: 4; first published 1873–74). There he argued that "as an *experimental* science, physiological psychology strives for a reform of psychological research which is not inferior in importance to the revolution which the introduction of experiment caused in the thinking in natural science." Furthermore, he held that this change in psychology would perhaps prove even more important than the revolution in the natural sciences "insofar as in the realm of the natural sciences exact observation is possible under favorable conditions even *without* experiment, while in the realm of psychology this is excluded."

We may also see the prominence of revolution in twentieth-century thought in a debate between two anthropologists in successive editions of the *Encyclopaedia Britannica*.[5] At issue is the question of whether culture evolved independently all over the globe, or whether it arose in or near Egypt and gradually diffused throughout the inhabited world. In an article on anthropology in the eleventh edition (1910–1911), Edward B. Tylor, an advocate of cultural evolution, concluded that anthropology was "grappling with the heavy task" of systematizing the accumulation of knowledge beginning with the discoveries concerning early man by Boucher de Perthes, Lartet, Christy, and their successors. "There have recently been no discoveries," he wrote, "to rival in novelty those which followed the exploration of the bone-caves and drift-gravels," which had "effected an instant revolution in all accepted theories of man's antiquity."

The twelfth edition (1922), which consisted of the twenty-nine volumes of the eleventh edition plus three supplementary volumes covering the period from 1910 to 1921, contained two articles on anthropology — the original one by Tylor, and a new one by Grafton Elliot Smith. Elliot Smith, an advocate of the 'diffusionist' theory of culture, took exception to Tylor's "remarkable statement" that "anthropology had . . . reached the limits of its discoveries." Almost every year since then, there has been a "rich harvest of anthropological data and a clearer vision of their significance." These years, he emphatically declared, "have witnessed a profound revolution in every branch of the study of man." Among the startling new facts, Elliot Smith introduced the recent discovery of 'Piltdown man', a curious choice (as Victor Hilts has remarked) in that the finger of suspicion has since been pointed at Elliot Smith himself as a possible perpetrator of this hoax. Despite their intellectual disagreement concerning independent cultural evolution versus diffusion, both of these anthropologists clearly conceived of the advancement of their science in terms of revolution.

Revolution in science was one of the central topics of discussion at a symposium on "American Morphology at the Turn of the Century" in 1981. At issue was a statement by Garland Allen (1978) that between 1890 and 1910 there had been a shift in American biology from morphology to experimentation.[6] This led to the more basic question, does science progress "in periodic spurts" or "revolutions to one degree or another" (Allen 1981, 172–174)? In the opening paper, Jane Maienschein observed that historians of biology 'have come increasingly to adopt the view that scientific change has been rapid and discontinuous," though she pointed out that such historians may not necessarily have endorsed "the Kuhnian idea of revolution" and may only mean "that individuals or groups rejected older ideas and hence speeded change" (p. 89). She countered: "I maintain that endorsing such a revolutionary view of how science changes risks distorting the facts in the effort to illustrate the expected patterns." She preferred to seek for "continua" and she concluded that to discuss the events in American biology in terms of revolution or evolution leads to a distortion. The problem of "whether to call scientific change in general evolutionary or revolutionary, continuous or discontinuous," is only "quibbling" (p. 112). Ronald Rainger, in another paper, was less concerned with revolutions in general than with an argument for "the persistence of a morphological tradition in paleontology" and hence "for continuity, not revolution, as the best means for understanding the historical development of the biological sciences during those years" (1981, 129–130).

In a reply to his critics, Garland Allen departed from the narrow theme in order to introduce his concept of revolution versus evolution in the sciences. He affirmed his belief that "an abrupt, or revolutionary, change (in the Kuhnian sense) occurred in the natural-history area of biology between 1890 and 1910," and he proposed "a model that is evolutionary," that is, "one in which evolutionary and revolutionary elements are both constantly at play" (1981, 172). Basically, Allen argued that "all revolutionary change is dependent upon antecedent evolutionary change" and that, "conversely, all evolutionary change leads . . . to revolutionary change" (p. 173). For him this meant that "quantitative" or "small, incremental, 'evolutionary'" changes lead to "qualitative" or "large, significantly different, 'revolutionary'" changes. He held that when such a transition from "quantitative to qualitative change" occurs slowly, it is an evolution, but when it occurs rapidly it is a revolution. He then compared the progress of science to the development of species in the model of 'punctuated equilibria', as proposed in paleobiology by Stephen Jay Gould and Niles Eldredge. There are "periods of rapid change, in which new species are formed or old ones die quickly, followed by stable periods of slow change, where adaptations are perfected." This introduction of a radical and not-yet-accepted position in evolutionary biology raises more problems than it solves, as Fred Churchill was quick to point out in his Epilogue. He found it "doubtful that the tide of human affairs possesses the analogous levels of organization and discrete units" (1981, 181).

This debate points to the fact that in very recent years a rather significant opposition to the view that revolution is a feature of scientific advance has been voiced. On more than one occasion a scientific colleague, having learned that I was writing a book on scientific revolutions, has written to me that he would like to have an argument about this topic. What astonished me in each case was the hostility aroused by the concept and word 'revolution' in the context of science, even when my correspondent had no idea of the line I was going to take in the proposed book. For a long time I was puzzled: What could there be about the concept of revolution that would so easily trigger such a negative response? I have since concluded that to some extent this attitude is a reaction to Kuhn's book. Obviously not all scientists agree with Peter Medawar (1979, 91) that "Kuhn's views have caught on — a sure sign that scientists find them illuminating because they haven't much time for what they think of as mere philosophizing." But although "Kuhn's views throw some light on the psychology of scientists" and are "an interesting comment on the history of science" (p. 92), they have one feature that could easily arouse the ire of many practicing scientists.[7] For it is an explicit feature of Kuhn's schema that most scientific research is a kind of "mop-up work," that this feature of ongoing science is not appreciated by "people who are not actually practitioners of a mature science" (1970, 24). Indeed, such "mopping-up operations are what engage most scientists throughout their careers." Although Kuhn says that this activity can be "fascinating . . . in the execution," many scientists must have felt that the very choice of expression was demeaning to the image of the scientist as a bold adventurer, charting new paths, making brilliant discoveries, and advancing the cause of truth.

26

The

Historians

Speak

In chapter 1 we saw that the historian of science George Sarton wrote, in 1937, of the progress of science as normally an incremental or cumulative activity rather than a succession of revolutions. This opinion was shared by a number of scientists and commentators on science, among them the chemist James B. Conant and the physicist Ernest Rutherford; and — as we saw in the last chapter — it is still held by a vocal minority today. But by the 1950s the concept of revolutions in science was gaining acceptance in the thinking of historians of science, spurred on primarily by three major works: Herbert Butterfield's *The Origins of Modern Science 1300 – 1800* (1949; rev. ed. 1957), A. Rupert Hall's *The Scientific Revolution 1500 – 1800* (1954; rev. ed. 1983), and Thomas S. Kuhn's *The Structure of Scientific Revolutions* (1962; rev. ed. 1970). Butterfield and Hall gave historical prominence to *the* Scientific Revolution, although Butterfield did introduce at least one other revolution in science. But the concept of revolutions in science as regularly occurring phenomena achieved general acceptance only after the publication of Kuhn's book.

It has been frequently said that Herbert Butterfield introduced the expression 'the Scientific Revolution' into historical discourse. When I once asked him about this, Butterfield — who had long been interested in the history of historiography — replied that he was fully aware of his role in giving currency to the name, but that he could claim no originality for its invention. And in fact, a careful reading of *The Origins of Modern Science*, which was based on a series of lectures given in 1948, shows that he did not once present himself as the originator of this designation.

Nevertheless, Butterfield was largely responsible for making the Scientific Revolution a central issue in the mind of every reader. With force and eloquence, Butterfield declared that the end product of the Scientific Revolution was "not only . . . the eclipse of scholastic philosophy but . . . the destruction of Aristotelian physics"; it "overturned the authority in science not only of the Middle Ages but of the ancient world." Accordingly, this revolution "outshines everything since the rise of Christianity and reduces the Renaissance and Reformation to the rank of mere episodes, mere internal displacements, within the system of medieval Christendom" (1949, vii). Coming from a general historian rather than a scientist or a specialist historian of science, Butterfield's dramatic conclusion was especially effective in convincing other general historians and philosophers — and even historians and philosophers of science — of the propriety of dealing with the emergence of modern science in the age of Galileo and Newton as a major revolution in history. Where Whitehead (1925, ch. 3) had encompassed the great scientific events associated with Galileo, Newton, and their contemporaries under the simple rubric of the "century of genius," Butterfield emphasized the revolutionary nature of the scientific thinking of that age in such phrases as "amongst the epic adventures" and "one of the great episodes in human experience" (1949, 179). Above all, Butterfield stressed the revolutionary consequences of what he called putting on "a different kind of thinking-cap" (ibid., 1) and he eschewed such easy explanations as the effects of the Reformation or of social and economic factors.

In *Origins* Butterfield not only gave prominence to the Scientific Revolution as possibly the most remarkable event in modern Western civilization but also wrote (ch. 11) of "The Postponed Scientific Revolution in Chemistry," thus indicating that he was aware of post-Newtonian revolutions in science. This title may have been intended to be no more than a simple variant of the long-current expression 'the chemical revolution', which ultimately goes back to an expression by Lavoisier, the chief architect of that revolution, and is a name that became widely used after the publication of Marcelin Berthelot's *La révolution chimique* (1890). I must confess that I have never been at all certain of exactly what it means to say that a revolution in science was 'postponed', a term that makes more sense for an outdoor event that is put off until a later date on account of rain.[1] Nor does Butterfield make it clear to his readers whether he was implying a real difference between his concept of a "scientific revolution in chemistry" and Lavoisier's "chemical revolution." Possibly all he intended was to show that the Scientific Revolution affected primarily mathematics, astronomy, and physics, but not chemistry, and that a comparable revolution in chemistry did not occur until the time of the French Revolution; that is, chemistry came within the fold of subjects altered by the Scientific Revolution about a century later than astronomy and physics.

Herbert Butterfield's influence was aggrandized by the fact that his book appeared just as the professional field of the history of science was exploding with activity and making its presence felt in many areas of intellectual concern: general history, philosophy, political science, economics, sociology. The many applications of science in World War II, the problems of international control of nuclear weapons, and the hopes and fears for a future conditioned by science and technology caused real concern among many scientists and nonscientists. This concern engendered an ever growing interest in the history of science, in revolutions in science, and in the making of modern science during scientific revolutions. Butterfield's stunning presentation of the first scientific revolution (the inaugural revolution of modern science) appeared at the right moment to exert a maximal influence. His book became almost at once the commonly used primer for both elementary and advanced students. Its ringing cadences carried the author's convictions to a generation of scholars and scientists.

Some Early Writers on Revolution

Herbert Butterfield was not the first historian of the twentieth century to write at length about scientific revolutions, however. A number of major authors cited by Butterfield had previously taken up the topic of revolutions in science and the Scientific Revolution. The earliest of these appears to have been Martha (Bronfenbrenner) Ornstein, whose

Columbia University doctoral dissertation on *The Rôle of the Scientific Societies in the Seventeenth Century*, published in 1913, is still a classic; it was reprinted in 1928 and again in 1975. In addition to writing about the Scientific Revolution as a single unified movement, Ornstein also applied the concept of revolution to particular events within the larger revolution. For instance, she referred to the telescope, which "had utterly revolutionized the science of astronomy," to "Linnaeus's revolutionary work," to "the revolutionary changes in optics," to "the revolution in the universities" (1928, 8, 13, 249, 262). Ornstein also wrote specifically about a change that occurred in the first half of the seventeenth century which "seems more like a 'mutation' than a normal gradual evolution from former times" (p. 21). She summed up her findings in a statement that during "the second half of the seventeenth century," the scientific societies were the carriers of culture, "much as the universities had been before the scientific revolution" (p. 262). And in memorable words, she concluded that there had been "a revolution in the established habits of thought and inquiry, compared to which most revolutions registered in history seem insignificant" (p. 21). The grandiloquence of this last quotatioh is on a par with the later expression by Butterfield. To us, what may be most notable is that at no point in her presentation can we discern even a bare hint that this concept of the Scientific Revolution, like the concept of revolutions in science, was being used in any other than a well-established mode of historical exposition and analysis.

An author of the 1920s who also exerted a significant influence on Butterfield and on many historians and philosophers of science was Edwin Arthur Burtt. His "historical and critical essay" on *The Metaphysical Foundations of Modern Physical Science* (1925) is still highly respected as a classic study of the philosophical underpinnings of science during the Scientific Revolution. A philosopher by training, Burtt later abandoned the study of the history and significance of early thought and pursued research in the philosophy of religion.

A little more than half of Burtt's book is devoted to Copernicus and Kepler, Galileo, Descartes, Gilbert, and Boyle; the remainder deals with the "metaphysics of Newton." Newton is introduced by Burtt in a reference to "the unprecedented intellectual revolution which he carried to such a decisive issue" (p. 203). This encomium is, however, accompanied by a statement of the "disappointment" that Burtt experienced on looking through Newton's pages for "a clear statement" by Newton "of the method used by his powerful mind in the accomplishment of his dazzling performances"; Burtt had searched through Newton's papers in vain for any "specific and illuminating directions for those less gifted."

In this very influential historical study, Burtt also wrote of "the revolutionizing of chemistry by Robert Boyle," who had applied to this subject the "method of specific hypothesis and experiment rather than that of geometrical reduction" (p. 200). For Burtt, the "exact mathematical movement in science" of the days of Kepler and Galileo brought "in its train" a "remarkable metaphysical revolution" (p. 156). There are also references (as on p. 118) to "the astronomical revolution." Galileo's "positive conception of causality" and the science that went with it are described as a "whole revolution" (p. 89), a fit phrase for Burtt's conception of "Galileo's revolutionary greatness" (p. 93) and one that has overtones of the older expression 'complete revolution', implying a motion that goes full circle so as to return to its starting point. In describing Galileo, Burtt referred to a "thought-revolution" (p. 84), an idea later developed more fully by Alexandre Koyré and Herbert Butterfield.[8] In Burtt's pages one can also encounter "the Copernican revolution" (p. 50), said to have been a "most radical revolution," one for which Burtt believed that the way had been paved by "the free speculations of such a thinker as Nicholas of Cusa" (p. 28). Burtt discusses Copernicus's expectation that the "simplicity" of his system could "rightly . . . decrease somewhat the prejudices which he knows his revolutionary view is certain to awake" (p. 27). And in the general introduction to the subject, in which Burtt lists all the radical innovations that occurred during the first two crucial centuries of modern science, he refers to "what was happening between the years 1500 and 1700" as "this revolution" (p. 16).

Burtt's book pioneered a new dimension of scientific thought — the metaphysical underpinnings and religious implications of physics from Copernicus to Newton. It is notable for displaying the degree to which this new science was "inseparable from the philosophical and religious currents of the sixteenth and seventeenth centuries" (Guerlac 1977, 63). But in the present context, this oft-reprinted work is of significance for its discussion of the Scientific Revolution and the "radical" Copernican revolution.

Another important writer who used the concept of revolution in relation to the science of the seventeenth century was the philosopher Alfred North Whitehead. Whitehead (1923, 165) remarked that the telescope "might have remained a toy," but in Galileo's hands "it created a revolution"; he then endeavored "to explain the main revolutionary ideas which Galileo impressed upon his contemporaries." In his *Science and the Modern World* (1925), based on the Lowell Lectures he gave in Boston in 1925, Whitehead referred again and again to what he called the "historical revolt" of the sixteenth century, which he conceived to have encompassed science, in which domain "it meant the appeal to experiment and the inductive method of reasoning" (p. 57). And although he did not here specifically use the word 'revolution' in relation to Galileo, there can be no doubt concerning his commitment to Galileo's revolutionary impact, producing "the most intimate change in outlook which the human race had yet encountered" (p. 3). Then, continuing in memorable and oft-quoted words, he said, "Since a babe was born in a manger, it may be doubted whether so great a thing has happened with so little stir." This may seem an odd metaphor for the Galilean revolution, since Galileo's own style was combative as he strove to establish the new philosophy, the new science, and the new astronomy, and to destroy the forces of reaction that he believed held his Church in thrall and committed it to error in science. But Whitehead viewed events of that time with historical aloofness, as was perhaps only fitting in a man for whom Paolo Sarpi's history of the Council of Trent was bedtime reading. It seems remarkable today that Whitehead was so far removed in his mind-set from the travails of the seventeenth century that he could write of Galileo's trial, abjuration, and sentence that Galileo suffered only "a mild reproof" and an "honorable detention" (p. 2). I assume that Whitehead's presentation was intended to convey his feeling that the revolutionary implications of Galilean science were not clearly manifest to his contemporaries and so did not really produce an immediate and violent effect on their thought. Hence the authorities merely issued a "mild reproof," much as one would punish a slightly naughty child.

Yet another popular book of the twenties was *The Making of the Modern Mind* by John Herman Randall, Jr. Published in 1926 (and reprinted in 1940, and in a fiftieth anniversary edition in 1976), this was the work of a young historian of philosophy in his twenties. Randall wrote that "neither Renaissance nor Reformation was the movement that produced the really great revolution from the medieval to the modern world; that was effected by the gradual development of science" (p. 164). "It was not humanism, and it was not the Reformation, that was destined to work the greatest revolution in the beliefs of men, however triumphant they seemed for centuries; it was science" (p. 203). Later on, in discussing the work of Copernicus, he declared that none of the ideas of Copernicus "was really revolutionary except in a negative sense" of his having introduced the thought that "the old authorities had been found in error" and that "even observation and common sense were fallible" (p. 230). Randall was sensitive to the notion that "the Copernican revolution [was] consummated by Galileo" (p. 235) and took the strong position that "even more significant" than the Copernican and Galilean revolution "was the Cartesian revolution which created a new physics." He not only saw in science a "revolution from the medieval to the modern universe" (p. 242), contrasting "the Copernican" and "the Cartesian revolution" (p. 244), but he coupled the revolution made by Spinoza with the one made by Descartes as the "two titanic revolutions in men's beliefs" (p. 247). Randall was also aware of later revolutions, quoting from Diderot that we are "on the point of a great revolution in the sciences" (p. 265). He also noted that "a present-day revolution bids fair to modify" the Newtonian system (p. 254). Newton and Locke "effected that outstanding revolution in beliefs and habits of thought" that is "fittingly styled the 'Age of Enlightenment and Reason'" (p. 253). Revolutions in science abound in *The Making of the Modern Mind*.

A special emphasis was given to the Scientific Revolution by the historian Preserved Smith, when he used this phrase as the title to a chapter in his *History of Modern Culture* (1930). Smith, like Butterfield, was a general historian and not a scientist nor even a historian of science; his professional reputation had been made by a scholarly treatise on Erasmus. Smith was a pioneer in envisioning science and its history to be an essential part of "modern culture." The prominence he gave to science is indicated by the subtitle of his first volume, "The Great Renewal 1543–1687," the dates being those of the publication of Copernicus's *De Revolutionibus* and of Newton's *Principia*. In stressing the Scientific Revolution, however, Smith may have been following the example of another general historian, James Harvey Robinson, whose *Mind in the Making* (1921) has a chapter called "The Scientific Revolution." Smith called the Scientific Revolution "the greatest revolution in history" and maintained — with a rhetorical flourish similar to Ornstein's and Butterfield's — that "the scientific achievements of that epoch surpass all that previously had been done in the whole life of man upon earth" (p. 144).

In 1939 the scientist J. D. Bernal published a challenging book called *The Social Function of Science*, a frontal Marxist attack on the established order in both science and in society at large. Considering the author's political stance, we are not surprised to encounter quite a few references to revolutions in science. These include, among others, "the first great revolution in human society" (p. 14) accompanying "the discovery of agriculture," the "Pneumatic Revolution" (p. 27) of the seventeenth century, the "revolutionary . . . new ideas on mechanics" (p. 167) "inspired by the flight of the cannon ball," the early twentieth-century "improvements in methods of communication and transport" (p. 170) that "were revolutionizing the possibilities of co-ordinating and directing the movements of millions of men at a time," the "great revolution in chemistry" (p. 335) "instituted by Lavoisier," and "the great quantum revolution of the twentieth century" (p. 368). Also an industrial revolution of the eighteenth and nineteenth century and a "second industrial revolution" (p. 392) in which "science is playing a much larger and more conscious part than the first," discoveries which have "their effect in revolutionizing the whole progress of science" (p. 343), and a future "fundamental revolution" (p. 361) in the techniques of mining. But Bernal did not seriously develop the theme of revolutions in science — past, present, and future — and although he wrote at length about the nature and effects of the Scientific Revolution, he did not make much of the concept of a general revolution in the seventeenth century. This theme appears only at the head of a single paragraph bearing the subtitle "The Scientific Revolution: The Role of Capitalism." But when Bernal expanded, revised, and completely rewrote this work in the post-Butterfield period of the 1950s, he made such extensive use of the concept of revolution (and the word itself) that the reader cannot help but gain the impression that revolutions in science had become part of the essential fabric of his historical thinking (Bernal 1954; 1969); volume two of Bernal's four-volume set *Science in History* was entitled *The Scientific and Industrial Revolutions*. Here is an illustration of the way that in the last thirty years or so the literature of the history of science has become saturated with references to revolutions in science and to the Scientific Revolution. The two stages of Bernal's book are thus characteristic of the times in which they were written: a pre-1950 recognition that such revolutions, including a Scientific Revolution, occur, versus a post-1950 widespread use of these concepts as essentials to our understanding of scientific change.

The Seminal Role of Alexandre Koyré

I shall conclude this pre-Butterfield survey with a discussion of Alexandre Koyré, who became, during the 1950s and 1960s, the most influential figure in the writing of the history of science. At least a decade before Butterfield, Koyré had made effective use of the notion of the Scientific Revolution as a central organizing principle. Koyré's *Galilean Studies* of 1939 is generally acknowledged as a seminal book in what has been called "a historiographic revolution in the study of science" (Kuhn 1962, 3). As a result of this revolution, historians of science no longer seek "the permanent contributions of an older science to our present vantage," but rather "attempt to display the historical integrity of that science in its own time." Thus, as Kuhn puts it (ibid.), the new historians of science "ask, for example, not about the relation of Galileo's views to those of modern science, but rather about the relationship between his views and those of his group, i.e., his teachers, contemporaries, and immediate successors in the sciences." Additionally, "they insist upon studying the opinions of that group and other similar ones from the viewpoint — usually very different from that of modern science — that gives those opinions the maximum internal coherence and the closest possible fit to nature."

This new approach — "perhaps best exemplified in the writings of Alexandre Koyré" (p. 3) — has centered on a new kind of conceptual analysis. Attention is paid not only to the thinking of the individual scientist but also to the scientific, philosophical, and even religious presuppositions of the age — including the canons of scientific acceptability or respectability according to the reigning or 'received' philosophy, or the 'themata' (see Holton 1977). Koyré's analysis brought to the fore certain big changes that occurred in seventeenth-century thinking, such as the destruction of the Aristotelian cosmos and the mathematicization of space: changes so fundamental in character as to suggest an intellectual revolution.

Koyré opened his celebrated *Galilean Studies* with a declaration of his aim to produce "a study of the evolution (and the revolutions) of scientific ideas." He referred to "the scientific revolution of the seventeenth century" as a "veritable 'mutation'" in human thought, in the special sense in which this term 'mutation' had been used by Gaston Bachelard. Koyré considered this to have been the most important such 'mutation' since the invention of the Cosmos by the Greeks. The Scientific Revolution was a "profound intellectual transformation of which modern physics (or, more precisely, classical physics) was both the expression and the fruit."[3] This mutation was based on a radical "geometrization of space," essentially a replacement of the "concrete cosmos" of Aristotle and Ptolemy by the "abstract space" of Euclid (1939, 5–6; 1978, 1, 39).

Believing that the revolutionary changes in science in the seventeenth century were "solely changes in the way the human mind reflected upon its natural environment," Koyré — as Rupert Hall has said (1970, 212) — "insisted again and again that the changes that had brought classical science into being were neither socio-economic nor technological, nor concerned with the methodology of science." Hall concludes, "Such an expression of the totality of intellectual change during the late Renaissance (which I find extremely convincing) compels an historian to see the scientific revolution as a great historical drama," one which "has its subplots and convolutions as all grand dramas do, one that works slowly towards the climax of accomplishment in the mid- to late seventeenth century" (p. 213).

Herbert Butterfield was deeply influenced by Koyré's work, and not only in relation to such particular themes as the Platonism of Galileo, the role of mathematics, the destruction of the Aristotelian cosmos, and the alleged minor importance of experimentation in the Scientific Revolution. He also took over and used effectively Koyré's notion of the essential changes in man's way of thinking about phenomena in the natural world.

Butterfield's Concept of the Scientific Revolution

Butterfield's conception of the Scientific Revolution differs rather markedly from the way revolutions in science and even the Scientific Revolution had been conceived in the eighteenth and nineteenth centuries, and even in the early decades of the twentieth. He did not have in mind a revolution patterned on the model of the French or the Russian revolutions. Rather, for him the Scientific Revolution became equated with the whole development of modern science from the days of Copernicus, or of Galileo and Kepler. Although Butterfield disarmingly states that he is not introducing a new concept, and carefully refers to "the so-called 'scientific revolution,'" or "what is called the scientific revolution," he implies that the Scientific Revolution was not merely a single series of historical events at the time of Galileo or of Galileo and Newton, as had been the case for Ornstein, Burtt, and Koyré. He conceived the revolution to be a continuing historical or history-making force acting right up to the present time.[4] The Scientific Revolution thus appears in his writings to have features similar to the Marxian 'permanent revolution', a point he often discussed in conversation but did not make in his book.

Accordingly, in Butterfield's chapter 10, "The Place of the Scientific Revolution in the History of Western Civilization," it becomes apparent that what makes the Scientific Revolution appear so significant is that "not only was a new factor introduced into history at this time among other factors." Rather, "it proved to be so capable of growth, and so many-sided in its operations, that it consciously assumed a directing rôle from the very first, and so to speak, began to take control of the other factors" (p. 179). In short, the Scientific Revolution not only marked a moment of great change but became institutionalized as modern science itself. Butterfield says explicitly (ibid.): "And when we speak of western civilisation being carried to an oriental country like Japan in recent generations, we do not mean Graeco-Roman philosophy and humanist ideals, we do not mean the Christianising of Japan, we mean the science, the modes of thought and all that apparatus of civilisation which were beginning to change the face of the West in the latter half of the seventeenth century."

Butterfield believed, furthermore, that "we are in a position to see its implications at the present day much more clearly than the men who flourished fifty or even twenty years before us" (p. 189). He makes it clear that "we" (as of 1949) are not "under an optical illusion," we are not "reading the present back into the past," for that which has "been revealed" in the 1940s and 1950s "merely brings out more vividly the vast importance of the turn which the world took three hundred years ago, in the days of the scientific revolution," in the days of Copernicus, Galileo, and Newton. Thus for Butterfield, the historical significance of the Scientific Revolution is augmented and clarified by the later and even recent developments of science. This helps to explain "why our predecessors were less conscious of the significance of the seventeenth century" and of the supreme importance of the Scientific Revolution, and "why they talk so much more of the Renaissance or the eighteenth-century Enlightenment." Western civilization got its modern characteristics in the seventeenth century, and specifically through or in association with the Scientific Revolution: "That is why, since the rise of Christianity, there is no landmark in history that is worthy to be compared with this" (pp. 189–190).

The Use of Revolution by Historians of Science

The foregoing examples show that the concepts of revolution in science and of the Scientific Revolution appeared prominently in the writings of a number of significant writers before the 1950s. Furthermore, such revolutions were discussed in the most widely used general textbook of the pre-Koyré or pre-Butterfield era, W. C. Dampier's oft-reprinted *History of Science*. In the first edition (1929), for instance, he discusses "a revolution" in astronomy "produced by the Copernican Theory" (p. 139), a "revolution in the intellectual outlook of mankind" (p. 189) during the age of Newton, a nineteenth-century "revolution" (p. 269) in man's "ways of thought" that arose in the biological sciences, "revolutionary physical discoveries" (p. 312), and a "revolutionary result [that] came from biology, when physiology and psychology examined the relations of mind and matter, and again when the theory of evolution was established by Darwin" (p. 312). Dampier also mentions a "revolution in psychology" (p. 326), and "a veritable revolution in thought" produced by twentieth-century mathematics and physics. Although he refers often to 'revolution' in this work, he does not display a clearly developed theory of scientific revolutions, nor does he use the concept of revolution in a significant way as an organizing principle. Furthermore, the Scientific Revolution is not a major theme.

Despite the frequent occurrence of the theme of revolution, it should not be concluded that, during the first half of the twentieth century, historians, historians of science, and scientists generally came to recognize the existence of the Scientific Revolution and to use it as an organizing principle, or that they all conceived of scientific change in terms of revolution, as is so commonly the case at the present time. In the 1930s a number of major studies in the history of science mention neither the concept of revolution in science nor of the Scientific Revolution. For example, the Scientific Revolution is conspicuously absent both as a technical term and as an idea in Robert K. Merton's classic study of 1938 on *Science, Technology and Society in Seventeenth Century England;* nor does Merton refer to revolutions in the sciences. Neither of these terms appears in the famous article of 1931 by the Russian Boris Hessen (1931; 1971), with its path-breaking Marxian analysis of "The Social and Economic Roots of Newton's 'Principia.'" And they are also not to be found in G. N. Clark's small book of 1937, in reply to Hessen, *Science and Social Welfare in the Age of Newton*. Finally, the Scientific Revolution is neither a main topic nor even a subheading in Henry Guerlac's pioneering syllabus of *Science in Western Civilization* of 1952, in which the only occurrence of the word 'revolution' in relation to science is a reference to the Chemical Revolution associated with Lavoisier.[5]

A major event in the historiography of the Scientific Revolution was the publication in 1954 of A. Rupert Hall's *The Scientific Revolution 1500–1800*. Subtitled *The Formation of the Modern Scientific Attitude,* this was the first book to be expressly devoted to the Scientific Revolution, though Hall acknowledged the significance of the work of Herbert Butterfield (pp. vii, 375). Hall stressed the "complementary" types of advance that began in the sixteenth century, "the distinct lines of conceptualization and factual discovery," a "twin advance" that "constantly occurs in science" (p. 37). He argued that "the scientific spirit of the [sixteenth] century developed naturally from the work and progress in the later middle ages." Drawing on a vast corpus of scholarship, and taking full advantage of the research and thinking of Alexandre Koyré, Hall took his readers back into the springs of Galileo's thought, the development of ideas leading up to his two great works. Hall observes that Galileo's *"Mathematical Discourses and Demonstrations Concerning Two New Sciences* (1638) does not, however, give a true picture of the revolution in dynamics as Galileo effected it, any more than the earlier series of *Dialogues* can now be considered as a balanced statement of the respective merits of the two cosmologies" (p. 77). He concluded this portion of his book with the important insight that, during the next century, "a major concern of science was the extent to which nature could be explained broadly in terms of Cartesian mechanism interpreted with the aid of Galileo's descriptive analysis of motion" (p. 101). In every chapter, Hall applied the full panoply of the new conceptual analysis. Gone was the succession of hero-images that characterized traditional textbooks in the history of science. In their place was a history in which ideas and facts, theories and experiments or observations, were blended into the general background of religious and philosophical ideas. Herbert Butterfield had burst upon the historical scene of the Scientific Revolution as a gifted and enthusiastic amateur; but Rupert Hall was decidedly a talented and knowledgeable professional.

Hall had previously published a technical study of *Ballistics in the Seventeenth Century* (1952), which presaged his later deep involvement in the history of technology. We are not surprised, accordingly, that one of his major innovatory chapters dealt with "Technical Factors in the Scientific Revolution," in which he displayed a real insight into the tradition of artisan-engineers and the interactions of science and technology. In retrospect, an especially impressive part of his book is the discussion of scientific instruments (pp. 237–243) — a topic introduced and illustrated in A. Wolf's *History of Science, Technology and Philosophy in the Sixteenth and Seventeenth Centuries* (1935) and in his similar book on the eighteenth century (1938). Hall has continued to write articles and books on science during the Scientific Revolution, and has prepared a completely rewritten version of his pioneering work (that appeared late in 1983). It may be noted that his proclamation of the Scientific Revolution as having occurred during the three centuries from 1500 to 1800 would make this the longest revolution in the historical record.

Uncertainty about the History of 'Scientific Revolution'

This brief examination shows that during the twentieth century, a number of historians of science and philosophers had been using the concepts of the Scientific Revolution and of revolution in science. Yet prior to about 1950, despite the occurrence of easily locatable individual statements about both the Scientific Revolution and particular revolutions in science, neither of these ideas was used in a significant way to organize the historical discussion. Nor was there much interest shown in the question of whether there are revolutions in science, or in the nature and structure of such revolutions if they actually occur. I have found that the works of historians and philosophers express little concern for these topics before 1950 or so, in contrast to the writings of scientists, especially those scientists who exclaimed against revolutions. The minimal role of revolution in science, especially of the Scientific Revolution, is shown by the general ignorance of historians of science concerning the early use of these terms. The current literature on the history of science contains a variety of scholarly opinions that include the following: Auguste "Comte was the first to conceive of, and to baptize, the Scientific Revolution"; the "catch phrase" Scientific Revolution "is of comparatively recent origin, dating from 1943 when Alexandre Koyré first used the term"; and the "term 'scientific revolution' to characterize this period, and above all to characterize the development of modern science . . . was, I think, first used by Herbert Butterfield in 1948." The only serious attempt I know to trace the origin of the concept of revolution in science concludes that "it was Denis Diderot who in 1755 introduced the concept of scientific revolution." These samples make it evident that the scholarly tradition does not include a continuous theme of revolution in science and of the Scientific Revolution running through the twentieth century.

The decade of the 1950s saw the spread of the idea of the Scientific Revolution, to a large degree an effect of the influence of Butterfield and Koyré. The latter's *Galilean Studies,* published in France in 1939, was a victim of the war and began its scholarly life only in the late 1940s and 1950s.[6] But neither Butterfield nor Koyré did much to advance the use of the concept of lesser revolutions, revolutions in the sciences, a topic not particularly prominent in their works. And so it remained for Thomas S. Kuhn to make historians, philosophers, and sociologists of science fully aware of this aspect of the development of science and to direct scholarly attention to the theme that revolutions not only do occur in the sciences but are a regular feature of the scientific enterprise as a whole. I have mentioned in earlier chapters that this achievement has turned out to be independent of the acceptance of Kuhn's particular proposed "structure of scientific revolutions." And I have also noted that a significant feature of his influence has been to shift the point of view of scholars away from the notion of a contest among competing ideas to that of a contest among individual scientists or groups of scientists holding ideas.[7]

Kuhn's influence may thus be described as transforming a growing scholarly concern for a single large-scale Scientific Revolution into a research program directed toward individual smaller-scale revolutions in the sciences. Even if we believe that there is a second, a third, or possibly a fourth Scientific Revolution, this is a mere handful compared to the enormously large number of revolutions in science. Kuhn, furthermore, sees a whole scale of revolutions occurring in the sciences, with the ones discussed in the present book (the Copernican, the Darwinian, the Einsteinian) at the large end of the scale.

Research on the Scientific Revolution

It is hardly surprising that the recognition of the existence of the Scientific Revolution has triggered many new studies of the nature of that revolution. One has resulted in the overturning of one of Alexandre Koyré's favorite themes, approved and echoed by Herbert Butterfield, that the role of experiment in the Scientific Revolution had been grossly exaggerated. In particular, Koyré insisted that the account of major experiments said to have been performed by such figures as Galileo and Pascal were philosophical romances, fictions invented to give their work an alleged empirical base. That Galileo, for example, did not perform the famous experiment of the inclined plane which he described in the *Two New Sciences* was proved for Koyré by Galileo's statement that there was an agreement among observations at different trials to one-tenth of a pulse beat. But when Thomas B. Settle constructed an apparatus similar to the one Galileo described and replicated the experiment, he found that it was easy to obtain this degree of accuracy. More recently, Stillman Drake has found new manuscript evidence that Galileo's early discoveries concerning the science of motion were made in the context of experiment. Of course, Koyré was correct in emphasizing the importance of a new way of thinking in Galileo's production of the new science of motion, but the new way of thinking needed experiment as an aid to discovery and as a test of the principles being discovered.

Another aspect of the Scientific Revolution currently being explored is the background of alchemy, Hermeticism, and the other parts of thought that have been ignored by scholars insisting on 'rational' science. A great pioneer in this area, wielding a tremendous influence, was the late Frances Yates. It is still too early for us to be able to evaluate the extent of the influence of these topics on the growth of science, or even the actual significance of such studies on a figure such as Newton. But at least we are now fully cognizant that Newton's studies of alchemy, of the Hermetic art and philosophy, and of prophecy were deep and continued over many years. It is a challenge to find out, if possible, exactly how and to what degree his intellectual activities in what we have too easily called nonscientific realms or nonrational thought may have influenced his science.

There has been a considerable study of the social structure of scientific activity, some of it quantitative, and a number of scholars have essayed interpretations of parts of the Scientific Revolution in terms of the influence of social factors. But thus far there has been a conspicuous and real neglect of the psychological study of scientific revolutionaries. Here is a promising area as yet unexplored, which may give historical studies of revolutions in science a wholly unexpected dimension that will mark a new era in the scholarly analysis of science and the activities of scientists.

27

Relativity

and

Quantum Theory

F or nonscientists and scientists, relativity symbolizes revolution in science in our century. But for those in the know, quantum theory (especially in its revised form as quantum mechanics) may have been an even greater revolution. We may find a measure of Albert Einstein's greatness as a scientist in his fundamental contribution to both revolutions.

In considering relativity, we must keep in mind that there are two different theories of relativity. One is the special theory (1905), which deals with space and time and with simultaneity, leading to the famous equation $E = mc^2$. The other is the general theory (1915), which deals with gravitation.[1] Although both theories have been revolutionary, most discussions of the relativity revolution center on issues arising from the special theory. But what brought the attention of the world to special relativity was the verification in 1919 of a prediction of the general theory, that starlight passing near the sun is bent by the sun's gravity. This verification, which occurred during a solar eclipse expedition, set into being the relativity craze that swept the world and overnight made Einstein a public figure.

Special Relativity

Einstein proposed the special theory of relativity in 1905, in a paper published in the *Annalen der Physik*.[2] It was followed by a supplement in that same year in which the first form of the mass–energy equivalence was stated for radiation. In 1907 Einstein put forth a comprehensive paper on relativity, containing the general result that $E = mc^2$. His radical paper set forth new concepts of mass, space, and time and challenged the apparently simple notion of simultaneity. At the outset Einstein displayed what he called "the 'Principle of Relativity,'" and introduced "another postulate," that "light is always propagated in empty space with a definite velocity c which is independent of the state of motion of the emitting body." Among the consequences of the theory of relativity was the abandonment of the ideas of 'absolute' space and time and of the concept of an all-pervading 'aether', then considered to be the medium for the transmission of light and all other forms of electromagnetic radiation.

In retrospect, the publication of Einstein's inaugural paper on relativity in the *Annalen der Physik* in June 1905 is a classic example of a revolution on paper. We have seen in chapter 2 how Max Born studied "the electrodynamics and optics of moving bodies" in Göttingen in 1905 and 1906 but never heard of Einstein or his work. It was the same in Cambridge, England, in 1906–07. Einstein, according to the recollections of his sister (Pais 1982, 150–151), "imagined that his publication in the renowned and much-read journal would draw immediate attention." Of course, he anticipated "sharp opposition and the severest criticism," but he was "very disappointed" by the lack of response, the "icy criticism." In time he received a letter from Max Planck asking questions on some obscure points in the paper, which was the cause of "especially great" joy since Planck was "one of the greatest physicists of that time." Planck's early and deep commitment to relativity was a major reason for the very rapid spread of interest in the new topic among physicists (Pais ibid.). In the year following the publication of Einstein's paper, Planck began to lecture on the theory of relativity in Berlin, not on Einstein's presentation but on Lorentz's theory of the electron. Planck's assistant, Max von Laue (later a Nobel Prize winner in physics), published a paper on relativity in 1907.

In September 1906 Planck gave an address on relativity to the German Physical Society (published in that same year), and in 1907 the first doctoral dissertation ever to be written on relativity was completed by

K. von Mosengeil under Planck's direction (Pais 1982, 150). Pais (p. 151) has indicated how a few physicists came into the fold, including Johann Jakob Laub of Würzburg and Rudolf Ladenburg of Breslau (Wroclaw). Von Laue went to Berne as a skeptic to visit Einstein and found it hard to believe that this "young man" could be the "father of the relativity theory." Years later, von Laue would write one of the very best technical introductions to the theory. In a letter to Einstein on 24 March 1917, von Laue expressed to Einstein his excitement about his own revolutionary work in physics: "It has happened! My revolutionary views [revolutionäre Ansichten] on wave optics are printed." At "this very moment," he added, they "undoubtedly excite the highest loathing of every peace-loving physicist"; but "I nevertheless persist in my reprehensible conviction."

In addition to Max Born's own account of how he first learned about relativity theory, we have an independent version by Leopold Infeld. Infeld (1950, 44) relates a story told him by his friend Professor Stanislaus Loria. Loria's teacher, "Professor Witkowski in Crakow (and a very great teacher he was!)," read Einstein's 1905 paper on relativity and "exclaimed to Loria, 'A new Copernicus has been born! Read Einstein's paper.'" Some time later (in 1907, according to Born), Loria met Born at a physics meeting, where he told him about Einstein, asking him if he had read the paper. "It turned out that neither Born nor anyone else there had heard about Einstein." According to Infeld's story, they then "went to the library, took from the bookshelves the seventeenth volume of *Annalen der Physik* and started to read Einstein's article." Immediately, says Infeld, "Max Born recognized its greatness and also the necessity for formal generalizations." It was Infeld's judgment that Born's own work on relativity theory, produced much later, "became one of the important early contributions to this field of science."

At first the number of physicists who were willing to accept Einsteinian special relativity was small — not as yet large enough on a world scale to produce a revolution in science, but including a sizeable fraction of German-speaking theoretical physicists. In July 1907 Planck wrote to Einstein that "the advocates of the relativity principle" form only "a modest-sized crowd"; hence he believed it "doubly important" that they "agree with one another" (Pais 1982, 151). The phrase 'relativity principle' could embrace Lorentz's theory, which Planck personally favored, as well as Einstein's. Yet Einstein's reputation was growing, albeit slowly. In the autumn of 1907 Johannes Stark (editor of the *Jahrbuch der Radioaktivität und Elektronik*) wrote to Einstein requesting a "review article" on relativity.[3] In 1906 Planck had used the name 'Relativtheorie', but by 1907 Einstein had adopted today's more familiar 'Relativitätstheorie' (relativity theory, or, theory of relativity; see Miller 1981, 88). The first reference in print to Einstein's relativity paper was in an article in 1905 by Walter Kaufmann, who considered Einstein's "research . . . as being 'formally identical' with Lorentz's" and no more than a useful generalization of the latter (see Miller 1981, 226). Kaufmann concluded (ibid., §7.4.2) that his own experimental data confuted the Einstein and the Lorentz theory of electrons. We shall explore the consequences of this in a moment.

Einstein's theory was the subject of a paper by Paul Ehrenfest in 1907 (for which see ibid., §7.4.4). In the following year, 1908, Hermann Minkowski published a radical transformation of Einstein's theory into mathematical form, beginning the "enormous formal simplification of special relativity" (Pais 1982, 152). By these stages, the revolution on paper was becoming a revolution in science. Pais (1982, 152) dates the rapid growth of Einstein's reputation and influence from 1908.

At this time Einstein's academic star was rising. From the humble post of patent examiner in the Swiss Patent Office at Berne he rose to gain, in the spring of 1909, an appointment as associate professor of theoretical physics at the University of Zürich, apparently for his work on quantum theory of solids. One of the faculty members who recommended him for the post wrote that Einstein "ranks among the most important theoretical physicists," that he had "been recognized rather generally as such since his work on the relativity principle" (Pais 1982, 185). On 8 July of that same year, he received an honorary degree from the University of Geneva, along with the chemist Wilhelm Ostwald and Marie Curie. He held his new post for only two years until in March 1911 he moved to Prague, where he became a full professor at the German university, the Karl-Ferdinand University. He stayed in this post for sixteen months, when he was succeeded by Philipp Frank. Einstein returned to Zürich, where he became professor of physics at the ETH, the Polytechnic Institute (Technische Hochschule).

The difficulties that lay in the way of accepting the special theory of relativity were largely conceptual, of course, but there was also a real experimental hurdle.[4] At the end of his foundational paper of 1905 Einstein had deduced a formula for the transverse mass of an electron. This was almost but not quite the same as the formula that appeared in the Lorentz theory, a discrepancy that was soon eliminated so as to make both theories give the same result. But the literature contained two publications by Walter Kaufmann, one of 1902 and the other of 1903, in which the results of experiment were quite different from the predictions of Lorentz's (and hence Einstein's) theory. Einstein ignored these results (see Miller 1981, 81 – 92, 333 – 334). In 1906 Kaufmann published an article in the *Annalen der Physik*, the same journal that a year earlier had carried Einstein's article on relativity, in which he gave an adequate summary of Einstein's ideas on space and time (Miller 1981, 343), referred to a Lorentz – Einstein theory of electrons, and concluded (ibid., §7.4.2) that the results of Kaufmann's own measurements were "not compatible" with "the fundamental assumptions" of the Lorentz – Einstein theory (see Holton 1973, 189 – 190, 234 – 235). Lorentz thereupon wrote a letter (discovered and printed by Miller 1981, 334 – 337; 1982, 20 – 21) stating that he was at his "wits' end" ("au bout de mon latin"). "Unfortunately," he wrote to Poincaré, his hypothesis was "in contradiction with M. Kaufmann's new experiments," and he believed himself "obliged to abandon it." But Einstein assumed that the existence of a "systematic deviation" of the data from the theory implied an "unnoticed source of error," and he had faith that additional and more exact experiments would vindicate the relativity theory. He proved to be right when new experiments, published in 1908 by A. H. Bucherer, agreed with the predictions of Lorentz and Einstein. Additional confirmation came from experiments by E. Hupka in 1910.[5] Decisive results were obtained from 1914 to 1916 and ever since that time there has been continuing abundant evidence of many kinds concerning the correctness of relativity theory.

While this experimental evidence was coming in, the theory itself was undergoing a fundamental reconstruction at the hands of Hermann Minkowski, a mathematics professor at Göttingen, who some years earlier had taught Einstein mathematics in Zürich. In 1908 Minkowski published a paper introducing the concept of a single four-dimensional 'space-time', to replace the separate conceptual entities of three-dimensional space and one-dimensional time. He also gave relativity theory its modern tensor form (requiring physicists thenceforth to learn a new mathematics recently developed by Ricci and Levi-Città), introduced new technical terms into relativity, and made it evident from the point of view of relativity that the traditional Newtonian theory of gravitation was not adequate (Pais 1982, 152). Einstein was apparently not an instant convert and even considered Minkowski's rewriting of the theory in tensors as mere "superfluous learnedness" (ibid.). But by 1912 he had been converted, and in 1916 he gracefully acknowledged the role of

Minkowski in his own progression from special to general relativity. Minikowksi's contribution was later stressed by Einstein (1961, 56–57), when he said that without it "the general theory of relativity . . . would perhaps have got no further than its swaddling clothes." This oft-quoted expression appears in translation as "no further than its longclothes"; although 'Windel' is the German word commonly used for 'diapers', the sense here is obviously that, but for Minkowski, the general theory would never have got beyond its infancy.

The first public presentation of Minkowski's ideas of space and time occurred in a lecture on 5 November 1907 entitled "The Principle of Relativity." It was not published until 1915, six years after Minkowski's death, but his ideas were circulated by means of two other papers, published in 1908 and 1909 (Galison 1979, 89). Fully aware of the significance of his contribution, he opened his 1907 lecture with these remarks: "Gentlemen! The views of space and time which I wish to lay before you . . . are radical [radikale] . . . Henceforth, space by itself, and time by itself, are doomed to fade away into mere shadows." Instead, in the first manuscript version of the lecture, Minkowski said that the "character" of his new views on space and time was "revolutionary," actually "mightily revolutionary" ("Ihr Charakter ist höchst/gewaltig revolutionär"; Galison 1979, 98). In the final printed draft, however, the reference to the "revolutionary" character was omitted.

Max Born's discussion of his first reading experience of Einstein's papers reminds us how difficult the Einsteinian conception was, even for those who had no problem with the mathematics. In 1907, when Loria directed him to Einstein's publications, Born had been a member of Hermann Minkowksi's university seminar, and so — he writes (1969, 104–105) — "quite familiar with relativistic ideas [presumably Lorentz's] and the Lorentz transformations." Even so, he recalled, when he read Einstein's papers, "Einstein's reasoning was a revelation to me." As an intellectual creation, Born found that "Einstein's theory was new and revolutionary." Einstein's presentation had "the audacity of challenging Isaac Newton's established philosophy," the traditional concepts of space and time. In retrospect Born recognized the force of Einstein's intellectual revolution and the revolution on paper, but he saw clearly that there was not yet a revolution in the sciences. The new ideas and the new way of thinking had still to be studied, accepted, applied, and made the basis of the common or shared beliefs of scientists. Born shows dramatically that Einstein's theory was in fact so radical — so "new and revolutionary," he wrote much later — that "an effort was needed to assimilate it." And, he reminds us, "not everybody was able or willing to do so"; this statement applied to Born himself at first. The Einsteinian revolution in science required a general accept-ance of Einstein's fundamentally new way of dealing with the physical world.

The problems of accepting Einstein's assumptions are shown clearly in a 1909 article by two Americans, Gilbert Lewis and Richard Tolman. They admitted that Einstein's principles "generalize a number of ex-perimental facts, and are inconsistent with none." Bucherer's experi-ments were cited as important evidence for the theory. But while they did not question one of the "laws" on which they said the theory rested, they were troubled by the other one. That is, there was no resistance to the general law that "absolute motion could not be detected," but they found it hard to accept the law that light has a constant velocity indepen-dent of the observer (see Miller 1981, 251–252). This second law would lead to "strange conclusions" about the relativity of length and time, they believed, which had to be "scientific fiction," that is, "in a certain sense psychological."

With each passing year, more and more physicists were won over, although many of them — while accepting Einstein's equations — preferred to believe in absolute time and simultaneity (as Lorentz did; see Miller 1981, 259), and accepted 'contraction' as a basis for the spatial problems arising from the constancy of the speed of light. A sensational element was added to the theory in April 1911, when the French physicist Paul Langevin addressed a congress of philosophers in Bologna. Lan-gevin was an able scientist; Einstein (p. 232) even said that if he himself had not discovered special relativity, Langevin would have done so. In discussing the relativity (or dilation) of time, Langevin replaced Ein-stein's abstract notion of clocks moving with respect to one another by human beings — thus giving birth to the 'twins paradox', a dramatic version of Einstein's 'clock paradox', which soon became one of the popular bizarreries of relativity theory. The time problem arises if one twin stays on earth while the other goes on a voyage; when the voyager returns home, it is found that the two twins are of different ages. In one of Langevin's examples, the traveling twin goes out on a straight line to a star, turns around and follows the same path home. If the speed of travel is high enough (though not quite the speed of light), the traveler could find that during *his* two years of travel two centuries would have elapsed on earth. The philosopher Henri Bergson later admitted that it was Langevin's talk in April 1911 that "first drew my attention to Einstein's ideas."

This clock (or twin) paradox rapidly became (and to some extent still remains) a source of perplexity and even of hostility to the theory of relativity. Writing about those who disagreed with the "intellectual content" of the theory, as opposed to its formulas or mathematical results, von Laue told Einstein in 1911 that a common basis of objection was "in particular, the relativity of time with its often paradoxical consequences" (quoted in Miller 1981, 257). Von Laue, in his textbook on relativity of 1912, the first ever to be written, held that such paradoxes and other similar problems about time had "great philosophical meaning" and "precisely for this reason" they should be treated only by "the methods of philosophy." In 1911, we may note, Einstein himself discussed such motion by "a living organism in a box" which could be sent on a "lengthy flight" and return "in a scarcely altered condition," while the organisms that had stayed at home would "already have long since given way to new generations."

While many people could not easily accept Eintein's radical reconstruction of basic physical thought, they did make use of his mathematical results — which, as von Laue (and others) had shown, were formally but not 'physically' identical with those of Lorentz's theory. Von Laue even went so far as to say (1911) that a "definite distinction" between the two theories "is not possible." But in time it was recognized that Einstein's theory was preferable, especially after general relativity gave special relativity a new importance.[6]

By about 1911 Einstein's special theory of relativity had gained enough adherents to constitute a revolution in science. In that year, Arnold Sommerfeld announced that the theory of relativity was "already so well established that it was no longer on the frontiers of physics" (Miller 1981, 257). Early in 1912 Wilhelm Wien, winner of the Nobel Prize in physics for 1911, proposed that Einstein and Lorentz be awarded this coveted prize. From a "logical point of view," he wrote (quoted in Pais 1982, 153), the principle of relativity "must be considered as one of the most significant accomplishments ever achieved in theoretical physics." There was now, he said, clear "confirmation of the theory by experiment." Lorentz, he concluded, was the first to find "the mathematical content of the relativity principle," but Einstein "succeeded in reducing it to a simple principle."

Of course all physicists did not accept the revolutionary new ideas. J. D. Van der Waals in 1912 argued that there was as yet no causal explanation why mass and length should vary with velocity (see Miller 1981, 258). In addition to a lingering concern about the relativity of time with its attendant paradoxes, there were more fundamental objections about dispensing with absolute length and time and mass, and it was hard to accept the 'relativity of simultaneity'. Even more difficult in some ways was the rejection of the aether. How could light and other electromagnetic waves exist in space if there were no medium there to undulate? The strength with which these objections were made may be taken as an index of the revolutionary character of the new doctrine.

Among the antirelativity arguments raised in 1911, Professor W. F. Magie of Princeton (1912, 293), in his presidential address delivered before the American Physical Society, said that "the principle of relativity" failed to meet the criterion that any "ultimate solution, . . . to be really serviceable, must be intelligible to everybody, to the common man as well as to the trained scholar." For him, the theory of relativity failed to be intelligible because it was not "expressed in terms of the primary concepts of force, space and time, as they are understood by the whole race of man." But he evidently was not aware how radical Newton's concepts of force and inertia had been in 1687. Nor did he apparently appreciate how few people, other than the small number who had studied academic physics, really understood the "primary concepts" of force and inertia.

Magie also asserted his "right to ask of those leaders of thought to whom we owe the development of the theory of relativity, that they recognize the limited and partial applicability of that theory and its inability to describe the universe in intelligible terms." He would also "exhort them to pursue their brilliant course until they succeed in explaining the principle of relativity by reducing it to a mode of action expressed in terms of the primary concepts of physics."

In a comment in *The Nation* (vol. 94, 11 April 1912, 370–371), Louis Trenchard More summarized Magie's published address and introduced the following remarks about revolutions in science:

Both Professor Einstein's theory of Relativity and Professor Planck's theory of Quanta are proclaimed somewhat noisily to be the greatest revolutions in scientific method since the time of Newton. That they are revolutionary there can be no doubt, in so far as they substitute mathematical symbols as the basis of science and deny that any concrete experience underlies these symbols, thus replacing an objective by a subjective universe. The question remains whether this change is a step forward or backward, into light or into obscurity. It is held, and apparently rightly, that the revolution effected by Galileo and Newton was to replace the metaphysical methods of the schoolman by the experimental methods of the scientist. Now the new methods might seem to be just the reversal of that step, so that, if there is here any revolution in thought, it is in reality a return to the scholastic methods of the Middle Ages.

A little more than two decades later, in his biography of Newton (1934, 333), More—now become dean of the graduate school of the University of Cincinnati—still expressed his abhorrence for "Professor Einstein's Generalised Theory of Relativity," which More called the "boldest attempt [yet] towards a philosophy of pure idealism"; such a "philosophy, which is merely a logical exercise of the active mind, and ignores the world of brute facts, may be interesting, but it ultimately evaporates into a scholasticism." And then he concluded that if relativistic physics (and its philosophy) persists, "it will cause the decadence of science as surely as the mediaeval scholasticism preceded the decadence of religion." The reader, accordingly, may not be surprised to find that More also decried the great advances being made in mathematical or symbolic logic, and wrote (ibid., 332): "It is a notable fact that these two works, probably the two most stupendous creations of the scientific brain, are now under attack,—the *Organon* by modern symbolists in logic, and the *Principia* by the relativists in physics." His final comment was that "Aristotle and Newton will be honoured and *used* when the modernists are long forgotten" (p. 332). Here we see examples of the way in which the profundity of a revolution in science can be gauged as much by the virulence of conservative attacks as by the radical changes in scientific thought it produces.

General Relativity

Einstein once remarked that even if he had never lived, the special theory of relativity would have come into being because "the times were ripe for it" (Infeld 1950, 46), but this was not equally true for general relativity. He doubted whether, if he had not produced this theory, "it would be known yet." General relativity has been called the "second

Einstein revolution" (ibid.). It was a tremendous leap forward that left many physicists behind, just at a time when a good number of them had been won over to special relativity. Planck, who had greeted special relativity with enthusiasm as one of its earliest supporters, said to Einstein: "Everything now is so nearly settled, why do you bother about these other problems?" Einstein did so because he was a genius, far ahead of his contemporaries. He knew that special relativity was incomplete, that it did not deal with accelerations and with gravity. He later wrote that the major idea that set him off ("the happiest idea in my life," he later recalled; quoted in Pais 1982, 178) occurred to him in November 1907, in the Patent Office in Berne, where he was employed. The idea was: "If a person falls freely he will not feel his own weight." This "simple thought," he said, impelled him toward a theory of gravitation. But it was not until 1915 that he was ready to present his general theory of relativity in a fully developed form; a year later he published what one of his biographers has called "the Authorized Version." This theory was built around what Infeld (1950, 55) denotes as "three themes": gravitation, invariance, and the relation between geometry and physics. Central to the theory were the new field laws for gravitation and the equations of the gravitational field. It has been said that Einstein did for the gravitational field what Maxwell had done for the electromagnetic field. One of the intellectually spectacular features of general relativity was to reduce Newtonian gravitational forces to aspects of the curvature of four-dimensional space-time. Writing in the *Encyclopaedia Britannica* in 1922 (12th ed., s.v. 'relativity'), James H. Jeans concluded that the new background to "the picture of the universe" is no longer "the varying agitation of a sea of aether in a three-dimensional space" but "a tangle of world lines in a four-dimensional space."

The new theory led to three testable predictions. The first was the advance of the perihelion of the planet Mercury, the phenomenon by which in its orbital motion the planet does not return to exactly the same point in space closest to the sun in every revolution but moves a little ahead. This fact had been known since the middle of the nineteenth century, but classical Newtonian celestial mechanics predicted the wrong amount of this advance. The second prediction was that light should be bent in a gravitational field. Hence, starlight passing near the body of the sun should be affected by the sun's gravity, with the result that the apparent position of such stars would be shifted. The only time when this phenomenon could be readily observed was during a total eclipse of the sun, since otherwise the bright light of the sun normally makes it impossible to see stars near the sun. (The elaboration of this phenomenon in its full quantitative details was made by the Swiss astronomer Martin Schwarzchild.) The third prediction was what is known as a 'red shift' of spectral lines that are part of the stellar radiation that passes close to the sun. Here were three tests that could be made of the theory. But it must be remembered that the year was 1915; the Great War occupied the leading scientific nations of the world. Einstein was in Berlin and no eclipse expeditions were possible.

But Einstein did not stop work. In 1917 he published a paper in the *Proceedings of the Prussian Academy* with the title "Cosmological Considerations in General Relativity." Though its conclusions are now rejected, this paper opened up a new subject in theoretical physical science. By showing that "general relativity can throw light" upon the "problem . . . of the structure of our universe," it began the study of scientific cosmology; it transformed cosmology from a branch of metaphysics to a part of physics and physical astronomy (Infeld 1950, 72; on Einstein and cosmology, see Pais 1982, §15*e*).

The English astronomer Arthur Eddington, who learned of Einstein's work during wartime (see chapter 25), became the chief agent for the dissemination and acceptance of Einstein's ideas. He was the author of the authoritative *Report on the Relativity Theory of Gravitation* (1918); of a technical work, *Mathematical Theory of Relativity* (1923); of two popular expositions, *Space, Time and Gravitation* (1920) and *The Nature of the Physical World* (1928; the Gifford Lectures for 1927); and of numerous other lectures, articles, and pamphlets. P. A. M. Dirac has recorded, as one example, how he first learned of relativity, while a student at Bristol University, by reading Eddington. Even more important, immediately after the cessation of hostilities, Eddington in 1919 organized a British eclipse expedition to test the predicted bending of light during a total eclipse of the sun. The confirmatory result electrified scientists and nonscientists alike all over the world.

It is difficult to recapture today the scientific excitement of 1919. Two expeditions had been mounted to test this prediction of Einstein's general relativity theory—one to Sobral in Brazil and the other (led by Eddington) to Principe Island, near the coast of what was then Spanish Guinea. In the autumn of 1919 the data had been analyzed and on 6 November 1919, at a joint meeting of the Royal Astronomical Society and the Royal Society, the Astronomer Royal announced that "light is deflected in accordance with Einstein's law of gravitation." This historic meeting is described in full in *The Observatory* (published by the

RAS) and in the *Proceedings of the Royal Society*. J. J. Thomson was in the chair and declared this to be "the most important result" in relation to "the theory of gravitation since Newton's day," the "highest achievement of human thought." On the next day, 7 November 1919, the usually staid London *Times* carried the headline "Revolution in Science," along with two subheads reading "New Theory of the Universe" and "Newtonian Ideas Overthrown." On November 8, the *Times* ran a second article on revolution, this one headed "The Revolution in Science": "Einstein v. Newton," "Views of Eminent Physicists." Readers were told that "the subject was a lively topic of conversation in the House of Commons," that Sir Joseph Larmor — a distinguished physicist, a Fellow of the Royal Society, a professor in Cambridge University which he represented as M.P. — had been "besieged by inquiries as to whether Newton had been cast down and Cambridge 'done in.'" The news was quickly printed in the Dutch newspapers. H. A. Lorentz published an article in a Rotterdam newspaper on 19 November, which was at once translated and printed in the *New York Times*. Max Born had an article in the *Frankfurter Allgemeine Zeitung* on 23 November, and on 14 December Einstein's picture appeared on the cover of the weekly *Berliner Illustrierte Zeitung*, with a caption which declared not only that Einstein has made "a complete revolution in our concepts of nature" but that he had produced insights comparable to those of Copernicus, Kepler, and Newton (Pais 1982, 308). On 4 December in *Nature,* E. Cunningham wrote that Einstein's "thought was revolutionary."

Abraham Pais (p. 309) has traced the headlines and stories in the *New York Times,* beginning with the issue for 9 November 1919. "Einstein's Theory Triumphs" was accompanied by "A Book for 12 Wise Men" (referring to Einstein's alleged comment that "No More In All The World Could Comprehend It"). There were not only stories but editorials, continuing in a steady stream well into December. From then on until Einstein died, Pais found, no year passed without the *New York Times* running at least one story about Einstein or having some reference to him. He had become a legend. When Einstein came to London in 1921, he was introduced at a lecture in King's College by Lord Haldane, at whose house he stayed. When Einstein entered Haldane's house and Haldane's daughter caught sight of the famous visitor, she "fainted from excitement" (Pais 1982, 312). In introducing Einstein to the lecture audience at King's, Haldane reported that before giving the lecture, Einstein had gone to Westminster Abbey "to gaze on the tomb of Newton."

From that day to ours, relativity—both the special and the general theory—and 'revolution' have been closely coupled in the writings of scientists and nonscientists, of historians and philosophers. In 1921, Haldane stated the theme of his book *The Reign of Relativity* as (ch. 4) "the revolution in our physical conception made by Einstein." For the philosopher Karl Popper (Whitrow 1967, 25), Einstein "revolutionized physics."[7] The physicists Max Born (1965, 2) and Silvio Bergia (French 1979, 82) have referred respectively to Einstein's "revolutionary notions of space and time" and "the Einsteinian revolution." Born (1965, 2) said that the "special theory of relativity of 1905" was an event that both signaled "the end of the classical period" of physics and "the beginning of a new era." Steven Weinberg (1979, 22) holds that Einstein's "most remarkable achievement" was that "he for the first time made space and time a part of physics and not of metaphysics." According to the mathematician Emile Borel (1960, 3), Einstein has "given us not only a new theory of physics but has also taught us a new manner in which to look at the world." As a result, "it will be impossible for those who have studied his theories to think as they would have thought had they not studied them." The Spanish philosopher José Ortega y Gasset (1961, 135) did not use the word 'revolution' explicitly, but he boldly declared that Einstein's "theory of relativity" is "the most important intellectual fact" of "the present time." Thus Einstein's two theories of relativity inaugurated a revolution in philosophy along with a revolution in physics.

There is no lack of evidence that general relativity, even more than special relativity, fulfills all the tests set forth in chapter 3 for a revolution. The general theory, however, has had a very different history from that of special relativity. For a long time the general theory was of interest to astronomers (and only astronomers concerned with cosmology or cosmogony) and not to physicists. Steven Weinberg (1981, 20) has pointed out that "all modern theories of physics that deal with matter at the most fundamental level" are based "ultimately on two great pillars," of which one is "special relativity," the other "quantum mechanics." But in reviewing the activities of physicists during the 1920s and 30s, Segrè (1976, 93) especially makes the point that "general relativity, as opposed to special relativity, was not at the forefront of physicists' interests." That is, the theory was not essential, as special relativity had become, to theories of matter and of radiation which were then the primary topics of research. For example, when I was doing graduate study in physics in the late 1930s, the special theory entered all courses — even elementary and intermediate ones—on atomic and nuclear physics and on quantum theory, but only a select few mathematicians (stimulated by G. D. Birkhoff) were concerned with general relativity. Furthermore, general relativity implied the essential falsity or inadequacy of one of the most successful theories physics had ever produced — the Newtonian theory of gravitation — and the introduction of a strange concept of curvature of four-dimensional space-time to account for gravitational forces. It must also be understood that the great eclipse test of 1919 was only a qualitative demonstration that light is affected by a gravitational field. Only later were such eclipse tests made more precise. But decades would pass before new modes of verification other than the three originally proposed by Einstein would be devised; only "forty years after Einstein developed the theory" (Weinberg 1981, 21) were new types of precise experiments conceived and carried out to confirm the general theory of relativity.

In the decades following World War II matters have changed considerably. Precise confirmatory tests in the laboratory have become possible. There has been a renewed interest in the nature of gravitation and in the relation of gravitational forces to the other forces of nature — electromagnetic forces and the weak and strong nuclear forces. A huge physics and astronomy 'industry' centering on general relativity and its applications to cosmology and cosmogony, as well as to other parts of physics, has grown up. One result, as Steven Weinberg suspects, may be that in order to "understand gravitation at extremely short distances," there may be required "another great leap to universal principles" (p. 24), another revolution of which we at present can have no idea. In short, today general relativity is an exciting subject for scientists — perhaps more than ever before.

The Beginnings of Quantum Theory: Planck and Einstein

The quantum theory differs from relativity in a number of major respects. Almost everyone has heard of relativity and its creator Albert Einstein, but only scientists and a small number of nonscientists (those who either are interested in science or have studied science) are aware of quantum theory. Contrariwise, almost everyone concerned with any aspect of the physical sciences — not just physicists but chemists, astronomers, biochemists, molecular biologists, metallurgists, and others — regularly make use of quantum theory or its results in their work, but much less frequently general relativity. In addition to permeating these areas of science, quantum theory has even — like relativity — produced a fundamental change in our scientific thinking and philosophy of science. The revolutionary quality of both relativity and quantum theory were recognized very early, but both long remained only revolutions on paper.

The quantum theory has had three major phases: the old quantum theory (Planck, Einstein, Bohr, Sommerfeld, Compton), quantum mechanics (de Broglie, Schrödinger, Heisenberg, Jordan, Born), and the newer relativistic quantum mechanics or quantum field theory. Each of the first two has been recognized as a revolution. In fact, physicists have difficulty in finding strong enough language to express the magnitude and profundity of the quantum revolution. For Victor Weisskopf (1973, 441), the "discovery of the quantum of action by Max Planck" was "the beginning of one of the most fruitful and also most revolutionary developments in science." Only rarely in history, he added, "have our views regarding the basis of the properties and behavior of matter been changed and expanded as profoundly as in the three decades following Planck's discovery." Paul Davies (1980, 9) writes of "the revolution that has taken place in science and philosophy since the inception of the quantum theory of matter at the beginning of the century." He finds it "remarkable that the greatest scientific revolution of all time has gone largely unnoticed by the general public," since he believes that "its implications" are so "shattering as to be almost beyond belief — even to the scientific revolutionaries themselves" (p. 11).

The date of the beginning of quantum theory is usually set at 1900, when Max Planck published his idea of a 'quantum of action'. Planck was not concerned, as Einstein would be five years later, with the quantum aspect of light or radiation itself or with the actual process by which matter and radiation are coupled in the production (or annihilation) of such light. Rather, he was solely concerned with the exchange of energy between the oscillating electrons in a container's walls and the enclosed radiation, which he found to occur in bursts, corresponding to an energy hv, where h is a new universal constant of nature introduced by Planck (Planck's constant), and v is the frequency of the exchanged energy. As Kuhn has shown, in 1900 Planck supposed only that the collection of oscillators (that is, of material bodies; not the ether vibrations) capable of vibrating at frequency v would have as a group a total energy proportional to their frequency v. This is a very limited assumption compared to the later concept of a quantum of light as an individual entity with the defining property of having an energy hv.

We may well understand why Planck did not even contemplate the more radical step of assuming that light 'consists' of discrete parcels or packets of energy. First, such an assumption was not necessary to his presentation. Second, and of even more importance, this idea would have run counter to one of the best established notions of physics — that light (and, following Maxwell and Hertz, all electromagnetic radiation) is a wave phenomenon, spreading out through space in a continuous undulation in which the concept of discrete units is inconceivably foreign. Indeed, when Einstein set forth his notion of light quanta five years later, a conceptual difficulty arose from the fact that the quantity of energy in a light-quantum is determined by the frequency of the light and measured by the wavelength — using the same techniques of 'interference' that had been responsible for the experimental base of the wave theory many decades earlier.

Planck later referred to his bold formulation of ideas as "an act of desperation" (Pais 1982, 370). His reasoning, according to Pais, "was mad," but it was the kind of "madness" that has a "divine quality" that "only the greatest transitional figures can bring to science." It led him to make "the first conceptual break" that has differentiated the physics of our century from all preceding ages, and it cast a deeply conservative thinker "into the role of a reluctant revolutionary."[8] Although Planck is often portrayed as one who was forced into the radical step of inaugurating quantum theory against his will, in many statements Planck shows that he truly appreciated the revolutionary character of his own work and of Einstein's. His boundless enthusiasm for Einstein's theory of relativity is manifest in his statement (quoted in Holton 1981, 14) that this "new way of thinking . . . well surpasses in daring everything that has been achieved in speculative scientific research, even in the theory of knowledge." For Planck, the "revolution in the physical *Weltanschauung* brought about by the relativity principle" could be "compared in scope and depth only with that caused by the introduction of the Copernican system." In the address which he gave when he received the Nobel Prize, Planck said, "Either the quantum of action is only a fictitious magnitude, and the whole deduction of the law of radiation more or less of an illusion, little more than a game played with formulae, or the deduction of the law is based upon a true physical idea." He then explained that if it is the latter, then the quantum of action "would have to play a fundamental role in physics." The reason is that the quantum "heralds the advent of a new state of things, destined, perhaps, to transform completely our physical concepts which — since the introduction of the infinitesimal calculus by Leibniz and Newton—have been founded on the assumption of the continuity of all causal chains of events." In this statement, the cautious Planck did not explicitly refer to his own discovery in terms of revolution. Planck's role as the inaugurator of a wholly new kind of physics was recognized by Einstein, who recommended him for the Nobel Prize in 1918 on the grounds that he had "laid the foundation of the quantum theory, the fertility of which for all of physics has become manifest in recent years" (Pais 1982, 371).

The state of intellectual limbo between 1900 and 1905 has been described for us by Max Born in his obituary notice of Planck for the Royal Society. Born had "no doubt" that Planck's "discovery of the quantum of action" was an event "comparable with the scientific revolutions brought about by Galileo and Newton, Faraday and Maxwell." He had earlier written that the "quantum theory dates from the year 1900, when Max Planck announced his revolutionary concept of energy atoms, or 'quanta'" (1962, 1). This event "was so decisive for the development of science," he declared, that it is "usually considered as the dividing point between *classical physics* and *modern* or *quantum physics.*" But Born (1948, 169, 171) warns us not to accept too readily the "generally acknowledged" opinion that "the year 1900 of Planck's discovery marks indeed the beginning of a new epoch in physics," since "during the first years of the new century very little happened." Born relates, "[This] was the time of my own student days, and I remember that Planck's idea was hardly mentioned in our lectures, and if so as a kind of preliminary 'working hypothesis' which ought of course to be eliminated." Born stresses the significance of Einstein's two papers (1905; 1907). But, although Born remarks that after 1900 "Planck himself turned to other fields of work," he "never forgot his quanta." This is shown by Planck's treatise on heat radiation of 1906, which "made a profound impression by the masterly presentation of the successive steps which led to the quantum hypothesis" (Born 1948, 171).

It is a measure of Einstein's greatness that at the time when he was inaugurating the revolution in relativity, he was also making fundamental contributions to quantum theory. He first mentioned quantum theory in a statistical paper of 1904, a topic to which he returned in 1906, when he founded what is now called the "quantum theory of the solid state." But chiefly it was his paper of March 1905 which marks the transformation of Planck's potentially revolutionary idea into a truly revolutionary one, although as yet only a revolution on paper.[9] This 1905 publication contains two fundamental postulates. One is that in its diffusion through space, light or 'pure' radiation is to be conceived as composed of separate and individual parcels or packets of energy (quanta); the other that the emission or absorption of light (or any form of electromagnetic radiation) by matter occurs in the same discrete quanta. These postulates not only went so far beyond Planck's hypothesis of 1900 as to constitute a radical transformation, but they also contravened the fundamentals of the received physics of that time. Pais (ibid.) finds this work to have been "Einstein's one revolutionary contribution to physics"; it "upset all existing ideas about the interaction between light and matter." It is the single one of Einstein's discoveries that we have seen Einstein himself to have described as "revolutionary."

Einstein's paper of March 1905 was entitled "Über einen die Erzeugung und Verwandlung des Lichtes betreffenden heuristischen Gesichtspunkt" ("On a Heuristic Viewpoint Concerning the Production and Transformation of Light"). The word 'heuristic' had not been commonly used in physics — it was a term from philosophy and education that meant something useful for discovering or explaining, but not necessarily to be considered as true. Einstein would use it again in a paper on relativity in 1907 and in his "popular exposition" of *Relativity, the Special and General Theory* (1917, ch. 14; Eng. trans. 1920). The reason that Einstein introduced this term is that he was proposing a particle-like concept that simply could not, in the form he gave it, account for most of the known phenomena of light. One of the greatest achievements of nineteenth-century physics had been the establishment of the wave theory of light and its validation by interference experiments. Einstein (Klein 1975, 118) was proposing nothing less than that "physicists consider abandoning the electromagnetic wave theory of light," which was "the greatest triumph of Maxwell's field theory and of all nineteenth-century physics." Therefore, Einstein was proposing a provisional hypothesis.

A wave is characterized by the fundamental properties of speed, wavelength, and frequency. In Einstein's concept of particle-like quanta of energy hv, the frequency v was derived from the wave equation, using the wavelength determined by techniques of 'interference'. But in the concept of the light quantum, the frequency — which is of such basic significance in the wave theory — could have no obvious physical meaning for a particle or a quantum of light. The contradiction between the continuous or wave properties and the discrete or particulate properties was so obvious that Einstein had to introduce into his paper the phrase "if our concept corresponds to reality." Planck continued to believe, as all other scientists did for a long time, that light and all other forms of electromagnetic radiation are composed of waves and so are continuously divisible: the discrete energy-units or quanta are only effects of the interaction of the continuous waves with matter, as in the absorption and emission of light, and are not features of the waves of light as such. According to Einstein's assumption of 1905, it is the light itself that is composed of discrete units or quanta: that is, light (and all forms of electromagnetic radiation) must have a corpuscular structure. The quanta, in Einstein's concept, are characteristic of the light itself and not merely of the process of interaction between light and matter. Although scientists and historians generally refer nowadays to 'Einstein's theory of photons', the concept of photon embraces the additional property of momentum, introduced long after 1905. Furthermore, until Einstein's dying day he insisted that this was — as he said in an interview about a week before he died — "not a theory," because it does not provide a complete account of optical phenomena.

Although Einstein's paper was hypothetical, heuristic, partial, and theoretical, it did contain one part which was significant in that it led to a direct experiment. This was Einstein's discussion of the photoelectric effect, discovered by Hertz in 1887, and of which many properties had been found by Philip Lenard in 1902. In this phenomenon, light incident on a metal surface causes an electron to be emitted or ejected. Experiments had shown that for an electron to be emitted, a certain minimal frequency is required in the incident light, a 'threshold' frequency that proves to be a characteristic of the particular metal. Assuming that the light is composed of discrete quanta, Einstein proposed "the simplest process conceivable," that one "quantum of light gives up all its energy to a single electron." If the light (or radiation) is monochromatic, of frequency v, each quantum has energy hv. This energy does two jobs: it does 'work' (P) against the forces that bind the electron to the metal, and it gives the electron the kinetic energy (E) with which it leaves the metal. In other words,

$$E + P = hv$$

or

$$E = hv - P.$$

Einstein's equation explained the laws of this effect. One law is that the kinetic energy E of the emitted electrons does not depend on the brightness or intensity of the light but only on the frequency. (Einstein's explanation is that the intensity is a measure of the number of photons, and hence the number of electrons emitted, not their energy.) The equation also shows how the energy E of the emitted electron is related to the frequency v of the incident light. Another law is that each metal requires a certain minimum frequency for the emission of photoelectrons. Einstein's equation explains that the photoelectric effect will occur only if v is large enough for hv to be greater than P.

Einstein's equation predicted that E should vary directly as v; that if the straight line graph of E and v were to be plotted, the slope would be Planck's constant h. Before long, experiments were under way, chiefly by one of J. J. Thomson's students, A. L. Hughes, which indicated the correctness of Einstein's equation. But the real test came from experiments made by R. A. Millikan; these not only perfectly confirmed Einstein's equation but also produced a new exact value for h (see Wheaton 1983).

Millikan's paper (1916) on these experiments is a strange one. Despite his statement that "in every case" Einstein's "photoelectric equation" appeared "to predict exactly the observed results" of the experiments, Millikan went on to state that the "semicorpuscular theory" used by Einstein to derive the equation "seems at present wholly untenable." Within a year he restated his position that Einstein's "hypothesis of an electromagnetic light corpuscle" was "bold" and actually "reckless." In his book on *The Electron* (1917), Millikan said that Einstein's equation was a "prediction as bold as the hypothesis which suggested it," but that there was no "basis" for Einstein's radical hypothesis. How astonishing, therefore, Millikan reported, that "this equation of Einstein's" could be "found to predict accurately all the facts which had been observed experimentally" by Millikan and others! In his book, Millikan, ever the foe of revolutions, did not tell his readers that in fact he had undertaken these experiments in order to prove the falsity of Einstein's equation and, by implication, the concepts on which it was based.[10] In 1949 he confessed that he had spent ten years of his life "testing that 1905 equation of Einstein's." Contrary "to all my expectations," he wrote, "I was compelled in 1915 to assert its unambiguous experimental verification in spite of its unreasonableness."

Millikan (1948, 344) clearly expressed the basis for his objection to Einstein's light quanta: They "seemed to violate everything that we knew about the interference of light," the experimental basis of the wave theory. In 1911 Einstein himself felt he must publicly "insist on the provisional character of this concept of light-quanta," because it "does not seem reconcilable with the experimentally verified consequences of the wave theory" (Pais 1982, 383). Pais has found that Einstein's caution was "almost hopefully mistaken for vacillation," a factor which explains why Einstein's admirer von Laue could write to Einstein in 1907 that he was pleased to learn that Einstein had "given up" his "light-quantum theory." He was not alone. In 1912 Sommerfeld said that Einstein no longer maintained "his original point of view [of 1905] in all its audacity," while Millikan said in 1913 that Einstein "gave . . . up, I believe," his idea of light quanta "some two years ago." In 1916 Millikan once again proclaimed that despite the experimental verification of Einstein's equation, the "physical theory" on which it was founded had proved "so untenable that Einstein himself, I believe, no longer holds to it." But Pais, who has had access to Einstein's papers and letters, finds "no evidence that he at any time withdrew any of his statements made in 1905." Roger Stuewer (1975, 75–77) gives convincing evidence that Einstein did not ever waver in his commitment to his light-quantum hypothesis, but actually "achieved deeper insights into it."

As late as 1918, Rutherford (quoted in Pais 1982, 386) argued that there "is at present no physical explanation possible" of the "remarkable connection between energy and frequency." Pais concludes his study of this episode by remarking that "even after Einstein's photoelectric law was accepted, almost no one but Einstein himself would have anything to do with light-quanta." As proof, Pais quotes the citation for Einstein's Nobel Prize in 1922, not for relativity, not for a quantum theory of light, but "for his services to theoretical physics and especially for his discovery of the law of the photo-electric effect." We can only conclude that Einstein's revolutionary innovation was still a mere revolution on paper, with practically no adherents.[11]

The example of Millikan's attempt to disprove Einstein's new idea should not be taken to imply the existence of a general opposition by physicists to Einstein's heuristic proposal; Einstein's theoretical suggestion tended to be ignored more than actively combatted. Millikan, a truly great physicist, was an exception to the rule. The general verdict of the physics community was expressed in 1913 in the formal document proposing that Einstein be elected a member of the Prussian Academy of Sciences. Signed by four great scientists and admirers of Einstein — Max Planck, Walther Nernst, Heinrich Rubens, and Emil Warburg — this document (published in Kahan 1962; cf. Pais 1982, 382) praises Einstein for his solid achievement, going so far as to say that "there is hardly one among the great problems in which modern physics is so rich to which Einstein has not made a remarkable contribution." Then, they felt the need to excuse Einstein for "sometimes . . . [having] missed the target in his speculations," as was the case "in his hypothesis of light-quanta." In offering an excuse for this lapse, they added that "it is not possible to introduce really new ideas even in the most exact sciences without sometimes taking a risk." Even a Homer may occasionally nod!

Quantum Theory and Spectra: Bohr's Model of the Atom

The line of development we have been following was not the only one for quantum theory. In 1912 a young Dane working in Rutherford's laboratory in Manchester proposed a revolutionary new model of the atom. Niels Bohr began with Rutherford's concept of the atom as a miniature solar system, with a central nucleus encircled by 'orbiting planetary' electrons. The revolutionary aspect of Bohr's model lay in the way in which he proposed that the "atom-model" could account for the emission and absorption of certain frequencies of light and no others. He used "Planck's theory of radiation," that there are "distinctly separated emissions" of energy v. He then said that "the general importance of Planck's theory for the discussion of the behaviour of atomic systems was originally pointed out by Einstein," and developed by other physicists. As is well known, Bohr postulated that in stable orbits the electron neither emits nor absorbs energy, but when it spontaneously 'jumps' from any such stable orbit to another orbit of lower energy, it emits a single quantum of light. Conversely, when an electron absorbs a single quantum, it 'jumps' to an orbit of higher energy. Bohr showed that on this basis he could derive several of the known laws of spectroscopy. This was the origin of the famous and revolutionary 'old' quantum theory.

It is difficult to judge how revolutionary Bohr initially considered his theory to be. He certainly tried, during the years from 1913 to 1924, to include as much of classical ideas as he could, to give his theory the appearance of belonging to the great tradition. Yet he also called his initial concept only a 'model' of the atom, in a manner that is reminiscent of Einstein's use of the qualifier 'heuristic' in his 1905 paper on the light quantum. By the early twenties there could have been little doubt that Bohr's theory was revolutionary, a fact recognized by numerous physicists. Its later developments included the extension from a one-electron atom (hydrogen) to an atom with two electrons (helium), and also the introduction of elliptical orbits. Many physicists contributed to this great theory, the major one — other than Bohr himself — having been Arnold Sommerfeld. Like all revolutionary ideas in science, Bohr's quantum theory did not gain immediate acceptance by the scientific community, despite the close numerical agreement he achieved with laws found by experiment. Perhaps the delay here was caused less by the revolutionary nature of Bohr's atom model and quantum theory of spectra than by the fact of the First World War. In the years following the war, however, almost every major physicist interested in fundamental issues participated in the development of quantum theory.

Bohr's theory is tied in with Einstein's because both supposed a one-to-one interaction between an electron and a light quantum. In the presentation of the photoelectric effect, Einstein had considered the case where the quantum would have enough energy to cause the electron that absorbed it to leave matter, but in Bohr's theory this was the extreme case (ionization); when the quantum energy was less, the electron would not leave the atom but only 'jump' to a higher orbit. What made Bohr's theory difficult to believe in was the idea of discrete and fixed states or orbits, with no intermediate states being possible. Furthermore, like Einstein he had supposed a direct contradiction of a fundamental principle of Maxwellian physics: that a charged body (that is, an electron) moving in an electric field (the field surrounding the positively charged nucleus) must radiate. According to all accepted principles, such an orbiting electron would constantly lose energy by radiation and hence would have to move in ever decreasing orbits until it would eventually fall into the nucleus. But Bohr postulated that an electron could move in a stable orbit around the nucleus without emitting or radiating energy. Here was a major obstacle to the acceptance of the theory. Max von Laue, for one, based his doubts about the Bohr theory on this flat rejection of Maxwellian physics.

Those who learned their physics, as I did, in the 1930s will recall that it was a feature of courses in quantum theory (and in many of the textbooks) to introduce the subject by a historical review, in which the students were taken step-by-step through the failures of the old theory of radiation (including the equipartition of energy) and the stages of development of quantum theory (by Planck and Einstein). Then, there was a discussion of the laws of spectroscopy and their explanation by Bohr's original theory, followed by its subsequent extension to elliptical orbits by Sommerfeld. Stress was given to such historically significant experiments as those of Millikan and of Frank and Hertz. Finally, the students would learn of the spin of the electron, quantum numbers, and the great Pauli exclusion principle. This historical rehearsal of the reasons why the quantum became accepted shows, in retrospect, that the professors and the authors of textbooks felt the need to take the students through all the experiences of a previous generation which had to be convinced about a radically new and imperfect foundation of physics. Here is an index to the revolutionary quality of quantum theory.

A careful examination of Bohr's writings from 1913 to 1923 shows that although he introduced Planck's constant and referred to Einstein's work on the photoelectric effect, he did not come out strongly for the quantum theory of light. That is, he was primarily concerned not with the nature of light itself nor with its transmission but with its absorption and emission when an electron changes its orbital (and therefore energy) state. In his original paper (1913) Bohr admitted that he had introduced a "quantity foreign to the laws of classical electrodynamics, i.e., Planck's constant" (see Heilbron and Kuhn 1969; Miller 1984). His theory seems in retrospect a strange combination of classical mechanics with quantization used in determining the stationary states plus discontinuities. Bohr (1963, 9) was aware that his "atomic model" was imperfect, and in its first form incomplete, since its "considerations conflict with the admirably coherent set of conceptions which have been rightly termed the classical theory of electrodynamics." In those days, 1910–1913, as Martin Klein has found, "such men as Max Planck and H. A. Lorentz had only sharply critical things to say about Einstein's light quanta" on the grounds that these quanta failed utterly to account for interference and diffraction (1970, 6). Bohr himself, in his inaugural presentation of 1913, wrote about the "*homogeneous* radiation" emitted from an atom and not light quanta (see Miller 1984). From 1913 to about 1920, Bohr tried to reconcile the classical wave theory of light with the atomic processes, eventually producing what he called the "correspondence principle." But Arnold Sommerfeld in 1922, in his influential treatise on *Atomic Structure and Spectral Lines,* could only hold it "astonishing" that so "much of the wave theory still remains, even in spectroscopic processes of a decidedly quantum character" (p. 254), and he concluded that "modern physics" is "for the present confronted with irreconcilable contradictions" (p. 56). Bohr himself even proposed the renouncement of what he called the "so-called hypothesis of light quanta" (see Miller 1984). The discussions of this era show not only the turmoil that arose in attempting to produce a satisfying quantum theory of spectra in relation to an atom model, but indicate the difficulties in combining a revolutionary new set of concepts with traditional physics. Sommerfeld (1922, 254) believed that modern physics simply must acknowledge the contradictions between the old and the new and "frankly confess its non-liquet," a point of view that elicited an enthusiastic approval from Wolfgang Pauli.

Bohr's theory passes all the tests for revolutions in science. For example, in a letter published in *Die Naturwissenschaften* in 1929 (p. 483), Rutherford said that "Professor Bohr's bold application of the quantum theory to explain the origin of spectra" constituted a revolution. Bohr's theory, he said, was "a direct development of Planck's hypothesis which has had such revolutionary consequences in Physics" and is even now "in process of revolutionizing our methods of thought and concepts of philosophy." In 1969, Sir John Cockcroft wrote of Bohr's "combination of classical mechanics with the quantum theory to describe motion in the electronic orbit" as a development which "revolutionized atomic theory." But like the Cartesian revolution, Bohr's revolution did not last. As with Descartes's work, some major elements of Bohr's theory were incorporated into the next revolution, quantum mechanics, of which it can be considered a first phase.

Toward Quantum Mechanics: The Great Quantum Revolution

In 1926 Einstein's light quantum was given the name 'photon'. This name was introduced by the physical chemist G. N. Lewis for a somewhat different concept; though the concept was rejected, the name rapidly became part of the standard vocabulary of physics (see Stuewer 1975, 325). The photon of the mid-1920s, however, differs from Einstein's original light quantum in having certain particle properties that Einstein himself had not at first taken into consideration, of which the chief one is momentum. But he did introduce momentum ($p = h\nu/c$) by 1916; this concept had actually appeared even earlier in a paper of 1909 by Johannes Stark (see Pais 1982, 409). The idea that a photon may have momentum was developed by P. Debye and Arthur H. Compton in 1923. Compton in fact made one of the most remarkable discoveries in modern physics, eponymously known today as the Compton effect. Compton proved, by unassailable experimental evidence, that "a radiation quantum carries with it directed momentum as well as energy"

(Stuewer 1975, 232). Roger Stuewer, tracing the history of this work, reveals that Compton's motivation — unlike Millikan's of a decade earlier — was not to test Einstein's predictions. Stuewer (p. 249) also found that the first use of the name 'Compton effect' occurred in a letter of congratulations written to Compton from Munich on 9 October 1923 by Arnold Sommerfeld, who reported that Compton's results had been the "chief topic of discussion" between him and Einstein in the previous August.

Although there was, at first, some dispute about Compton's results (for which see Stuewer 1975, ch. 6), it was soon recognized (for example, by Heisenberg) that the Compton effect was a turning point not only in the quantum theory of radiation but in the whole of physics (p. 287). Compton himself very early recognized the revolutionary character of his work. In a lecture to the American Association for the Advancement of Science in 1923, printed in 1924 in the *Journal of The Franklin Institute,* he declared straightforwardly that his discovery presented "to us a revolutionary change in our ideas regarding the process of scattering of electromagnetic waves." Yet he also wrote in 1923 in the *Proceedings of the National Academy of Sciences* (9: 350–362) that "the present quantum conception of diffraction is far from being in conflict" with classical wave theory. Einstein saw his own idea at last vindicated and announced that there were now two different theories of light, wave and corpuscular, "both indispensable and — as one has to admit today in spite of twenty years of immense efforts by theoretical physicists — without any logical connection."

At about this same time, Louis de Broglie was encouraged by Compton's work to develop his own idea of matter waves. In a paper on this topic in 1923, he cited "the recent results of A. H. Compton," along with the photoelectric effect and Bohr's theory, as sources of his conviction that the dualism of wave and particle that marked the Einsteinian conception of the light quantum was "absolutely general"; the work of Einstein, Bohr, and Compton reinforced his acceptance of the "actual reality of light quanta."

De Broglie did not account physically for the duality of light, but he argued that this duality was a general aspect of nature, that even ordinary matter (an electron, for example) should exhibit wave as well as particle properties. This revolutionary idea was developed by de Broglie in his doctoral dissertation at the Sorbonne (submitted 25 November 1924), and carried further by Einstein, who was responsible for bringing it to the attention of Schrödinger (see Wheaton 1983). De Broglie's concept led to experimental confirmation by Davisson and Germer in America and by G. P. Thomson (J. J.'s son) in England. More importantly, it was the prelude to the new quantum mechanics, which we associate in the first instance with Schrödinger and Heisenberg (see Klein 1964; Jammer 1966; Raman and Forman 1969; Stuewer 1975; and, for a dissenting view, Miller 1984). The consequences of this new revolution, notably after the introduction of probability considerations by Max Born, are at the very core of the physics and the philosophy of nature of the last half-century.

It is a well known fact of record that, beginning in the 1920s, Einstein broke with the whole of the physics community by refusing to accept quantum mechanics as more than a 'provisional' account of nature. What he objected to was the absence of classical causality and determinism in the new physics, the introduction of probability as the foundation for physical events, and the consequent (as it seemed to him) incomplete description of nature. Nevertheless, he recognized that, however provisional, quantum mechanics was a great step forward in physics, and he recommended to the Nobel Prize committee that its co-inventors Schrödinger and Heisenberg be awarded that coveted prize (see Pais 1982, 515). And paradoxically, Einstein himself had made an important contribution to the statistical basis of quantum mechanics.[12]

The history of the quantum mechanics revolution, or the second quantum revolution, and its rapid successive transformations from revolution in itself to revolution on paper and to revolution in science, could easily have become the subject of a chapter in this book. The revolutionary implications of quantum mechanics for physics are manifest in the history of that subject during the past half-century, and the importance of these developments for science and for thought in general are apparent in almost any work on philosophy of science in recent decades (see, notably, Born 1949; Davies 1980; Feynman 1965; Jammer 1974; and Suppe 1977).

The "Last Stand" of the Original Quantum Theory

We may conclude this presentation with a curious episode that indicates the truly revolutionary character of Einstein's concept of light quanta. In 1924 Bohr (together with H. A. Kramers and J. C. Slater) published an article that essentially rejected the concept soon to be named the photon. The time was a year after the first announcement of the Compton effect. Bohr had used the concept of quanta in his theory of the atom, which had won universal acceptance and had revolutionized that branch of physics — although there were many thorny problems which would not be enodated until the advent of quantum mechanics a few years later.[13] But Bohr's theory, like Planck's original quantum theory, did not concern itself with 'the free radiation field', that is, with the quantization of light or of electromagnetic radiation in space. Like most physicists who adopted the quantum theory during the two decades after Einstein's paper of 1905, Bohr accepted the quantization of emission and absorption but not of light itself. We must not forget that many experiments (from interference and diffraction phenomena) gave what seemed to be unassailable evidence for the continuous wave transmission of light.

The Bohr – Kramers – Slater proposal was a last step in Bohr's basic denial of the quantum theory as a general description of light. His own 'correspondence principle', he believed, could bridge the gap between the quantum theory of emission and absorption and the accepted electromagnetic wave theory of transmission. In 1919 and in succeeding years he even professed a willingness to take the very extreme step of abandoning the principle of conservation of energy if that were necessary in order to preserve "our ordinary ideas of radiation" (see Stuewer 1975, 222). This topic arose in his Nobel Prize lecture on 11 December 1922, when he explained that "the predictions of Einstein's theory have received . . . exact experimental confirmation in recent years." Nevertheless, he was quick to add, "in spite of its heuristic value," Einstein's "hypothesis of light-quanta" is "quite irreconcilable with so-called interference phenomena" and "is not able to throw light on the nature of radiation." This theme dominates the Bohr – Kramers – Slater paper of 1924, the aim of which is stated as follows: the cause of the properties of radiation is not being sought "in any departure from the electrodynamic theory of light as regards the laws of propagation in free space," but rather "in the peculiarities of the interaction between the virtual field of radiation and the illuminated atoms." In this paper the authors declare that in processes on the level of single atoms, they would "abandon . . . a direct application of the principles of conservation of energy and momentum," which would be laws only for an average taken over a large aggregate, but not laws for individual atoms. Two years earlier Sommerfeld had said that to abandon the conservation of energy would be the "mildest cure" for the disease of the duality of a wave and quantum theory of light (Pais 1982, 419). A few years later, reviewing the history of those days, Heisenberg (1929) described the Bohr – Kramers – Slater theory as "representing the height of the crisis in the old quantum theory" (Pais 1982, 419); it was, according to Pais, the "last stand of the old quantum theory."[14]

In a letter to B. L. van der Waerden, Slater later wrote that the "idea of statistical conservation of energy and momentum" had been "put into the theory by Bohr and Kramers, quite against my better judgment" (Stuewer 1975, 292). Bohr and Kramers, Slater pointed out, were able to maintain "that no phenomenon at that time demanded the existence of corpuscles" of light (or quanta) in space. Slater "became persuaded that the simplicity of mechanism obtained by rejecting a corpuscular theory more than made up for the loss involved in discarding conservation of energy and rational causation."[15]

Negative reactions to this theory "abounded" (Stuewer 1975, §7B4). The true answer, however, did not come from discussions on the theoretical level but rather from direct experiments, illustrating what we have seen Huxley call "a beautiful hypothesis killed by an ugly fact." An experiment proved that the conservation laws actually are valid on the single-encounter level. The experiment used the technique of the Compton effect. The first results were obtained in Berlin by W. Bothe and H. Geiger, and then (even more conclusively) in Chicago by A. H. Compton, working with A. W. Simon. When Bohr heard the news on 21 April 1925, he wrote at once that "there is nothing else to do than to give our revolutionary efforts as honourable a funeral as possible" (Stuewer 1975, 301; Pais 1982, 421). In July of the same year, he once again wrote of revolution in an article in the *Zeitschrift für Physik*. "One must be prepared," he said, "for the fact that the required generalization of the classical electrodynamic theory demands a profound revolution in the concepts on which the description of nature has until now been founded."[16] This episode, and the discussion of it by Bohr, may show how advances in quantum theory led men to invoke the language of revolution.

28

Einstein

on Revolution

in Science

Although the relativity revolution has become the archetypical revolution in science for many historians, philosophers, sociologists, and scientists, Einstein himself argued that his intellectual creation should be considered as part of an evolutionary rather than a revolutionary development of physics. Einstein never wrote a major essay on the topic of revolution versus evolution, but he did express himself rather forcibly on a number of occasions.

In evaluating Einstein's stated views on revolution in science, we must keep in mind that before he gained worldwide notoriety, Einstein may have had different views about revolutions than he did later on. This would account for the fact that in March of 1905, in a letter to Conrad Habicht, he characterized his own idea about the light quantum as "very revolutionary [sehr revolutionär]" (Seelig 1954, 89), whereas in 1947 he came out strongly against the notion that science progresses by a steady stream of revolutions. The letter to Habicht is, so far as I know, the only occasion on which Einstein used the term 'revolutionary' to characterize any aspect of his own work or, for that matter, any part of the physics of his own century. Einstein's other recorded statements about revolutions in science were either parts of a letter or a speech or an essay on an aspect of his own science or on the achievement of a particular scientist, so that each opinion expressed must be interpreted in the particular context in which it is embedded. I have seen no evidence that Einstein ever devoted serious thought to the mode of progress of the sciences and worked up a true theory of the way the sciences develop. I should add that there is additionally the problem of language — Einstein's original German phrases and their translations.

A year after Einstein had published his "very revolutionary" concept of light, his theory of relativity, and his development of Brownian motion, he evidently was concerned about his doldrums and feared that perhaps he would never regain the creativity impulse that had produced those works. Was the great creative phase over? On 3 May 1906 he wrote to Maurice Solovine about his sadness in no longer producing new science of any consequence. "Soon," he wrote, "I will arrive at the stationary and sterile age where one laments over the revolutionary mentality of the young [revolutionäre Gesinnung der Jungen]" (Einstein 1956, 5; see Feuer 1971, 297; 1974). This statement is somewhat ambiguous, but I read it as, in part, suggesting that a young creative scientist tends to have a "revolutionary mentality" and thus implying that such a person is apt to produce a "very revolutionary" concept. I see no reason to suppose that in these two letters of 1905/1906 the word 'revolutionary' had any sense other than the common one then current among scientists. That is, Einstein was especially stressing that his concept of light quanta embodied a strong element of discontinuity, a revolutionary break in the progress of physical science.

The invocations of revolutionary science of 1905–1906 may be contrasted with Einstein's remarks in 1947. On 30 January the *New York Times* carried the news: "Einstein's Theory Reported Widened." At issue was Erwin Schrödinger's claim to have "solved a thirty-year-old problem: the competent generalization of Einstein's great theory of 1915." The *Times* reported that Schrödinger alleged that he had extended general relativity beyond the confines of gravitation to the domain of electromagnetic phenomena. This research, he was said to have immodestly declared, was "the kind of thing we scientists should be doing instead of creating atomic bombs." Einstein was approached by the *Times* to give a statement. In a report printed with the news story about Schrödinger's self-declared achievement, Einstein was quoted merely to the effect that he was unable to "make any comment at this time.""I have no first-hand information," Einstein said, and he had only "limited correspondence about scientific matters" with Schrödinger.

But although Einstein did not issue a public comment for the press, he did compose one. The text, in an English version, has been quoted by Martin Klein (1975, 113). "The reader," Einstein said, "gets the impression that every five minutes there is a revolution in science, somewhat like the *coups d' état* in some of the smaller unstable republics." Einstein believed that (according to Klein's summary) "overusing the term scientific revolution produced a false image of the process by which science develops." This "process of development," Einstein actually wrote, is one "in which the best brains of successive generations add by untiring labor," a process which "slowly leads to a deeper conception of the laws of nature." It should be observed that in such proclamations Einstein, though stressing the cumulative aspect of the progress of science, does not wholly rule out the occasional revolution.[1]

Klein observes that when Einstein did "occasionally refer to scientific revolutions," it was "only when the transformations involved were, so to speak, on the scale [for science] of the French or Russian Revolutions" (ibid.). We have seen that Einstein again and again referred to the Maxwellian revolution (or the revolution of Faraday, Maxwell, and Hertz). In his "Autobiographical Notes" (1949, 37), Einstein stated that "the revolution begun by the introduction of the [electromagnetic] field was by no means finished." Klein (1975, 118–119) has given a perceptive analysis of the way in which Einstein had not actually created the new "theory of light quanta," but had only proposed "a hypothesis" that could serve as "a heuristic guide in constructing the new theory that was needed." Klein also argues that in his proposal of the principle of relativity, Einstein "did not claim to have developed the new fundamental theory that was needed." Hence, in his 1907 paper (and again in his book of 1915), Einstein could legitimately argue that special relativity was no more than "a heuristic principle." It follows that for Einstein, relativity could not constitute a revolution.

Although Einstein applied the word 'revolutionary' to only one of his three great contributions of 1905, his fellow scientists, students, collaborators, and biographers have tended to agree with historians of science on the revolutionary quality of all three: special relativity, light quanta, and Brownian motion. The last of these is least well known to the general public, but it is revolutionary in its own right, since it deals in a radically new way with a fundamental problem of molecular motion, in the course of which Einstein set forth "the first serious treatment of statistical fluctuations ever given" (Klein 1975, 116). Developed simultaneously and independently by the Polish physicist Marian von Smoluchowki, this theory was considered revolutionary by "many of their contemporaries," especially after its experimental confirmation by Jean Perrin, The Svedberg, and others. But Einstein did not consider this work to be revolutionary because "it was only a fully consequent working out of the implications of the mechanical world view" (Klein 1975, 116).

The revolutionary implications of Einstein's 1905 paper on the quantum of light are discussed in Chapter 27 above, but here we may note that Einstein actually used the word 'heuristic' in the title. It was not a theory that he was presenting, but rather a means of accounting for diverse phenomena on the basis of a principle whose truth or falsity was not in question, a principle serving only as a basis for the explanation. Until the end of his life Einstein would not use the word 'theory' in relation to the quantum hypothesis of light. In an interview given a little more than a week before he died, Einstein corrected his visitor who had referred to Einstein's "quantum theory of light." No, he said emphatically, "not a theory." Because unlike relativity, which Einstein saw as a logical and evolutionary development of earlier physics, the quantum hypothesis was irreconcilable with established principles, Einstein viewed his idea about light as strange and perhaps even ultimately untenable. Possibly his use of the word 'revolutionary' in referring to the paper on light was an indication of this quality of not fitting and even not being true, rather than simply being novel.

It is well known that Einstein devoted much of his mature scientific life to the unsuccessful struggle to create a 'unified field theory' that would unite gravitation and the other forces of matter in a causal way to produce a true and complete description of the physical world. Martin Klein sees Einstein's later opinions on revolutions in science to be part of Einstein's belief in a coming revolution that would restore to physics some of the qualities it had lost as a result of the twentieth-century onslaughts. "When Einstein reacted skeptically to claims that this or that new discovery or new theory had revolutionized physics," Klein (1975, 120) writes, "he did so in the name of the True Revolution. The old regime of the Newtonian world view had been abandoned, but its genuine successor must provide a comprehensive, consistent, and unified picture of physical reality to replace what was gone. Provisional structures of thought that offered less could be valued for what they were, but Einstein refused to grant their claims of a revolution accomplished."

Let me now turn to Einstein's discussion of Galileo, who was one of Einstein's heroes, along with Kepler and Newton. Einstein admired Galileo not only for his scientific achievement but also for what Einstein saw as the "leitmotif" of his work: "the passionate fight against any kind of dogma based on authority." Einstein lauded Galileo for accepting only "experience and careful reflection" as "criteria of truth," and he commented that in those days such an attitude was "sinister and revolutionary [*unheimlich und revolutionär*]." This phrase occurs in a preface Einstein wrote for Stillman Drake's translation of Galileo's *Dialogue* (1953), which has not thus far been incorporated into discussions of Einstein and revolution.

Einstein's preface to Galileo was printed in both German and in what is claimed to be an "authorized translation by Sonja Bargmann." Whereas in this example the same word ('revolutionary'/ 'revolutionär') occurs in both languages, a quite different word, 'bahnbrechend' (literally 'path-breaking'), appears in a German passage which reads (in the "authorized translation") as the "revolutionary factual content [of] the Dialogue." In this example Einstein is comparing Galileo to a political revolutionary. Galileo, according to Einstein, discarded the authority of ancient savants and dogmas and put his trust in his own reason. Since at that time there were few men who possessed "the passionate will, the intelligence, and the courage" to stand up against "the host of those" who relied "on the ignorance of the people and the indolence of teachers in priest's and scholar's garb," in order to "maintain and defend their positions of authority," Einstein conceived that Galileo's position was 'path-breaking' or 'revolutionary'. But he did not refer to a 'Galilean revolution' as such. He was aware that even without a Galileo, the seventeenth century would have witnessed a break from "the fetters of an obsolete intellectual tradition," and he was wary lest he himself should fall in with "the general weakness of those who, intoxicated with devotion, exaggerate the stature of their heroes."

Now Einstein had an excellent command of English, both written and spoken, even though he preferred to write in German. We do not know how carefully he checked the "authorized translation," but I do not believe he would have passed the rendering of 'bahnbrechend' by 'revolutionary' if it had not expressed his intent. Would the translator, then working closely with Einstein, have distorted his meaning? In any event, he had used the word 'revolutionär' a few lines earlier, and the context of 'revolution' is, accordingly, unambiguous. Four years earlier, in his "Autobiographical Notes" (1949, 53), Einstein had used this same word 'bahnbrechend' when speaking of the time "after Planck's trailblazing work" ("nach Plancks bahnbrechender Arbeit"), but on this occasion he did not say of Planck, as he did of Galileo, that this work was 'revolutionary'. He discussed the "fundamental crisis"—a crisis "the seriousness of which was suddenly recognized because of Max Planck's investigations into heat radiation (1900)" (p. 37).

In Einstein's discussion of the revolutionary ("revolutionär") quality of Maxwell's theory in his "Autobiographical Notes" (1949, 32–35), he compared "the pair Faraday-Maxwell" with "the pair Galileo-Newton," implying that the first member of each pair grasped "relations intuitively," whereas the second one formulated "those relations exactly" and applied "them quantitatively." I believe that anyone who reads these "Autobiographical Notes" in conjunction with the Galileo preface cannot help but conclude that Einstein was implying that there had been a pair of comparable great revolutions. The first was the Newtonian revolution, based on the Galilean revolution, in which the concepts of mass and acceleration became coupled with the new idea of force, a force acting at a distance. The second was the Maxwellian revolution, to a degree based on Faraday's ideas, in which Newtonian "action at a distance is replaced by the field," which—Einstein quite properly stressed—"thus also describes the radiation" (1949, 35).

In a commemorative essay on Newton, published in *Die Naturwissenschaften* in 1927, Einstein wrote of "the revolution in electrodynamics and optics brought about by Faraday and Maxwell." It was, he said, the "first great fundamental advance in theoretical physics since Newton." Again, in context, it would seem that Einstein was treating—by implication—of a Newtonian revolution. On this occasion, Einstein did not use the word 'Revolution' as in other essays, but rather wrote of "die Faraday-Maxwellsche Umwälzung der Elektrodynamik und Optik." We have seen that 'Umwälzung' is a commonly used synonym for revolution.

Later on in Einstein's essay on Newton (1927; 1954, 260), he concludes with an expression of his belief that the "general theory of relativity formed the last step in the development of the program of the field-theory." He then said, "Quantitatively it modified Newton's theory only slightly, but for that all the more profoundly qualitatively." This is a typical statement of Einstein's belief in the evolutionary quality of general relativity. The phrase 'modified Newton's theory' is an expression of Einstein's inner conviction that his creation was a transformation rather than the creation of something wholly new. We have seen that recognizing such a process of transformation in no way necessarily lessens our evaluation of the revolutionary changes that the new ideas may introduce. In this same essay, Einstein said that the "theory of Maxwell and Lorentz led inevitably to the special theory of relativity, which, since it abandoned the notion of absolute simultaneity, excluded the existence of forces acting at a distance." He wanted his readers to conceive of special relativity as a stage in evolution, even though we may see that the implications of such a transformation, however evolutionary, are of such magnitude as to be revolutionary. In this essay, Einstein actually spelled out the implications of both special and general relativity which — to most historical observers — must seem not only revolutionary but even revolutionary to the highest degree.

This evolutionary theme was developed by Einstein in many articles. In a popular presentation written for the London *Times* (28 Nov. 1919; Einstein 1954, 230), he said that the "special theory of relativity" was "simply a systematic development of the electrodynamics of Maxwell and Lorentz." In a lecture at King's College, London, in 1921, he amplified this idea, saying that "the theory of relativity . . . put a sort of finishing touch to the mighty intellectual edifice of Maxwell and Lorentz," seeking "to extend field physics to all phenomena, gravitation included" (ibid., 246). And then he unambiguously declared: "We have here no revolutionary act but the natural continuation of a line that can be traced through centuries." We shall explore below the possibility that such a public statement may have been a response to journalistic extravagance. But we may note that this evolutionary theme appears also in other and later writings, such as the essay on Newton (p. 261), where Einstein discusses the "evolution of our ideas about the processes of nature." Yet the difficulty in trying to force Einstein's opinions into a simple mold is seen in the fact that in the same essay he invoked a quite different image of the development of science: "a revolution in our fundamental notions has gradually occurred since the end of the nineteenth century." The latter appears in German as "ein Umschwung der Grund-Anschauungen," which has been translated by Sonja Bargmann (Einstein 1954, 257) as a gradual "shift in our fundamental notions." We may, however, gain a gloss on this phrase by referring to Einstein's essay on Maxwell (see chapter 20, above). There Einstein wrote of "der grosse Umschwung, der mit den Namen Faraday, Maxwell, Hertz für alle Zeiten verbunden bleiben wird" ("the great change [or revolution] which will be associated for all time with the names of Faraday, Maxwell, and Hertz"). But in the very next sentence, Einstein referred to this as "Revolution," making it clear that he was using 'Umschwung' and

'Revolution' synonymously. The first translator of the essay on Newton (Einstein 1934*a*, 53) rendered Einstein's 'Umschwung' by 'revolution' — the first sense of the word given in many dictionaries (for example, Wildhagen-Héraucourt 1972) — but he changed the author's grammar by making this "a gradual revolution in our fundamental notions." Perhaps this introduces a new concept of historical change in the sciences that in fact is logically self-contradictory. But whichever sense of the word we accept, there can be no doubt that Einstein appreciated that great revolutionary alterations of science could and did occur, but they were rarely (if ever) sudden, dramatic, and unexpected changes that had no logical links of development with previous thought. He himself, however, never either in public or in private said that relativity theory constituted such a revolution.

In an article on Einstein in 1981, Gerald Holton discusses Einstein's "ideas on the way scientific theory develops by evolution" (p. 14). He stresses Einstein's statement that "the most beautiful fate of a physical theory is to point the way to the establishment of a more inclusive theory, in which it lives on as a limiting case." Particularly convincing is a statement made by Einstein when he first came to America (*New York Times*, 4 April 1921; quoted in Holton 1981, 15):

> There has been a false opinion widely spread among the general public that the theory of relativity is to be taken as differing radically from the previous developments in physics from the time of Galileo and Newton, that it is violently opposed to their deductions. The contrary is true. Without the discoveries of every one of the great men of physics, those who laid down the preceding laws, relativity would have been impossible to conceive and there would have been no basis for it. Psychologically it is impossible to come to such a theory at once, without the work which must be done before. The men who have laid the foundation of physics on which I have been able to construct my theory are Galileo, Newton, Maxwell, and Lorentz.

Michael Pupin must have been consciously paraphrasing Einstein's own statement when — at Columbia University — Pupin introduced him as the discoverer of a theory which is "an evolution, not a revolution of the science of dynamics."

The foregoing presentation of Einstein's opinions shows how difficult it is to make a simple statement that Einstein either did or did not believe revolutions occur in science. He was certainly aware that most people — scientists and nonscientists alike — considered relativity to be a revolution, and he went out of his way (on several occasions) to show that relativity was a logical and evolutionary step and not a direct break with the past. Einstein said more than once that there was a Maxwellian revolution, and he unambiguously used the word 'revolutionary' in the context of Galileo's *Dialogue* in 1953, much as he had done in relation to his own 'heuristic' concept of light a half-century earlier.

In discussing Einstein's expressed opinions about revolution and evolution in the sciences, we must not forget that Einstein never wrote an essay on this subject, nor was this subject discussed in full in any recorded conversation or in any letter that is available to us. Furthermore, we know that Einstein was in many ways a modest man, and hence he would have reacted strongly against the newspaper headlines about his having made a revolution. In his most direct statement on this topic, he chiefly attacked the way in which journalists had given the impression of a revolution in science "every five minutes," but it is to be noted that even in that riposte to the announcement of Schrödinger's accomplishment, Einstein did not rule out the possibility of all revolutions in science. This combination of natural modesty and reaction to journalistic practice could have been a major factor in his referring to his revolution as 'evolution'.

Additionally, in 1905 and 1906 the word 'revolution' had a very different connotation for young idealistic intellectuals than it did after 1917. Einstein's major statements about his own work as evolutionary and not revolutionary were made in the aftermath of the revolution in Russia in 1917 and the abortive revolutions throughout all of central Europe in the period immediately following the end of World War I. There was even bloody fighting in the streets of Berlin. Later on, in the 1920s and 1950s, as we have seen, Einstein was willing to write about a Galilean (and possibly a Newtonian) revolution and on a number of occasions he wrote about the Maxwellian revolution. It is significant, I believe, that a strong and simple declaration of a Maxwellian revolution appears so prominently and strikingly in Einstein's autobiography, written in the 1940s. When Einstein referred to the Maxwellian revolution as a revolution due to Faraday, Maxwell, and Hertz (adding that Maxwell had the "lion's share"), he was no doubt stressing the magnitude of the conceptual change without giving consideration to the time span. Since Faraday's papers were published in the 1830s and Hertz's in the 1890s, this revolution would have required more than half a century. This example would indicate that for Einstein the great revolution in science was not strictly analogous to those sudden and dramatic political events which change the form of government.

Einstein's former assistant, Banesh Hoffmann, who has written several major books about Einstein and about modern physics, tells me that he never heard Einstein say anything against the occurrence of revolutions in science. In the book which Hoffmann wrote about Einstein in collaboration with Einstein's long-term secretary and companion, he found no inconsistency with Einstein's own views about science in the abundance of references to Einstein and revolution (for example, pp. 70, 74, 78, 79, 83). He simply applied to Einstein's science the ideas Einstein had expressed about Galileo's and Maxwell's science, but not about his own. In his book about Einstein and relativity, Leopold Infeld, a close associate of Einstein's and co-author with Einstein of *The Evolution of Physics* (1938), calls the special theory "The First Einstein Revolution" and the general theory "The Second Einstein Revolution" (1950, 23, 40). Infeld evaluates Einstein's contribution to quantum theory as "revolutionary but at the same time conciliatory," a major step in "the great unfinished revolution" of quantum theory (p. 102). The journalist Alexander Moszkowski, who reported the fruits of many extended conversations with Einstein, said that the special theory embodied "revolutionary changes in physical thought" (1921, 113), that the general theory required "ideas of a revolutionary nature" (p. 6), and that "very few of us are aware of the further inner revolution that awaits us along the line of development of Einsteinian ideas" (p. 141). Max Planck, "a conservative person in thought and expression," apparently had no difficulty in proclaiming the extreme revolutionary character of Einstein's achievement (Holton 1981, 14):

> This new way of thinking about time makes extraordinary demands on the physicist's ability to abstract, and on his imaginative faculty. It well surpasses in daring everything that has been achieved in speculative scientific research, even in the theory of knowledge . . . This revolution in the physical *Weltanschauung*, brought about by the relativity principle, is to be compared in scope and depth only with that caused by the introduction of the Copernican system of the world.

But Denis Sciama (1969, ix) observes that "Newton's laws of motion are logically incomplete by themselves, and the problems they raise lead one step by step to the full complexity of General Relativity." There is support among many scientists and historians for Einstein's view of relativity as an extension and transformation of existing scientific ideas, just as there is no want of testimony that the theory of relativity has constituted one of the greatest revolutions of our century and a major revolution in science.

Einstein's opinions about evolution and revolution as factors in scientific change, which appear in numerous essays and in his autobiographical statement, have been the subject of two recent studies. Gerald Holton (1981) concentrates on Einstein's own expressions of the evolutionary quality of relativity and his statement against revolutions apropos of Schrödinger's claims in 1947. Accordingly, he mentions but does not discuss the letter to Habicht about quantum theory and does not explore Einstein's many statements about the Maxwellian revolution. Martin Klein (1975), on the other hand, while aware of Einstein's belief that relativity was part of an evolutionary process, has studied Einstein's expressions about revolution — in relation to Maxwell, Schrödinger, and Einstein's own hypothesis concerning light quanta.

29

Continental Drift
and Plate Tectonics:
A Revolution
in Earth Science

————◆————

The recent revolution in earth science is notable for a number of features that illuminate the nature of all revolutions in science. But this revolution also exhibits some novel aspects that are peculiarly characteristic of the science of our times. Essentially, this revolution consists in the rejection of the traditional view that continents were formed or have grown or developed in fixed places, and the adoption of the radical notion of a 'drift' of continents, with respect to one another and the ocean basins, on the surface of the earth. A feature of this revolution is the concept of plate tectonics: that the earth is divided into a set of rigid plates, containing continents and portions of the ocean floors, which rotate very slowly relative to each other.

The hypothesis of continental drift was put forth as a serious scientific proposal by Alfred Wegener (1880–1930) in 1912 and advanced in a major published monograph a few years later (Wegener 1915). Recognized almost at once as potentially revolutionary, requiring a drastic revision of the very fundamentals of geologic science, the idea of continental mobility was widely discussed by geologists during the 1920s and 1930s and was almost completely rejected. Wegener's proposal of continental drift thus produced only what I have called a revolution on paper until some time in the mid-1950s, when new kinds of evidence began to favor the possibility of a movement of continents. But it was not until the 60s that a revolution in science actually occurred.

Historical analysis reveals that the revolution in science that ended the half-century status of revolution on paper was not merely a belated acceptance of a somewhat dormant or rejected set of ideas or theory. The revolution in science followed the introduction of new techniques for studying the earth and the broadening of knowledge by using new sources of information. Not only did many earth scientists begin to think along nontraditional lines, but there was an important input made by scientists trained in physics rather than geology. Hence the eventual revolution in earth science was not so much a mere revival of a long-rejected theory of continental drift as it was a radical transformation of the older idea, encompassing a newly conceived theory of plate tectonics to describe the motion. In a sense, Wegener's original theory never produced a revolution in science, but the eventual revolution in science did embody the central concept of continental mobility and the idea of two types of domains (continents and ocean floors) of Wegener's theory.

An outstanding feature of the recent revolution has been a general consciousness among working geologists that they have been living through a revolutionary moment in their science. A number of earth scientists have written articles or monographs stressing the revolutionary nature of the changes in thinking about continents and the earth, and have produced books with such titles as *A Revolution in the Earth Sciences: From Continental Drift to Plate Tectonics* (Hallam 1973) or *Critical Years of the Revolution in Earth Science* (Glen 1982). This stress on revolution has not only been a feature of later historical or review articles and books but of working papers during the stages of revolution. For instance, a seminal technical paper in *Science* (Opdyke et al. 1966), entitled "Paleomagnetic Study of Antarctic Deep-Sea Cores," bore the subtitle "Revolutionary Method of Dating Events in Earth's History." In 1970, during a discussion of "a new class of faults," J. Tuzo Wilson remarked that recent discoveries concerning reversals of terrestrial magnetism constituted a "revolution" in earth science. In the final report (1972) of the Upper Mantle Project (of the International Council of Scientific Unions), the "emergence of a unifying concept of plate tectonics" during the "period of the U.M.P." was said to be a "revolution" in the development of the earth sciences (Sullivan 1974, 343).

A consciousness of revolution in the historical reviews and summaries written during the early 1970s (primarily by English-speaking scientists) in some measure reflects the fact that the acceptance of continental motion and plate tectonics in the 1960s coincided with the considerable attention given to Thomas S. Kuhn's seminal tract, *The Structure of Scientific Revolutions*, published in 1962. Thus Allan Cox (1973), Anthony Hallam (1973), Ursula Marvin (1973), and J. Tuzo Wilson (1973, 1976) all refer specifically to Kuhn in reviewing or discussing the recent changes in theories about the continents. This revolution in science is also notable for having produced—in a little more than a decade after its occurrence—a series of well-documented histories or historical articles, many written by earth scientists, some of whom had themselves made fundamental contributions to the revolution.

As a result of these recent historical investigations, we now know that Francis Bacon was not the originator of the idea that continents have moved (Marvin 1973). Rather, he merely pointed out that there is a general conformance of the west coastal outlines of Africa and Peru. Nor did Alexander von Humboldt, almost two centuries later, advance from a recognition of the congruence of the facing coasts across the Atlantic to the suggestion that the two continents had once been joined and had then moved apart. But a breaking-up of a protocontinent and the subsequent separation of its parts was suggested in 1859 in a marginally scientific book entitled *Creation and Its Mysteries Unveiled*, written in French by Antonio Snider-Pellegrini, an American living in Paris. Claims have also been advanced (wrongly, as pointed out by Ursula Marvin 1973, 58) for the Austrian geologist Eduard Suess as an early proponent of the idea of continental drift. But Suess did put forth, at the beginning of the twentieth century, the hypothesis that there had originally been two large Paleozoic continents: "Atlantis" (located in the North Atlantic Ocean) and Gondwana-Land (in the South). The latter he named for Gondwana, a district in central India (home of the Gonds). Suess—like some of his nineteenth-century predecessors—believed that our present continents were remnants of larger primitive ones, pieces of which had foundered into the ocean basins; but he did not propose that this process entailed a continental drift as we would understand that term today (Marvin 1973, 58).

A better case can be made for an American geologist, F. B. Taylor, who published in 1910 "a lengthy paper giving the first logically worked out and coherent hypothesis involving what we would now recognize as continental drift" (Hallam 1973, 3). First presented in a pamphlet in 1898, Taylor's theory was astronomical, not based on geography or geology. He hypothesized that a long time ago the earth had captured a comet which became our moon, an event which increased the earth's speed of rotation and generated a large tidal force; the combination of these two effects tended to pull the continents away from the poles. In his paper of 1910, and in later publications, Taylor elaborated his argument for continental movements by means of geological evidence (Aldrich 1976, 271), but these never attracted much attention in the geological community (Marvin 1973, 63–64). In 1911 another American, Howard B. Baker, proposed that there had been a displacement of the continents, caused by astronomical forces, among them planetary perturbations in the solar system (p. 65). When Wegener published his treatise, he summarized the work of a number of possible predecessors, including a lengthy paragraph discussing Taylor's contributions (1924, 8–10). But Wegener twice asserted that he "became acquainted with all these works only when the displacement theory in its main outlines had already been worked out" (pp. 8, 10). In the last edition of his book, Wegener (1962, 3) added some new names to the historical register. He now said: "I also discovered ideas very similar to my own in a work of F. B. Taylor which appeared in 1910."[1]

Wegener's Theory of Continental Motion

Serious discussion of the hypothesis of continental drift by geologists and geophysicists began with the publications of Alfred Wegener. He was by training and profession not a geologist but rather an astronomer and meteorologist (whose doctoral dissertation was in the domain of the history of astronomy).[2] Wegener's academic career embraced, first, a post in astronomy and meteorology at Marburg and, later, a professorship of meteorology plus geophysics at Graz (1924–1930). In his twenties and thirties Wegener went on meteorological expeditions to Greenland; he lost his life on a third such expedition in 1930. According to Lauge Koch, who went with Wegener on the first of these expeditions, the idea of continental drift arose while Wegener was observing slabs of ice split apart in the sea. But Wegener himself says only that at about

Christmastime in 1910 he was struck by the way the coastlines on the two sides of the Atlantic fit together and that this suggested to him the possibility of a lateral motion of the continents.

Apparently, Wegener did not take his own idea too seriously, dismissing it as "improbable" (Wegener 1924, 5; 1962, 1). But he began to develop the hypothesis of continental motion in the autumn of the following year, he relates, when "quite accidently" he came upon "a compendium of references describing the faunal similarities of Paleozoic strata in Africa and Brazil" (Marvin 1973, 66). In the compendium, the existence of identical or near-identical fossils of animals on both sides of the Atlantic in the distant past was used in the then-traditional manner to argue for the existence of ancient land bridges between Africa and Brazil. Snails, for example, simply could not have swum across the vastness of the Atlantic Ocean; hence to find the same or very similar fossil snails on both sides of the South Atlantic carries a strong probability that there must have been some connection by land between South America and Africa long ago. The alternative would be to assume a similar but independent evolutionary development on a large scale in the two regions—a highly improbable occurrence.

Wegener was much impressed by the fossil evidence, but he rejected the hypothesis that the continents had once been connected by some kind of land bridge or now-sunken continent; this hypothesis required the further supposition of the sinking or disintegration of these land connections, for which there was no scientific evidence. Of course there are land bridges between continents, for example the isthmus of Panama and a former bridge across the Bering Strait, but there was no real evidence for a hypothesized land bridge spanning the South Atlantic in ancient times. As an alternative theory, Wegener revived his earlier ideas on the possibility of continental drift, transforming them from what, as he confessed, was only a "fantastic and impractical" idea—no more than a jigsaw puzzle solution of no real significance for earth science—into a working scientific concept. Wegener developed his hypothesis, adducing various kinds of supporting evidence, and summarized his results in 1912 at a meeting of geologists. His first two reports were published later in that year, and in 1915 he published a monograph, *Die Entstehung der Kontinente und Ozeane (The Origins of Continents and Oceans)*, in which he marshaled all the evidence he had found to support his ideas.[3] This book appeared in revised versions in 1920, 1922, and 1929 and was translated into English, French, Spanish, and Russian. In the English version (1924), based on the third German edition of 1922, Wegener's expression "Die Verschiebung der Kontinente" was accurately rendered as 'continental displacement', but almost at once the phrase generally used was 'continental drift'.[4]

Wegener based his argument on geological and paleontological evidence, not mere pattern-fitting. He put great stress on aspects of geologic similarity on both sides of the ocean. In the last edition of his book, he adduced supporting evidence from the field of paleoclimatology, a subject on which he had written a book in 1924 (in collaboration with W. Köppen), inferring a wandering of the poles of rotation of the earth. (For a concise summary of Wegener's detailed geographical, geological, palaeoclimatic, and palaeontological and biological arguments and evidence, see Hallam 1973, ch. 2.) Wegener postulated that in the Mesozoic era, and continuing up to the fairly recent past, there had been a huge supercontinent or protocontinent, which he named Pangaea, that had broken apart. The fragments of Pangaea separated and moved away from one another, producing the individual continents we know today. Two possible causes of a drift or motion or displacement of continents were put forth: tidal forces produced by the moon, and a force of "flight from the poles" ('Pohlflucht'), that is, a kind of centrifugal effect produced by the rotation of the earth. But Wegener was aware that the solution to the problem of the cause of continental movement still eluded him. "The Newton of drift theory," he wrote (1962, 166), "has not yet appeared," thus echoing the sentiments of Cuvier, van't Hoff, and others. "It is probable," he admitted, "that the complete solution of the problem of the driving forces will still be a long time coming." In retrospect, Wegener's most fundamental and original contribution was his concept that the continents and ocean floors were two distinct layers of the earth's crust, differing from each other both in rock composition and in altitude. Most scientists at that time believed that the oceans, with the exception of the Pacific, had Sialic floors. Wegener's basic ideas were validated by plate tectonics.

Though Wegener's theory of continental drift was only a revolution on paper for some time, this does not mean that his ideas attracted no attention and no followers. Far from it! The decade of the 1920s was marked by a series of violent international controversies. An unsigned review of the second edition of Wegener's book (1920), in the influential journal *Nature* on 16 February 1922 (vol. 109, p. 202), gave an adequate summary of the main points of the theory and expressed the hope that an English version might soon appear. Noting that there was "strong opposition from a good many geologists," the reviewer concluded that if Wegener's theory could be substantiated, a "revolution in thought" would occur similar to "the change in astronomical ideas at the time of Copernicus" (p. 203). A generally favorable report on a lecture by Wegener appeared in 1921 in the most important German scientific journal, *Die Naturwissenschaften* (1921, 219–220). The reporter, O. Baschin, said that those who heard the lecture, at the Geographical Society of Berlin, were "extraordinarily convinced" and that the theory aroused "general approval," although there were some objections and warnings of caution in the ensuing discussion. Baschin concluded that there "is no factual proof against Wegener's theory," but that "solid proof must still be found before it can be unreservedly accepted."

Quite different in tone was the review in the British *Geological Magazine* for August 1922, in which Philip Lake stated bluntly that Wegener "is not seeking truth; he is advocating a cause and is blind to every fact and argument that tells against it." In America, in the October 1922 issue of *Geographical Review*, Harry Fielding Reid delivered what he saw as the coup de grâce to both continental drift and polar wandering. In the autumn of that same year, the continental drift hypothesis was the subject of debate and discussion at the annual meeting of the British Association for the Advancement of Science; the published report (by W. B. Wright) described the event as "lively but inconclusive." But a favorable presentation of "The Displacement of Continents: A New Theory" was published by Professor F. E. Weiss in the *Manchester Guardian* on Thursday, 16 March 1922. Weiss said Wegener's theory was "of fundamental importance to the sciences of geography and geology" and "of great interest to the biological sciences," and he concluded that the theory "constitutes a good working hypothesis" which will "greatly stimulate further inquiry."

A major scientific event of the 1920s was the debate in Tulsa, Oklahoma, in 1926, convened by the American Association of Petroleum Geologists and published under the title *Theory of Continental Drift: A Symposium on the Origin and Movement of Land Masses . . . as Proposed by Alfred Wegener* (van der Gracht 1928). Among those in attendance were Wegener himself and Frank B. Taylor. The other eleven participants included eight Americans and three Europeans. The chairman, W. A. J. M. van Waterschoot van der Gracht (a Dutch geologist, vice president of Marland Oil Co.), contributed to the published proceedings a lengthy introduction supporting continental drift and a concluding rebuttal to the opponents; these two papers occupy more than half of the volume. Some of the participants (Chester Longwell of Yale, John

Joly of Dublin, G. A. F. Molengraaff of Delft, J. W. Gregory of Glasgow, Joseph T. Singewald, Jr., of Johns Hopkins) were somewhat tolerant but highly skeptical, while others (Bailey Willis of Stanford, Rollin T. Chamberlin of Chicago, William Bowie of the U.S. Coast and Geodetic Survey, Edward W. Berry of Johns Hopkins) larded their geological counterevidence with sarcastic remarks about the faulty science and defective method which they claimed characterized Wegener's thinking and writing. Looking back from today's vantage point, what is most impressive is the rancor and virulence in these criticisms.[5] It is obvious that Wegener had launched what seemed to be a frontal attack on the foundations of earth science and of sound belief.

The Wegener hypothesis aroused hostility on a number of grounds. First, it went directly counter to the mind-set of almost all geologists and geophysicists, who had been conditioned from their earliest days to think of the continents as essentially stable, of the earth as terra firma. To suggest that continents might have a lateral motion with respect to one another was as "heretical" and "absurd" in a laic sense as the Copernican hypothesis had been in the time of Galileo.[6] Second, the new hypothesis supposed that the earth is not as rigid as had been believed and as seemed evident even to the most superficial observer. Hence, there seemed to be required some imagined forces of tremendous magnitude, by far exceeding—as geophysicists like Harold Jeffreys were quick to point out—the two proposed by Wegener. The argument could be boiled down to the impossibility of having what has been called "a weak continental ship" sail "through an unyielding oceanic crust" (see Glen 1982, 5).

It is unfortunately common to attempt to dismiss a proposed revolution in science by making *ad hominem* comments about the proponent of the new theory. Wegener not only was attacked for his method but was denied the right to discuss geology because he lacked credentials, being a meteorologist rather than a geologist—and a German meteorologist at that. Charles Schuchert, emeritus professor of paleontology at Yale (1928, 140), referred to continental drift as a "German theory," and he quoted with evident approval the remarks of P. Termier (director of the Geological Survey of France) that Wegener's theory was only "a beautiful dream," the "dream of a great poet"; when one "tries to embrace it," one "finds that he has in his arms but a little vapor or smoke." Wegener, furthermore, according to Schuchert, "generalizes too easily" and "pays little or no attention to historical geology" (p. 139). He was an outsider, who had never done any actual field work in paleontology or geology; it "is wrong," Schuchert concluded, "for a stranger to the facts he handles to generalize from them to other generalizations."

That Wegener was rejected—at least in part—because he was not a member of 'the club' is borne out by the literature. In his attack on Wegener's theory, evidence, and method of scientific procedure, Harold Jeffreys (1952, 345) declared that "Wegener was primarily a meteorologist." In 1944, in an article in the *American Journal of Science* (vol. 242, p. 229), Chester R. Longwell wrote condescendingly that "charitable commentators" suggest that Wegener's inconsistencies and omissions "may be overlooked on the ground that [he] . . . was not a geologist." And, as late as 1978, George Gaylord Simpson (1978, 272) repeated his earlier opinion that "most of Wegener's supposed paleontological and biological evidence" was either "equivocal" or "simply wrong"; he criticized Wegener (a "German meteorologist") for daring to venture into fields with which he "had no firsthand acquaintance."

Most geologists in the 1930s and 40s would have agreed with Jeffreys when he said in the third edition of his influential treatise, *The Earth* (1952, 348), that "the advocates of continental drift have not produced in thirty years an explanation that will bear inspection." Orthodox geologists and paleontologists even used the idea of continental drift to "supply comic relief" in their classroom lectures. Percy E. Raymond, professor of paleontology at Harvard, told his students how "half of a Devonian pelecypod" had been found in Newfoundland and another half in Ireland. The two "matched perfectly" and so must have been "the two halves of the same pelecypod, which had been wrenched apart by Wegener's hypothesis in the late Pleistocene" (Marvin 1973, 106).

The twenties and thirties were not wholly devoid of Wegener supporters, however. Reginald A. Daly of Harvard favored the general idea of continental mobility, although he was far from being a strict Wegenerian, having once himself referred to Wegener as "a German meteorologist" (1925). Daly produced his own version of the motion of continents, which, in retrospect, began "an approach to the current model of plate tectonics" (Marvin 1973, 99). On the title page of his book, aptly called *Our Mobile Earth* (1926), Daly placed the words "E pur si muove!" a version of the statement allegedly made by Galileo after recanting his Copernican allegiance to the theory of a moving earth ("Eppur si muove" = "Nevertheless it does move").[7]

One of the primary supporters of Wegener's ideas during the 1920s was Emile Argand, founder and director of the Geological Institute at Neuchâtel, Switzerland. In 1922, at the first post-World War I international congress of geologists, Argand delivered a spirited defense of Wegener's fundamental concept in relation to "Tectonics of Asia." Argand not only gathered and organized an impressive mass of evidence supporting Wegener, but also performed a valuable function in making a distinction between a Wegenerian kind of "mobilisme" and orthodox "fixisme." He declared that "fixisme" "is not a theory but a negative element common to several theories" (Argand-Carozzi 1977, 125). Although he made an affirmative exhibition of the variety and details of the arguments favoring "mobilisme," Argand could not help admitting in his conclusion that "almost nothing is known about the forces responsible for continental drift" (p. 162).

Wegener's two chief supporters during the 1930s were Arthur Holmes, regarded by many "as the greatest British geologist of this century" (Hallam 1973, 26), and the South African geologist Alexander du Toit. Holmes was a convert by the time of the appearance of the American symposium on continental drift, a volume he reviewed in *Nature* (Sept. 1928). Here he noted that "all the adverse criticism is . . . mainly directed against Wegener" rather than against Wegener's ideas in general, that "when all has been said, there remains a far stronger case for continental drift than either Taylor or Wegener has yet put forward." Although he accepted the general idea of the motion of continents and became Britain's major advocate of continental drift, Holmes proposed a new mechanism for producing this effect, according to which convection currents in the earth's mantle (the part of the earth immediately below the crust) would tend to produce mountain-building and continental drift (see Marvin 1973, 103; Hallam 1973, 26ff.). Du Toit lived in Johannesburg, "in the heart of ancient Gondwana," as Ursula Marvin reminds us (1973, 107), "where the evidence of continental drift is most compelling." He summed up his ideas in a work called *Our Wandering Continents* (1937), subtitled "an hypothesis of continental drifting." It was dedicated "To the memory of Alfred Wegener for his distinguished services in connection with the geological interpretation of Our Earth." In this book, du Toit proposed a theory of the earth that differed somewhat from Wegener's (see Marvin 1973, 107–110; Hallam 1973, 30–36). For instance, instead of Wegener's original single continent, Pangaea, du Toit proposed two—Laurasia in the northern hemisphere and Gondwanaland in the southern hemisphere.

Du Toit attributed the rejection of Wegener's hypothesis to two factors: (1) the lack of a satisfactory mechanism for producing the drift, and (2) the "deep conservatism" which he found to be characteristic of the whole history of geology. Yet du Toit was fully aware that the acceptance of continental drift would "involve the rewriting of our numerous text-books, not only of Geology, but Palaeogeography, Palaeoclimatology and Geophysics" (p. 5). He had no doubt, he said, that "a great and fundamental truth" was embodied in the "postulated Drift" and that Taylor and Wegener had proposed a "revolutionary Hypothesis" (p. vii).

Du Toit was not the only person to view Wegener's work as revolutionary. The reporter in *Die Naturwissenschaften* in 1921, the anonymous reviewer in *Nature* in 1922, F. E. Weiss in 1922, van der Gracht and others in 1926 — friends and foes alike — used the term. Daly (1926, 260) characterized continental drift as a "new, startling explanation" which many "geologists have found . . . bizarre" and even "shocking," a "revolutionary conception." The same revolutionary character of "ideas so novel as those of Wegener" was manifest in Philip Lake's pronouncement that a "moving continent is as strange to us as a moving earth was to our ancestors" (1922, 338). In reviewing the proceedings of the Tulsa symposium of 1926 (van der Gracht et al. 1928) in *Geological Magazine* (1928, vol. 65, pp. 422–424), Lake explicitly referred to Wegener's "revolutionary theory."

Wegener himself was fully aware of the revolutionary potential of his new ideas. Writing in 1911 to W. Köppen (a colleague and teacher), a year before he made any public statement of his new ideas in lectures or in print, Wegener asked why we "should hesitate to throw the old view overboard?" He continued: "Why should one hold this idea back for 10 or even 30 years? Is it perhaps revolutionary? [Ist sie etwa revolutionär?]" He appended a bold and simple answer to his rhetorical question: "I do not believe the old ideas have more than ten years to live."[8]

Because of its revolutionary qualities, stronger evidence than usual would be necessary in order for this theory to win support from the scientific community. For any fundamental and radical change to be accepted by scientists, there must either be unassailable or compelling evidence, or there must be a whole set of obvious advantages over existing theories. It seems clear that in the 1920s and 30s, neither of these conditions had as yet obtained. In fact, such unassailable evidence did not become available until the 1950s and after. And in the meantime the radical restructuring of the whole science of geology which would be required by the acceptance of Wegener's ideas did not seem to be at all attractive in the absence of more compelling facts. The Chicago geologist R. T. Chamberlin reported at the 1926 symposium of the American

Association of Petroleum Geologists that at an earlier meeting of the Geological Society of America (held in Ann Arbor in 1922) he had heard it said that "if we are to believe Wegener's hypothesis, we must forget everything which has been learned in the last 70 years and start all over again." In retrospect, this has proved to be perfectly true. Chamberlin's remarks, we may observe, were restated in a somewhat different context some forty years later when J. Tuzo Wilson (1968a, 22) wrote, "If indeed the Earth is, in its own slow way, a very dynamic body, and we have regarded it as essentially static, we need to discard most of our old theories and books and start again with a new viewpoint and a new science."

As for Wegener's failure to provide a satisfactory mechanism for continental drift — generally agreed to be "the greatest stumbling-block to acceptance of Wegener's hypothesis" — Hallam (1973, 110) tries to counter such an argument by pointing out that "gravity, geomagnetism, and electricity were all fully accepted long before they were adequately 'explained.'" He adds that in geology the lack of any "general agreement" about the "underlying cause" has not prevented the universal acceptance of the "existence of former ice ages." And J. Tuzo Wilson (1964, 4) argued, has not man always been willing to accept the existence of phenomena (for example, the earth's magnetic field) or of processes (such as thunderstorms) "long before he could account for them?" At this point, however, a clarification is required. Rachel Laudan has wisely pointed out that the problem of *cause* of *'mechanism'* with respect to continental drift differs greatly from the situation with respect to "gravity, geomagnetism, and electricity," or "the existence of thunderstorms." In the case of continental drift, the problem "was not simply that there was no *known* mechanism or cause," but rather that "any *conceivable* mechanism would conflict with physical theory" (R. Laudan 1978 = Gutting 1980, 288). Furthermore, there existed unifying theories of the nature of the earth and its interior which explained most observations tolerably well.

S. K. Runcorn (1980, 193) observes that whereas in the 1950s and earlier "the absence of a mechanism" was "widely held to be a most weighty objection to accepting the geological or palaeomagnetic evidence for continental drift," today "the 'theory' of plate tectonics is [becoming] accepted without any consensus about the physical mechanism responsible." He sees, as of 1980, "the question of the nature of the forces that move the plates" as "the most important challenge facing geophysicists today."

Transformations of Wegener's Theory

In retrospect, now that we have witnessed the revolution, there appear to have been two radical innovations that separate the age of the revolution on paper from the present. The first is the accumulation of new and convincing evidence that continents and ocean floors are indeed real entities that have moved with respect to one another — evidence that by several orders of magnitude transcends the fitting of coastlines and even the matching of transoceanic geological or ecological features and of fossil flora and fauna. The second is a radical restatement of the theory that has so altered the fundamental concepts as to make it problematic whether the revolution that was ultimately achieved can legitimately be identified as the attempted revolution which failed for almost half a century. This situation has many points of similarity with that of the so-called Copernican revolution. The eventual revolution, set into motion by Galileo and Kepler and achieved by Newton, kept only the most general Copernican cosmological idea, that the earth moves and the sun stands still, while rejecting the essential features of Copernicus's astronomy. Similarly, the revolution in earth science kept only Wegener's most general idea, that there may be a motion of continents with respect to one another, while rejecting the essential features of Wegener's theory — that continents (made of Sial) move on or through the oceanic crust (Sima) as individual or separate entities, while the denser envelope of Sima remains fixed in location.

The current idea is rather that large blocks or plates covering the earth's surface move and that some of these may carry continents or parts of continents, together with ocean floor. Hence a theory of individual continents in motion has been replaced by a different theory in which the motion of the continents is but the visible part of a more fundamental motion. In the process, Wegener's hypothesizing of a 'Pohlfluchtkraft' and a tidal force became irrelevant.

The new evidence from the 1950s came, in the first instance, from research on paleomagnetism and on the magnetism of the earth. Paleomagnetism is the study of 'remanent' rock magnetism, the magnetism that remains in a rock sample as it solidifies from a molten state. This magnetism is imprinted into iron oxides in the rock as an effect of the earth's magnetic field. Studies made by P. M. S. Blackett in London and by S. K. Runcorn in Newcastle (later in Cambridge) and others showed that the magnetic field of the earth has not been constant but has changed — even undergoing reversals — according to a time pattern that can be determined. (These studies were made possible by a sensitive new instrument, the magnetometer, of which Blackett was the primary inventor.) When the path of the position of the magnetic poles had been carefully plotted, it was found that the movement of the pole differed from one place to another, suggesting that each of these land masses had moved independently. The evidence pointed toward a time when the southern continents had been joined together in the south polar region as a protocontinent, Gondwanaland, so that there must have been lateral displacements of its parts—our present continents (see McKenzie 1977, 114–117).

These first results did not at once convince the community of earth scientists that there had been a continental displacement; no doubt, there were too many unsolved problems about details of the history of the earth's magnetic field, while the argument from magnetism was of a "complicated and specialist nature" with many untested assumptions (MacKenzie 1977, 116). But sufficient interest was aroused, chiefly among geophysicists, that in 1956 a symposium was held on the subject of continental drift. E. Irving of the Australian National University reviewed the magnetic studies of the past years and concluded that "the balance of the evidence favours the idea that . . . the Earth's axis [has] changed its position relative to the Earth as a whole [and] that also the continents have 'drifted' with respect to each other" (Irving 1953; 1958; see Marvin 1973, 150–152).

A second line of research that tended to advance the revival of Wegener's basic idea, although not Wegener's theory, was the study of mid-ocean ridges. The oceans and the inland seas cover about 70 percent of the surface of the earth; therefore, since knowledge of the nature of the regions below the oceans was rudimentary in the 1930s and 40s, we can understand why the pre-war debate concerning continental drift had turned out to be so inconclusive (Hallam 1973, 37). Yet the mid-Atlantic ridge had been mapped and in 1916 F. B. Taylor said it appeared as though the Atlantic continents had crept away from the ridge to either side. Wegener, furthermore, did write out the evidence that the ocean floors were basaltic—arguing from density, magnetism, and compositions, and so on—but nobody paid attention. The direct key to our current knowledge of the motion of continents came from studies of the earth beneath the oceans. During the International Geophysical Year (1957–58), new techniques were introduced for measuring gravity and for combining seismic and gravity data. Geophysicists found ways to determine the rate at which heat flows through the floor of the ocean. The implication of these studies was that huge blocks of oceanic crust could "apparently be displaced large distances with respect to each other" (Hallam 1973, 52). The congruence of these findings with those obtained from magnetic studies gave strong support to the notion that continents had undergone displacement with respect to one another. At this time, a mode of continental drift was put forth in the now widely accepted concept of plate tectonics,[9] a system of architecture according to which the earth's crust is analogous to a "mosaic of large plates — akin, on a mega-scale, to ice floes or paving stones." These plates move as separate rigid units, and experience deformation at their borders with other plates. Ursula Marvin especially stresses the fact that these "moving plates are not the continents, as postulated by Alfred Wegener, nor are they the individual ocean floors" (1973, 165). Since each of these plates may include both continents and ocean floors, their motion is very different from Wegener's conception of moving continents. Hence, the original term 'continental drift', implying a motion of individual continents, is no longer strictly accurate (Hallam 1973, 74). In 1968, it was proposed that six large plates and twelve subplates cover the earth. Since then, additional subdivisions have been suggested.

Sea-Floor Spreading

The theory of plate tectonics is bound up with the concept of 'sea-floor spreading' to explain the instability and changes in the earth's crust. Proposed originally in 1960 by Harry H. Hess of Princeton, sea-floor spreading postulates a process whereby the ocean floor is constantly being pushed out on both sides of the ridge that runs through the major oceans.[10] Hess first circulated his ideas in 1960 in the draft of a chapter for a book on oceans which was not published until 1962. His central concept was so radical that he introduced it by saying that he considered his chapter to be "an essay in geopoetry." He proposed that the great ridges running down the center of oceans are outlets for the up-pouring of molten material from the earth's mantle, the region below the crust. This matter spreads out equally on both sides of the ridge, where it cools and solidifies, so as to become part of the old crust to which it becomes attached. As the crust grows in this way alongside the ridge, the material (a giant plate) moves laterally away from the ridge. Since the earth does not expand, it follows that this plate also cannot simply expand with the addition of new matter. Hence there must be a region away from the ridge where the plate disintegrates. That is, at the boundary furthest from the ridge, the plate slides beneath another plate, descending down into the mantle in an oceanic trench where it releases its water and becomes molten once again. This process has been linked to a kind of convection 'conveyor belt', bringing up matter from the mantle at a ridge, forcing it forward or away from the edge, and eventually taking the matter of the mantle back into the mantle itself at a distant trench.

There is thus a continual enormous pressure that pushes away from the mid-Atlantic ridge the two plates that carry Africa and South America. About 180 million years ago, South America and Africa were joined, forming the continent of Gondwanaland. The line of cleavage between the continents coincides with the still-active ridge that causes the spreading. This line is marked by the occurrence of earthquakes and is today roughly equidistant from the Atlantic boundaries of South America and Africa.

Hess thus provided what is generally considered to be a major element in the subsequent development of plate tectonics and our understanding of continental drift by showing the manner in which the ocean floor is constantly being created on one side of a continent and destroyed on the other (McKenzie 1977, 117). The speed with which the floor of the Atlantic moves out from the ridge is about four centimeters (an inch and a half) per year, so that the time required for the ocean floor to well up at the midocean ridge, travel across the ocean, and finally descend back into the mantle is about two hundred million years. This number at once explained a number of mysteries. For instance, fossils from borings in the ocean floor were never older than some two hundred million years (the age of the Mesozoic era), even though marine fossils dug up on land showed that marine life dated back to at least ten times that age. Again, if the ocean beds were as old as the continents, then the normal rate of sedimentation would have had to produce deep layers of sediment, but dredging expeditions had found that there is little sediment occurring on the ocean floors. In short, during the several billions of years of existence of the oceans (Uyeda 1978, 63), the ocean floors have not been constant but continually changing, continually moving.

When Hess's ideas are combined with the general notion of plate tectonics, another process may be envisioned whereby the addition of new matter at the boundary of a plate does not increase its size. The plate can be continually shortened by compression, which manifests itself in the formation or alteration of mountain ranges where two plates collide.

In presenting his ideas on sea-floor spreading, Harry Hess was explicit on the fact that his theory was "not exactly the same as continental drift" (1962, 617). According to all continental-drift thinking, "continents . . . flow through oceanic crust impelled by unknown forces," but he put forth the idea that the continents "rather . . . ride passively on mantle material as it comes to the surface at the crest of the ridge and then move laterally away from it."

I have mentioned the general agreement that the initial purely paleomagnetic evidence did not fully convince most earth scientists of the need to abandon the concept of 'fixism'. The final evidence came from new magnetic studies that dramatically confirmed the hypothesis of sea-floor spreading. Ship-born magnetometers revealed that there are areas of the ocean floor which are magnetized in stripes (see Hurley 1959, 61). If Hess's explanation was correct, then there ought to exist a symmetry with respect to the magnetic stripes on both sides of an oceanic ridge. This was the test proposed by F. J. Vine, then a research student at Cambridge University, and his supervisor, D. H. Matthews. Measurements soon proved that there was just that predicted symmetry.

According to theory, as hot molten material spreads out on both sides of an oceanic ridge and solidifies, it acquires the current magnetism of the earth. Though it is pushed out from the ridge by new matter, it retains the magnetism it acquired while cooling. Thus each successive stripe of matter should carry a magnetic imprint of the date of its formation into a solid mass, and this should be the same in both directions, on both sides of the ridge. This important hypothesis was published by Vine and Matthews in *Nature* in 1963 and successfully passed the test of experiment in the following years. In fact, the history of the earth's magnetism shows not only minor changes but definite reversals at now-known dates, all of which were actually found in successive stripes.

Though this hypothesis may sound very logical and unexciting today, it was in fact radical and daring at the time. Vine recalls that when he first mentioned his idea to Maurice Hill, a Cambridge geophysicist, Hill "thought I was totally mad," according to Vine, although he "was polite enough not to say anything; he just looked at me and went on to talk about something else" (Glen 1982, 279). Vine also talked about his hypothesis with Sir Edward Bullard, who "was much more encouraging," even though "he realized it was a bit of a long shot." Vine was "quite keen to publish the idea with Teddy Bullard," thinking "it would look great: 'Bullard and Vine.' Teddy Bullard quite rightly said 'no way.' He didn't want his name on the paper." Bullard was a major innovator in geophysics who made important contributions to the theory of heat flow in the earth. Always receptive to new ideas, he "received the hypothesis with enthusiasm and spoke favorably of it to many," despite his unwillingness to accept Vine's initial offer of co-authorship (p. 358).

The hypothesis proposed by Vine and Matthews (and, independently, by Lawrence Morley in Canada; for details, see Glen 1982, 271ff.) was "equal in importance to any formulated in the geological sciences in this century" (p. 271). In addition to independently confirming Hess's idea of sea-floor spreading, it could enable the speed of spreading to be calculated on the basis of an independent, accurately determined time scale for the reversals of the earth's magnetic field. There seems to be agreement that the confirmation of the Vine–Matthews–Morley hypothesis "triggered" the "revolution" in the earth sciences (p. 269). The next step was the "formulation of a new theory of global tectonics" (Hallam 1973, 67) and the reconstruction of our knowledge of the earth.

Acceptance of the Revolution

Whoever examines the history of the past twenty years of earth science, as presented in such works as Sullivan (1974) and especially Glen (1982), will recognize how much significant research was required in order to complete the revolution (including the heroic contributions, which I have scarcely mentioned, of such scientists as Edward Bullard, J. Tuzo Wilson, Maurice Ewing, and others).[11] Furthermore, many eminent geophysicists long refused to accept not only the theory of plate tectonics but even the concept of sea-floor spreading. "Not a single aspect of the ocean-floor spreading hypothesis can stand up to criticism," wrote Vladimir V. Beloussov in 1970 (Sullivan 1974, 105), a man described in *Nature* as "the Soviet Union's most distinguished geophysicist." A dozen years later, in December 1982, in the *Geophysical Journal of the Royal Astronomical Society* (vol. 71, 555–556), Sir Harold Jeffreys, aged 91, for six decades the leading arch-foe of continental drift, still scorned the theory and likened the "subduction of oceanic plates" to "cutting butter with a knife also made of butter."

The conservatism of earth scientists, to which du Toit referred, is further illustrated by the article on Wegener in the authoritative *Dictionary of Scientific Biography* (*14*: 214–217), published in 1976. Written by K. E. Bullen of the University of Sydney, this account concludes with a grudging presentation of the evidence (from paleomagnetism and studies of the ocean floor) that has tended to support the rebirth of Wegener's ideas, followed by a listing of the arguments — old and new — "against continental drift": while "ad hoc answers to all these questions have been put forward by protagonists of the theory, yet again their answers have been questioned" (p. 216). In 1976, three years after the publication of Hallam's and Marvin's histories (both of which proclaimed the success of the revolution and analyzed its structure), the final judgment in the current biography of Wegener is that the "enthusiasms of a considerable number of earth scientists lead them to assert, sometimes with a religious fervor, that continental drift is now established" (p. 217).

Bullen's reference to "religious fervor" is especially apt because the language of the changes in opinion during the 1960s carried strong overtones of religious metaphor, especially in relation to conversion. This is a rather general feature of scientific revolutions, as T. S. Kuhn has observed. A typical example is the experience of J. Tuzo Wilson who, in 1959, was a primary antagonist of the theory of continental drift. But within a few years, Wilson had undergone conversion and referred to himself as "a reformed anti-drifter" (see Wilson 1966, 3–9, for a discussion of his subsequent conversion). He not only then developed some of the fundamental geological evidence favoring continental drift but became the major herald of the revolution. In 1963, at a Symposium on the Upper Mantle Project in Berkeley during the XIIIth General Assembly of the International Union of Geology and Geophysics, Wilson boldly announced that "earth science is ripe for a major scientific revolution" (Takeuchi 1970, 244). The present situation in earth science, he said, was "like that of astronomers before the ideas of Copernicus and Galileo were accepted, like that of chemistry before the ideas of atoms and molecules were introduced, like that of biology before evolution, like that of physics before quantum mechanics."

In the late 1960s and early 70s, there were a large number of symposia on continental drift or on allied topics such as paleomagnetism and geomagnetism. One of these, part of the Annual General Meeting of the American Philosophical Society in April 1968, was entitled "Gondwanaland Revisited: New Evidence for Continental Drift." Here, in a paper entitled "Static or Mobile Earth: The Current Scientific Revolution," Wilson suggested (1968, 309, 317) that this "major scientific revolution in our own time" should "be called the Wegenerian revolution" in honor of its "chief advocate." Fellow earth scientists have generally agreed with Wilson that there has been a revolution and that Wegener should be honored as a first major articulator of the concept of continental drift. But this eponymous honor has not as yet been given to Wegener as it has to Copernicus, Galileo, Newton, Lavoisier, Darwin, and Einstein.

As we have seen, many writers have compared the revolution initiated by Wegener to the Copernican revolution. And it is similar, in that the eventual revolution in earth science was as far removed from the original theory of Wegener as the system of Kepler, Galileo, and Newton was from what was actually proposed by Copernicus. Just as there was no revolution in astronomy until more than a half-century after the publication of Copernicus's book in 1543, so there was no revolution in geological science until about a half-century after the publication of Wegener's original paper and his book. The eventual revolution that has been called Copernican was really a Newtonian revolution, based upon the achievements of Galileo and Kepler, and the 'Copernican system' that was at the foundation of the revolution was the system of Kepler. In like fashion, the eventual revolution that occurred in the earth sciences in the 1960s does not embody Wegener's theory but only his basic idea that the continents have not been fixed in their present place throughout all the earth's history and that they once clustered around the poles. Wegener's main contribution, the idea of mobilism as opposed to stabilism, played a role much like that of Copernicus's main contribution, the vision that it is possible to build a system of the world upon the concept of a moving rather than a stationary earth.

The general shift in earth science from stabilism to mobilism — specifically to ideas of continental drift and plate tectonics — is undoubtedly a revolution according to the four major tests presented in chapter 3. First, this change in accepted geological opinion has been recognized as a revolution by major observers *at the time*, including practitioners in this subject area. I take this to be a primary index of the occurrence of a revolution in science. Second, an examination of the contents of the science before 1912 and after 1970 shows the magnitude of the change to be sufficient for a revolution. Third, critical historians have concluded that the changes in the framework of geological thinking are of sufficient magnitude to constitute a revolution. Admittedly, this is a somewhat subjective judgment, but it serves as a corroboration of the conclusions of geologists and geophysicists participating in the revolution. We have seen that long before there was a successful revolution (that is, before there was a revolution in science, as opposed to a revolution on paper), a number of geologists—even those who opposed Wegener's ideas—were aware of the revolutionary character of the concept of continental drift and understood the revolutionary implications that its acceptance would have for the whole of geological science. Fourth, earth scientists today generally agree that a revolution has occurred in their discipline.

But what is the magnitude of the revolution? Is it a major revolution, to be compared to the Darwinian revolution, the revolution of quantum mechanics and relativity, or the Newtonian revolution? Or is it a revolution on a smaller scale, more akin to the Chemical Revolution? We have seen that George Gaylord Simpson called it a "major sub-revolution." D. P. McKenzie (1977, 120–121), in the conclusion of an article, "On the Relation of Plate Tectonics to the 'Evolution' of Ideas in Geology," contrasted the "impact of plate tectonics on the geological sciences" with the impact of "the discovery of the structure of DNA in the biological sciences." He concluded that "plate tectonics was less fundamental a revolution than the discoveries which began molecular biology" and noted that for this reason "the new ideas have . . . been assimilated and exploited faster in the geological sciences." But to any outside observer, aware of the radical change in our notions about the history of the earth, the magnitude of the shift in ideas would suggest a great revolution. Only because of the complete lack of an ideological component would it seem to be anything less.[12]

30

Conclusion:

Conversion as a Feature of

Revolutions in Science

M any aspects of revolution
have not been explored in this book — the process of creation and the
role of the individual scientist in the formulation and spread of revolutionary scientific ideas, the personalities of scientific revolutionaries,
and the effect of changing technologies and methods of scientific communication on revolutions in science. I have barely touched on the
many degrees and levels of interaction between revolutions in science
and their social, political, institutional, and perhaps even economic
matrices, and I have only indicated by examples the possible points of
contact or of succession between revolutions in science and social and
political revolutions.

But there is one variety of experience in scientific revolutions that
occurs again and again in the primary and secondary literature, and I
would like to give it some consideration here. That phenomenon is
conversion. Max Planck (1949, 33–34) is often quoted to the effect that
"new scientific truth does not triumph by convincing its opponents and
making them see the light, but rather because its opponents eventually

die, and a new generation grows up that is familiar with it." A similar sentiment was expressed a half-century earlier by Harvard's Professor Joseph Lovering, when he told his students that there are two theories of light, the wave and the corpuscular. Today, he is said to have remarked, everyone believes in the wave theory; the reason is that all those who believed in the corpuscular theory are dead. There is a measure of truth in such statements, as we all know, and yet a new scientific idea does win adherents, and even convinces some opponents, as has been seen in many examples throughout this book. Planck himself actually witnessed the acceptance, modification, and application of his fundamental concept by his fellow scientists. This feature of scientific revolutions—the winning over of working scientists—is common enough that I have used its magnitude as a mark of the transition from a revolution on paper to a revolution in science.

Such a change in belief may be a traumatic event. The adoption of radically new ideas almost always causes a rethinking of fundamentals — space and time, simultaneity, fixity of species, indivisibility of the atom, existence of genes, incompatibility of particle and wave, causality, predictability. Additionally, the new ideas cast previously accepted beliefs in a wholly different light or perspective. It is no wonder that scientists use such expressions as 'having seen the light' or write of a 'conversion', consciously or unconsciously comparing their experience to the classical drama of religious experience.

In his *Structure of Scientific Revolutions* (1962), T. S. Kuhn evoked this phenomenon with two phrases: an irreversible "gestalt switch," and a "conversion experience." He explicitly discussed the shift of allegiance from one paradigm to another as an act similar to religious conversion. While this was not a major feature of his schema, it does appear prominently, although Kuhn does not document it by examples. But whoever reads the literature of scientific revolution cannot help being struck by the ubiquity of references to conversion. Sometimes the introduction of conversion by a scientist is merely the use of the image and word, as when in 1796 Joseph Priestley (1796 [1929], 19–20) wrote about how "Dr. Black in Edinburgh, and as far as I hear all the Scots have declared themselves converts" to Lavoisier's new system of chemistry. Two centuries later, the physicist A. Pais (1982, 150) used the same language in relation to the new physics. And in the previous chapter we have seen the case of J. Tuzo Wilson, one of the architects of the theory of plate tectonics and an early scientific propagandist for the general theory of continental drift.

About a hundred years ago, T. H. Huxley wrote that the "one act of faith in the convert to science, is the confession of the universality of order and of the absolute validity in all times and under all circumstances, of the law of causation." The occasion of this remark was an answer to the charge that Darwin had "attempted to reinstate the old pagan goddess, Chance." Darwin's detractors, according to Huxley, have said that Darwin "supposes variations to come about 'by chance,'" and that the fittest survive the 'chances' of the struggle for existence." Thus they argue that in Darwin's theory "'chance' is substituted for providential design." In replying to Darwin's conservative critics, Huxley indicated that those who have talked thus of 'chance' are the "inheritors of antique superstition and ignorance, . . . whose minds have never been illumined by a ray of scientific thought." They are the unconverted, they have not as yet been converted to science; they have not made the "confession," to which Huxley referred, "of the law of causation." This "confession," Huxley explained, "is an act of faith." The reason is that "by the nature of the case, the truth of such propositions is not susceptible of proof." Yet such faith differs from other faiths because it "is not blind, but reasonable." This faith "is invariably confirmed by experience, and constitutes the sole trustworthy foundation for all action." Huxley not only went to great lengths to confute the opponents of Darwin; he did so by what must seem to us today to be an extravagant use of the analogy of religion. He carried this even further than I have indicated, referring to "people in whom the chance-worship of our remoter ancestors thus strangely survives" (Darwin 1887, 2: 199–200).

Conversion figured prominently in Darwin's correspondence. Here are some samples from 1858 and 1859:

[To A. R. Wallace, 25 Jan. 1859:] You ask about Lyell's frame of mind. I think he is somewhat staggered, but does not give in, and speaks with horror, often to me, of what a thing it would be, and what a job it would be for the next edition of 'The Principles,' if he were "*per*verted." But he is most candid and honest, and I think will end by being *per*verted. Dr. Hooker has become almost as heterodox as you or I, and I look at Hooker as *by far* the most capable judge in Europe.

[To C. Lyell, 20 Sept. 1859:] Your previously felt doubts on the immutability of species, may have more influence in converting you (if you be converted) than my book . . . I cannot too strongly express my conviction of the general truth of my doctrines, and God knows I have never shirked a difficulty. I am foolishly anxious for your verdict, not that I shall be disappointed if you are not converted; for I remember the long years it took me to come round; but I shall be most deeply delighted if you do come round, especially if I have a fair share in the conversion.

[To W. D. Fox, 23 Sept. 1859:] I am not so silly as to expect to convert you.

[To J. D. Hooker, 15 Oct. 1859:] Lyell is going to reread my book, and I yet entertain hopes that he will be converted, or perverted, as he calls it.

[To T. H. Huxley, 15 Oct. 1859:] I am very far from expecting to convert you to many of my heresies.

[To Asa Gray, 11 Nov. 1859:] Lyell . . . is nearly a convert to my views.

[To A. R. Wallace, 13 Nov. 1859:] Hooker thinks [Lyell] a complete convert.

Lyell later discussed this topic in a letter to Hooker (Darwin 1887, 2: 193n): "I find I am half converting not a few who were in arms against Darwin, and are even now against Huxley." He had to abandon "old and long cherished ideas, which constituted the charm to me of the theoretical part of the science in my earlier days, when I believed with Pascal in the theory, as Hallam terms it, of 'the archangel ruined.'" In these extracts, we may not only notice the use of the word 'convert' and other religious terms, but the expression by Darwin of the "long years" it took to convert himself. This is a common theme among scientists. In his autobiography, J. J. Thomson wrote about how difficult it was to convince himself that atoms are composite.

Darwin's correspondence also gives us some insight into the actual experience of conversion. On 21 November 1859, H. C. Watson, who called Darwin "the greatest revolutionist in natural history," wrote that "natural selection" embodies "the characteristics of all great natural truths": it clarifies "what was obscure," simplifies "what was intricate," and adds "greatly to previous knowledge" (Darwin 1887 2: 226). In his account of the reception of the *Origin of Species*, Huxley (1888), explained the effect of the new theory of evolution: like "the flash of light, which to a man who has lost himself in a dark night, suddenly reveals a road which, whether it takes him straight home or not, certainly goes his way." Then, using a religious metaphor, Huxley said that "Darwin and Wallace dispelled the darkness, and the beacon-fire of the 'Origin' guided the benighted" (Darwin 1887, 2: 197).

One of the most notable accounts of scientific conversion was written by the chemist Lothar Meyer. Years after the event, Meyer was recalling a striking event that occurred at the end of the Karlsruhe Congress in 1860, "one of the most important congresses in the history of chemistry" (van Spronsen 1969, 42), convened by the great organic chemist August Kekulé. This was the first time that an international assembly of scientists attempted to solve a pressing internal problem of science. At issue was the confusion between competing and rather different systems of atomic weights. So great was the uncertainty that many chemists turned from atomic weights to combining weights. The differences in atomic-weight systems arose from the divergence in atomic and molecular concepts; for example, could 'atoms' of the same chemical element combine to form 'molecules' (as the Italian chemist Amadeo Avogadro believed), or were chemical bonds only formed between 'atoms' of different elements (as held by John Dalton, founder of the atomic theory)? The congress was called to solve "once and for all" this thorny question, on which all structural formulas for organic chemistry had ultimately to depend (see de Milt 1948).

The chemists from all over the world, as may not be surprising, did not conclude their deliberations with a simple, universally accepted solution. But this congress did have positive results: Near the end, Stanislas Cannizzaro, a professor of chemistry at the University of Genoa, distributed a pamphlet in which he presented what is now the accepted solution of the problem for which the congress had been assembled. Drawing primarily on the work of Amadeo Avogadro, but also on the ideas of Charles Frédéric Gerhardt, Cannizzaro had been regularly teaching to his students the solution of the problem that baffled the whole scientific community. Lothar Meyer was 'converted' to Cannizzaro's system as soon as he read the pamphlet, and he went on to become one of the discoverers of the periodic table (or system) of the elements, along with several other scientists in different countries. Two years after the congress, Cannizzaro published his analysis in the *Jahresbericht über die Fortschritte der Chemie,* and Meyer, writing an introduction to a reprint some decades later, described his own conversion experience. This is so classic an account that it is worth quoting (translated into English) in full:

> After the close of the congress, at the author's behest, a friend, Angelo Pavesi, distributed a short, apparently insignificant work, Cannizzaro's *Sunto* . . . , reproduced here in translation. It had appeared a few years earlier but had not become well known. I, too, received a copy, which I pocketed in order to read it on the way home. I also read it over and over at home and was astonished at the light which the little work shed upon the most important points of controversy. Scales as it were fell from my eyes, doubts evaporated, and a feeling of the most tranquil certainty took their place. If a few years later I was able to contribute something to clarifying the situation and calming down overheated spirits, this is in no small part thanks to Cannizzaro's paper. Something similar must have happened to many other participants in the congress. The towering waves of the controversy began to smooth out. (Cannizzaro, 1891, 59–60)

The reader may note the reference to Saul of Tarsus (Acts 8:11) in the phrase "scales . . . fell from my eyes," as well as other language from religious experience. Meyer obviously believed that there were strong common elements in scientific and religious conversion.

In a historical study of revolutions, one cannot discuss conversion without having in mind that in classical times 'conversio' meant a revolution in the old cyclical sense, and even in religion 'conversion' retains some of the ancient sense of spiritual rebirth. But the modern use of this term, especially in science, implies radical change and the acceptance of a wholly different point of view. And so with these transformations of the concept we come full circle. Though analysts of science do not approve of a religious word such as 'conversion' in a discussion of scientific change, the historian's primary task in studying 'conversion'—as in studying 'revolution'—is not to make judgments concerning the actions and utterances of the past, but to record and to analyze them.

Supplements

A Note on Citations and References

Notes

References

Index

———◆———

Supplements

———◆———

1.1. Comparisons of Scientific and Political Revolutions

From the late seventeenth century onward, scientists and historians of science have used the language of political revolution (or political warfare) to describe the progress of science. An early example occurs in Walter Charleton's presentation of the atomic philosophy of the ancients and of Gassendi in *Physiologia Epicuro-Gassendo-Charltoniana* (1654). In the opening pages Charleton uses such expressions as "a Despot in Physiologie," "the dishonourable tyranny of that Usurper, Autority," "no Monarchy in Philosophy, besides that of Truth," "Assertors of Philosophical Liberty," and he speaks later of Galileo's "subversion . . . of Aristotles Doctrine" (pp. 435, 453).

In the eighteenth century, comparisons were frequently made between scientific and political revolutions. An early analyst who used political analogies in discussing revolutions in science was the French astronomer and historian Jean-Sylvain Bailly (for whom see 7.4 and 3.1, below). Bailly's *History of Astronomy* was published in 1775–1782, years of the American Revolution and the prelude to the French Revolution. Bailly often adopted the full panoply of political metaphor, as when he stated that "Newton accomplished, but in milder and fairer ways, what conquerors who usurped the throne sometimes attempted in Asia"

(1785, 2: 560). Bailly then explained that these usurpers "wished to efface the memory of the preceding reigns, so that their reign would serve as an epoch, so that everything would begin with them." Bailly (1: 337) also wrote about Copernicus as a man with a "seditious mind" ("esprit séditieux"), and of the need "to overthrow the throne of Ptolemy" ("renverser le trône de Ptolémée"). He also introduced the work of Newton by a declaration that "Aristotle and Descartes still shared the empire and were the preceptors of Europe" (2: 560).

Another eighteenth-century scientist and historian who invoked the political analogy of revolutions in science was Joseph Priestley. In 1796, seven years after the outbreak of the French Revolution, Priestley addressed a reply to the surviving chemists who had advocated a new system of chemistry; in an introduction to one of their works on that subject Lavoisier had predicted a revolution (see chapter 14, above). Priestley's essay was entitled *Considerations on the Doctrine of Phlogiston, and the Decomposition of Water* (1796). In his introduction (Priestley 1929), he acknowledged that there have been few, if indeed any, revolutions in science "so great, so sudden, and so general" as the change accomplished by "the Antiphlogistians," and he referred to his own time as "this age of revolutions, philosophical [i.e., scientific] as well as civil" (p. 20). Developing the metaphor of political revolution, Priestley declared that no man "ought to surrender his own judgment to any mere *authority*, however respectable" (p. 17). Then he invoked a direct image of the French Revolution: "As you would not, I am persuaded, have your reign to resemble that of *Robespierre*, few as we are who remain disaffected, we hope you had rather gain us by persuasion, than silence us by power." A little later, he referred again to the French Revolution. He pointed out that the advocates of the new chemistry and the new nomenclature had achieved a victory over Richard Kirwan, an ardent phlogistonist, as Priestley still was, perhaps the last surviving believer in the old chemistry. If they were to "gain as much" by their eventual reply to this pamphlet of Priestley's as they had done by their attack on Kirwan, Priestley said, then "your power will be universally established" and "there will be no *Vendée* in your dominions" (p. 18).

The theme of political and scientific revolutions occurs frequently during the nineteenth century. An outstanding example is found in the writings of the mathematician, philosopher, and historian of science William Whewell, Master of Trinity College, Cambridge. In his *Philosophy of the Inductive Sciences* (1847, 2: 255–257), Whewell castigated Descartes for his "endeavour to revive the method of obtaining knowledge by reasoning from our own ideas only, and to erect it in opposition to the method of observation and experiment." The Cartesian philosophy thus seemed to Whewell "an attempt at a counter-revolution." This introduction of political metaphor is very striking, especially in the context of "the great reform of scientific method which was going on in his [Descartes's] time" (p. 257). Whewell continued to use the language of politics as he described how "Descartes came to be considered as the great hero of the overthrow of the Aristotelian dogmatism" (p. 258). Whewell held that Descartes's "physical doctrines" were of value, however, because of "their being arrived at in a way inconsistent with his own professed method, namely, by a reference to observation" (p.260). But Whewell insisted that "the actual progress of physical science" in the seventeenth century was based on experiment or observation and induction and thus constituted a "revolution from the philosophy of tradition to the philosophy of experience" (p. 263). This was the "revolution in the method of scientific research" (cf. pp. 204, 208, 228).

In a hitherto unpublished letter from the anthropologist and brain surgeon Paul Broca to Carl Vogt (Paris, 8 March 1865; this letter, in the Bibliothèque Publique et Universitaire de Genève, was brought to my attention by Joy Harvey) Broca lamented the recent death of the physiologist Louis Pierre Gratiolet, who—according to Broca—"believed unreasonably in species and was consequently anti-Darwinian." (Gratiolet, however, was "resolutely polygenist" and considered man to be an animal.) Broca combined his socialist politics with his scientific position in order to describe Gratiolet's place "in our science." In science he was representative of "a party comparable to the one which in the old political assemblies used to be called the left center." Then Broca expressed a general philosophical point of view: "It must be admitted that if in the history of progress, movement is provoked by the extreme left, it is gradually realized by men who are less advanced, less logical, but more in harmony with the masses."

In our own century, echoes of political revolution are also found to spill over into discussions of scientific revolution. For example, in a famous debate over behaviorism, William McDougall accused John B. Watson, the founder of behavioristic psychology, of propounding ideas that would appeal only to those who were sympathetic to bolshevism (see chapter 25, above). In his Gifford Lectures for 1927 on *The Nature of the Physical World*, Sir Arthur Eddington took account of the charges of bolshevism raised against the new physics, notably relativity. We shall see, in the discussion of reactions to behaviorism and relativity, the suggestion that relativity (bolshevism in science) was a symptom of the same malaise of the world that had enabled bolshevism to triumph in Russia. This sentiment was most vocally expressed by Professor Charles Lane Poor (see chapter 25, above), abetted by Professor Thomas Chrowder Chamberlin, an emeritus professor of earth science at the University of Chicago. Introducing Poor's book (1922), Chamberlin admitted that he was deeply concerned by the "present loosening of ties and venturesome drift," which he believed was "not confined to the strenuous affairs of life just now distraught by extraordinary conditions." No, he found (p. viii) that this "reaches down into the fundamentals of thought and touches the intellectual instincts inherited from the great past." In particular, "the basal concepts of space and time, the very framework of thought, are being called in question." He found a terrible anarchy, in which "the very roots of the thinking of the ages are being digged about with little show of care whether it will promote growth or lead to withering." What bothered Chamberlin was not only that it was being "urged that the geometry of Euclid, the dynamics of Galileo, and the celestial mechanics of Newton are basally defective," but that "these revolutionary claims are put forth, of course, in the name of progress." Poor himself stated in his book, as in his articles and his lectures, that there is a worldwide "mental as well as physical" "state of unrest." He saw the "physical aspects of this unrest" in "the strikes, the socialistic uprisings, the war"; and he found evidence for "the deep mental disturbances" in "the widespread interest in social problems," the "futuristic movements in art," and the "easy way in which many cast aside the well tested theories of finance and government in favor of radical and untried experiments." Relativity theory indicated that "the same spirit of unrest has invaded science." And what made matters worse was that "this radical, this destroying theory has been accepted as lightly and as easily as one accepts a correction to the estimated height of a mountain in Asia, or to the source of a river in equatorial Africa" (p. v).

A somewhat different theme was expressed at about the same time by the astronomer James H. Jeans, like Eddington a leading, high-level popularizer of relativity and quantum physics. Writing on relativity in the twelfth edition of the *Encyclopaedia Britannica* (1922, 32: 262), Jeans observed how the language of political action had entered the discussion of science. He had been describing the experiments of A. A. Michelson and in particular the Michelson–Morley experiment, which had had a "*null*" result." It appeared "that if the earth moved through the aether this motion was concealed by a universal shrinkage of matter," that is, a Lorentz–Fitzgerald 'contraction'. This "shrinkage," however, "was in turn concealed by some other agency or agencies whose wit, so far, appeared to be greater than that of man." Jeans's description of the then-current state of mind is redolent of an anthropomorphism that seems not fitting in a discussion of physics. But Jeans next introduced the parallelism between political and scientific discourse that he found physicists to be using. "At this time," he wrote, "the word 'conspiracy' found its way into the technical language of science." It was supposed that there was "a conspiracy on the part of the various agencies of nature to prevent man from measuring his velocity of motion in space." He concluded, "The conspiracy, if such there was, appeared to have been perfectly organized."

On first reading this paragraph, I thought that Jeans was exercising his well-known caustic wit, but the next paragraph quickly dispelled any such notion. "A perfectly organized conspiracy of this kind," he wrote, "differs only in name from a law of nature." He gave, as example, the case of an inventor who would try to produce a perpetual-motion machine. It would appear to him that "the forces of nature have joined in a conspiracy to prevent his machine from working." A deeper knowledge of physics, however, would show him that "he is in conflict not with a conspiracy, but with a law of nature—the conservation of energy." Jeans hailed Einstein's theory of 1905, in which Einstein "propounded the hypothesis that the apparent conspiracy might be in effect a law of nature." He then proceeded to analyze and to discuss the consequences of Einstein's "hypothesis of relativity" in a remarkably lucid presentation of special relativity, followed by an equally brilliant account of general relativity.

The German physicist Johannes Stark was one of the first scientists to recognize the significance of relativity theory, and he was certainly the first to request an article from Einstein on this subject. Writing in 1922, in a book entitled *The Current Crisis in German Physics (Die gegenwärtige Krisis in der deutschen Physik*, pp. 14–15), Stark linked political and social revolution with scientific revolution. He explained how seditious "propaganda found fruitful ground in the time of political and social revolution [Revolution]" after the first World War, when supporters of Einstein's theory of relativity "spoke of the revolution [Umsturz = revolution, upheaval] in our hitherto held views of space and time and of a theory embracing the world."

A final example is provided by the discussion of relativity by the Spanish philosopher José Ortega y Gasset in his *The Modern Theme* (1923; Eng. trans. 1961). Ortega compared science and politics in a contrast of the intellectual positions of Einstein and Lorentz. Here he saw displayed the very differing "intellectual temperaments" of these two physicists. Lorentz represented "the old rationalism," which obliged him "to conclude that it is matter which yields and contracts." This theory of contraction seemed to Ortega to be an "example of utopianism." It was, he wrote, "the Oath of the Tennis Court transferred to physics." The "contrary solution" was adopted by Einstein: "Geometry must yield, pure space is to bow to observation, to curve, in fact." This led Ortega to conclude: "In the political sphere, supposing the analogy to be a perfect one, Lorentz would say: Nations may perish, provided we keep our principles. Einstein, on the other hand, would maintain: We must look for such principles as will preserve nations, because that is what principles are for." He ended his political analogy by referring to a change in the whole course of physical science imposed by Einstein, who broke the centuries-old "undisputed dictatorship" of the "*role* of geometry, of pure reason," to which he gave a "much more modest" *rôle*. After Einstein, reason descended "from dictatorship to the status of a humble instrument, which has, in every case, first to prove its efficacy."

4.1. Matthew Wren and the English Revolution

Possibly the earliest writer to use the word 'revolution' in relation to the English Revolution was Matthew Wren, the son of Dr. Matthew Wren, Bishop of Ely. Wren studied at Cambridge, and afterward was a private student in Oxford. After the Restoration, he became secretary to the Earl of Clarendon, served in Parliament, and was secretary to the Duke of of York until his death in 1672. He was one of the founding Fellows of the Royal Society and a charter member of the Council. In a tract of 1659, entitled *Monarchy Asserted, or the State of Monarchicall & Popular Government in Vindication of the Considerations upon Mr Harrington's Oceana,* Wren stressed the fact that he and Harrington had drawn very different conclusions from the same natural causes. He drew on the science of optics for an analogy, referring to the different colors of objects seen in different lights. On page 19, he discusses "those two waies of *Naturall and Violent Revolution* by which Mr Harrington imagines *Propriety may come to have a being before Empire.*"

He was particularly concerned with the "Ballance" between the power of money and of land and (in ch. 3) discussed the question: "Whether the Ballance of Dominion in Land be the Naturall Cause of Empire." He concluded (in ch. 6) that "the Solution of this Question may give Birth to a New One, Whether Mr *Harrington* be the better *Civill Lawyer* or *Mathematician?*" In chapter 3, Wren produced two corollaries, of which the second reads: "And that the Conformity of the Ballance to the establish't Government does not necessarily secure a State from Changes and Revolutions" (p. 33). It is not at all clear as to whether he was thus making an antithesis between ordinary mutations and cyclical sequences of events or 'revolutions', or was using 'revolution' to indicate a change of considerable magnitude. It has been suggested that by 'revolution' Wren intended "to convey to his reader the notion of a completed political cycle." It is more likely, however, that Wren was using 'revolution' here in the sense of a great change, a noncyclical event that would be 'revolutionary' in the sense in which this word is generally used today.

For more precise information, we may turn to another tract, not published until 1781, entitled *Of the Origin and Progress of the Revolutions in England*. This work ends with the entrance of the victorious rebel army which "had no hindrance from marching to *London;* where they turned all the Lords, and the greatest part of the Commons out of doors, and began to set things in order for the King's trial" (p. 252). Wren's stated aim was to present "an exact view of the Beginning, and Progress, of those Mischiefs that have devoured the Church and Crown of *England*" (p. 228). Wren used such words as 'revolt' and 'rebellion', but he also wrote of 'revolution'. He said he would "not make it my business to relate" how the "Parliament . . . by accusations of treason deprived the King of his most considerable Councellors, and deterred the rest; . . . how by tumults they got the Bishops out of the Lords house, and drove the King from *Whitehall;* . . . and finally how the war was begun, and carried on." "The world is full both of books, and pamphlets," he wrote, "who have nothing to do but to teach their readers these events." His own design was rather "summarily to treat of the most general causes of those strange revolutions we have seen" (p. 242). I find no implication here of a cyclic sequence of events, but rather the notion of extraordinary occurrences or unusual changes.

And the same is true of the only other occurrence of 'revolution' in the text of the tract. Having sketched out the main background of the Rebellion—and having delineated the causes of the "actions, which disposed the matter, and (as it were) burnt all to tinder for that great flame, which has devoured three Kingdoms"—Wren turned to the events of the Civil War leading up to the capture of the King. But first he referred to the "hands that endeavoured to strike fire" into the tinder, to those who "were (not to speak of seditious preachers and libellers) a pack of discontented Noblemen and Gentlemen of this and the *Scottish* nation, who, either by ambition, revenge, or avarice, were engaged to labour a revolution of affairs." This, again, reads as if Wren had in mind an overthrow of the established system, an extraordinary event, an occurrence very much like that which we would call a revolution today.

4.2. Cromwell and Revolution

The writings of Oliver Cromwell typify the difficulties in interpreting the mid-seventeenth-century usage of 'revolution':

> They that shall attribute to this or that person the contrivances and production of those mighty things God hath wrought in the midst of us, and [say] that they have not been the revolutions of Christ himself, upon whose shoulders the government is laid, they speak against God, and they fall under his hand without a Mediator.
>
> Therefore whatsoever you may judge men for, and say, this man is cunning, and politic, and subtle, take heed, again I say, how you judge of his revolutions, as the products of men's inventions.
>
> And I say this, not only to this assembly, but to the world, that that man liveth not, that can come to me, and charge me that I have in these great revolutions made necessities; I challenge even all that fear God. And as God hath said, *my glory I will not give unto another;* let men take heed and be twice advised, how they call his revolutions, the things of God, and his working of things from one period to another, how, I say, they call them necessities of men's creations; for by so doing they do vilify and lessen the works of God, and rob him of his glory, which he hath said he *will not give unto another,* nor suffer to be taken from him. We know what God did to Herod, when he was applauded, and did not acknowledge God. And God knoweth what he will do with men, when they shall call his revolutions human designs, and so detract from his glory.

In each of these five occurrences of 'revolutions', the sense is clear insofar as Cromwell is saying that men should not take "the revolutions of Christ" or "his [i.e. God's] revolutions" as "the products of men's inventions" or as the "necessities of men's creations"; but it is far from clear whether these revolutions are cyclical events of some sort or simply moments of great change with no antecedents (see Cromwell 1945, 591, 592).

In this context, it may be of interest to note that the Great Seal of 1651 expressed the spirit of those days in these words: "Freedom by God's Blessing restored." Hence the English Revolution (as Hannah Arendt 1965, 43, observed) "was officially understood as a restoration."

4.3. The Reformation as Revolution

Whoever contemplates the great changes that have occurred in modern times must give primacy of place to the Reformation — a real revolution of the sixteenth century, more of a revolution than the almost contemporaneous one alleged by historians to have been begun by Copernicus. The Reformation fulfills all of the usual requirements for a revolution—the seizure of power, the overthrow of one system and its replacement by a radically different one. The Reformation, furthermore, was marked by violence and religious wars that are analogous to the civil wars subsequent to revolutions within a single state. Why, then, it may be asked, do we not refer to the Reformation as the 'Protestant Revolution'?

The answer to this question has two parts. First, these momentous events have been and still are sometimes called the 'Protestant Revolution'. For example, in a current college textbook on European history (Burns et al. 1982) a major section of chapter 19 — "The Age of the Reformation (1517–c.1600)" — is entitled plainly "The Protestant Revolution."

The second part of the answer to the question of the Reformation as a 'Protestant Revolution' is the matter of date. The *Oxford English Dictionary* gives evidence that in England the name 'Reformation' was used as early as 1563 and 1588 and in the mid-seventeenth century, well before the concept and name of 'revolution' had entered general discourse to denote violent radical changes in which a new system was introduced. Furthermore, there had been a long tradition, going back to the fifteenth as well as the sixteenth century, of using the word 'reformation' to indicate what the *O.E.D.* defines as: "Improvement of (or in) an existing state of things, institutions, practice, etc.; a radical change for the better effected in political, religious, or social affairs." In French two words are used for the Reformation: 'réformation' was originally "the general term to designate the religious revolution of the XVIth century" and 'réforme' was applied "restrictively to the work of Zwingli and Calvin" (Littré 1881, 1546). Nowadays, however, these "two terms are used indifferently, one for the other," although *réforme* and not *réformation* is used in speaking of monasteries.

In the seventeenth century Thomas Sprat discussed the Reformation in his *History of the Royal Society* (1667, 371). Comparing the reformation in the church with the design of the newly founded Royal Society, Sprat wrote that both the church and the society forsook the "Ancient Traditions and ventured on novelties." Both thought that their ancestors may have made mistakes. By the end of the eighteenth century, the designation 'revolution' was frequently used for the Reformation, especially in Germany.

I have not attempted to trace the history of the conception of the Reformation as revolution, but I did find that this was not uncommon in the late eighteenth and early nineteenth centuries. In 1791 the German philosopher and poet Karl Heinrich Heydenreich referred to the Reformation as "that revolution [*Revolution*] which was so beneficial theologically" (Buonafede 1791, pt. 1, pref.). On 18 April 1802 the Institut de France announced a prize competition for an essay on the subject: "What was the influence of Luther's Reformation on the political situation of the different states of Europe and on intellectual progress?" The prize went to Charles de Villers, with honorable mention to the Marquis de Maleville (who wrote from a Catholic viewpoint) and the writer Jean-Jacques Leuliette (Wittmer 1908, 194–253). In the published essays (1804), all three wrote of the Reformation as a 'révolution'. In 1820, in a series of lectures on the philosophy of Kant, published almost forty years later (1857), Victor Cousin wrote of the Reformation that what was wanted at first was "only a reform, as formerly at Constance and Basle," but that the "rejected reform engendered a revolution" (p. 11). Cousin, in lectures of 1829, again referred to the Reformation as one of the two components in the revolutions by which the modern era threw off the Middle Ages and attained liberty. This change was achieved in part by the religious revolution (the Reformation) of the sixteenth century and in part by the political revolution (the English Glorious Revolution) of the seventeenth century and accomplished fully by the French Revolution of the eighteenth century. He also wrote of a concomitant revolution in philosophy (Cousin 1841).

5.1. Revolution, Newness, and Improvement

Anyone who was conversant with the activity of scientists in the seventeenth and eighteenth centuries would be aware of a succession of new discoveries and the invention of new instruments like the telescope and the microscope, which were enlarging the mode of inquiry into the operations of nature. A constant theme among writers about the sciences is this feature of newness, which tended to be seen as an improvement rather than a revolution in knowledge. The breaking of the bounds of tradition of science was symbolized in the emblematic design of the title-page of Francis Bacon's *Novum Organum* of 1620, setting forth the "new tool" of the sciences: the method of induction. Here was displayed a ship symbolically sailing through the Pillars of Hercules, at the western end of the Mediterranean, which in ancient times had marked the end of the known world. Thus Bacon was implying that the traditional motto *Ne plus ultra* (No further!) was to be altered to *Plus ultra* (Further!). This interpretation appears in Joseph Glanvill's choice of title, *Plus Ultra*, for a book of 1668, subtitled *The Progress and Advancement of Knowledge since the Days of Aristotle*.

Bacon declared the newness of his method for the sciences when he sent a presentation copy to the University of Cambridge; he evidently felt the need to excuse himself for the boldness of his creation. Accordingly, he sent an accompanying letter (Mullinger 1884, 2: 573) saying, "Let it not trouble you that the way is new." And he added the comment, "For such innovations must be brought into being through the revolutions of the ages and the centuries." ("Nec vos moveat quod via nova sit. *Necesse est enim talia per aetatum et saeculorum circuitus evenire*.")

Glanvill's *Plus Ultra* is typical of the writings of his age. In the preface he refers to "the late *Improvements of Science*" and in the beginning of his text (p. 6) mentions the "many *Arts, Instruments, Observations, Inventions* and *Improvements*" that "have been disclosed to the World since the days of *Aristotle*." Chapter two stresses the "ways whereby *knowledge* may be *advanced*," or the "*Ways of improv-*

ing Useful Knowledge," exemplified in two subjects in which the moderns had by
far (and obviously) surpassed the ancients: chemistry and anatomy. Anatomy
(pp. 12, 13) has undergone *"vast Improvements,"* even *"wonderful* Improvements,"
of which "the Noblest" (p. 15) is Harvey's discovery of the circulation of the
blood. Chapter three introduces "the Improvements of *Mathematicks,"* followed by chapter four on *"Improvements* in *Geometry* by *Des Cartes, Vieta,* and *Dr.
Wallis."* The theme of newness goes on throughout the book, but the result of
innovations is always improvement rather than revolution. Thus there are *"late
Improvements* of *Astronomy"* (ch. 5), *"Improvements* of *Opticks* and *Geography"*
(ch. 6), and *"Improvements* with *Arts* that search into the *recesses* and *intrigues* of
Nature, with which *latter Ages* have assisted *Philosophical* Inquiries," that is,
telescopes, microscopes, thermometers, barometers, and air-pumps. A whole
chapter (13) is devoted to Boyle, lauded for his "promotion of *Useful* Knowledge," who not only "cultivated" science but "improved it."

Glanvill, like many others of the seventeenth century, did not assign a major
role in the development of astronomy to Copernicus. Whereas almost two pages
are devoted to "the Noble *Ticho* Brahe," a mere sentence (sandwiched between
the work of Sacrobosco, Thābit, and Regiomontanus and of Joachimus [Coper-
nicus's pupil Rheticus] and Clavius) suffices for Copernicus. The latter merely
"restored the *Hypothesis* of *Pythagoras* and *Philolaus,"* and "gave far more *neat*
and consistent *Accounts* of the *Phaenomena."*

Not only did Glanvill fail to conceive of improvements in science as revolutions; he argued explicitly that progress in science is gradual and incremental.
The *"true* knowledge of *general Nature,"* he said (p. 91), is "like *Nature* it self." It
"must proceed *slowly,* by degrees almost insensible." That is, "what *one* Age can
do in so *immense* an Undertaking" is "little *more* than to remove the *Rubbish,* lay
in *Materials,* and *put* things in *order* for the *Building."* He asserted that "the
business of the *Experimental Philosophers"* is to *"seek* and *gather, observe* and
examine, and *lay up* in *Bank* for the Ages that come after."

The theme of improvement rather than revolution is a feature of the many
books of the late seventeenth century in England and in France that discuss the
achievements of the 'moderns' or compare and contrast the 'ancients' and the
'moderns' or that — in England — defend or attack the new Royal Society. In
France the two chief proponents of the moderns in *The Quarrel of the Ancients
and the Moderns* were Charles Perrault (1688) and Bernard de Fontenelle (1688).
In England the literature is greater and is usually grouped together under the
name of Swift's satire, *The Battle of the Books* (1704). Perrault left the reader in no
doubt of his belief that knowledge had increased enormously since the beginning of the seventeenth century and especially since the founding of the Royal
Society and the Paris Academy of Sciences. Among the improvements introduced by the moderns were (pref.; 1964, 97) telescopes and microscopes and of
course (1688, 97–8; 1964, 125) the circulation of the blood. Not only did Perrault
come out for progress and improvement and not revolution, but he too expressed a belief (1688, 70; 1964, 118) that "the progress of the Arts and of the

Sciences, . . . although very considerable, . . . is made only bit by bit and in an imperceptible manner."

Fontenelle is kinder to the ancients than many of the other writers of such comparisons and contrasts. In his *Digression on the Ancients and the Moderns* he argues that since science is cumulative and progressive, it follows that more recent scientists must know more than their predecessors. He developed this theme most notably in his *New Dialogues of the Dead* (1683), a popular work which was twice translated into English (by R. Bentley, London, 1685; by John Hughes, London, 1708, 1730). Scientists appear in many of the dialogues; among them Erasistratus, Raymond Lull, Paracelsus, Harvey, Galileo, Descartes. Although Fontenelle lauds the accomplishments of the moderns, he often points out that their achievements and improvements have not fully solved the fundamental problems. Thus, we have seen that, in the dialogue between Harvey and Erasistratus, Fontenelle has Harvey admit that doctors are still unable to cure diseases and prevent deaths. In neither these dialogues nor in his book on the ancients and moderns does Fontenelle indicate that there may have been a revolution in the sciences. Moreover, even in a famous essay on the utility of mathematics, written at about the same time as his declarations that the calculus constituted a revolution, Fontenelle does not invoke the notion of a revolutionary change. Rather (1699) he wrote of "improvements in geometry" and other useful sciences. But in this essay, he did use the concept and term 'revolution' in the traditional sense of "the perpetual Revolutions of Human Affairs, of the Beginning and Fall of Empires." This same sense of revolutions occurs in one of the *Dialogues of the Dead* in which the protagonists are Paracelsus and Molière (author of *Le médecin malgré lui*). Molière is made to refer to the "Revolutions [that] may happen in the Empire of Letters" and also "the Vertues of Numbers, the Properties of Planets, and Fatalities linked to certain Periods, or certain Revolutions" (Hughes trans., 1730, 209–214). But neither Molière nor Paracelsus introduces the concept of revolution as a notable change in medicine or in science.

Like Fontenelle, Thomas Sprat used the word 'revolution' in a nonscientific context. In his *History of the Royal Society* (1667, 58), he wrote about the political events ("Twenty years Melancholy" which belongs "to another kind of History") in terms of "those dreadful revolutions" ending with "the *King's* Return." Another occurrence of 'revolution' appears in Sprat's discussion of America. Referring to the inaccessibility of the lands under Spanish domain, Sprat expressed the hope that some day knowledge of this part of the world would become "more familiar to *Europe*, either by a *free Trade*, or by *Conquest*, or by any other *Revolution* in its Civil affairs." The choice of language leaves no doubt that for Sprat such a 'revolution' would be merely a great change.

The participants in the quarrel of the ancients and the moderns or the battle of the books were part of a struggle to establish the idea of progress. This theme has been the subject of John B. Bury's celebrated monograph of 1920, which elaborated on a grand scale a theme discussed by Ferdinand Brunetière in 1893.

In 1945 Edgar Zilsel added a new dimension to such studies in his analysis of "The Genesis of the Concept of Scientific Progress." As a result of such studies we are now aware that men of the sixteenth and seventeenth centuries were divided in their attitudes toward novelties. Shakespeare (*1 Henry IV*, V, 1) railed at "hurly burly innovations," but Sprat (1667, 321) said that he was not "frightened at that, which is wont to be objected in this case [that is, the new method of experimentation], the hazard of alteration and novelty." The acceptance of the new knowledge rather than the ancient verities was put very clearly by William Watts (for whom see Jones 1961, 72) when he wrote: "Of this one thing I am confident: that you are all so *rationall* and *ingenuous* as to preferre *Truth* before *Authority: Amicus Plato, Amicus Aristoteles,* but *Magis amica veritas.*" This sentiment (see Guerlac 1981, ch. 1) was invoked by Watts in relation to the way in which accounts of voyages "conduce to the improvement of Philosophy."

In 1638 John Wilkins (see Jones 1961, 77) published *Discovery of a New World,* in which his chief purpose is to show that the moon is inhabited. He also set forth the principle of "the Growth of Sciences." That is, he asserted that all discoveries have not been made. There is much "left to Discovery," he wrote, "and it cannot be any Inconveniency for us to maintain a New Truth, or rectifie an ancient Error." Wilkins attacked the "superstitious" and "lazy Opinion" that Aristotle's works are "the Bounds and Limits of all Human Invention," beyond which there can be "no possibility of reaching."

Although (in a work published a year later), Wilkins referred to Bacon as "our *English Aristotle*," he stressed the aspect of progress in Bacon's philosophy, referring with admiration to "that Work truly stiled the *Advancement of Learning*" (ibid., 78). Wilkins held that Bacon had not merely devised a new system replacing the old one of Aristotle, but the new Baconian philosophy embodied a method that would produce progress in all fields of learning, including science.

In his *Ancients and Moderns* (1936; 1961), R. F. Jones has stressed the significance of Francis Bacon's disparagement of the ancients and has especially called attention to George Hakewill's *Apologie, . . . the Common Errour touching Nature's Perpetuall and Universal Decay* (1627; 3rd ed., 1635), an influential work that primarily addressed a central question in discussions of ancients and moderns: has nature (the universe, the earth, and all of living nature) been subject to decay since ancient times? Hakewill (Jones 1961, 36) argued against a "corruption of nature" on the basis of his belief in a cyclical process of successive decays and ascendencies. But he recognized that if there were no constancies in the qualities of nature and of men's minds, there would be no basis of comparing and contrasting ancient and modern knowledge. Apparently, Hakewill was the first to introduce into this controversy "the new science which was destined later to play an overwhelming part in it."

In the present context the importance of Hakewill is his recognition that there have been new discoveries in science (he points to the work of Vesalius, Fallopius, and other anatomists) and new inventions. But he was not very conversant with the new science. Out of some six hundred pages only about eight are devoted to anatomy and medicine (bk. 3, ch. 5, §6; ch. 7, §4–5) and about four to astronomy, geometry, and physics. But a whole chapter (bk. 3, ch. 10) deals with the three great inventions: printing with movable type, gunpowder, and the magnetic mariner's compass. In the discussions of anatomy he tends to list names of anatomists rather than their specific achievements. But in the field of astronomy and mathematics, he was aware enough of his ignorance to turn for help to his "learned friend Mr *Brigges*, Professour of *Geometrie* at Oxford." Briggs's essay prints to almost a page (p. 301) and is in Latin.

Briggs explained the main features of Copernican astronomy, by which, he said, it is possible to account for planetary phenomena "much more easily and more accurately than by the Epicycles of Ptolemy or any ancient writer or by other Hypotheses." Briggs noted that "the recently discovered Optical tube" enables us "to see that Venus and Mercury move in orbit round the sun" and he devoted a whole paragraph to the "four Medicean stars," the planets which "revolve about the star Jupiter." This name, he observed, was given to them "by the Florentine Galileo Galilei, who was the first to discover them through an Optical Tube."

At the very start of the book, Hakewill sets forth his theory of cycles, "though there be many changes and variations in the World yet all things come about one time or another to the same points again." Hence he believes that there are "revolutions and returnes" in "*Arts* and *Sciences*" as there are "revolutions and returnes of *vertues* and *vices.*" This usage of the word 'revolution' occurs frequently, as in "the revolution of a few ages" (p. 39). But despite his initial avowal that "there is nothing new under the *Sunne*," Hakewill did find new things coming into being. For example, he declared that there was "a *new* kinde of *Physicke* professed by a *new sect* of *Physitians*," which was "never heard of in the world before" (p. 275). It was "altogether differing from the *Ancients*" in "*name* and *tearmes of art*," in "*matter*, and *methode* and *manner* of proceeding," in both "*doctrine* [and] . . . *practise*." This new medicine had been founded ("if wee may credit himself") by Paracelsus.

Although Hakewill believed in "a kinde of *circular progress*," he also held that "the *Arts* and *Sciences*" had in recent times "either been *revived* from *decay*, or *reduced* to *use*, or *brought forward* to *perfection*" (p. 312). With each new flourishing, after a failing and fading away, there were significant novelties and improvements. For Hakewill, such steps toward perfection were most discernible in knowledge (that is, science) and in arts or inventions.

5.2. Humanism and Revolution in Science in the Sixteenth Century: Vesalius and Copernicus

Men of science of the sixteenth and seventeenth centuries tended to view their own achievements from a special point of view that was characteristic of Renaissance humanism. It used to be thought that the primary, even the defining, characteristic of humanists was that they "devoted themselves to the study of the language, literature, and antiquities of Rome, and afterwards of Greece"*(O.E.D.)*. But the newer interpretation of the humanist movement would stress, in addition to a respect for antiquity, a set of aesthetic, historical, philosophical, and scientific ideals that set them apart from contemporaries (see Burdach 1963; Cassirer 1963; Kristeller 1943; Edelstein 1943). It was characteristic of such humanists that they would not present their science in either the vernacular tongues or the scholarly Latin of their day; they refined and purified the vocabulary and cast aside the simplified or colloquial prose style of Latin in favor of "the terminology of a time long past." Thus Vesalius (as Olschki 1922, 2: 81ff., 98ff., was the first to note) consciously adopted the 'Kunstprosa' (artistic prose) of Cicero, apparently the "first anatomist to do so" (Edelstein 1943, 548). He chose his language and terminology from the Roman Celsus, said to have been the "only ancient author to apply the rules of Ciceronian 'Kunstprosa' to the composition of medical writings" (ibid., n. 3). Vesalius adopted the "stately and rhythmic style" and "rhetorical word order" of Cicero consciously. This may be seen clearly in his *Letter on Bloodletting* of 1539, described by Edelstein as "a masterpiece, embodying all those qualities which Vesalius admired in the ancients." In his edition of Rhazes (published in 1537), Vesalius included a prefatory letter (trans. in Cushing 1943) disparaging the coarse Arabic style and displaying his own "appreciation of the classical style, its beauty, harmony, elegance, and conciseness." Whoever examines Vesalius's *De Fabrica* (1543) is immediately aware of the author's concern for philology and insistence on proper technical terms. A careful study has shown how the second edition of the *De Fabrica* embodied changes in style, grammar, and punctuation (Roth 1892, 76, 98f., 216). Although Vesalius's *De Fabrica* is devoted to anatomy, there are two asides in which Erasmus is criticized for faulty Latin (Roth 1892, 149, 232; Edelstein 1943, 549, n. 2).

This excursus on Vesalius's language and style is important not only to place him directly and consciously in the humanists' camp, but also to make precise his role as a possible revolutionary. Vesalius had made the text containing his great empirical discoveries difficult for the doctors and anatomists of his day to read, writing it in a style and using a vocabulary that was foreign to them. This aspect of his presentation (as Olschki 1922, 2: 99, suggested) explains why a German translation of Vesalius's "complicated Latin" text was made speedily and published in 1543, the same year as the original publication. There is no other large scientific treatise of this period, or even of the next century, which was at once so translated from the Latin and published in a vernacular language. If I seem to be stressing unduly the humanist stylistic quality of Vesalius's text, the reason is that it is necessary to understand the degree of Vesalius's commitment to the humanist creed in order fully to appreciate his sentiments about ancient knowledge and his self-image as a revolutionary.

The word 'revolutionary' is apt to leap into the judgment of any scholar who examines Vesalius's achievement today. What he accomplished was radical and new. He set anatomy on a modern foundation, based on careful empirical studies, on the results of actual dissections of human bodies, and he turned the profession away from reliance on older and faulty books. From that day to this, doctors have been introduced to medicine by performing or observing actual dissections of human bodies. But in his great work "on the construction of the human body," Vesalius does not say that he has produced something new and unheard-of, as men of science would tend to declare in the next century, and as Paracelsus, Telesio, and Cardano were apt to do in Vesalius's own time. Rather, Vesalius's mode of presenting his magnificent achievement is one that has been characterized as "only the revival of the work of ancient anatomists" (Edelstein 1943, 551). Vesalius himself leaves no doubt on this score. Anatomy, he unambiguously says, "should be recalled from the dead, so that if it did not achieve with us a greater perfection than at any other place or time among the old teachers of anatomy, it might at least reach such a point that one could with confidence assert that our modern science of anatomy was equal to that of the old." Vesalius hoped that in the age in which he lived "anatomy was unique both in the level to which it had sunk and in the completeness of its subsequent restoration." He found grounds to rejoice that "there is hope that anatomy will ere long be cultivated in all our academies as it was of old in Alexandria" (trans. Farrington 1932, 138iff.). We "are witnessing," he wrote in the preface to *De Fabrica*, "the fortunate rebirth of medicine."

Vesalius was not merely indulging in fancies and hypotheses when he wrote of the long-past knowledge of anatomy. Cicero and Celsus had written about dissections of the human body by the ancients; in *De Fabrica* Vesalius, in four different passages, on the authority of Galen, gives the names of great anatomists of Hellenistic times (Roth 1892, 148 n. 5). Even Vesalius's interest in the anatomy of the living body (a particular feature of the illustrations), and in the possibility of vivisection (*De Fabrica*, §7, ch. 19; cf. Roth 1892, 152), has been traced back to a passage in the preface to Celsus's *De medicina* concerning the superiority of vivisection to the dissection of a corpse as a means of learning about the structure and function of the human body.

Also Vesalius's argument concerning the cause of the decay of medicine and anatomy—in that physicians had stopped performing surgical operations and doing actual dissections—was to be found in ancient writings (see chapter 11, above). Galen had used almost identical phrases in relation to doctors of his own day who were, he believed, endangering the state of medicine (Edelstein 1943, 553). Vesalius fixed the time of decline in the West "after the Gothic deluge" ("post Gotthorum illuviem") and in the East "after the reign of Mansor at Bokhara in Persia" ("Mansorumque Bocharae in Persia regem"). But if Vesalius conceived that he was effecting a 'renaissance', and only a 'revolution' in the sense of a return to an antecedent condition (perhaps with 'improvements'), how could this be squared with the prosecution of original research which contradicted the authority of Galen? Ludwig Edelstein (1943, 557) suggests a simple answer, consonant with the aims of humanists: that "Vesalius proves himself to be a true humanist" precisely because he "recommends the study of nature" and asks men "to employ their own faculties." Edelstein points to the example of the fifteenth-century painter Cimabue, who was praised (for example, by Ghiberti) for having painted in the 'maniera greca' and thus having been "the reviver of art." Ghiberti did not thereby imply that Cimabue had imitated or even copied Greek paintings, since no Greek paintings were available to him. Rather Cimabue, and all the others who followed in the tradition he established, "studied and followed Nature." Nature herself had been the preceptress of Greek thinkers and artists and it was to act in the Greek manner to be guided by and to put one's trust in one's own powers of observation and reasoning. For Vesalius as well, there were no extant Hellenistic books on anatomy, and so he read only "the book of the human body which cannot lie" (*De Fabrica*, 600, l. 42). In this proud declaration he used, apparently for the first time (Roth 1892, 68 n. 5, 117 n. 5, 125 n. 3), the image of the human body as a book to be read, thereby transforming the existing image of a 'book of nature'. Vesalius not only called attention to the loss of the writings of ancient anatomists who had actually performed dissections of cadavers, he even advanced a theory concerning their destruction, that they had been willfully destroyed by envious people who were so envious that they begrudged later generations the enjoyment of truth (*De Fabrica*, pref., f. 3ʳ).

Copernicus, whose book *De Revolutionibus* was published in the same year, 1543, as Vesalius's great work on the human body, was also a deeply committed humanist. He published a translation into Latin of a Greek text by the seventh-century author Theophylactus Simocatta. In the preface to his major treatise, he states how he came to be dissatisfied with the received Ptolemaic astronomy, of which the central feature was the orbital motion of the sun and planets about a fixed earth. He then turned to ancient authors in order to find out whether in any of their writings they might have proposed alternative doctrines to Ptolemy's. During this study, he said, he encountered the ideas of the Pythagoreans concerning the motion of the earth. It was only then, assured by a tradition of antiquity, that in humanist fashion he began to consider the astronomical consequences of the earth's moving in orbit, since he knew that "others before me had been given the same liberty" ("quia sciebam aliis ante me hanc concessam libertatem"). In comparing this episode with the experience of Vesalius, Edelstein (1943, 557) has noted that "the works of the mathematicians with whose theories Copernicus concerned himself were lost, as were the books of the ancient anatomists whom Vesalius admired."

In this supplement I have drawn heavily on the important study (1943) by the late Ludwig Edelstein, with whom I discussed these issues on many occasions.

5.3. Revolutions in Mathematics

Reference has been made to the recognition by Fontenelle that the invention of the calculus constituted a revolution in mathematics. This very early instance of the perception of a revolution occurred early in the seventeenth century, at a time when Newton and Leibniz (co-inventors of the modern calculus) were still alive and making contributions to the subject. It is considerably easier to discern that a revolution has taken place in mathematics than in the physical and biological sciences, a factor which may account for the fact that this revolution was perceived almost at once.

In the present work, such revolutions in mathematics are not generally discussed, although some attention is paid to the Cartesian revolution in mathematics and mention is made of the revolutionary aspects of nineteenth-century probability theory and statistics and the radically new set theory of Georg Cantor. The topic of revolutions in mathematics must, however, be considered here because a number of mathematicians and historians of mathematics have denied that revolutions can ever occur in mathematics. For example, in a presentation of ten 'laws' of the history of mathematics, Michael Crowe (1975, 165) declares that "revolutions never occur in mathematics." In favor of this proposition, he cites J. B. Fourier and Hermann Hankel, both of whom held that mathematics grows by accretion, preserving "every principle it has once acquired." Crowe cites with special approval the opinion of Hankel (see §9.2, below) that mathematics differs from "most sciences," because in mathematics each generation does not have to destroy the structures erected by its predecessors, but rather "builds a new story to the old structure." Crowe's third witness against revolutions is Clifford Truesdell, a great master of both mathematics and its history, who has unequivocally stated that "while 'imagination, fancy and invention' are the soul of mathematical research, in mathematics there has never yet been a revolution" (1968, foreword).

Crowe emphasizes that his 'law' that there are no revolutions in mathematics depends on "the minimal stipulation that a necessary characteristic of a revolution is that some previously existing entity (be it king, constitution, or theory) must be overthrown and irrevocably discarded." He holds that in mathematics "new areas are 'formed' or created without the overthrow of previous doctrines." An example he gives is Euclidean geometry, which "was not deposed by, but reigns along with, the various non-Euclidean geometries." (For a discussion of Crowe's paper see Mehrtens 1976.) This same nonrevolutionary character of the history of mathematics is embodied in the title of Raymond Wilder's book, *Evolution of Mathematical Concepts* (1968). Another historian of mathematics who came out flatly against revolutions is Carl B. Boyer (see §9.2, below).

It would be foreign to my purpose in this book to try to settle the question of whether or not revolutions have actually occurred in mathematics. But it is of some interest to explore the ways in which some major innovations in mathematics have been considered as revolutions. What I find particularly striking about the history of mathematics in this regard is the very early recognition by Fontenelle that a mathematical revolution was taking place. His opinion may be contrasted with that of D. J. Struik (1954, 132), who prefers to conceive of a "gradual evolution of the calculus."

From the early eighteenth century, when Fontenelle was applying the concept of 'revolution', to our own day, there have been mathematicians who have seen the history of their subject as a succession of revolutions. A little less than a century after Fontenelle discerned a revolution in the calculus, the philosopher Immanuel Kant wrote about a revolution in geometry. The occasion was the preface to the second edition (1787; see ch. 15, above) of his *Critique of Pure Reason*. The revolution, according to Kant, consisted of transforming geometry from an empirical to a logical subject. This occurred in ancient Greece, Kant wrote, and the "history of this intellectual revolution" and of "its fortunate author has not been preserved."

The nineteenth century, according to a number of historians of mathematics, produced a non-Euclidean revolution, or an anti-Euclidean revolution. In *The Development of Mathematics* (1945, 329), the mathematician E. T. Bell refers to non-Euclidean geometry as the "revolutionary geometry" and says it was "one of the major revolutions in all thought" (p. 330): "To exhibit another [revolution] comparable to it in far-reaching significance, we have to go back to Copernicus; and even this comparison is inadequate in some respects. For non-Euclidean geometry and abstract algebra were to change the whole outlook on deductive reasoning, and not merely enlarge or modify particular divisions of science and mathematics."

This comparison of this revolution and the Copernican revolution may appear to be less than positive, however, when read together with Bell's earlier and more popular account of Lobachevsky in his *Men of Mathematics* (1937, ch. 16). Bell entitled his chapter on Lobachevsky "The Copernicus of Geometry," a phrase taken from William Kingdom Clifford. But Bell began his chapter: "Granting that the commonly accepted estimate of the importance of what Copernicus did is correct, we shall have to admit that it is either the highest praise or the severest condemnation humanly possible to call another man the 'Copernicus' of anything." Yet, in the conclusion, Bell wrote: "The full impact of the Lobatchewskian method of challenging axioms has probably yet to be felt. It is no exaggeration to call Lobatchewsky the Copernicus of Geometry, for geometry is only a part of the vaster domain which he renovated; it might even be just to designate him as a Copernicus of all thought."

D. J. Struik (1954, 253) did not use the actual word 'revolution', but the sense is clear in his statement that Nikolai Ivanovich Lobachevsky and János Bolyai were "the first to challenge openly the authority of two millennia and to construct a non-euclidean geometry." The "first in time to publish his idea was Lobachevsky." Morris Kline (1972, 879), however, uses the word 'revolutionary' to express the significance of non-Euclidean geometry.

At the very end of the nineteenth century Georg Cantor evaluated his own invention of infinite set theory as revolutionary, an opinion seconded by the Swedish mathematician Magnus G. Mittag-Leffler (see chapter 18). In analyzing Cantor's "revolutionary advance in mathematics," Joseph Dauben (in an unpublished paper of October 1974) argues for 'scientific resolution' as a key to understanding the development of Cantor's set theory, and concludes that there is an "inherent structure of logic" which "determines the structure of mathematical evolution," the latter being "necessarily cumulative." But it was

Dauben who called our attention to a letter from Cantor to Paul Tannery, of 8 December 1895 (Tannery 1934, 304), referring to his "introduction of *transfinite cardinal numbers* and the transfinite order-types" as "(in my eyes) the *most important* and *most revolutionary* innovation [*wichtigste* und *revolutionärste* Neuerung] within my 'Théorie des ensembles.'"

In our own time, there has occurred what has been called a second non-Euclidean revolution. This expression has been invented by Benoît Mandelbrot (1983) in relation to a new geometry which he has pioneered and to which he has applied the name 'fractal'. The origins of the second revolution have been traced by Mandelbrot to a letter of 1877, written by Georg Cantor to his friend and confidant, the mathematician Richard Dedekind. In this letter Cantor disclosed that he had apparently proved that a square does not contain more points than any of its sides. It was in the subsequent exchange of letters that Mandelbrot (1982, 232) would see the first step toward what he calls "the second anti-Euclidean revolution." Mandelbrot is thus a member of that group of scientists and mathematicians who have seen the revolutionary quality of their own contributions. This feature of revolution was particularly remarked by other mathematicians and scientists, notably by J. Freeman Dyson in an article on Mandelbrot and fractals in *Science* in 1979.

Evidently, there is a difference among mathematicians concerning the occurrence of revolutions in the domain of mathematics. Opinions on this subject are not limited to mathematicians and historians of mathematics. In his celebrated *Introduction to the Study of Experimental Medicine* (1927, 41), Claude Bernard explained why mathematics differs from the experimental sciences with respect to the occurrence of revolutions. Because "mathematical truths are immutable and absolute," he wrote, it follows that "the science of mathematics grows by simple successive juxtaposition of all acquired truths." But since, on the contrary, the experimental sciences "are only relative," they "can move forward only by revolutions and by recasting old truths in a new scientific form."

7.1. Copernicus's Revolutionary Advocacy of Scientific Realism

In the history of thought, and even in the history of astronomy, the Copernican revolution is apt to be cited as the archetypal revolution in the sciences. There is a sense in which Copernican astronomy seeded the Scientific Revolution, since that revolution reached its climax in Newtonian physics, which gave a dynamical basis for the astronomy generally called Copernican. We shall see that Newton's starting point was the Keplerian astronomy, the Galilean science of motion, and the Cartesian physics—all of which were either related to or modifications of Copernican science.

Even though there was no revolutionary effect on the technical development of the sciences for more than half a century, the basic Copernican idea—that the sun stands still at (or near) the center of the universe, while the earth rotates and revolves—was bold and challenging, even if it was to some degree a restatement of a doctrine set forth, though not worked up in any detail, in Greek antiquity. At the very least, Copernicus drew attention to yet another system of the world, in addition to the recognized ones: the Aristotelian and the Ptolemaic. One necessary consequence of his system was the position that the literal interpretation of Scripture cannot be the ultimate test for scientific explanations of the observed phenomena of the world of nature around us. Like it or not, *De Revolutionibus* could not avoid constituting a challenge to authority. A significant feature of the Scientific Revolution was to base knowledge on experiment and observation and to disdain any authorities other than nature herself. The motto of the Royal Society, founded a little over a century after the publication of *De Revolutionibus,* was "Nullius in verba" (On the word of no man). Whether or not Copernicus was actually a major figure in this revolutionary tilt of knowledge away from authority, he has come to symbolize the first mover in this direction of science and it is an honorable role.

Copernicus himself did advocate an unorthodox position with regard to scientific knowledge. It was just the opposite of the guidelines given to the reader in Osiander's anonymous preface to *De Revolutionibus.* As Pierre Duhem pointed out long ago (1908 [1940]), Copernicus went against a centuries-old tradition of astronomers who were not so much concerned with questions of reality as with the devising of computing schemes that would "save the phenomena," that would accurately predict astronomical events. In arguing for the 'reality' of his own system, and in not going along with those for whom 'reality' was not a central question, Copernicus was certainly a rebel. It is even reasonable to call him a revolutionary.

In order to evaluate Copernicus's position, it is necessary to take cognizance of two very different traditions in science from antiquity to the sixteenth century. Pierre Duhem (1954, 40) has differentiated these two traditions as follows: "The Greeks clearly distinguished, in the discussion of a theory about the motion of the stars, what belongs to the physicist—we should say today the metaphysician—and to the astronomer. It belonged to the physicist to decide, by reasons drawn from cosmology, what the real motions of the stars are. The astronomer, on the other hand, must not be concerned whether the motions he represented were real or fictitious; their sole object was to represent the *relative* displacements of the heavenly bodies." The doctrine that the goal of astronomers was merely to devise schemes for "saving the phenomena," and not to establish the physicists' "reality" was widely held in the Middle Ages and apparently led to the adoption of Ptolemaic astronomy rather than the Aristotelian physicalist system of imbedded spheres. Whether or not there were two wholly separate and distinct traditions, as Duhem argued, there seems to be no question that at the time of Copernicus it was not the generally accepted position to argue for the reality of cosmological systems as Copernicus did. Rather, there was an adherence to the kind of instrumentalism advocated in the preface to *De Revolutionibus,* written by Osiander but thought to have come from the pen of Copernicus.

In an article on "Copernicus and the Scientific Revolution," Edward Grant has argued (1962, 197) that "Copernicus is really the initiator of a very basic attitude which came to be held in some form or other by most of the great figures of the Scientific Revolution—namely, that fundamental principles . . . about the universe must be physically true." Grant sees Copernicus as having restricted the principle of saving the phenomena to "true hypotheses," and, to that extent, standing apart from "the ancient and medieval tradition" in which "truth and falsity were not at issue." Since Copernicus held that his astronomical "revolutionary hypotheses were true," it follows that "time-honored physical notions about the earth had now to be drastically altered" (p. 215). Hence Grant sees Copernicus's doctrine that "physics must follow the basic requirements of a true astronomy" to have constituted "a momentous break with an almost sanctified tradition." In his final assessment of Copernicus's "quest for reality," Grant asks us to consider Copernicus as "the first great figure of the Scientific Revolution." It was "essentially his attitude," Grant concludes, that "came to prevail with Kepler, Galileo, Descartes, and Newton" (p. 216).

Kepler strongly endorsed the Copernican position in a number of writings (for which see Duhem 1954, 42), in particular in a letter announcing that the preface to *De Revolutionibus* which contains the traditional opinion was not written by Copernicus at all but by Osiander. In this letter (trans. in Rosen 1971, 24), Kepler declared it to be "a most absurd fiction" that "the phenomena of nature can be demonstrated by false causes." This fiction, he added, "is not in Copernicus," who "thought his hypotheses were true"; and "he did not merely think so but he proves that they were true." Galileo expressed a similar opinion in his account of the state of celestial science in Copernicus's day. In his *Dialogue Concerning the Two Chief World Systems*, Galileo says that an "astronomer might be satisfied merely as a calculator," but "there was no satisfaction and peace for the astronomer as a scientist" (Drake trans., 1953, 341). Although Copernicus was aware that "the celestial appearances might be saved by means of assumptions essentially false in nature," Galileo insists that Copernicus knew "it would be very much better if he could derive them from true suppositions." When he did so, by adopting the concept of a moving earth and a stationary sun, Copernicus found that "the whole then corresponded to its parts with wonderful simplicity," and so "he embraced the new arrangement and in it he found peace of mind." In many passages, Descartes also adopted a realist position. In a letter to Mersenne on 11 March 1640 (1971, *3*: 39), he wrote: "As to physics, I should think I knew nothing about it if I could only say how things may be without demonstrating that they cannot be otherwise." (On Descartes's views concerning the literal and physical truth of his astronomical and physical principles or hypotheses, see Duhem 1954, 46; also Blake, Ducasse, Madden 1960, 89–97.) As these examples indicate, to the degree that "the quest for physical reality cannot be overestimated as a turning point in the history of science" (Grant 1962, 220), Copernicus "mapped the new path and inspired the Scientific Revolution by bequeathing to it his own ardent desire for knowledge of physical realities."

In a somewhat different presentation of these same issues, Michael Heidel-

berger (1976, 332) separates Copernicus's astronomy and computational
methods. The astronomy, he observes, could be and was considered apart from
the physics. That is, Ptolemeans could simply consider Copernicus's own preface and the chapter of *De Revolutionibus* (bk. 1, ch. 10) dealing with the heliocentric worldview and planetary distances as "philosophical blunders which were
not to be taken seriously." Heidelberger compares this to the situation today,
when "prefaces to books in physics are often viewed as speculative." Heidelberger would thus see Copernicus's mathematical theory of the planetary
movements — as opposed to his worldview and his insistence on the reality of
astronomical hypotheses — to be not at all revolutionary but rather within the
tradition, so that astronomers could use some of Copernicus's astronomy and
yet not accept his radical new philosophy or physics. They could, in fact, even
consider the astronomy to represent "the completion and restoration of the
tradition." Hence Heidelberger considers that the essential part of "the Copernican revolution" was the transition "from instrumentalism to the realistic
paradigm constitutive of nature" (p. 334).

Anyone who is aware of the tremendous labor Copernicus expended on the
calculations for *De Revolutionibus* will not find it difficult to suppose that his
motivation in some good measure must have been associated with his conviction
that he was working to delineate the true system of the world and not merely a
feigned computing scheme. And this is even truer of Kepler, who complains in
print of the fatiguing arduousness of the computation he has undertaken and
who was sustained by his zeal that his "new astronomy" was at last a mathematical and physical delineation of reality, of the actual and not a hypothetical
universe. In his zeal for reality, he was perhaps speaking in a sense that was far
more Copernican than in the actual formulation of his final system of astronomy. Galileo too was in this sense Copernican when he declared for the fixed
sun and moving earth as the true and only system; he was actually forced to
admit that since God is omnipotent, he could have made the universe in a
different fashion. In his long struggle to establish the new astronomy, Galileo
was certainly sustained, if not actually motivated, by a Copernican zeal for what
he conceived to be the true and real system of the world. Newton's commitment
to reality was equally great. In the conclusion to his *Principia,* he argued
straightforwardly that universal gravity "really exists" ("revera existat"). And
so perhaps there is a ground on which the Scientific Revolution may be said to
have been Copernican, one which may even constitute a more fundamental
contribution to developing science than the alleged Copernican revolution in
astronomy.

7.2. Some Seventeenth-Century Judgments on
Copernicus and Astronomy

In the later seventeenth century, following the publication of Kepler's *Astronomia Nova* and the *Epitome Astronomiae Copernicanae* and *Tabulae Rudolphinae,*

many 'Copernican' presentations of astronomy followed Kepler to the extent of using elliptical orbits rather than the complex Copernican schemes. From that day to ours, the phrase 'Copernican system' has come to mean only that the sun is stationary, that the earth rotates on its axis while it revolves around a sun fixed in position, as the other planets do, and that the fixed stars are immobile (save for some slight 'proper motion' with respect to each other).

In a general history of astronomy published in 1675, a decade before the *Principia*, it is said (Sherburne 1675, 49):

> [In Copernicus's] *De Revolutionibus Orbium Coelestium*, in Six Books, . . . he not only revived, but most happily united, and formed into an *Hypothesis* of his own, the several Opinions of *Philolaus, Heraclides Ponticus*, and *Ecphantus Pythagoreus*. For according to the opinion of *Philolaus*, he made the Earth to move about the Sun, as the Center, whence its *Annual* Motion; and with *Heraclides* and *Ecphantus*, he likewise gave it a Motion like that of a Wheel about its own Axis, whence its *Diurnal* Motion; an Hypothesis so near the Truth, that like that when persecuted, maugre all Opposition,
>
> > *Per damna, per caedes, ab ipso*
> > *Sumit opes animumque ferro;*
>
> As *Ricciolus* (though a Dissenter from it) observes.

The two lines, from Horace, have been rendered into English as follows:

> Through wounds, through losses, no decay can feel,
> Collecting strength and spirit from the steel.

Sherburne's presentation, we may observe, is limited entirely to the basic simple (or simplified) Copernican cosmological ideas and does not even mention any of the astronomical features of *De Revolutionibus*. Indeed, Sherburne does not even go so far as to say that this system had any real advantages over the older Ptolemaic system. He does say that those major ideas of Copernican astronomy were revived—that is, not invented—by Copernicus, which was a common opinion in the seventeenth century.

In 1694 William Wotton published his *Reflections upon Ancient and Modern Learning*, in which he included a seven-page epitome of ancient and modern astronomy, commissioned by Wotton from Edmond Halley. After describing ancient astronomy, Halley turned to the main assignment, "to compare the Ancients with the Moderns" (p. 279). In order to "make a Parallel as just as may be," Halley wrote, "I oppose the Noble *Tycho Brahe* or *Hevelius* to *Hipparchus*, and *John Kepler* to *Claudius Ptolemee*"—a comparison that Halley could "suppose no one acquainted with the Stars will doubt." He concluded with a reference to "that elegant Theory of the planetary Motions, first invented by the acute Diligence of *Kepler*, and now lately demonstrated by that excellent Geometer Mr. *Newton*." It is significant that Halley chose to omit Copernicus altogether.

We shall see that Newton's *Principia* (1687), the work that set the seal to the Scientific Revolution, was based directly on the Keplerian system, not the Copernican. That is, by the time of Newton, Copernican astronomy (as expounded in *De Revolutionibus*) was all but forgotten, save as a historical curiosity. Only occasionally was the name of Copernicus even introduced into astronomical discussion. In the *Principia* Newton referred only twice to Copernicus: he mentioned Copernicus's value for the mean distance of the moon from the earth (bk. 3, prop. 4), and he referred to "the Copernican hypothesis," by which he meant the Keplerian system, since the hypothesis supposes that the planets move in orbit around the sun in ellipses according to the law of areas (bk. 2, final schol.). Earlier, in the preliminary version of book 3 of the *Principia*, the *Essay on the System of the Universe* (see Cohen 1971, 241 n.8), Newton referred to "Hypothesis, quam Flamstedius sequitur, nempe Keplero-Copernicaea," that is, "the Hypothesis which Flamsteed follows, namely the Keplero-Copernican." On 28 April 1686, when Newton's *Principia* was presented to the Royal Society, it was described as containing "a mathematical demonstration of the Copernican hypothesis as proposed by Kepler" (Cohen 1971, 130). Just a little less than a decade earlier, Robert Plot (1677, 225) referred to the Bishop of Sarum, Seth Ward, who had "first *Geometrically* demonstrated the *Copernico-Elliptical Hypothesis* to be the most *genuine, simple,* and *uniform.*"

7.3. Luther, Melanchthon, and Donne

One reason for supposing that the basic cosmological idea of Copernicus was revolutionary may be the fact that it was vehemently attacked. For example, Martin Luther once referred to Copernicus as a "fool," an "upstart astrologer" who wishes to turn the "whole science of astronomy upside down" by having "the earth revolve rather than the heavens or the firmament, sun and moon" (see Norlind 1953, 275). Luther's abhorrence for the basic Copernican idea of a moving earth was neither philosophical nor scientific. He was not apparently concerned about the philosophical niceties of the abode of man being itself unique and, accordingly, different from the planets and hence the body fixed in space and motionless. Much less was Luther interested in astronomy, or in science at large; in fact he never read either the *De Revolutionibus* or the preliminary *Commentariolus,* nor did he read Rheticus's *Narratio Prima,* the first printed account of the Copernican system (1540). In fact, Luther made his famous declaration in 1539, before Rheticus's tract was published and four years before the printing of *De Revolutionibus* (1543).

Luther's sole concern was the plain and literal interpretation of the Bible. He said that Copernicus was putting things upside down because "the Holy Scripture tells us that Joshua commanded the sun and not the earth to stand still" (Dreyer 1906, 352).

The case of Melanchthon is even more interesting. As the astronomer and historian J. L. E. Dreyer remarked (1953, 352), it is not surprising that Luther should have been opposed to Copernican doctrines: "Luther had always been a stranger to humanism." But Dreyer found it "remarkable that the highly cultured Melanchthon should give vent to more than one sweeping condemnation of Copernicus."

Two years before the publication of the book of Copernicus, Melanchthon wrote to a correspondent that wise rulers ought to coerce such unbridled license of mind. (The wise rulers of Rome did that in 1633, so Protestants have no right to blame them.) And in his *Initia doctrinae physicae*, published in 1549, he goes fully into the matter in a chapter headed: "Quis est motus mundi?" First he appeals to the testimony of our senses. Then he serves up the passages of the Old Testament in which the earth is spoken of as resting or the sun as moving. Finally he tries his hand at "physical arguments," of which the following is a specimen: "When a circle revolves the centre remains unmoved; but the earth is the centre of the world, therefore it is unmoved." A beautiful proof. It would have been wiser if he had stuck to his Scriptural arguments or to the *argumenta ad hominem* which he advocated in 1541.

The degree of abhorrence expressed by Luther and Melanchthon is a gauge of their commitment as Protestants to the literal interpretation of Scripture, rather than a gauge of the revolutionary quality of the Copernican cosmological doctrine.

Melanchthon's early condemnation of Copernicus occurred in a letter to Mithobius (Burkard Mithob) of 16 October 1541. A year earlier he had received the first sheets of Rheticus's *Narratio Prima* (1st ed., 1540; 2nd ed., 1541), sent to him with Rheticus's compliments (see Rosen 1971, 394). His reaction in the letter may have been based on an immediate hostility to the radical idea of the earth's being in motion, rather than a study of the technical aspects of Copernican astronomy. Melanchthon, unlike Luther, actually had the skill to read Copernicus. He taught and wrote about Ptolemaic astronomy (see Westman 1975, 401) and was on friendly terms with Rheticus. On Rheticus's return from his famous visit to Copernicus, Melanchthon recommended his appointment to the Deanship of the Faculty of Arts and Sciences. In 1549, six years after the appearance of Copernicus's *De Revolutionibus*, Melanchthon (Prowe 1883, *1* (2): 232–233) published the *Initia Doctrinae Physicae*, setting forth three grounds for rejecting the Copernican doctrine: the evidence of the senses, the thousand-year consensus of men of science, and the authority of the Bible. In the second edition, in 1550, as Emil Wohlwill found (1904, 261), Melanchthon eliminated some of his extreme statements, such as characterizing Copernicus as one of those who argue for the motion of the earth "either from love of novelty or from the desire to appear clever."

Those who wish to indicate the profound or revolutionary effect of Copernican astronomy (or, really, Copernican cosmology) cite the following lines of

John Donne, as if they characterized the general thought of an age:

> And new Philosophy cals all in doubt,
> The Element of fire is quite put out;
> The Sun is lost, and th'earth, and no mans wit
> Can well direct him, where to looke for it . . .
> 'Tis all in pieces, all cohaerence gone;
> All just supply, and all Relation.

This extract, from the *First Anniversarie,* composed in 1611, was strongly influenced by the new discoveries made by Galileo and published by him in the *Sidereus Nuncius* in 1610. As Marjorie Nicolson has shown (1976, §5), Donne had been deeply impressed by Galilean astronomy and the revelations of the telescope. The above lines, like most of John Donne's poetry and thought, are highly idiosyncratic and reflect the torment of a single very learned mind to a greater degree than they do the general anxiety of his age. It has even been argued (Carey 1981, 250) that this "glum account" is not even to be taken seriously "as Donne's 'real' response" to the new astronomy, much less be considered an index of a general state of cultural reaction to a new system of the world. We are not to imagine that thousands or even hundreds of men and women felt abandoned or lost in a Copernican world in which their earthly abode had no fixed place. In any event, the point about Donne's expressions is that they are reflections of the new Galilean astronomy of his own day and not of the Copernican cosmology of a half-century and more earlier.

7.4. J.-S. Bailly and the Historian's Invention of a Revolution

We shall see (§13.1) that the eighteenth-century historian of astronomy Jean-Sylvain Bailly developed a complete theory of scientific revolutions.

Bailly does not use the actual expression 'Copernican Revolution' ('révolution copernicienne'), but he leaves no doubt that one of the major revolutions in science was inaugurated (if not, however, accomplished) by Copernicus. Copernicus, according to Bailly, was "the renovator of physical astronomy" responsible for the introduction of "the true system of the world" (1785, *1:* 337). Bailly said that a radical step had to be taken at the time of Copernicus, for it was necessary "to forget the motion which we see, in order to believe in that which we do not experience. One man dares to propose it . . . That is not all: it was necessary to destroy an accepted system . . . and to overthrow the throne of Ptolemy . . . A seditious mind gives the signal, and the revolution takes place. Copernicus . . . dared to shake off the yoke of authority, and he rid humanity of a longstanding prejudice that had impeded all progress" (ibid.).

Copernicus thus fulfilled the two necessary conditions that—according to Bailly's implied standards—made his work qualify as a revolution. He undermined the authority of the old or accepted system, and he set up a better one in its place. It made little difference to Bailly that the Copernican system itself might have been a revival of an older system of Aristarchus (p. 23); what mattered was only that Copernicus overthrew the yoke of authority and established a system of the universe different from the one that "had received the homage of fourteen centuries" (p. 337).

Bailly's concept of a two-stage revolution is especially pronounced in another of his presentations of the work of Copernicus. Bailly had been describing briefly the transition of astronomy from the Greeks to the Arabs, and from the Arabs to the Europeans, who began to cultivate this science (3: 320): "Walther and Regiomontanus in Germany constructed instruments and made new observations. In each new domicile science was subjected to a new scrutiny and the transmitted knowledge was verified, but at this period there was a great revolution which changed everything. The genius of Europe was revealed and heralded in Copernicus." Declaring, furthermore, that "Copernicus had taken a big step toward the truth," Bailly pointed out that the "destruction of the Ptolemaic system was an indispensable preliminary, and this first revolution had to precede all the others" (p. 321).

Although Bailly does not expressly say here that Copernicus created or started a revolution, there is no doubt from his text that this was the thrust of his argument. Bailly's is not, however, the earliest reference I have found to a revolution associated with Copernicus. A predecessor was the historian of mathematics, Jean Étienne Montucla. The latter wrote of Copernicus's "bold step," which "was as it were the signal for the successful revolution which philosophy underwent a short time later" (1758, *1*: 507). Quite clearly, Copernicus's "sublime discovery" did not — of and by itself — constitute a revolution, but was only a sign of a revolution to come in the seventeenth century, as Montucla indicates in subsequent pages (521–522). In this way, Montucla — like others — was able to harmonize the significance of the Copernican system (as developed by Kepler and Newton) with the limitations in the work of Copernicus himself. But even in the revised edition of Montucla's history, published after the French Revolution, neither the general concept of revolutions in science nor the concept of a Copernican revolution is fully developed, at least not in the sense that this had been done by Bailly.

Ever since the time of Bailly and Montucla, scientists and historians of science have — often unthinkingly — written about a Copernican revolution, usually confusing the revolutionary or daring steps of shifting the center of the universe with an achieved revolution in astronomy. The continuing dilemma of scholars concerning Copernicus's astronomy and the so-called Copernican revolution is illustrated by the treatment of this topic in many textbooks. In a widely used textbook that is of the highest qualities of accuracy, information, and perceptiveness, the opening statement presents the traditionally accepted opinion that there was a Copernican revolution, a revolution attendant on the publication of Copernicus's book in 1543. The author then proceeds through a sequence of qualifications and denials until it becomes evident that the revolution occurred only after the contributions of Kepler and of Galileo and that this revolution was not strictly Copernican at all. The first sentence reads: "When the 17th century dawned, the Copernican revolution in astronomy was over fifty years old." The first qualification: "Perhaps one should say rather that Copernicus' book, *De revolutionibus orbium coelestium* (1543), was over fifty years old." The denial: "Whether the book would initiate a revolution had yet to be determined." The second denial: "Two men who had scarcely passed the thresholds of their scientific careers in 1600 were to be the primary agents in assuring that it would [that Copernicus' book would initiate a revolution]."

The next sentence says, however, that Copernicus had already begun a revolution: "Kepler . . . and Galileo . . . acknowledged Copernicus as their master; both devoted their careers to confirming the revolution in astronomical theory he had begun." Then it is said that Kepler and Galileo radically altered astronomy and "modified Copernicanism in a way the master might not have accepted." In short, the revolution—made by Kepler and Galileo—was not strictly Copernican at all. The conclusion is that Copernicus had done no more than propose a "limited reformation": "By the time Kepler and Galileo were done, the limited reformation had become a radical revolution."

9.1. Francis Bacon's "Revolutio Scientiarum"

The earliest example I have found of the coupling of 'revolution' and 'science' occurs in Francis Bacon's *Novum Organum*, first published in 1620. It occurs in aphorism 78, book 1, and reads as follows (Bacon, 1857–1874, *4*: 77; 1905, 279; 1889, 272–273):

> Out of the five and twenty centuries over which the memory and learning of men extends, you can hardly pick out six that were fertile in sciences or favourable to their development [quae scientiarum feraces earumve proventui utiles fuerunt]. In times no less than in regions there are wastes and deserts. For only three revolutions and periods of learning can properly be reckoned [Tres enim tantum doctrinarum revolutiones et periodi recte numerari possunt]; one among the Greeks, the second among the Romans, and the last among us, that is to say, the nations of Western Europe, and to each of these hardly two centuries can justly be assigned. The intervening ages of the world, in respect of any rich or flourishing growth of the sciences [quoad scientiarum segetem uberem aut laetam], were unprosperous. For neither the Arabians nor the Schoolmen need be mentioned; who in the intermediate times rather crushed the sciences with a multitude of treatises [qui per intermedia tempora scientias potius contriverunt numerosis tractatibus], than increased their weight. And therefore the first cause of so meagre a progress in the sciences [prima causa tam pusilli in

scientiis profectus] is duly and orderly referred to the narrow limits of the time that has been favourable to them.

In trying to make precise the meaning of this extraordinary passage, the first question to be answered is what Bacon meant by the word 'scientia' ("scientiarum") or science. That Bacon did not simply mean what we do today by 'science' is evident from the following extract, in which he discusses science in relation to 'philosophia naturalis' or natural philosophy. Bacon says that during "those very ages in which the wits and learning of men have flourished most, or indeed flourished at all, the least part of their diligence was given to natural philosophy. Yet this very philosophy it is that ought to be esteemed the great mother of the sciences. For all arts and all sciences, if torn from this root, though they may be polished and shaped and made fit for use, yet they will hardly grow" (aph. 79). In his learned commentary on this latter passage, Fowler (Bacon 1889, 273 n.19) admits the difficulty in attempting "to reconcile Bacon's conception of Natural Philosophy, so far as we can discover it from this and the next Aphorism, with the account of it given in De Augmentis, bk. iii." But Bacon "seems here to have conceived of Natural Philosophy as the one general science with which the more special sciences should be brought into connexion"; thus it "should deal with the laws of nature and man in a more comprehensive manner than it was possible for the more special sciences to do." Fowler thus sees Bacon's natural philosophy as providing "a general survey of nature," which "the more special sciences" then "follow out in detail" (p. 277 n. 25). Bacon himself complains that "natural philosophy . . . has been made merely a passage and bridge to something else." That is, it has had "to attend on the business of medicine or mathematics" (aph. 80).

A few aphorisms earlier, Bacon had expressly declared that the "sciences which we possess come for the most part from the Greeks" (aph. 71). In this context he says that whatever "has been added by Roman, Arabic, or later writers is not much nor of much importance." There is no need to dwell on Bacon's ignorance or prejudice in thus equating the magnitude and significance of the Roman and Arabic contributions to science. But there is an obvious contradiction with aphorism 78, where Bacon says that Roman times constitute one of the three periods favorable to "learning" that were "fertile in sciences or favourable to their development." Here, in aphorism 71, he characterizes this same period as one in which "not much" has been added to the stock of scientific knowledge, nothing "of much importance [having been] produced." Additionally, in aphorism 79, he complains that during the Roman period, "the meditations and labours of philosophers were principally employed and consumed on moral philosophy." Since this was "to the Heathen . . . as theology to us," we can see why the sciences as such did not flourish. In aphorism 71, furthermore, Bacon—while contending that whatever the Romans or Arabs had added to the sciences "is built on the foundation of the Greek discoveries" — does not even refer to those whom we would consider to have been the outstanding Greek scientists and mathematicians. That is, Bacon lists "the elder of the Greek philosophers, Empedocles, Anaxagoras, Leucippus, Democritus, Parmenides, Heraclitus, Xenophanes, Philolaus, and the rest," omitting "Pythagoras as a mystic." But Bacon does not even mention Eudoxus, Aristarchus, Ptolemy, Theophrastus, Hippocrates, Hipparchus, Archimedes, Galen, or even Euclid.

A second ambiguity in Bacon's statement about revolutions is whether these are revolutions in the sense of an ebb and flow or periods of sharp and radical change. Bacon's three revolutions ('revolutiones') are also "periods of learning." They comprise the six out of twenty-five centuries of recorded history that were either "fertile in sciences" or "favourable" to the development of sciences. They are all bounded at one terminus by long periods or ages that were "wastes and deserts," or "unprosperous" for "any rich or flourishing growth of the sciences" (aph. 78). Hence the historical image is that of a sequence: barren early antiquity, followed by a "period of learning" among the Greeks; barrenness again, followed by a period of fruitfulness among the Romans; then barrenness once more, followed by the recent century and a half. Bacon does not make it absolutely clear whether there is such a barren period between the Greeks and Romans or whether these two past ages were temporally contiguous, but he probably thought there was an age of separation between them. It has been suggested (by Fowler in Bacon 1889, 272 n.17) that Bacon apparently envisaged "the Greek period [as extending] from Thales to Plato, the Roman period from Cicero or Lucretius to Marcus Aurelius," and the "modern period from the beginning of the revolt against Aristotle, or, perhaps, from the invention of Printing, to Bacon's own time." Thus it would seem that Bacon's three revolutions partake of the character of a tidal ebb and flow of knowledge.

In reading an English translation (for example, Bacon 1911, 414), one must exercise caution, because of the phrase "three revolutions and epochs of philosophy." This would introduce the word 'epoch', which was then considered to be a signal for the sense of revolution as a real break in continuity rather than a cyclical succession. But the Latin word for 'epoch' would have been 'epocha', whereas the expression used by Bacon is "tres . . . doctrinarum revolutiones et periodi" (Bacon 1889, 272). Nor is "epochs of philosophy" an exact translation of "doctrinarum . . . periodi."

From the foregoing discussion, it would seem that in aphorism 78 Bacon had in mind the image of three successive periods favorable to the sciences following long intervals of barrenness, in the sense of an alternation or an ebb and flow, that here the idea of radical change (in the form of a break with the past) was not his primary meaning of revolution. Additionally, although he does discuss "scientiae" or sciences, his context of revolution is not revolutions in sciences so much as "revolutions in learning" ("doctrinarum revolutiones").

I have mentioned the additional ambiguity of the meaning of 'scientia' for Bacon. It is obvious that the Roman period from Cicero to Marcus Aurelius was hardly a flourishing period for the sciences as we know them today or even as they were known in Bacon's day: astronomy, mathematics, physics, kinematics, botany, zoology, anatomy, physiology. There were two major scientists during this period, both Greeks, Galen and Ptolemy. In the *Novum Organum*, however, Bacon does not once mention either Ptolemy or Galen. It is thus clear, I believe, that Bacon's "doctrinarum revolutiones" are not scientific revolutions in the sense in which we commonly use this expression today, since they did not refer to 'science' in our present meaning and since the concept of revolution is not that of a noncyclical sudden and dramatic change.

In any event, this reference to revolutions is an isolated example of the use of this concept by Bacon as a means of describing the course of intellectual, philosophical, or scientific development—a subject of major concern for him. Aphorism 78, in point of fact, was intended to reveal what Bacon called "the first cause of so meagre a progress in the sciences," namely "the narrow limits of the time that has been favourable to them." The second is that even during such "ages in which the wits and learning of men have flourished most, or indeed flourished at all, the least part of their diligence was given to natural philosophy" (aph. 79). While men's minds were thus "diverted . . . from the philosophy of nature," and even "during those three periods [in which] natural philosophy was in a great degree either neglected or hindered, it is no wonder if men made but small advance in that which they were not attending." Additionally (aph. 80):

> [Let] no man look for much progress in the sciences—especially in the practical part of them—unless natural philosophy be carried on and applied to particular sciences, and particular sciences be carried back again to natural philosophy. For want of this, astronomy, optics, music, a number of mechanical arts, medicine itself—nay, what one might more wonder at, moral and political philosophy, and the logical sciences—altogether lack profoundness, and merely glide along the surface and variety of things. Because after these particular sciences have been once distributed and established, they are no more nourished by natural philosophy, which might have drawn out of the true contemplation of motions, rays, sounds, texture and configuration of bodies, affections, and intellectual perceptions, the means of imparting to them fresh strength and growth. And therefore it is nothing strange if the sciences grow not, seeing they are parted from their roots.

This leads Bacon to a major conclusion: that "the sciences have made but little progress" because the goal has not been correctly chosen (aph. 81). And then he states what we have come to consider the Baconian point of view: "Now the true and lawful goal of the sciences is none other than this: that human life be endowed with new discoveries and powers." This familiar doctrine of the association of truth and utility was tempered by the further statement that "fruits and works [that is, practical applications or embodiments of scientific truths or discoveries] are as it were sponsors and sureties for the truth of philosophies [natural philosophy or science]" (aph. 73).

Other factors that have kept the sciences from their full development are poverty of method (aph. 82) and lack of experiment (aph. 83). Another reason for the lack of rapid progress is "reverence for antiquity" (aph. 84). Also when one looks at "the variety and the beauty of the provision which the mechanical arts have brought together for men's use," there is a tendency to be "satisfied with the discoveries already made," to "be more inclined to admire the wealth of man than to feel his wants" and so be driven to make new discoveries (aph. 85). In every age, furthermore, "Natural Philosophy has had a troublesome adversary . . . , namely, superstition, and the blind and immoderate zeal of religion" (aph. 89). In schools, academies, colleges, and "similar bodies destined for the abode of learned men and the cultivation of learning," Bacon found "everything . . . adverse to the progress of science." There are no "prizes and rewards" for those who make advances in the sciences (aph. 91), and, lastly, "by far the greatest obstacle to the progress of science and to the undertaking of new tasks and provinces therein" is that "men despair of and think things impossible." And this leads Bacon to discuss the ebbs and flows of the sciences (aph. 92).

Whoever studies Bacon's *Novum Organum* is aided by an impressive monument of nineteenth-century scholarship, Thomas Fowler's edition (1889), in which the text is in Latin, but the extensive notes (a running commentary) and the general introduction are in English. The index, covering both text and notes, has entries in Latin and English, the preponderance being in Latin. There is only one entry for revolution. It reads: "Revolutiones scientiarum, 297." On first encountering this reference I thought that in some unaccountable manner the concept of revolutions in science might have originated decades earlier than all my research had indicated. But I was a little puzzled by the syntax, since 'scientiarum' is in the genitive plural, as is 'orbium' in Copernicus's *De Revolutionibus Orbium Coelestium*. That is, the index is apparently pointing to "revolutions *of* the sciences," in the same sense of Copernicus's "revolutions *of* the celestial spheres." The Latin text, and its English version, read as follows (bk. 1, aph. 92):

Itaque [viri prudentes et severi] existimant esse quosdam scientiarum, per temporum et aetatum mundi revolutiones, fluxus et refluxus; cum aliis temporibus crescant et floreant, aliis declinent et jaceant . . .

[Bacon 1889, 297]

And so [wise and serious men] think that there are what might be called ebbs and flows of the sciences in the course of the revolutions of the times and ages of the world since at some times [the sciences] grow and flourish and at other times they decline and die . . .

It may be seen that despite the index, it is a set of revolutions "temporum et aetatum mundi" that Bacon specifically invokes, and not "revolutiones scientiarum." Here "scientiarum . . . fluxus et refluxus," or tidal "ebbs and flows of the sciences," are revolution only in the old cyclical sense and as such might somewhat anachronistically be so classified in a nineteenth-century index. Fowler does have an entry for "Fluxus et refluxus maris" (the ebb and flow of the sea). He would accordingly, have been more consistent if he had listed aphorism 92 only under "Fluxus et refluxus scientiarum." But I doubt whether many readers who were concerned about the sciences in general would have found the entry for "Fluxus et refluxus scientiarum," whereas at least this reader did find the entry under "Revolutiones scientiarum."

This passage from aphorism 92 may serve as a gloss on aphorism 78. It links "the revolutions of the times and ages of the world" with the "ebbs and flows" of the sciences, thereby confirming the impression that, similarly, in aphorism 78, Bacon had in mind a cyclical ebb and flow and not a sudden and dramatic change producing something wholly new.

Bacon also discussed revolutions in an English essay (58: "Of Vicissitude of Things"). His main topic here is "the vicissitude of sects and religions" (1857–1874, 6: 514). While "true religion" is "built upon the rock," the other "sects and religions" are "tossed upon the waves of time." They are temporary, they rise and fall like the ebbs and flows of the tides in the seas. In short, the cyclical succession of these vicissitudes is almost beyond human control; but Bacon does offer "some counsel concerning them, as far as the weakness of human judgment can give stay to so great revolutions."

9.2. Descartes and Revolutions in Mathematics

In an analysis of revolutions in mathematics, T. W. Hawkins suggests that the conservative character of mathematics compared to the natural sciences is not quite so absolute as historians of mathematics such as Hermann Hankel (1884) have asserted. "In mathematics alone," Hankel stated, it is not the case as it is for "most sciences," that "one generation tears down what another has built, and what one has established another undoes." Hawkins argues that mathematics differs from the sciences in that mathematicians are primarily engaged in solving problems, which are mathematical problems and not "the problems or puzzles provided the scientist by the 'external world.'" Hawkins finds a revolution in mathematics occurring when "the methods of solving mathematical problems are radically changed on a large scale." In this sense, a revolution occurred in mathematics in the seventeenth century — the principal figures in this revolution were François Viète, René Descartes, Pierre de Fermat, Isaac Newton, and G. G. Leibniz. Of course, as Hawkins points out, their collective endeavor "did not involve a 'rejection' of ancient mathematics in the sense that, for example, Euclid's *Elements* were declared 'false.'" But their work "did involve a rejection of the methods by which the ancients solved problems" and introduced "new methods," which were devised on the basis "of the premise that mathematical problems should be reduced to the symbolic form of 'equations' and the equations used to effect the resolution." Ultimately, the introduction of these new methods altered "the nature of the problems posed and ultimately radically altered the scope and content of mathematics." The

"Galois theory of algebraic equations, differential and integral equations, bilinear forms, tensors, manifolds, covariant differentiation, functions, Fourier analysis, complex function theory, matrices, hypercomplex numbers, and so on were all by-products of the revolution." Furthermore, "without the revolution in mathematics which altered the objects and ultimately the problems of mathematics, these concepts and theories are inconceivable." For Hawkins, the "central figure in initiating this revolution in mathematics was René Descartes." Hawkins thus believes it legitimate to regard Descartes "as the 'Copernicus' of the revolution." Hawkins finds that with respect to content and influence, Descartes's treatise *Geometry* (1637) was the major work in "the transition from ancient to modern mathematics."

A different point of view was taken by Carl Boyer in his *History of Mathematics* (1968, 369), where he says that the "philosophy and science of Descartes were almost revolutionary in their break with the past," but his mathematics "was linked with earlier traditions." Boyer would see this contrast as "the natural result of the fact [!] that the growth of mathematics is more cumulatively progressive than is the development of other branches of learning." For Boyer, "Mathematics grows by accretion, with very little need to slough off irrelevancies, whereas science grows largely through substitutions when better replacements are found." So he would see "Descartes' chief contribution to mathematics, the foundation of analytic geometry," to have been "motivated by an attempt to return to the past." Hawkins disagrees, finding it "very misleading to characterize the goal of Descartes' *Geometry*" in Boyer's terms as "generally a geometric construction." Rather than finding Descartes seeking to return to the past, Hawkins finds the true goal of Descartes's *Geometry* to be to establish the superiority of his method, "the key to which is the setting up of a symbolical algebraic equation, from which the solution of the problem is then derived." Descartes approached the solution of construction problems by transforming the traditional methods into the setting up of an algebraic equation in one unknown, and then by "specifying canonical construction procedures for the roots of equations, Descartes sought to reduce construction problems completely to their algebraic component."

Of course, Descartes had been anticipated in the general method of the solving of construction problems by a form of 'algebraic analysis' by François Viète, but Viète's work was limited to equations in one unknown whereas Descartes went beyond this; furthermore, Viète used a cumbersome method of algebraic symbolism. Descartes not only proceeded from equations in one unknown to equations in two unknowns, but he essentially introduced the symbolic language of modern mathematics. Boyer (1968, 371) quite correctly says that Descartes's *Geometry* is "the earliest mathematical text that a present-day student of algebra can follow without encountering difficulties in notation." Except for an archaic symbol for equality, Descartes uses letters near the beginning of the alphabet for parameters (a, b, c, . . .) and those near the end of the alphabet (. . . , x, y, z) for the unknown quantities. Except for *aa* (which, as

D. J. Struik points out, was still used in the nineteenth century by Gauss), he generally uses exponentials, and he also uses "the Germanic symbols + and −," all of which makes "Descartes' algebraic notation look like ours, for of course we took ours from him" (Boyer 1968, 371).

That Descartes did not consider his method primarily a return to the past is shown by his arguing for the power of his new method, in which the solution was found by means of a fundamental algebraic equation. Descartes, it must be remembered, introduced his *Geometry* as one of the three examples of the new method which he had discovered, and therefore presented this work as one of the three treatises which were supplementary to his famous *Discourse on Method* (1637). Descartes stressed the fact that the problem of the "five line locus" indicated the special character of the method, since according to Pappus the Greeks could not solve this problem by using their method. Hawkins would see Descartes's "most convincing proof of his thesis" that "the equation is the source of knowledge regarding a curve" in his method for determining the normal (and therefore the tangent) to a curve. Descartes's legacy, therefore, was "a new mathematical method which, in fact, implied a new conception of mathematics," in which "the principal objects of mathematics were symbolical equations from which, and in terms of which, mathematical problems were to be resolved." Whoever reads carefully the first volume of D. T. Whiteside's edition of *The Mathematical Papers of Isaac Newton* cannot help but be impressed by the fact that Newton's ideas were formed by his studies of Descartes's mathematics, in an edition which contained commentaries by a number of other mathematicians including Frans van Schooten. It can be argued legitimately that in many ways Descartes's contemporary Fermat was the more modern mathematician, that Fermat came closer to our conception of analytic geometry than Descartes did. But the fact remains that it was from Descartes, and the commentators on Descartes, that Newton (and also Leibniz) took their start.

D. T. Whiteside, in his survey of "Mathematical Thought in the Later Seventeenth Century," has neatly summed up his views as follows (1961, 290): Descartes, he writes, elevated "Greek coordinate systems with the analytical power of the free variable" and "laid the foundations of an analytic study of geometrical forms." It was Descartes's "*Géométrie* which rapidly became standard in the new university mathematical courses in Western Europe from the middle of the century." And Whiteside adds the comment, "Nor did any contemporary mathematician—and least of all the great geometers Newton and Huygens— deny that fact." The revolution in mathematics recognized by Fontenelle, the final formulation of the differential and integral calculus by Newton and by Leibniz, was in a direct line of succession that goes back to the revolutionary innovations of Descartes.

[This discussion of Descartes and mathematical revolutions draws heavily on conversations with Thomas W. Hawkins and from the notes he allowed me to read of a lecture on "Descartes and the Mathematical Revolution of the Seventeenth Century," part of his course on the history of mathematics.]

11.1. John Freind on Harvey and Medicine

One of the most complete discussions ever written about the practical or medical consequences of Harvey's discovery is to be found in John Freind's *History of Physick* (1723; 4th ed., 1750). Harvey, according to Freind, "had thoughts of composing . . . a Work himself, to shew the advantages of this doctrine, in relation to practice," but "sickness and death" prevented him from achieving this aim, which would have indicated what "improvements" in medical practice, "even in the cure of distempers, might be made" (1750, *1*: 237ff.). In the absence of such a work, Freind himself wrote up some hints "of what use a perfect knowledge of the circulation may be to us, if rightly applied, in the practical part of our profession."

Freind's first example was "tying up the arteries in *Amputations*," which was much to be preferred to "that old painful and cruel" practice "of stopping the blood by *Cauteries, Caustics,* or *Escharotics* alone," that is, by the method of burning with a hot iron or caustics. There was not only the question of "extreme torment" to the patient, but — as a result of Harvey's discovery — it was now understood that the blood, "by the laws of its motion," would exert "such a force" against the "*Eschar*" (or dead tissue sealing off the artery) that only a ligature could resist it. Yet, as Freind admitted, the practice of ligatures in place of cautery and the use of caustics or escharotics had been introduced in the sixteenth century by Ambroise Paré. Harvey's discovery had only served to "convince men of the usefulness of it" (*1*: 239). Had Harvey not discovered the circulation of the blood, Freind concluded, Paré's procedure would never "have been so much in vogue." To prove his point, he contrasted English practice with that of the Germans (who "are but little acquainted with it") and of the Dutch (who "intirely reject it").

Freind's second example was also an explanation and again was related to amputation. When "the trunk of the artery is cut off" during an amputation, the "course of the blood is nevertheless preserved." The Harveyan doctrine accounts readily for this phenomenon, by showing how "the lesser arterial branches in this case supply the defect." By "distending themselves gradually to a greater dimension," these lesser arteries "are able to furnish . . . what is necessary for motion and nourishment."

These first two examples indicate how Harvey's ideas affected practice by explaining and justifying existing procedures; the second two showed how practice had been changed. When an aneurysm (a dilation or enlargement of an artery) occurs after "a puncture," the "true method" of treatment follows "at first sight" from Harvey's doctrine. Doctors should not use compression ("which seldom stops the current in the artery") but ought rather, "after having made proper ligatures, to divide the vessel." The artery should not only be tied "above the puncture," but below the puncture as well, "as in the case of a *Varix*" (an abnormally dilated vein), "in order to hinder any supply of blood from other branches, which everywhere almost in the body communicate with one another." The British, thanks to Harvey, follow "the true method" of treating such aneurysms, whereas "the practice of another nation is very defective in this point." The final example was the Harveyan solution of "as warm a controversy as ever was in Physick": "whether in a *Pleurisy*, a vein should be opened on the *same* or on the *opposite* side" (*1* : 241). From the standpoint of the circulation, it turns out that "the difference is so minute, that one wou'd wonder there ever cou'd have been any dispute about it." Freind concluded with some remarks about the practice of "bleeding in general." The "Circulation," he declared, "has quite confounded and superseded all those rules, which had been before with so much pains and formality laid down, as to opening, in particular cases, this or that vein."

Whoever reads Freind's arguments some two and half centuries later cannot but be struck by the fact that Harvey's discovery had made no real difference to the Galenical base of the practice of medicine as a balancing of the humors. At best, bloodletters now could "have this advantage at least from the Circulation, of knowing exactly how indifferent it often is, which vein is made choice of," and "if there be any preference, of judging without any hesitation, which vein to choose" (*1* : 243). Additionally, it will seem as surprising to readers as it did to me to find no mention made of the one obvious implication of the circulation: the choice of which side of a wound a tourniquet should be applied to in relation to an artery or a vein. Be that as it may, Freind certainly has made no case for a revolution in medical practice attendant on Harvey's doctrine of the circulation of the blood. His list of the medical improvements provided by knowledge of the circulation thus confirms the opinion of William Temple and Bernard de Fontenelle that the significance of this great discovery was intellectual and scientific and that, radical as the new ideas were, their revolutionary effect in their own time (and for a century thereafter) was confined to the realms of biological understanding.

11.2. Paracelsus and Revolution

So great was the force of Paracelsus's personality and the power of his scientific and medical ideas that as late as the eighteenth century both the man and his ideas could still arouse virulent hostility. In 1750 the fourth edition of John Freind's *History of Physick* continued the disparagement of Paracelsus which had been so marked a feature of the first edition in 1723. This took the form of an attack on Daniel Le Clerc, author of a renowned work on *History of Medicine* that ended with the age of Galen. A new edition had appeared in 1723, containing a "sketch" of a "plan" (1750, *1* : 2) for extending the history up to the middle of the sixteenth century, featuring—as its conclusion—the work of Paracelsus.

Freind not only censured Le Clerc in the text for "the long detail" he "gives of *Paracelsus's* idle system" (1750, 2: 336) but wrote a special introduction, dated 10 May 1723, still printed at the head of the first volume in the fourth edition in 1750, which was largely an animadversion of Le Clerc's sketch. Le Clerc's proposal, according to Freind, would have been unsatisfactory even if "he had not filled half of it with relating all the obscure jargon and nonsense of that illiterate Enthusiast, *Paracelsus.*"

(Freind's strictures of Le Clerc for his enthusiasm for Paracelsus may be contrasted with his expressed admiration for Le Clerc's scholarship in ancient medicine. According to Freind (1750, *1*: 1–2), Le Clerc has "amply and clearly represented all the Philosophy, the Theory, and Practice of the ancient Physicians"; "there is scarce a Notion, a Distemper, a Medicine, or even the name of an Author, to be met with among them, of which he has not given a full and exact account." Furthermore, Freind asserted (1750, *1*: 227), "How much the art of Physick was improv'd, polish'd, and perfected by the *Greeks,* has been accurately explain'd by Mr. *le Clerc.*")

Two rejoinders to Freind's attack on Le Clerc's sketch were published: an anonymous article (said to have been written by Le Clerc himself) in the *Bibliothèque Ancienne & Moderne* (1727, 27) and a pamphlet "By C.W., M.D." (London, 1726). "C.W." is generally considered to have been William Cockburn, M.D., who published a translation of the French article, under the title *An Answer to What Dr. Freind has written . . . Concerning Several Mistakes, which he pretends to have found in a Short Work of Dr. Le Clerc, intituled, An Essay of a Plan, &c.* (London, 1728). The title announced that there was a preface by W. Cockburn, M.D. The text (p. 9) refers to Paracelsus as an "Innovator; the most famous that ever was in Physick" —a precise translation of "Novateur, le plus fameux qu'il y ait jamais eu dans la Médecine" (1727, 27: 394). But in the preface, Cockburn praised Paracelsus extravagantly for "introducing the method of curing Diseases by chymical Medicines, to the total Overthrow of *Hippocrates* and *Galen*" (p. xi). In this way, he said, Paracelsus "made one of the greatest Revolutions in Physick." This statement carries the new sense of a 'revolution' as an overthrow of a long-lived system of medical thought and treatment and its replacement by a wholly new and different system. But Cockburn went a step further. He said that this revolution was not transitory, but rather had lasted (or, the new medicine that had triumphed in the revolution had lasted) for what he said was a "considerable period of time."

Cockburn's words imply a belief that there had been other revolutions in medicine, and in fact that some of them had been great revolutions. But whereas this pamphlet of 1728 contains the theme of revolutions in medicine, an earlier work of Cockburn's—a preface to James Harvey's *Praesagium Medicum* (1720)—is more conventional in its image and language, using the more traditional concept of improvements. It is anyone's guess whether this difference between the two prefaces reflects the growing use of the concept and term 'revolution' during the first quarter of the eighteenth century or arises from a difference in subject matter.

13.1. J.-S. Bailly's Theory of Scientific Revolutions

We have seen that in his presentation of Copernican astronomy, Bailly introduced his notion of a two-stage revolution. This appears also with respect to the Newtonian natural philosophy — celestial mechanics based on universal gravity — which Bailly called a 'revolution' in more than one chapter of his history. A notable passage which particularly praises Newton for his modesty (a-propos of the preface to the first edition of the *Principia*) says: "Newton overturned or changed all ideas. Aristotle and Descartes still shared the empire; they were the preceptors of Europe. The English philosopher destroyed almost all their teachings and propounded a new [natural] philosophy; this philosophy effected a revolution" (1785, 2: 560). Newton, in other words, both established a new science and showed the falsity or inadequacy of current ideas and systems. An example of the latter occurs at the end of book two of the *Principia,* where Newton proved that Cartesian vortices cannot be reconciled with Kepler's laws.

In this passage on Newton, Bailly shows clearly his belief that a revolution in science is a two-stage action. These two phases or stages occur in Bailly's presentations of the grand revolutions associated with Copernicus and Newton but not of the revolution associated with the micrometer nor other innovations such as those he predicted in his history. It would seem as if the two-stage revolution was a requirement only for revolutions on a large scale, such as the introduction of a new system of the world (Copernicus), or a new natural philosophy of dynamics and celestial mechanics (Newton).

Bailly alerted his readers to the time scales of revolutions in science. Newton's "book on the *Mathematical Principles of Natural Philosophy,*" he wrote, "was destined to produce a revolution in astronomy." But, he noted, "this revolution did not occur all at once" (2: 579).

Bailly did not universally apply the designation 'revolution' to the major radical innovations in astronomy. Two outstanding examples of innovators of the first rank who seem not to have quite merited the accolade of 'revolution' were Kepler and Galileo. Kepler apparently fulfilled the qualifications of the two stages, since he had first to get rid of "all the epicycles which Copernicus had allowed to remain" in astronomy, before introducing his own concepts of elliptical orbits and motion according to the law of areas. As to his significance: "The privilege of great men is to change received ideas and to announce truths which propagate their influence on the centuries to come. With respect to both of these activities," according to Bailly (2: 2–5), "Kepler merits being considered as one of the greatest men who have ever appeared on the earth." He is, in fact, "the true founder of modern astronomy." For all that, however, Bailly does not ever evaluate Kepler's work as having constituted a 'revolution'. And the same is true of Galileo, who had first to destroy the accepted Aristotelian notions of motion — including the artificial distinction between natural and violent motion, and the "ridiculous" distinction between naturally light and naturally heavy bodies — before he introduced his own laws of accelerated motion and falling bodies, and the resolution and composition of motion (so as to find the parabolic path of projectiles). But this too apparently did not merit the designation of 'revolution' (2: 79).

I do not intend to imply that Bailly necessarily gave serious thought to the question, or even posed the specific question, of whether Kepler and Galileo did or did not actually produce a revolution. But it is significant that in a history of astronomy in which the concept and actual term figure so prominently, 'revolution' is applied again and again to Newton and Copernicus and not to Kepler and Galileo. We may guess that Bailly, who fully understood the technical aspects of Newton's completion of the work of Kepler and Galileo, naturally considered that Kepler and Galileo had only taken the great steps that made a revolution necessary. He would not, in this event, have deemed the achievement of either Kepler or Galileo revolutionary in character, to have been a 'revolution'.

13.2. Symmer's and Marat's Self-Proclaimed Revolutions

The earliest known example of a scientist referring to his work as revolutionary occurs in a letter written by Robert Symmer in 1760 (as noted by John Heilbron [1979, 433]). Symmer, a Scot who found employment in the British financial administration in London, is not generally well known either to scientists or to historians. His major contribution arose as a result of his curious habit of wearing two pairs of silk stockings on the same leg—one white for warmth and one black for mourning. When he took off his stockings at night he discovered that they would strongly attract one another. This led him to the study of static electricity. It was unfortunate for his reputation, as John Heilbron has observed (1976, 224), that "he continued to use his stockings to generate the electricity for his experiments." This was, he wrote to a friend in 1761, "enough to disgust the Delicacy of more than one Philosopher" (Brit. Libr., MS Add. 6839, 220–221).

Symmer's observations were as follows: If the stockings were removed together, they would exhibit almost no ability (or none at all) of attracting electrically; when separated, each stocking would stretch out as if there were still a leg inside of it and they would then strongly attract each other and also attract small bodies at a distance of six feet. When the stockings came together, they would collapse so as to lie flat on one another, and their attractive power in this combined state would extend only to a couple of inches; on separation once again, the stockings would regain their strong electrical power. In other experiments he threw the stretched-out or expanded socks against the wall, to which they would adhere, dancing to the movement of any breeze. Later, Symmer charged up 'Leyden jars' (condensers or capacitors) with his socks. No wonder that Symmer wrote that "the frequent mention of pulling on, and putting off, of *stockings*" was a "circumstance . . . so little philosophical [that is, scientific], and so apt to excite ludicrous Ideas, that I was not surprised to find it the Occasion of many a joke, among a sarcastical set of minute Philosophers" (Brit. Libr., MS Add. 6839, 182–183; see Heilbron 1979, 432–433).

These experiments led Symmer to conclude that Benjamin Franklin's theory of electrical action needed major emendation. Franklin had explained the occurrence of two states or varieties of electrification — for which he coined the expressions 'negative' and 'positive' or 'minus' and 'plus' electrification — as arising from the loss (minus) or gain (plus) of a single electrical 'fluid'.

Symmer, however, decided that his experiments with silk stockings implied "that the contrary electricities arose from *two distinct positive principles,* perhaps materialized as two different counterbalancing fluids. In this, as he rightly believed, he differed from all his predecessors" (Heilbron 1979, 431). Franklin, who was in London, helped Symmer in some experiments, even though they were intended to demonstrate the double flux of the alleged two different electricities. It was characteristic of Franklin's generosity (his "great civility," as Symmer put it) in scientific matters and his devotion to advancing the truth that he helped Symmer even though Symmer's proposal was designed as a direct challenge to Franklin's own theory of electrical action (ibid., 432).

Symmer's new theory was not generally accepted, although some of his experiments proved to be of notable importance, especially for Continental 'electricians'; it was only at the century's end that the 'two-fluid' theory of electricity, espoused by Augustin Coulomb (who found by experiment the mathematical law of electrical force), began to take over and to replace the Franklinian 'one-fluid' theory. But Symmer was fully aware that his ideas were radical: "In establishing the Principle of *Two Counteracting Powers,*" he wrote, "I was to differ with all who had ever written upon the subject." Yet he hoped that his "Revolution in the System of Electricity may possibly make some little Eclat among the Philosophers abroad, who are generally more alert in those matters, than we are at home" (Brit. Libr., MS Add. 6839, f.183). The "revolution" would extend far beyond the confines of electrical science as such, he wrote, since

> besides accounting for the Phenomena of Electricity, it may turn out to be a matter of very great Importance, in many useful Branches of Human Knowledge. For my Part, I am really of Opinion, it may be found hereafter to account for Magnetism, for Gravity, and prove to be the genuine Principle of the Newtonian Philosophy; and may likewise throw a Light upon the Principles of Chymistry, Vegetation, and Animal Life.

Another scientist who claimed that his work was revolution-making was Jean-Paul Marat, "l'ami du peuple." Although Marat's political career is well known, as is his murder by Charlotte Corday while he was in the bath, his scientific career remains obscure even to scholars. For example, was he a medical doctor or a veterinarian? Heilbron (1979, 429) calls him "doctor to the grooms, or perhaps the horses, of the Duc d'Artois." Was he truly a scientist? The authoritative *Dictionary of Scientific Biography,* recently completed in 15 huge volumes, does not give him even the mere pittance of a single column — although he published books based on his scientific researches, "découvertes" concerning fire, electricity, and light (1779), *Découvertes sur la lumière* (1780), and a treatise on electricity (1782); he also published a new French translation of Newton's *Opticks*

(1787). Benjamin Franklin came to see his electrical experiments when he was in Paris as the American 'ambassador' during the Revolution, but he did not push Marat's candidacy for membership in the French Academy of Sciences. Most recently, Heilbron (1979, 429–430) refuses even to give the reader the stated principles of Marat's theory, accepting the rating that Marat was no more than a crank (the opinion of Volta and Lichtenberg).

But in his own day there were many who held his work in high esteem; in particular, because of his attacks on Newton, he was even hailed as the dethroner of Newton (see Fayet 1960, 30), as one who, according to Brissot de Warville, had been "endowed by nature with a genius for observation and an indefatigable ardor for the search for truth," a scientist who believed "only in experiment" and not at all in respect for great names in science, a man who "courageously knocked down the idol of the academic cult and substituted well-proved . . . facts for the errors of Newton, with respect to light" (pp. 30–31). Marat began his publications with the word 'discoveries' or 'new discoveries'. The astronomer Lalande said that "no one was more prolific in discoveries than Marat. The greatest men [of science] have only made two or three discoveries during a long life. But Marat made them by the hundreds."

In the first volume of the *Journal de Littérature, des Sciences et des Arts* (1781, 371), an article appeared on Marat's researches in optics as follows:

> The revolution [*La révolution*] that M. Marat has just made in optics has created such a sensation among the Physicists who cultivate this science that they still have not recovered from their astonishment. Those who actually are most inclined to accept innovations cannot deny that, since the publication of [Marat's] book of *Discoveries Concerning Light*, Newton has lost the greatest jewel in his crown.

Joseph Fayet (1960, 31 n.17) has indicated that there is very convincing evidence that the author of this encomium must have been none other than Marat himself.

For information concerning Symmer's researches and their influence, see Heilbron (1976; 1976a; 1979, ch. 18). Symmer published in 1759 a four-part report on his research, *Phil. Trans.* 51: 340–389. See also his correspondence with Andrew Mitchell (Brit. Libr., MS Add. 6839).

For Franklin's theory of electrical action, see Cohen (1956 and rev. ed.; 1972) and Heilbron (1979, chs. 14–15). On Marat and science, see Fayet (1960, 26–49) and Dauben (1969).

14.1. Engels' Comparison of Lavoisier's Revolution in Chemistry and Marx's Revolution in Economics

On two major occasions Friedrich Engels essayed a measure of the greatness of Karl Marx's achievement by a comparison with a revolutionary moment in the sciences. He did not, as might be supposed, hail Marx as a 'Newton'; his Hegelian bias was far too strong for a direct Newtonian image to come to mind. Rather, he invoked the Chemical Revolution of Lavoisier and the recent Darwinian revolution in natural history.

When Marx died, Engels delivered a graveside speech in English at Highgate Cemetery (London, 17 March 1883). Here he said:

> Just as Darwin discovered the law of development of organic nature, so Marx discovered the law of development of human history . . . [: that] the production of the immediate material means of subsistence and consequently the degree of economic development attained by a given people or during a given epoch form the foundation upon which the state institutions, the legal conceptions, art, and even the ideas on religion, of the people concerned have been evolved, and in the light of which they must, therefore, be explained.

Then, without referring to Newton by name, Engels declared that "Marx also discovered the special law of motion governing the present-day capitalist mode of production and the bourgeois society that this mode of production has created." The invocation of a Marxian "law of motion" implied a comparison to Newton and his system of dynamics based on three "axioms" or "laws of motion." "Such," concluded Engels, "was the man of science," but "this was not even half the man" because "Marx was before all else a revolutionist."

One year later, in 1884, Engels wrote a preface for the second volume of Marx's *Capital,* of which the first edition (in German) appeared in 1885. Marx had left this volume in fragments, some of which had been revised, but only a small portion of which had been "made ready for the press" by the author. Engels not only produced a coherent text out of Marx's various manuscripts but did a considerable amount of editorial work, for instance, translating quoted extracts (chiefly from English) into German. In the preface to the second volume, Engels devoted a considerable amount of space to a refutation of the charge that Marx had "plagiarised the work of Rodbertus" (the economist Karl Rodbertus-Jugetzow). He concluded not only in a vindication of Marx, but in the observation that a certain "officially and ceremoniously installed professor" had "forgotten his classical Political Economy to such an extent that he seriously charges Marx with having purloined things from Rodbertus which may be found even in Adam Smith and Ricardo." And then he asked,

> But what is there new in Marx's utterances on surplus-value? How is it that Marx's theory of surplus-value struck home like a thunderbolt out of a clear sky, and that in all civilised countries, while the theories of all his socialist predecessors, Rodbertus included, vanished without having produced any effect?

In reply to his rhetorical question, he says, "The history of chemistry offers an illustration which explains this."

> We know that late in the past century the phlogistic theory still prevailed. It assumed that combustion consisted essentially in this: that a certain hypo-

thetical substance, an absolute combustible named phlogiston, separated from the burning body. This theory sufficed to explain most of the chemical phenomena then known, although it had to be considerably strained in some cases. But in 1774 Priestley produced a certain kind of air "which he found to be so pure, or so free from phlogiston, that common air seemed adulterated in comparison with it." He called it "dephlogisticated air." Shortly after him Scheele obtained the same kind of air in Sweden and demonstrated its existence in the atmosphere. He also found that this kind of air disappeared whenever some body was burned in it or in ordinary air and therefore he called it "fire-air." "From these facts he drew the conclusion that the combination arising from the union of phlogiston with one of the components of the atmosphere" (that is to say, from combustion) "was nothing but fire or heat which escaped through the glass."

Priestley and Scheele had produced oxygen without knowing what they had laid their hands on. They "remained prisoners of the" phlogistic "categories as they came down to them." The element which was destined to upset all phlogistic views and to revolutionise chemistry remained barren in their hands. But Priestley had immediately communicated his discovery to Lavoisier in Paris, and Lavoisier, by means of this discovery, now analysed the entire phlogistic chemistry and came to the conclusion that this new kind of air was a new chemical element, and that combustion was not a case of the mysterious phlogiston *departing* from the burning body, but of this new element *combining* with that body. Thus he was the first to place all chemistry, which in its phlogistic form had stood on its head, squarely on its feet. And although he did not produce oxygen simultaneously and independently of the other two, as he claimed later on, he nevertheless is the real *discoverer* of oxygen vis-à-vis the others who had only *produced* it without knowing *what* they had produced.

Marx stands in the same relation to his predecessors in the theory of surplus-value as Lavoisier stood to Priestley and Scheele. The *existence* of that part of the value of products which we now call surplus-value had been ascertained long before Marx. It had also been stated with more or less precision what it consisted of, namely, of the product of the labour for which its appropriator had not given any equivalent. But one did not get any further . . .

Now Marx appeared upon the scene. And he took a view directly opposite to that of all his predecessors. What they had regarded as a *solution*, he considered but a *problem*. He saw that he had to deal neither with dephlogisticated air nor with fire-air, but with oxygen—that here it was not simply a matter of stating an economic fact or of pointing out the conflict between this fact and eternal justice and true morality, but of explaining a fact which was destined to revolutionise all economics, and which offered to him who knew how to use it the key to an understanding of all capitalist production. With this fact as his starting-point he examined all the economic categories which he found at hand, just as Lavoisier proceeding from oxygen had examined the categories of phlogistic chemistry which he found at hand.

The quoted extracts in these paragraphs were taken by Engels from "Roscoe and Schorlemmer, *Ausführliches Lehrbuch der Chemie*, Braunschweig, 1887, I, pp. 13, 18."

14.2. Lichtenberg on Revolutions in Science and Lavoisier's Chemical Revolution

Georg Christoph Lichtenberg was one of the most interesting figures on the eighteenth-century scene. Professor of mathematics and physics at the University of Göttingen, he had a wide range of interests and "became the leading German expert in a number of scientific fields, including geodesy, geophysics, meteorology, astronomy, chemistry, statistics, and geometry" (Bilaniuk 1973, 321); his major specialty was experimental physics. Among his many contributions to science, perhaps the most curious is his pioneering in the "electrostatic recording process," which brings him credit for discovery (1777) of "the basic process of xerographic copying." His contribution to this area of science and the technology based upon it are memorialized in the name 'Lichtenberg figures'.

Like others of his century (see ch. 13, above), Lichtenberg wrote (1794) on "die physischen Revolutionen der Erde" or "The Physical Revolutions of the Earth" (1800, 2: 25 ff.), a theme which he developed by making a contrast with political revolutions. Lichtenberg wrote extensively about revolutions and revolutions in science, and he even developed a theory of 'paradigmata'. Among those whom he most admired, Copernicus, Newton, and Kant figure prominently.

He early on took a strong position against the new French chemistry or the chemistry of Lavoisier. Unlike Joseph Priestley and Karl Scheele (co-discoverers of oxygen), he did eventually give up the old doctrine of phlogiston, which he found to be faulty, but he remained strongly critical of some aspects of the new system of chemistry of Lavoisier and other French chemists. He was a great believer in simplicity and apparently could not bring himself to accept the new system of chemistry with its great proliferation of 'elements'. On this point he was quite explicit (1800, 9: 187; see Stern 1959, 38n; Mautner and Miller 1952, 227): "I cannot say that I like these discoveries of new kinds of matter [Erden]. Such conglomerations of new substances [Körper] remind me of the epicycles in astronomy. What would those astronomers have done with their epicycles if they had known of the aberration of the fixed stars? They might have shown a great deal of ingenuity, as for instance Copernicus did in his mistakes." Lichtenberg concluded that "if chemistry doesn't soon get its Kepler, it will be smothered by the sheer number of epicycles." Then, "inertia, rather than understanding (which would have done it better), will at last simplify it." He could not help but believe that "there *must* [es muss und muss] be a standpoint from which everything looks simpler."

Lichtenberg's comments on the new chemistry are articulated in his lengthy introduction ('Vorrede') to the sixth edition of Johann Christian Polycarp Erxleben's *Anfangsgründe der Naturlehre* (1794). In the first paragraph he turns to "the French or new chemistry" (p. xxi). His criticism of the "antiphlogistic theory" is general, based on a view of physics (or natural philosophy) at large and is in no sense a support of phlogiston. He states that "this revolution in chemistry, which has had so great an influence on natural science," is in fact "a kind of masterpiece" (pp. xxi–xxii). As an aside, he observes that the initial reception of the new theory in Germany reflected "the character of the nation from which it arose." He concludes: "France is not the country from which the German is accustomed to expect lasting scientific principles." "Where is Cartesian physics now," he asks, which was so strongly defended by the Bernoullis? (p. xxii). On one occasion, Lichtenberg even referred to the new chemistry as a "französische désordre" (Hahn 1927, 60).

I shall not discuss the basis of Lichtenberg's quarrel with the new chemistry, which is of real significance for a study of resistance to scientific change. But it is important in the present context to note that in the pages of Lichtenberg's arguments with their marshaling of counterevidence, he does admit that there has been "eine Revolution in der Chemie" (for example, pp. xxii, xxiii). In one of his discussions of this revolution he introduces a satirical note in relation to a reported ceremony which was "childish" and "genuinely French," in which "Mme Lavoisier, garbed as a priestess, burned phlogiston in a gathering." Lichetenberg said that the only comment on this was that "If Newton had been able to have his wife—if he had a wife—burn the Cartestian vortices, he would not possibly have been able to write the *Principia*" (p. xxiii).

One of Lichtenberg's notebooks (1967, *1*: 826–827) contains an observation that is of interest because it links political revolution with scientific and philosophical progress. In the *Allgemeine Litteratur-Zeitung* (1793, nr. 78, p. 622), he wrote, "it was once observed that the statues in Paris should have been locked up in order to prevent the barbarism of their destruction" during the Revolution; and in another place (nr. 85, p. 675) that "much should not have happened with such violence in the Revolution." But, Lichtenberg commented, this is "as if nature would hand over the realization of her plans to metaphysics." He added (referring to a dreadful earthquake in Calabria in 1793, in which some 40,000 people were said to have been killed): "It also probably would have been good if the towns in Calabria had been kept safe long enough for nature to complete the building of the cellar which she had in process—that's the way it is. I should think that if in spite of all of the repairs and all the propping, the building finally does collapse, then the better arrangement of the whole was not a part of the rebuilding plan and the law of its progress."

This led him to comment on religion, philosophy, and science: "From a Catholicism steadily improved, but improved according to its own ground rules, Protestantism could never arise, nor Kantian philosophy from an improved popular philosophy. From a gradually improved Cartesian physics the true Newtonian physics could never develop. The greatest mathematicians [that is, the Bernoullis] have turned and bent the vortices in order to make them work. But it was all to no avail, for these vortices had to be pushed aside while universal gravity ascended the throne and now reigns from the Milky Way to the Sun, and will reign until the end of time." Such sentiments about scientific change have led some commentators (such as Hans-Werner Schütt 1979) to consider Lichtenberg to have been a kind of Kuhnian.

Lichtenberg, in fact, was in a sense a real precursor of Kuhn, in that he developed a schema of scientific change based on 'paradigmata'. Unlike Kuhn, however, Lichtenberg saw his paradigms as a grammatical analog to the sciences and even wrote of "paradigmata according to which to *decline* in the [various] sciences" (see Stern 1959, 97, 103, 126). He evidently had in mind a variety of "formulas of procedure embodied in sensible form and relatable (by means of 'Transcendentmachung') from one science to another, and across the natural sciences to philosophy." Thus J. P. Stern sees Lichtenberg's paradigms as "archetypal configurations or Goethean 'Urphänomene'; issuing somewhere between facts and laws, these 'paradigms' would in themselves be actual parts of natural science (as grammatical paradigms are parts of natural language)" (p. 103). Lichtenberg did not fully develop the concept of paradigm. It was a time when our modern grammatical analysis was being formalized and this term was coming into general use to denote the inflections in standard form of Latin and Greek nouns, verbs, adjectives, and pronouns. Lichtenberg was arguing for primitive or self-explanatory elements of scientific knowledge that could be considered the analog of the grammatical standards, and to which more complex and puzzling phenomena could be referred (see Toulmin 1972, 106).

J. P. Stern wrote about Lichtenbergian paradigms in 1959, three years before the publication of *The Structure of Scientific Revolutions*, and so did not relate this eighteenth-century idea of paradigm with its twentieth century successor. But in Stephen Toulmin's *Human Understanding* (1972, 106), Lichtenberg was cited as a predecessor of Kuhn and it was pointed out, furthermore, that in the twentieth century the term 'paradigm' had been used before Kuhn by Ludwig Wittgenstein, W. H. Watson, N. R. Hanson, and S. Toulmin. (The general history of 'paradigm' in relation to the understanding of science and its growth has been explored by Daniel Goldman Cedarbaum.) A careful analysis of all these usages makes abundantly clear the high degree of originality with which both Lichtenberg and Kuhn used the concept of paradigm in the context of science.

J. P. Stern's *Lichtenberg* (1959) is an excellent introduction, replete with extracts; there is a very useful classified bibliography. See also the article by O. M. Bilaniuk in the *D.S.B.*

15.1. Hume's Alleged Copernican Revolution

Kant is not the only philosopher of the eighteenth century who is alleged to have said that his own thoughts constituted a Copernican revolution in philosophy.

In a discussion of David Hume's "psychological speculation" concerning "our misconceptions about necessities in things," Anthony Flew (1971, 254) says that Hume's presentation of this topic, in relation to determinism, "was modelled on the story about secondary qualities which he found in Newton and Galileo, and both appeared to him to be trophies of his Copernican revolution in reverse." This notion "of a sort of Copernican revolution in reverse" (Flew 1971, 88) is said to be a "key concept" of Hume's *Treatise of Human Nature*:

> Whereas the astronomy of Copernicus (1473–1543) had shifted the earth — and hence, apparently, man — from the centre of the universe, the new moral science being opened up by Hume promised to restore humanity to the middle of the map of knowledge: " . . . all the sciences have a relation, greater or less, to human nature . . . It is impossible to tell what changes and improvements we might make in these sciences were we thoroughly acquainted with the extent and force of human understanding, and could explain the nature of the ideas we employ, and of the operations we perform in our reasonings." [*Treatise,* Introduction]

In a similar vein, John Passmore, one of the most scholarly and perceptive writers on historical problems in philosophy of our age, says simply: "Hume had hoped to initiate a Copernican revolution in the theory of passions" (1952, 126).

Both Flew and Passmore refer in this connection to Hume's *Treatise.* One might very well have expected Hume to introduce a comparison with philosophy and the sciences in this work, since it is subtitled "An Attempt to Introduce the Experimental Method of Reasoning into Moral Subjects." In the introduction to the *Treatise* Hume does not refer to the (or a) Copernican revolution, nor even to Copernicus. In fact, no scientist is mentioned by name, although there are references to Thales and Bacon. In this portion of the introduction, however, Hume merely says that "*Mathematics, Natural Philosophy, and Natural Religion, are in some measure dependent on the science of* MAN" (1888, p. xix). Hume does, however, refer once to Copernicus in this work. The passage in question occurs at the conclusion of section 3 of part 1 of book 2, "Of the Passions." Here Hume says explicitly that "moral philosophy is in the same condition as natural, with regard to astronomy before the time of *Copernicus.*" We have seen that Flew interprets Hume's Copernican revolution as "reverse," in that it restored man to "the middle of the map of knowledge." This is hardly surprising in a work devoted to what would be called today "human studies," covering "all investigations of man and society" (Flew, 1971, 88) — implying that the main feature of the original Copernican revolution would have been to dethrone man from the center of the physical universe. But Hume's own view of what happened in astronomy is quite different. Here is the whole paragraph (1888, 282):

> The antients, tho' sensible of that maxim, *that nature does nothing in vain,* contriv'd such intricate systems of the heavens, as seem'd inconsistent with true philosophy, and gave place at last to something more simple and

natural. To invent without scruple a new principle to every new phaeno-
menon, instead of adapting it to the old; to overload our hypotheses with a
variety of this kind; are certain proofs, that none of these principles is the
just one, and that we only desire, by a number of falsehoods, to cover our
ignorance of the truth.

Hume was not at all referring to the problem of whether the sun or the earth is
at rest in the middle of the world, nor whether man is at the center, nor to any
of the features commonly associated with a Copernican revolution. He was
concerned simply and primarily with the question of whether "intricate systems" or those "more simple and natural" are more consistent with true
science. He considered Copernicus's achievement to have been a simplification
of the system of the world.

Hume, furthermore, does not necessarily imply that the foregoing change
was a revolution rather than an improvement. He does not, in the *Treatise,*
mention Newton, Galileo, Harvey, or any other scientist except Copernicus,
and he gives no clue whether he believes science advances by sudden great leaps
forward or by steady and gradual progress. Elsewhere in the *Treatise* (bk.3, §10),
and in quite another context, however, he does show that he believed that
revolutions occur in the political realm. It is here that he discusses the Glorious
Revolution, "that famous *revolution,* which has had such a happy influence on
our constitution" (1888, 563).

Hume, like Kant, did suggest that philosophy could be considered as in a state
similar to that of astronomy before Copernicus. But he did not tell us whether
he believed that revolutions occur in science, much less that there had been a
'Copernican revolution' as opposed to a Copernican improvement or simplification. But even if Hume did not refer (as alleged) to a Copernican revolution,
his introduction of the comparison of philosophy with pre-Copernican astronomy may have been of real significance. For we know that Kant read Hume, and
it is possible that his own reference to Copernicus may ultimately have been
derived from Hume.

18.1. Some Nineteenth-Century Theorists of
Revolution in Science

Antoine François Claude, Comte Ferrand (1751–1825)

The publication in 1817 of a four-volume work on *Theory of Revolutions* is a sign of
the early nineteenth-century recognition that revolutions are a normal part of
the regular course of history. This philosophical, historical, and analytical study
of revolutions is, so far as I am aware, the only study in which physical revolutions in nature are associated on a grand scale with revolutions in human affairs.
The author was a French statesman and political writer. An émigré, who returned to Paris in 1801, he became minister of state and postmaster-general

under the restored royal government in 1814. Author of many literary, historical, and political works, he was made a member of the French Academy in 1816.

Ferrand's general thesis (expounded in the introduction) is bipartite. First, he held, there are cyclical movements in the worlds both of physical nature and of political action, in which there is a well-defined point of departure, a point of furthest extension, and an inevitable return (as in the orbital motions of planets, comets, and stars). Second, there is a 'genre' of revolutions which are not predictable, although they may be subject to causal analysis from the vantage point of the historian. Examples are the physical revolutions of the earth, such as are produced by the actions of subterranean fires, floods, or earthquakes. These are said to be "metaphorically and concretely" related to "political and moral revolutions." Thus he calls political revolutions the "volcanoes of humanity." One of his major aims is to show how the revolutionary upheavals in human history are linked to the catastrophic revolutions of our globe through a network of cause and effect. He concludes that the analysis of any single revolution demands an examination of the events which precede and follow it as well as a study of the links between that revolution and revolutions of other 'orders'. Although in the introduction he divides all revolutions into the above-mentioned three 'orders' — political, moral, and physical — in book 1 he states that all revolutions fall into two classes, physical and social, with a subdivision of social revolutions into political and religious revolutions.

Ferrand was strongly convinced that a complete explanation of revolutions requires taking account of the role of divine providence in physical and human events. The most terrible physical revolution in human memory was the Noachian Flood, because it was causally linked to political and religious (that is, social) revolutions. Thus the Flood annihilated whole societies with their laws and customs. Ferrand also discussed physical revolutions in which human action becomes the destructive agent for the divine providence. The movement of populations in northeastern Europe had causal origins in the 'physical' needs of those who migrate (ch. 3); social revolutions were produced in lands where new tribes established themselves, introducing new customs and laws, and in bordering lands in which the displaced people established themselves with their customs and laws. A similar line of reasoning was applied to the "discovery of the New World," which "has produced and will still produce great political revolutions in the Old World." This discovery was "actually a physical revolution" since it was equivalent to "the sudden creation of a quarter of the earth" (ch. 4).

One of the most original parts of the book is the discussion of the "Relations of Physical Revolutions and the Social Order." After referring to the great advances in our knowledge of the revolutions of the earth (made by Bergmann, Buffon, Werner, Dolomieu, Deluc, and Cuvier), he asks directly what influence on political revolutions was exercised by these "great upheavals of nature." His primary example is the way in which the Dutch people have sought to maintain a physical existence in their struggle with the sea ever since the twelfth and thirteenth centuries, which has "strongly influenced their political life, that of

Europe, and that of other parts of the world." The formation of Zeeland gave the Dutch a new maritime province and provided facilities to "augment their naval forces and their maritime commerce." Hence "the physical revolutions which continually threatened the Dutch have made the whole world dependent on their commercial industry, because that industry alone could maintain — but at great cost — the physical existence which it itself had created."

We may be especially interested in Ferrand's presentation of the revolutions in England, because he was among the pioneers who believed that a revolution had occurred in the mid-seventeenth century — and perhaps influenced later writers such as Guizot. Ferrand's analysis begins with the age of William the Conqueror and concludes with the reign of William of Orange and the Glorious Revolution. Primarily the revolutions in England (ch. 2 of bk. 2, "Des révolutions sociales") have involved disputes over succession, the deposing of kings, and the fortunes of ruling houses or families. A major exception to these political or dynastic revolutions is Henry VIII's "revolutionary act" in breaking away from the authority of Rome. Among the revolutions he cites are the Civil War and execution of Charles I plus the Protectorate of Cromwell and the Glorious Revolution of 1688. Ferrand drew attention to many parallels between the French Revolution and the English Revolution, as Guizot would do again at greater length a few years later. "The English people, who began a revolution without cause, made a revolution that was needlessly unrestrained," he wrote (bk. 5, ch. 1). He lamented that "this unhappy people, without Parliament and without king, prostituted themselves before the first brigand who despised them sufficiently to make them his slaves."

Although Ferrand produced one of the lengthiest studies of revolution ever written, and although he sought links between physical revolutions and social and political revolutions, he did not invoke the notion of revolution in science. In discussing geology (for example, in bk. 1, ch. 5), he referred to revolutions undergone by the earth at the same time that he mentioned great successive advances in the science of geology. He predicted that "from century to century" new discoveries would be made "in a science that seems inexhaustible" (vol. 1, p. 11). But he did not even so much as hint that any or all of these "new discoveries" might constitute a revolution.

John Playfair (1748–1819)

Playfair's professional career was spent as professor of mathematics at Edinburgh. His reputation as a scientist does not come from mathematics (although he did propose what is often known as Playfair's axioms in plane geometry) but rather his celebrated exposition of James Hutton's *Theory of the Earth* in readable language. Although Playfair wrote extensively about the history of science, these efforts did not merit so much as a single line in the account of Playfair in the recent *Dictionary of Scientific Biography.* Playfair's chief historical work is *A General View of the Progress of Mathematical and Physical Science since the Revival of Letters in Europe,* written as one of the supplements to the second and fourth editions of the *Encyclopaedia Britannica*. Playfair also wrote on the astronomy and trigonometry of the Brahmins (1822, *3*: 91ff., 255ff.) and on the history of porisms (pp. 179ff.).

In a "Review of Le Compte rendu par l'Institut de France," published in the *Edinburgh Review* in 1809, Playfair (1822, *4*: 333ff.) quoted, in English translation, several extracts on revolutions in science. One is from the astronomer J.-B. Delambre (ibid., 355) on "the revolution recently brought about in chemistry." Another, from Cuvier's report (ibid., 365), states that "the theory of chemical affinities has undergone an entire revolution in the hands of Berthollet, who denies the existence of elective attractions and absolute decomposition." (As reported by Cuvier, in Playfair's translation (ibid., 365), Berthollet "has undertaken to prove, that in all the compositions and decompositions made by what is called elective attraction, there takes place a division of the substance combined between two other substances that act upon it with opposite forces." The "proportion in which this division is made, is determined, not only by the absolute energy with which these substances act, but also by their quantity.") A great change had occurred in ideas about heat, constituting "a body of science, of which the philosophers and chemists of the first half of the eighteenth century had not the most distant conception" (p. 365). According to Cuvier, "The discoveries of Black in Scotland, and Wilke in Sweden, led the way to this revolution." Playfair concluded (ibid., 379) with his own expression of belief that science was moving forward in "a revolution which no individual is sufficiently powerful to effect." In a "Review of Kater on the Pendulum" (p. 502), Playfair invoked the image of a "great revolution" which had taken "place in the construction of instruments intended for the measurement of angles."

Although these examples indicate a usage of 'revolution' that is distinctly post-1789, for the establishment of something new and better, Playfair also used the word in a somewhat more traditional sense. In his *Illustrations of the Huttonian Theory of the Earth* (1802), he wrote of advancements of knowledge as a sequence of steps, one following another — an "ancient physical system" giving way to "Aristotelian physics," the latter being replaced by "the physics of Descartes," Cartesian physics finally yielding to the Newtonian natural philosophy. Should one conclude, he asks (p. 512), "that in the revolutions of science, what has happened must continue to happen"? The reply is negative, because these earlier changes (or revolutions) are but "terms of a series that must end when the real laws of nature are discovered." Because "the Newtonian system" is true, it will not be supplanted. The Newtonian revolution is thus the final revolution. Playfair concludes, "It seems certain, therefore, that the rise and fall of theories in times past, does not argue, that the same will happen in the time that is to come."

Playfair's most extensive discussion of revolution occurs in his *General View of the Progress of Mathematical and Physical Science* (1820, pt. 1). This work opens with a statement of Playfair's philosophy, which is in embryo much like that propounded later in the century by Auguste Comte. Playfair organized his book on the principle of "the subserviency of one science to the progress of another, and to the consequent priority of the former in the order of regular study" (p. 1). Hence he began with "the history of the pure Mathematicks."

Playfair held that Copernicus was "destined to make . . . an entire revolution" in astronomy (p. 125). (We may note, in passing, that his library contained the histories by Bailly and Montucla.) It was Galileo, according to Playfair, who first "made a great revolution in the physical sciences" (p. 103). For him, Kepler did not produce a revolution but only a "reformation" of Copernican astronomy, which went beyond "the reformations of Copernicus" (p. 135).

Playfair's presentation is of special interest because it contains an early version of Planck's famous dictum. Explaining the delay in the complete acceptance of the Copernican system after the work of Kepler and of Galileo, Playfair said that many of "the adherents of the old system" had been "too long habituated to it to change their views"; "but as they disappeared from the scene, they were replaced by young astronomers, not under the influence of the same prejudices, and eager to follow doctrines which seemed to offer so many new subjects of investigation" (p. 144).

The second part of Playfair's history was entitled *Sketch of the Progress of Natural Philosophy from the Revival of Learning to the Present Time*. The first revolution mentioned by Playfair (1820, pt. 2, p. 23) is Descartes's "great revolution in the mathematical sciences" (Playfair had previously only given this the plaudit, pt. 1, p. 29, of 'epoch-like'). Playfair says explicitly that the invention of the calculus was also a revolution; he calls it "the greatest discovery ever made in the mathematical sciences" (p. 21). By the introduction "of the *infinitesimal analysis*," furthermore, the "domain of the mathematical sciences was incredibly enlarged in every direction" (p. 26). This great "revolution in science, . . . made by the introduction of the new analysis" (p. 40), moreover, aroused some opposition. In "every society," Playfair observed, "there are some who think themselves interested to maintain things in the condition wherein they have found them."

In introducing the subject of mechanics, Playfair pointed out that although there are some obvious revolutions, the "developement of truth is often so gradual, that it is impossible to assign the time when certain principles have been first introduced into science" (p. 45). An example was "the principle of *Virtual Velocities*" (p. 45).

Not unexpectedly, for Playfair, the "*Principia Mathematica* of Newton, published also in 1687, marks a great era in the history of human knowledge" (p. 46). This work "had the merit of effecting an almost entire revolution in mechanics, by giving new powers and a new direction to its researches." In his opinion, Newton effected "a greater revolution in the state of our knowledge of nature than any individual had yet done, and greater, perhaps, than any individual is ever destined to bring about" (p. 87).

Playfair had a number of interesting and significant ideas about the history of science. For example, he was very much aware that a major revolution needed preparation and could not occur unless there had been preliminary accomplishments. This was true even of Newton (p. 116): "The condition of knowledge at the time when Newton appeared, was favourable to great exertions; it was a moment when things might be said to be prepared for a revolution in the mathematical and physical sciences." But for Copernicus, Kepler, and Galileo, there would have been no Newton. He could not have done it all by himself (p. 117):

> In the progress of knowledge, a moment had arrived more favourable to the developement of talent than any other, either later or earlier, and in which it might produce the greatest possible effect. But, let it not be supposed, while I thus admit the influence of external circumstances on the exertions of intellectual power, that I am lessening the merit of this last, or taking any thing from the admiration that is due to it. I am, in truth, only distinguishing between what is possible, and what it is impossible, for the human mind to effect.

> With all the aid that circumstances could give, it required the highest degree of intellectual power to accomplish what Newton performed. We have here a memorable, perhaps a singular instance, of the highest degree of intellectual power, united to the most favourable condition of things for its exertion.

Playfair was aware of the ways in which revolutions in one part of science affect other parts. Writing of the Copernican revolution, he pointed out that "it was not astronomy alone which was benefited by this revolution, and the discussions to which it had given rise" (pt. 1, p. 145). Additionally, a "new light . . . was thrown on the physical world," and there was revealed the principle of "inertia of body" and the "composition of forces":

> To reconcile the real motion of the earth with its appearance of rest, and with our feeling of its immobility, required such an examination of the nature of motion, as discovered, if not its essence, at least its most general and fundamental properties. The whole science of rational mechanics profited, therefore, essentially by the discovery of the earth's motion.

Playfair is essentially a modern. While he invokes the concept of revolution, he also writes of improvement and progress. Although he is willing to praise the achievements of ancient mathematicians, he finds their physics and astronomy to be defective and erroneous. He makes the same kind of distinction between improvement, advance, and revolution that we still do. He recognized the importance of mathematics for physics as well as astronomy, and he stressed the innovatory quality of experiment as a feature of the new science. Bacon was one of his heroes and he had no praise for Descartes, except as a mathematician. Above all, while he saw the great advances and revolutions in terms of the heroic creations of single individuals, Playfair was very modern in recognizing that even a Newton could not have achieved his greatness otherwise than by living at the right time.

Sir John Leslie (1766–1832)

Playfair's history ends just before Newton. The continuation was written by Leslie, another Scot, who taught mathematics and natural philosophy at Edinburgh. His main field of scientific endeavor was heat. He is known eponymously for his invention of the Leslie's cube.

Leslie's continuation of Playfair's history is less philosophical. There are very few references to revolution. But he does say that Descartes "effected a revolution in scientific procedure" (1835, 594). It is of some interest to note that Leslie, unlike Playfair, mentioned modern periods devoid of revolution. One was the eighteenth century with respect to algebra, which "still continued to advance" and to acquire "greater perfection in all its details," although "it underwent no revolution" (p. 595).

I have always been fond of Leslie's remarks about finding precursors, once a great discovery has been made. In discussing Galvani's discoveries in electricity, Leslie mentioned that the "Germans lay claim to the origin of Galvanism" (p. 623n). He then referred briefly to observations by Sulzer and Swammerdam. "Such facts," according to Leslie, "are curious, and deserve attention." He concluded, however, that "every honourable mind must pity or scorn that invidious spirit with which some unhappy jackals hunt after imperfect and neglected anticipations, with a view of detracting from the merit of full discovery."

Louis Figuier (1819–1894)

Figuier was a medical doctor and a professor at l'Ecole de pharmacie in Paris. He taught for only a few years before deciding to devote himself to writing and was the author of many books on scientific subjects.

Figuier became famous as a popularizer of science, what the French call honorably a master of 'haute vulgarisation'. He was a derivative writer, not a scholar or original thinker and researcher. But we may admire his honesty in usually acknowledging his sources, many of whom are sound. He is of interest to us because he had a clear notion of revolution in science and was fully aware that there had been a general revolution in the sciences in the sixteenth and seventeenth centuries. He serves to indicate for us the general state of thinking about revolution in science during the middle and late nineteenth century. His books were evidently popular. The *Vie des savants illustres*, a five-volume work, had at least three editions in the 1870s and 80s. He also wrote popular books on astronomy and other branches of science.

In the preface to his book on the Renaissance, Figuier states that this was the era of "the great moral revolution." In the sciences or in philosophy, "one sees, at this epoch, only revolution or reform." For him, Copernicus "revolutionized astronomy" (p. ii) at about the same time that the Reformation was an "immense revolution" from the "political, philosophical, and social points of view" (p. 4). Vesalius, like Copernicus, accomplished a revolution, "a genuine revolution in the natural sciences" (p. 290). Figuier apparently took this opinion directly from Henri de Blainville, who had said in his history that Vesalius's treatise of 1543 "brought about a genuine revolution in science" (1845, 2: 187), one of the rare examples of Blainville's conceiving that there had been a revolution in science.

Figuier stated unequivocally (in the opening of his preface to the book on the seventeenth century; 1876, p. i) that "the seventeenth century is the time when the modern sciences were definitively created." Four men, he held, established the "new philosophy": Kepler, Descartes, Galileo, and Bacon. He hailed Harvey's discovery of the circulation of the blood as a "profound revolution" (p. ii). Later he said that discoveries such as Harvey's had the quality to "revolutionize a whole era in science" (p. 37).

He began the book with a statement of theory: "In order for a scientific revolution to spread and develop without obstacle, it is not enough that its principles be clearly expressed by men of genius. The generation to whom the revolution is proposed must also be ready to receive it" (pp. 1–2). Figuier is thus another proponent of the now-familiar generational theory.

In astronomy, Figuier praised Kepler for having brought to a head "the revolution begun by Copernicus" (p. 49). Descartes produced a "radical revolution in philosophy" (p. 152). Bacon's importance "as reformer of science" had been greatly inflated by the "British self-esteem" (p. 249), but he did inaugurate the era of the "renewal of the sciences." Figuier's discussion of Boyle is less interesting for what it says about Boyle or chemistry than for its flat declaration that there had in fact been a "scientific revolution" ("révolution scientifique") inaugurated by Galileo, Kepler, Bacon, and Descartes (p. 406). Boyle was for Figuier the "apostle of the scientific revolution." This is the earliest reference that I have found to a discussion of *the* Scientific Revolution in a general work.

In the volume on the eighteenth century, Figuier introduced the subject of Newton (p. 41) with a discussion of the general revolution in scientific method of the seventeenth century — "as it was called then, in the *science of sciences.*" Figuier evidently did not believe Newton to be a true innovator of the caliber of Galileo, Kepler, Bacon, and Descartes (1879, 19):

> His only merit consists in having made a mathematical demonstration and a marvelous generalization. No doubt a powerful genius was required to reach that result; but it was not the genius of invention. A genius of calculation and reasoning sufficed for the task. It is sure that if Newton had not existed, another would have been there, a bit later, to capture the glory that fell to the English philosopher.

What is perhaps most interesting about Figuier's chapter on Newton is his characterization of the reply of the Cartesians to the Newtonian natural philosophy as "this counter-revolution" ("cette contre-révolution"; p. 43).

Figuier saw Leibniz's philosophy as revolutionary (p. 52), and he applied the word 'révolution' to d'Alembert (p. 85) and Haller (pp. 284, 296), following the style of Condorcet's *éloges.* Lavoisier was the "creator of modern chemistry" (p. 444) and produced a revolution.

William Whewell (1794–1866)

Whewell was one of the most influential figures in nineteenth-century thought. Master of Trinity College, Cambridge, he was named vice-chancellor of the

university in 1842 and again in 1855. His chief contributions were in the history
and philosophy of science, but he also was known for his work in physical astronomy and his efforts to reform mineralogy. He wrote an influential report on the
state of theories of electricity, and in particular studied the tides. A three-volume *History of the Inductive Sciences* was published in 1837, his *Philosophy of the
Inductive Sciences* in 1840. In 1838 he was appointed professor of moral science (=
philosophy) at Cambridge.

William Whewell is the major figure among nineteenth-century writers on
the history of science to use the concept and name 'revolution in science' in a
significant way. Although it has recently been asserted that Whewell "made
virtually no use of the concept of scientific revolution," his *History of the Inductive
Sciences* and other works are replete with references to revolution and display a
veritable philosophy of scientific development in relation to revolutions.

Very near the opening of volume 1, it is announced, under the heading
"Terms record Discoveries" (1837, *1*: 11): "We shall frequently have to notice the
manner in which great discoveries . . . stamp their impress upon the terms of
a science; and, like great political revolutions, are recorded by the change of the
current coin which has accompanied them." Here Whewell has only made a
comparison between "great discoveries" and "great political revolutions"; he
has not actually said that such discoveries are or may be revolutions. The
unwary reader who goes no further than the opening pages of the introduction
to volume 1 might thence conclude that Whewell did not approve of the concept
of revolution in application to the sciences. And this impression might easily be
strengthened by an admonition given by Whewell on the previous page to the
effect that "the history of each science, which may . . . appear like a succession of revolutions, is, in reality, a series of developements"; the reason is that
"earlier truths are not expelled but absorbed, not contradicted but extended"
(p. 10). But shortly afterwards Whewell refers to the "great changes
which . . . take place in the history of science" as "the revolutions of the
intellectual world" (p. 11).

Whewell introduced the work of Lavoisier with this observation (ibid., *3*: 128):
"Few revolutions in science have immediately excited so much general notice as
the introduction of the theory of oxygen." He also contrasted Lavoisier with
those "who had, in earlier ages, produced revolutions in science," noting that
Lavoisier "saw his theory accepted by all the most eminent men of his time, and
established over a great part of Europe within a few years from its first promul-
gation" (p. 136). By this time (1837), such references to Lavoisier's chemical
'revolution' were not uncommon. Of more interest is the fact that in his discussion of Lavoisier, Whewell seems to have lost sight of the caution he had advocated earlier with respect to 'revolutions' in science. And he introduced 'revolution' in an even more striking way in his account of the *Genera Plantarum* (1789),
which the younger Jussieu had published after he "had employed himself for
nineteen years upon botany." This work, said Whewell, "was not received with
any enthusiasm; indeed, at that time, the revolutions of states absorbed the
thoughts of all Europe, and left men little leisure to attend to the revolutions of
science" (1837, *3*: 337).

Reference had earlier been made to the philosophical work of Francis Bacon, concerning whom Whewell wrote: "He announced a New Method, not merely a correction of special current errors; he thus converted the Insurrection into a Revolution, and established a new philosophical Dynasty." This quotation is taken from "the third edition, with additions" of Whewell's *History* (1865, *1*: 339), where it occurs in a lengthy extract, printed within square brackets, of a supplement added by Whewell in the "2nd Ed." Between the first and second editions of the *History,* Whewell published his *Philosophy of the Inductive Sciences* (see below), in which he discussed more fully the role of Bacon as a revolutionary.

A couple of years later, in his *Philosophy of the Inductive Sciences, Founded upon Their History* (first published in 1840), Whewell particularly evoked the metaphor of revolution in relation to the sciences, especially in book 12, "Review of Opinions on the Nature of Knowledge, and the Means of Seeking It." There, in discussing "the revolutions" in opinions as to the methods whereby the physical sciences might progress, he referred to two "movements" — "the Insurrection against Authority and the Appeal to Experience." In language more befitting political revolution than the rise of science, Whewell (1840, 2: 319–320) wrote about the "submission of the mind" that had been demanded by "the Scholastic System" and related that "the natural love of freedom in man's bosom, and the speculative tendencies of his intellect, rose in rebellion . . . against the ruling oppression." His aim was "to trace historically the views which have prevailed respecting such methods, at various periods of man's intellectual progress." He found that "among the most conspicuous of the revolutions which opinions on this subject have undergone, is the transition from an implicit trust in the internal powers of man's mind to a professed dependence upon external observation; and from an unbounded reverence for the wisdom of the past, to a fervid expectation of change and improvement" (p. 318).

Whewell presented certain presages of a "philosophical revolution" and introduced the work of some sixteenth-century philosophers, again in the language of political revolution, writing that "these insurrections against the authority of the established dogmas, although they did not directly substitute a better positive system in the place of that which they assailed, shook the authority of the Aristotelian system, and led to its overthrow" (p.352). Then Whewell turned his attention to "the practical reformers of science," who had a "greater share in bringing about the change from stationary to progressive knowledge, than those writers who so loudly announced the revolution." The list includes Leonardo da Vinci, Copernicus, Fabricius, Maurolycus, Benedetti, Tycho, Gilbert, Galileo, and Kepler.

Introducing Copernicus, Whewell notes that "even in those practical discoverers to whom, in reality, the revolution in science, and consequently in the philosophy of science, was due," we do not always find a "prompt and vigorous recognition of the supreme authority of observation" and a "bold estimate of the probable worthlessness of traditional knowledge" (p. 370). Later, after describing the work of these practitioners of science, Whewell announced the coming on the scene of Francis Bacon in these words (p. 388):

> A revolution was not only at hand, but had really taken place, in the great body of real cultivators of science. The occasion now required that this revolution should be formally recognized;—that the new intellectual power should be clothed with the forms of government;—that the new philosophical republic should be acknowledged as a sister state by the ancient dynasties of Aristotle and Plato. There was needed some great Theoretical Reformer, to speak in the name of the Experimental Philosophy; to lay before the world a declaration of its rights and a scheme of its laws. And thus our eyes are turned to Francis Bacon, and others who like him attempted this great office.

Political rhetoric abounds in Whewell's presentation of Bacon's philosophy of science. He is "not only one of the Founders, but the supreme Legislator of the modern Republic of Science" (p. 389); he is "the Hercules who slew the monsters that obstructed the earlier traveller"; and he is "the Solon who established a constitution fitted for all future time." But above all, Bacon has been properly hailed "as the leader of the revolution" (p. 392), that is, the leader of the "revolution in the method of scientific research" that had been going on in the sixteenth century, and of which he became the articulator and chief spokesman (p. 391). "If we must select some one philosopher as the Hero of the revolution in scientific method," Whewell concluded, then "beyond all doubt Francis Bacon must occupy the place of honour" (p. 392). But Whewell also said that "Bacon was by no means the first mover or principal author of the revolution in the method of philosophizing which took place in his time" (p. 414).

Over and over again in the *Philosophy of the Inductive Sciences,* Whewell invokes the concept of revolution. Thus Harvey's "reflections on the method of pursuing science" are said to be "strongly marked with the character of the revolution that was taking place" (p. 414). It is in this context that the Cartesian philosophy—held to be "an endeavour to revive the method of obtaining knowledge by reasoning from our own ideas only" and to erect a "philosophical system" in "opposition to the method of observation and experiment"—is called "an attempt at a counter-revolution" (p. 415).

Thus far, I have indicated some of the chief ways in which Whewell referred to revolutions in the sciences, including revolutions in the methods or philosophy of science. I shall conclude with a few words on Whewell's conception of a great revolution in philosophy. In a "Review of Opinions on Knowledge," Whewell remarked that the "mode of research by experiment and observation, which had, a little time ago, been a strange, and to many, an unwelcome innovation, was now become the habitual course of philosophers." And this led him to conclude, "The revolution from the philosophy of tradition to the philosophy of experience was completed" (pp. 424–425). A whole chapter (ch. 14 of bk. 12) is devoted to "Locke and His French Followers." Here Whewell observed that Locke "came at a period when the reign of Ideas was tottering to its fall" (1840, 2: 458). Locke, "by putting himself at the head of the assault, became the hero of his day: and his name has been used as the watchword of those who adhere to the philosophy of the senses up to our own times." Whewell also observed, in a somewhat unexpected fashion for one who had sounded the clarion so vehemently for Bacon (p. 426):

> Perhaps too, as was natural in so great a revolution, the writers of this time, especially the second-rate ones, were somewhat too prone to disparage the labours and talents of Aristotle and the ancients in general, and to overlook the ideal element of our knowledge, in their zealous study of phenomena. They urged, sometimes in an exaggerated manner, the superiority of modern times in all that regards science, and the supreme and sole importance of facts in scientific investigations.

This led Whewell to the following discussion of revolutions and revolutionaries (p. 458):

> Locke himself did not assert the exclusive authority of the senses in the extreme unmitigated manner in which some who call themselves his disciples have done. But this is the common lot of the leaders of revolutions, for they are usually bound by some ties of affection and habit to the previous state of things, and would not destroy all traces of that condition: while their followers attend, not to their inconsistent wishes, but to the meaning of the revolution itself; and carry out, to their genuine and complete results, the principles which won the victory, and which have been brought out more sharply from the conflict.

This analysis shows that Whewell was not merely using 'revolution' as metaphor but had a rather fully developed theory of revolutions in thought, one in which he did not hesitate to introduce metaphors and expressions that had come into general usage in discourses about political revolutions.

18.2. Revolution and Evolution: Marx, Poincaré, Boltzmann, Mach

Since the middle of the nineteenth century, as we have seen, many scientists, and nonscientists, have been committed to the belief that science does not advance by revolution but by evolution or an incremental process that does not provide for large quantum jumps. Although Albert Einstein believed in a Maxwellian revolution and a Galilean revolution, he generally tended to view scientific advance as an evolutionary process (see Holton 1981) and even insisted that the theory of relativity — generally held to be the prototypical revolution in science of the twentieth century — should be understood in terms of evolution and not revolution. Many scientists and historians of science have tended to downgrade or even deny the occurrence of revolutions in science.

Karl Marx

I do not know who was the first to advocate a Darwinian position with respect to the sciences, but it was apparently Karl Marx who first applied Darwinian concepts to the development of technology. The reasoning was a footnote on "The Development of Machinery" which appeared for the first time in the second edition (1873) of *Das Kapital*. Discussing the want of a history of technology, particularly of tools and machines, Marx (1954, vol. 1, ch. 15, §1, p.372) said:

> A critical history of technology would show how little any of the inventions of the 18th century are the work of a single individual. Hitherto there is no such book. Darwin has interested us in the history of Nature's Technology, i.e., in the formation of the organs of plants and animals, which organs serve as instruments of production for sustaining life. Does not the history of the productive organs of man, of organs that are the material basis of all social organisation, deserve equal attention? And would not such a history be easier to compile, since, as Vico says, human history differs from natural history in this, that we have made the former, but not the latter? Technology discloses man's mode of dealing with Nature, the process of production by which he sustains his life, and thereby also lays bare the mode of formation of his social relations, and of the mental conceptions that flow from them.

This is one of two references to Darwin in volume 1 of *Das Kapital*. The other (in a note to ch. 14, §2, p. 341) refers to Darwin's "epoch-making work on the origin of species," and discusses Darwin's ideas about organs of animals and manmade tools, such as knives. I do not know of any such early attempts to apply Darwinian evolution to the history of science.

Lucien Poincaré

Among scientists who have adopted an evolutionist view of the history of science, three physicists may claim our attention: Lucien Poincaré, Ludwig Boltzmann, and Ernst Mach. Lucien Poincaré (1862–1920) was a member of the famous family that included the mathematician and philosopher Henri Poincaré, who referred to the Maxwellian revolution and to revolutions going on in the science of his day. Lucien, a physicist and educator, was a brother of Raymond Poincaré, the ninth president of the third French Republic. He published a general book of 'haute vulgarisation' entitled *The New Physics and Its Evolution*, which was awarded a prize by the French Academy of Sciences. The theme of the opening chapter was summarized in the table of contents as follows: "Revolutionary change in modern Physics only apparent: evolution not revolution the rule in Physical Theory." As he developed his analysis of recent physics, he was aware that "the evolution of the different parts of physics does not . . . take place with equal speed" (1907, 7). Sometimes questions are forgotten until "some accidental circumstance suddenly brings them new life" and—as in the case "of the X rays"—then "they become the object of manifold labours, engross public attention, and invade nearly the whole domain of science." Lucien Poincaré alleged that the "discovery of the X rays" was "no doubt" considered by physicists "as the logical outcome of researches long pursued by a few scholars working in silence and obscurity on an otherwise much neglected subject." Nevertheless, "the extraordinary scientific movement provoked by Röntgen's sensational experiments" was a "lucky chance" that has "hastened an evolution already taking place"; "theories previously outlined have received a singular development." Lucien Poincaré saw no singularities in "the continuous march of physics," no discontinuities that might merit the name of 'revolution'. He was echoing a long French tradition of physicists such as Jules Violle, who in 1884 wrote of the "necessary evolution" of the "science of nature" and of physics in particular, or Alfred Cornu, who in 1896 referred to the "successive progressing of physics."

There is a curious footnote to Lucien Poincaré's insistence that science advances by evolution and not by revolution. F. Legge, editor of the series in which the English translation of Poincaré's book appeared in 1907, pointed out that the author's official position as Inspector-General of Public Instruction caused him to visit universities and *lycées* in order to report on the state of studies in the sciences. "Hence," the editor concluded, effectively denying the validity of Poincaré's views on the way science develops, the author "is in an excellent position to appreciate at its proper value the extraordinary change which has lately revolutionized physical science" (p. v).

Ludwig Boltzmann

More interesting is the case of Ludwig Boltzmann (1844–1906), famous for his contributions to the theory of gases and a founder of the science of statistical mechanics. He had the "profound intuition" that "thermodynamic phenomena were the macroscopic reflection of atomic phenomena regulated by mechanical laws and the play of chance" (J. W. Herivel in Williams 1969). He is considered to be as important in the philosophy of science as in theoretical physics. In Boltzmann's writings we find combined a belief that there are revolutions in science and that science progresses according to the principles of Darwinian evolution. In 1896 he wrote that theoretical physics, like every science, has "experienced the most varied revolutions" ("die mannigfachsten Umwälzungen"; 1905, 104). Three years later (22 Sept. 1899) in a lecture "On the Development of the Methods of Theoretical Physics in Recent Times," Boltzmann referred to the great changes that had occurred in physics during his active lifetime, during the second half of the nineteenth century. He felt as if he were "a monument of ancient scientific memories," he said, as he looked back "on all these developments and revolutions" ("diese Entwicklungen und Umwälzungen"; 1905, 205; 1974, 82). In particular he found that the "rapidly growing importance of the principle of energy has led to an attempted revolution [Umwälzung] involving the whole of physics" (1905, 216; 1974, 91); and he warned his hearers that (1905, 224; 1974, 97) "history often shows unforeseen revolutions" ("unvorhergesehene Umwälzungen").

In a lecture (1904) on statistical mechanics he again introduced the theme of revolutions in science. He "might almost say" that "theoretical physics . . . is in the course of a revolution" ("Umwälzung"; 1905, 346–347; 1974, 160–161). By contrast, every experimental result, every "securely ascertained fact remains for ever immutable"—it can be extended or even complemented but never "entirely overthrown." Hence, he believed, there is an important difference between theoretical and experimental physics. The "development of experimental physics proceeds continuously," he said, and does not have "sudden" leaps; that is, experimental physics "is never visited by great revolutions or commotions" ("von grossen Umwälzungen und Erschütterungen"). Apparently, Boltzmann did not seriously take into account the revolutionizing effects of new instruments and new methods of observation.

Boltzmann also referred to "powerful revolutions [mächtige Umwälzungen]," which may even affect "theories that proudly describe themselves as free from hypotheses" (1905, 348; 1974, 161). With a pointed dig at W. F. Ostwald and his school, he then said that no one could "doubt that the theory known by the name of energetics must radically change its garb if it is to go on existing." He developed this notion further as follows (1905, 348; 1974, 161):

> Of course we can keep our hypotheses fairly indefinite or even frame them in the form of mathematical formulae or in words that express a thought equivalent to them. Then we can check step by step that there is agreement with what is given; a total overthrow of what was previously built is of course not to be excluded even then, if for example the law of conservation of energy turned out at last to be false after all. However such a revolution will be extremely rare and in some cases so improbable as to be inconceivable.

It is to be noted that Boltzmann has used the word 'Umwälzung' for 'revolution' except in the final example, where the word 'Umsturz' occurs twice in the same paragraph; in the English version 'ein gänzlicher Umsturz' is translated as 'a total overthrow' but in the next sentence 'ein solcher Umsturz' becomes 'such a revolution'.

These statements concerning revolutions are in striking contrast to Boltzmann's pronouncements concerning evolution. We are reminded by S. R. de Groot (Boltzmann 1974, p. xii) that "Boltzmann had a tremendous admiration for Darwin and . . . wished to extend Darwinism from biological to cultural evolution," even considering "biological and cultural evolution as one and the same thing." And, indeed, after describing the tremendous technological innovations of the nineteenth century, Boltzmann (1974) said that future generations would not characterize this period as "the century of iron, or steam, or electricity," but that it will without any question "be named the century of the mechanical view of nature, of Darwin" (1974, 15). He added that a "mighty upswing" had taken place in such sciences as geology and physiology "since the general acceptance of the ideas of Darwin" (p.16). In listing "the great theories" of modern science, he mentioned "those of Copernicus, atomism, the mechanical theory of weightless media, Darwinism" (p. 34; see also p. 69), thus joining the ranks of those for whom Darwin was the only biologist who had precipitated a revolution in science.

A few years later (in 1899) Boltzmann gave four lectures at Clark University, in Worcester, Massachusetts. The first was "On the Fundamental Principles and Equations of Mechanics," in which Boltzmann gave a prominent place to "that most splendid mechanical theory in the field of biology, namely the doctrine of Darwin" (1974, 133). He discussed such purely biological and psychological issues as adaptation and heredity, memory, and perception. And he pointed out that "Darwin's theory explains by no means merely the appropriate character of human and animal bodily organs, but also gives an account why often inappropriate and rudimentary organs or even errors of organization could and must occur" (p. 136). He discussed a number of biological questions, referring to "Darwin's theory" and to the "Darwinian point of view," which, he claimed, enables us to "grasp . . . what is the relation of animal instinct to human intellect" (p. 138).

Boltzmann's strongest expression of Darwinism occurs in "An Inaugural Lecture on Natural Philosophy," first published in 1903. Here he referred (1974, 153) to the writings of Ernst Mach, who had "ingeniously" proposed that no theory is either "absolutely true" or "absolutely false," but that each theory "must gradually be perfected, as organisms must according to Darwin's theory" (1974, 153). Boltzmann then asserted that in establishing his own philosophical position, "I absorbed Darwin's theory," a theory which he said had a profound effect on his thinking (p. 157).

In this lecture Boltzmann discussed "logic" and "the laws of thought," remarking that Kant had explained that both the case for and that against matter being "continuous" or being "atomistically constituted . . . can be proved by strict logic" (pp. 164–165). Kant was also aware that it is "strictly provable that the divisibility of matter can have no limit and yet infinite divisibility [of matter] contradicts the laws of logic." At "every step," according to Boltzmann, we find that "philosophical thought becomes enmeshed in contradictions."

These laws of thought have evolved according to the same laws of evolution as the optical apparatus of the eye, the acoustic machinery of the ear and the pumping device of the heart. In the course of mankind's development everything inappropriate was shed and thus arose the unity and perfection that can give the illusion of infallibility. So too the perfection of the eye, the ear and the arrangement of the heart calls forth our admiration yet we could not assert these organs to be absolutely perfect. Just as little must the laws of thought be taken as absolutely infallible. Indeed it is precisely they that have evolved for the sake of grasping what is necessary for life and practically useful. [p. 165]

Then, invoking the specifically Darwinian doctrine of adaptation, in relation to "inherited and acquired ideas," Boltzmann announced that "our task cannot be to summon data to the judgment-throne of our laws of thought but rather to adapt our thoughts, ideas and concepts to what is given." Another Darwinian exercise occurs a few pages later when he discussed the development of thought processes from "the simplest living beings" to "more highly organised beings" (p. 167).

Toward the conclusion, Boltzmann derived "the concept of happiness . . . from Darwin's theory" (p. 176), a rare introduction into his lectures of the concept of "natural selection." He concluded with the bald assertion that "in my view all salvation for philosophy may be expected to come from Darwin's theory" (p. 193). And he then proceeded to account for "the position of the so-called laws of thought in logic" in "the light of Darwin's theory," stressing "heredity" and defining "the will" as "the inherited striving towards intervening in the world of phenomena in a way that is helpful to us," which "has resulted in our ideas becoming gradually perfected" (pp. 194–195).

Boltzmann's aim was to use Darwinian concepts to account for man's intellectual development and, in particular, the development of "laws of thought." Yet despite his laudation of Darwin and his claim that Darwin's theory was the key to understanding the basic problems of philosophy and to finding the source of the "concept of happiness," he did not really use the key ideas of Darwinian theory: variation, struggle for existence, competition, and natural selection. Nor did he discuss evolution in relation to the survival of types most suited to their environment. He expressed only a general belief in a process of evolving. He declared it a "matter of indifference" as to how life began on earth, whether over a long period of time "the first protoplasm developed 'by chance'" or whether "egg cells, spores or some other germs in the form of dust or embedded in meteorites" came to the earth from space (p. 196). Ruling out creation, he held that it was just absurd to imagine "highly developed organisms" falling "out of the skies." He then supposed that such "simple cells or particles of protoplasm" would be in constant motion, the "so-called Brownian molecular motion," and would grow by absorption and multiply by cell-division in a manner "explicable by purely mechanical means." Then, in a more Darwinian manner, he explained that of course these motions would be affected by the physical environment. Some "particles" would experience changes in a way that would permit them to move to "regions where there were better materials [that is, food] to absorb" and they would be thereby "better able to grow and propagate so as soon to overrun all the others." This was "a simple process that is readily understood mechanically" and in it "we have heredity, natural selection, sense perception, reason, will, pleasure and pain all together in a nutshell." The same principle, changed only in its quantitative degree, enables us to form a progression "via the whole world of plants and animals to mankind with all its thinking, feeling, willing and acting, its pleasure and pain, its artistic creation and scientific research, its nobility and vices." This led him to his most Darwinian expression, the Darwinian derivation of the concept of happiness (pp. 176–177):

Cells that had come together into rather large collections in which there occurred division of labour and hived off, by division, cells with similar tendencies, had greater opportunities in the struggle for existence, especially if certain cells when subjected to harmful influences did not rest until the working cells had removed such inroads as much as possible (pain). The action of these cells was particularly effective if, so long as the harmful influences were not completely eliminated, it continued and left behind a tension that diminished only very slowly, thus stressing the memory cells and inciting the motor cells to even more vigorous and circumspect collaboration when similar.harmful circumstances recurred. This state is called lasting displeasure, a feeling of unhappiness. The opposite, complete freedom from such vexatious after-effects and a warning to the memory cells that the motor cells are to act in the same way again if similar circumstances supervene, is called lasting pleasure, a feeling of happiness.

It is to be noted, however, that Boltzmann's evolutionism was never applied in an organized and systematic manner to the development of scientific theories and concepts, although he was concerned with the order of development of ideas in each individual. Thus he held that "the only immediately given" for an individual "consists of his own mental phenomena," while "atoms, forces, and energy forms are mental concepts constructed much later in order to represent the law-like features of perceptions" (p. 177).

Ernst Mach

How different from all this is Mach's Darwinism! Ernst Mach (1836–1916) was a very influential figure in science and philosophy and a true founder of the modern discipline of history of science. He is the father-figure for contemporary positivism, in its form of 'logical positivism', and his analysis of the concepts of space and time and force (Mach 1960) had a profound influence on the young Albert Einstein (Schilpp, 1949, 21) and other physicists. His own scientific work was in the area of perceptual psychology (Mach 1914) and aerodynamics (where his contribution is commemorated in the eponymous 'mach numbers'). His extreme positivist position led him to reject the concept of atoms (Mach 1911). But his idea that scientific laws are no more than an expression of the 'economy of thought' has played a significant role in recent philosophy of science (see Losee 1972).

Mach's ideas on the Darwinian development of science occur primarily in his inaugural address as rector of the University of Prague on 18 October 1883, "On Transformation and Adaptation in Scientific Thought." In the English version of this essay (Mach 1943, 214–215), Mach is made to refer to Galileo's "mighty revolution"— "And how great and how far-reaching that revolution was!" But in the original German (1896, 237–238) the "mighty revolution" is "die gewaltige Veränderung" and the second reference reads "Und gross genug war diese Veränderung!" The word 'Veränderung' (unlike 'Umwälzung') does not, however, carry the force of 'revolution', but merely means 'change' or 'alteration' and can even signify 'variation' or 'fluctuation'. The concept of revolution in the sciences would be quite out of place in Mach's evolutionary analysis of scientific thought.

Mach (1943, 215–218) began his essay by comparing Darwin and Galileo, even showing a parallel between the analysis of motion and the problems of heredity and adaptation. Darwin's ideas, he said (p. 217), were — only thirty years after Darwin's publication — "firmly rooted in every branch of human thought, however remote": "Everywhere, in history, in philosophy, even in the physical sciences, we hear the watchwords: heredity, adaptation, selection. We speak of the struggle for existence among the heavenly bodies and of the struggles for existence in the world of molecules." Then Mach announced his intention "to consider the growth of natural *knowledge* in the light of the theory of evolution." He reasoned that "knowledge . . . is a product of organic nature"; hence (p. 218), "although violent comparisons should be avoided, still, if Darwin reasoned rightly, the general imprint of evolution and transformation must be noticeable in ideas also." Mach's presentation of the modifications and transformation of scientific ideas still provides an exciting and fruitful experience for anyone interested in the conceptual development of science (see Cohen 1980, 283ff.). The following is a sample of Mach's use of Darwinian concepts to express his views on the core of the history of science, the growth of ideas (Mach 1943, 63):

> Slowly, gradually, and laboriously one thought is transformed into a different thought, as in all likelihood one animal species is gradually transformed into new species. Many ideas arise simultaneously. They fight the battle for existence not otherwise than do the Ichthyosaurus, the Brahman, and the horse.
>
> A few remain to spread rapidly over all fields of knowledge, to be redeveloped, to be again split up, to begin again the struggle from the start. As many animal species long since conquered, the relicts of ages past, still live in remote regions where their enemies cannot reach them, so also we find conquered ideas still living on in the minds of many men. Whoever will look carefully into his own soul will acknowledge that thoughts battle as obstinately for existence as animals. Who will gainsay that many vanquished modes of thought still haunt obscure crannies of his brain, too fainthearted to step out into the clear light of reason? What inquirer does not know that the hardest battle, in the transformation of his ideas, is fought with himself.

Mach concluded that the "transformation of ideas thus appears as a part of the general evolution of life, as a part of its adaptation to a constantly widening sphere of action" (p. 233).

Not all thinkers would go along with this personification of ideas, as if in the evolutionary process they had lives of their own. The Spanish philosopher Unamuno rebutted the notion that ideas develop and that ideas may be in conflict (Cohen 1980, 327). He could not conceive that ideas could have any life of their own, apart from the men who conceived them. "No hay opiniones," he wrote, "sino opinantes" ("There are no ideas, only holders of ideas"). This thought, he said, he had adapted from "some doctors" who maintained "No hay enfermedades sino enfermos" ("There are no illnesses, only people who are ill"; Unamuno 1951, 2: 271; cf. Marichal 1970, 104). In fairness to Mach, it should be pointed out that he was fully aware that "ideas, as such, do not comport themselves in all respects like independent organic individuals" so that "violent comparison should be avoided" (1943, 218).

Among scientists, Mach's specific use of Darwinian concepts in describing scientific change was the exception rather than the rule. Most scientists who wrote of the 'evolution' of their disciplines tended to refer to 'evolution' only in a general and not specifically Darwinian sense. Sometimes a scientist would even write of 'revolution' and 'evolution' in different paragraphs of the same article. These points are well illustrated by one of the presentations given in 1904 at the Congress of Arts and Science held at the Universal Exposition in St. Louis. In reviewing "The Progress and Development of Chemistry during the Nineteenth Century," F. W. Clarke, Chief Chemist of the U.S. Geological Survey, wrote of the second half of the century as witness to a "great but silent revolution," one "whose magnitude will be better appreciated by posterity than by ourselves" (1905, *4*: 226). He believed that even if there had been no more than the replacement of "supposition by experiment, and chance discovery by orderly research" — in which chemistry "surely played one of the leading parts" — "the revolution would still have been one of the greatest in the history of mankind." Four paragraphs later, however, he was describing the discovery of the periodic system as the result of a "process of evolution" (p. 227) and he characterized the advance of "any science" as an "evolution" (p. 234).

Comparing Darwin and Galileo, Mach wrote (1943, 215–216):

> Of scarcely less importance, it seems, was that movement which was prepared for by the illustrious biologists of the hundred years just past, and formally begun by the late Mr. Darwin. Galileo quickened the sense for the simpler phenomena of *inorganic* nature. And with the same simplicity and frankness that marked the efforts of Galileo, and without the aid of technical or scientific instruments, without physical or chemical experiment, but solely by the power of thought and observation, Darwin grasps a new property of *organic* nature—which we may briefly call its *plasticity*. With the same directness of purpose, Darwin, too, pursues his way. With the same candor and love of truth, he points out the strength and the weakness of his demonstrations. With masterly equanimity he holds aloof from the discussion of irrelevant subjects and wins alike the admiration of his adherents and of his adversaries.

18.3. Evolution vs. Revolution in Claude Bernard's Writings about the Development of Science

Claude Bernard is of exceptional interest in a study of the concept of revolution in science because he wrote extensively of both revolution and evolution as modes of scientific advance. His ideas about the ways in which science progresses have a real significance because he was one of the giants of nineteenth-century science, often described as the "founder of experimental medicine." Chief among his achievements is his great discovery of the glycogenic function of the liver, the process whereby the liver produces a substance he named glycogen, which it metabolizes into sugar; this discovery was related to Bernard's introduction of the general concept of glands of "internal secretion." Bernard is also noted for finding the role of the pancreas (or pancreatic juice) in the digestion of fat and the conversion of starch into maltose, the existence of vasomotor nerves and their effect in regulating the flow of blood, the independent contractility of muscle, the mode of carrying oxygen by the blood, and the theory of the constancy of the "internal environment."

Bernard's goal was to make medicine scientific by transforming medicine from a collection of empirical treatments into an "experimental science." Toward this end he did experimental work that actually pushed medicine into scientific paths and that raised physiology to a new level; but he also wrote extensively about the experimental method. Convinced that "method by itself produces nothing," he did not set forth a series of rules, but rather discussed all aspects of the scientific process from the initial intuition of the experimenter to the development of a theory and its ultimate confirmation or rejection by experimental test. Bernard's ideas about science were written down in various notebooks, were incorporated into his lectures (which were published), and were organized into a general philosophical and methodological work published in 1865 as the *Introduction à l'étude de la médecine expérimentale (Introduction to the Study of Experimental Medicine)*. The latter was originally intended to be the preface to a large-scale work on the principles of experimental medicine, and it contains a clear expression of Bernard's views on the nature of science, on fact and theory, the method of experiment, the progress of science, and much else besides.

One of Bernard's theses, prominently displayed in the *Introduction*, is that there has been a fundamental revolution in the sciences. "The revolution which the experimental method has effected in the sciences is this," he wrote (1927, pt. 1, ch. 2, §4): "it has put a scientific criterion in the place of personal authority." This idea is also developed in other writings of Bernard. The experimental method is "dependent only on itself" for the reason that "it includes within itself its criterion," namely, experience — observation and experiment. In this, science differs from other areas, in which a personal authority dominates. Experimental science "forces" this "impersonal authority" even on "great men of science." Science is thus the direct opposite of the activity of "scholastics," who sought "to prove from texts that they are infallible and that they have seen, said, or thought everything discovered after them." In science the "fundamental precept" is "non-submission to authority" and "in the experimental sciences, great men are never the promoters of absolute and immutable truths." The great revolution, for Bernard, was to establish the experimental method (in the age of Galileo) and to free science from the yoke of theological and philosophical dogmas. To avoid any "mistake," he hastened to add, "I mean to speak here of the evolution of science only." In art and letters, "personality dominates everything." Although Bernard had been discussing personal authority, and the revolution that has overthrown such authority and has established the experimental method, he apparently saw this kind of event as part of a general evolution of the sciences.

Once the revolution had occurred, and science had become transformed by the introduction of the experimental method, science developed in a way that Bernard called evolution. Physics and chemistry, "as established sciences, display the independence and impersonality which the experimental method demands." But he found that medicine is "still in the shades of empiricism" and is "still more or less mingled with religion and with the supernatural." Since "medical personality is placed above science by physicians themselves," there is clear proof "that the experimental method has by no means come into its own in medicine." Hence there is need for a revolution.

In this context Bernard seems to have envisaged two separate processes. One was the large-scale dramatic revolution, the introduction of the experimental method. This had occurred in physics in the seventeenth century, in chemistry in the eighteenth, and had not yet happened in the life sciences. The second was "science in evolution," primarily an incremental process having two aspects: using knowledge acquired already, in which all scientists are "more or less equal," and attaining new knowledge, "the darker regions of science," where "great men . . . are marked by ideas which light up phenomena hitherto obscure and carry science forward." This kind of evolution was for Bernard simply a mode of advance, without specific Darwinian overtones. But Bernard did not believe that the forward progress of experimental science is always and necessarily gradual and incremental. We have seen him to say that this is a feature only of mathematics. Since "truths in the experimental sciences . . . are only relative," he wrote, "these sciences can move forward only by revolution and by recasting old truths in a new scientific form" (ibid.). Thus Bernard envisaged that the evolution of the experimental sciences is punctuated by occasional revolutions. In other places, he associated such revolutions with Lavoisier and Bichat. In the *Introduction* (pt. 2, ch. 2, §10) he also said that Vesalius (along with Mundinus) "contradicted Galen by confronting his opinions with animal dissections" and that they were accordingly "considered innovators and revolutionaries."

Bernard (ibid.) considered it an error to confuse the history of science with the history of the men who made it. "The logical and didactic evolution of experimental science," he said, "is by no means expressed in the chronological history of the men concerned with it." He also argued that "experimental science must be the goal" of the "very evolution of human knowledge," and that "this evolution requires that the earlier sciences of classification shall lose their importance to the degree that the experimental sciences develop." Toward the end of the *Introduction*, he gave young doctors some advice on how to conduct themselves in dealing with their patients even though their medicine is not yet scientific. They must respect the ideals of experimental science but they must accept the fact that much of their practice will have to depend on "compassion and blind empiricism," which are "the prime movers of medicine." Bernard saw a process of "medical evolution" (pt. 3, ch.4, §3), in which empiricism leads to reflection and doubt and later to experimental test. "This medical evolution," he wrote, "can still be verified around us every day, for every man goes on learning, as does all humanity." Bernard's vision of the new "experimental medicine" was, he repeated, "nothing but the consequence of the wholly natural evolution of scientific medicine."

The earliest reference to revolution in science that I have found in Bernard's writings occurs in a notebook of the 1850s, often called the *Cahier rouge* (1942) or the *Cahier des notes* (1965), translated into English as *The Cahier Rouge* (1967). Here, Bernard explained that there are two ways in which science expands: "by adding new facts and by simplifying what already exists" (p. 163). His own work, he said, was of the first sort. But in a later passage under "Ideas to develop," he proposed an "Opening Lecture" on this topic: "Science moves forward by revolutions and not by addition pure and simple" (p. 165). In other words, the advance of science depends on new ideas and not on the mere accumulation of facts. Advance by revolutions relates "to theories, which are always successive," as in the "striking example of chemistry." He had in mind the overthrow of the phlogiston theory, "which explained very well, etc." In another note, which he also marked as an idea to develop, he held that "Science does not grow successively and regularly; it moves by bounds and by revolutions" (p. 196). He added, "It is the changes in theories that mark the bounds. Science is revolutionary. Develop the idea."

In his published writings, Bernard referred frequently to the evolution of the sciences, in particular of physiology or experimental medicine. But he tended to emphasize that such evolution could be hastened by favorable circumstances. In a celebrated *Rapport sur les progrès de la physiologie* (1867) he said explicitly, "The experimental sciences can develop in a country only to the degree of support given to their means of work." In France there were "museums with their collections" for "the natural sciences, geology, zoology, botany, etc.," but "the experimental science of living beings, that is to say physiology, has still none of the needed laboratories." Bernard stressed the need not only for laboratory facilities, but also for improved education and training of young scientists, and a clarification of the aims and methods of experimental biology (1867, 139–140, 146–149, 233; 1872, 398–401, 469, 582; 1927, 217–218; 1963, 50–55, 174, 448; 1966, 12). He said (1867, 8) that there are two paths in the advancement of physiology: "the force of discoveries and of new ideas," and "the power of the means of research and of scientific developments." In "the evolution of the sciences," he declared, "invention is, without contradiction, the essential part." Nevertheless, "new ideas and discoveries are like seeds; it does not suffice to give them birth and to plant them; they must also be nourished and developed by a scientific environment."

He concluded (ibid., 149) that "to make advances in physiology, as in the other sciences," two things are needed: genius and facilities for research. The occurrence of the first is beyond our control, but we can affect the latter. Hence, since the pace of evolution in the sciences depends on the support of research, the future is to a high degree determined by our own actions.

Many of Bernard's sentiments on evolution and revolution appear in shorter and longer writings, most notably in his *Leçons de pathologie expérimentale,* based on lectures given in the Collège de France and first published in an English version by Benjamin Ball and then translated back into French (see Olmsted 1938, 57). In this work Bernard discusses the revolution by which science threw off the yoke of authority and became experimental and he also writes of a general evolution of the sciences and particular revolutions (1872, 401–405).

In *La science expérimentale* (1878), a collection of shorter works that had been published separately, Bernard discussed "the evolution of experimental medicine" and "the natural progress" (p.61) that comes from the applications of "the experimental method" — "le *sentiment,* la *raison,* l'*expérience*" (p. 80); he also mentioned the evolution of human knowledge (pp. 79, 92). He wrote of "the natural sciences" contributing to "the evolution" of physiology (p. 102), a theme developed again in terms of "experimental physiology" considered "as the fruit of the total evolution of the other biological sciences" (pp. 143–144). In this work he stated that "the ideas of Bichat produced a profound and universal revolution in physiology and in medicine" (p. 162).

Bernard was familiar with Charles Darwin's ideas. For example, in his *Leçons sur les phénomènes de la vie communs aux animaux et aux végétaux* (1966), he refers to evolution and in particular to "the general law of the struggle for existence" (pp. 147, 149) and "selection" (p. 342). It is in this work that he makes the famous statement that "it matters little whether one be *Cuviériste* or *Darwiniste*" (p. 332). These names merely signify "two different ways of understanding the history of the past and the establishment of the present regime"; neither one can furnish a means of determining the future. (On Bernard and Darwin, see Schiller 1967.)

A first-rate introduction to Bernard's scientific achievement and philosophical thought is given by M. D. Grmek in the *Dictionary of Scientific Biography* (1970, 2: 24–34, with bibliography). Works in English include J. M. D. Olmsted's *Claude Bernard, Physiologist* (New York, 1938), J. M. D. Olmsted and E. Harris Olmsted's *Claude Bernard and the Experimental Method in Medicine* (New York, 1952), F. L. Holmes's *Claude Bernard and Animal Chemistry: The Emergence of a Scientist* (Cambridge, 1974), and the volume edited by F. Grande and M. B. Visscher on *Claude Bernard and Experimental Medicine* (Cambridge, 1976), which contains an English translation of Bernard's *Cahier rouge.* Bernard's *Introduction to . . . Experimental Medicine* is available in an English version by H. C. Greene (New York, 1927). A good general work is Joseph Schiller's *Claude Bernard et les problèmes scientifiques de son temps* (Paris, 1967).

19.1. Darwin's Awareness of Revolution

It is difficult for us today to conceive of the state of revolution and the awareness of revolution in the late 1840s and the 1850s. Priscilla Robertson (1952, vii) has observed that no one "has ever numbered the revolutions which broke out in Europe in 1848," but she reckons that — by counting "those in the small German states, the Italian states, and the provinces of the Austrian Empire" — there were well over fifty. Reports on revolution and on particular revolutions appeared during the 1840s and 1850s in issue after issue of the *Quarterly Review*, a journal read by Darwin in the decades before composing the *Origin of Species* (1859); in the 1840s there were also such articles in the *Westminster Review*, but not quite so many.

Revolutions were being discussed during the 1840s and 1850s not only in relation to political events but also in the context of science. Robert Chambers, the anonymous author of *Vestiges of the Natural History of Creation* (1844), introduced the concept of revolution in a *Sequel* to the *Vestiges* (1845, 149), boldly declaring that if the origin of the organic kingdom should "ever be cleared up in a way that leaves no doubt of a natural origin of plants and animals," there would have to be "a complete revolution in the view which is generally taken of our relation to the Father of our being." This statement was widely circulated. In 1847, Darwin wrote to the botanist J. D. Hooker (Darwin 1887, *1*: 355–356) that he had "just received a presentation copy of the sixth edition of the 'Vestiges'" from Robert Chambers, who — Darwin was convinced — "is the author." Darwin read Chambers's *Vestiges* very carefully and filled his copy (which did not include the *Sequel*) with copious annotations. (Darwin's annotated copy of the *Vestiges* is preserved among other books from his personal library in the Darwin collection in the University Library, Cambridge.) Darwin not only read Chambers's book carefully (see 1887, *1*: 333n) but was "much flattered and unflattered" that the work had been "by some" attributed to him. He found the author's "geology . . . bad, and his zoology far worse." The book, he said, was "strange, unphilosophical, but capitally written." In the "Historical Sketch" which Darwin added to later editions of the *Origin*, Darwin (1971, 10–11) referred to "the tenth and much improved edition (1853)" (which likewise does not include the *Sequel*). Darwin praised the "powerful and brilliant style," but deplored the author's "great want of scientific caution" and his lack of "accurate knowledge." On the whole, he concluded, the book "has done excellent service" in "calling attention to the subject, in removing prejudice, and in thus preparing the ground for the reception of analogous views."

It would seem highly unlikely that Darwin, who was interested enough in Chambers to write about two different editions of the *Vestiges* (1847, 1853), would have paid no attention to the *Sequel*, which had been published in two London editions (1845, 1846). In it he would have encountered Chambers's statement that the discovery of "a natural origin of plants and animals" would constitute "a complete revolution." This particular statement by Chambers was prominently displayed (and printed in italic type) in a review of the New York edition (1846) of the *Sequel*, written by Darwin's friend and correspondent the Harvard botanist Asa Gray. Gray's discussion of the *Sequel* was published anonymously in the *North American Review* (1846, 62: 465–506). Commenting on this passage, Gray enlarged Chambers's idea of a "complete revolution." He evidently assumed, as Mill had done in contemplating his own statement about the revolution of introducing new technology into under-developed countries, that a complete revolution implied a returning, as well as an introduction of something wholly novel. "Yes," he wrote, "the revolution would be complete," and, "for aught that we can see, would take us back again to the days of Democritus and Epicurus."

It is not likely that Darwin read this review, since it was published in an American nonscientific journal; although Darwin and Gray had met briefly in 1839, their real association did not begin until 1851 (Dupree 1959, 129). But in the 1840s Darwin would have encountered a discussion of revolution in science in a lengthy article in the *Quarterly Review* by Sir Henry Holland (1848). The occasion was a review of Thomas Graham's *Elements of Chemistry* (2nd ed. 1847) and of a new edition of Edward Turner's *Elements of Chemistry* (8th ed. 1847) "by Baron Liebig and Professor Gregory." In the course of the review Holland referred to a number of revolutions in chemistry. For example, he spoke of the "discovery" of the "great laws of chemical combination" as effecting "as great a revolution in Chemistry, as did the Newtonian law of gravitation in Astronomy" (pp. 53–54).

An author who, we have seen, wrote extensively and incisively about revolutions in science was William Whewell, whose *History of the Inductive Sciences* was published in three volumes in 1837. His companion work, *The Philosophy of the Inductive Sciences, Founded upon their History*, appeared in two volumes in 1840. Darwin knew Whewell personally, and in his autobiography (1958, 66) he described how "on several occasions I walked home with him at night." Darwin had high praise for Whewell (p. 104). We know from Darwin's comments in the species notebooks the care with which he read and reread Whewell's *History* (see Ruse 1975; 1979, 176). Darwin thought so highly of Whewell that he placed an extract from Whewell's *Bridgewater Treatise* on the page facing the title page of the first edition of the *Origin of Species*, along with a quotation from Bacon's *Advancement of Learning*. Whewell wrote to Darwin on 2 January 1860, saying that he had not been won over to the new theory. "But there is so much of thought and of fact in what you have written," he added, "that it is not to be contradicted without careful selection of the ground and manner of the dissent." Darwin's son Francis (1887, 2: 261n) appended to the transcript of this letter a wry comment that "Dr Whewell dissented in a practical manner for some years," in not permitting the Library of Trinity College, of which he was Master, to have a copy of the *Origin* on its shelves.

John Stuart Mill's *System of Logic* (1843, p. iii; 1973, 7: cxi) opened with a reference to revolution. In the preface he declared that he made "no pretence of giving to the world a new theory of our [*subsequently changed to* the] intellectual operations" and then expressed his views on the state of science and of revolutions: "In the existing state of the cultivation of the sciences, there would be a very strong presumption against any one who should imagine himself to have [*later changed to* that he had] effected a revolution in the theory of the investigation of truth, or added any fundamental new process to the practice of it." This statement, in addition to its intrinsic interest as an expression of Mill's attitude toward his own contribution, is noteworthy in the present context because Mill exhibits a belief that at other times such revolutions may be, and have been, effected. Since Darwin read and admired the works of Mill, he almost certainly would have encountered Mill's expressions concerning revolutions.

In the 1840s Darwin might also have encountered a reference to revolutions in the social sciences and philosophy, since this was a time when the increasing use of the concept of revolution in the sciences was mirrored to some degree in the social sciences. A single example will suffice. In 1844 Thomas De Quincey published a work entitled *The Logic of Political Economy*. In the preface he lamented that "Political Economy does not advance." In fact, he wrote, "Since the revolution effected in that science by Ricardo (1817), upon the whole it has been stationary." In a review of this book in the *Westminster Review* (June 1845, *43*: 319–331), John Stuart Mill lauded De Quincey's "early and consistent appreciation of Ricardo, the true founder of the abstract science of political economy," and prominently quoted an extract from De Quincey containing this sentence on Ricardo and revolution. Mill did not take issue with De Quincey's reference to Ricardo's revolution, but he did strongly "dissent from the opinion that political economy does not advance" and even suggested that this subject was "in a state of most rapid progression."

21.1. Vogt: The Scientist as Quondam Political Revolutionary

The career of Carl (or Karl) Vogt as scientist and revolutionary both resembles and differs from that of Virchow. Vogt began his career as a chemist, studying under Liebig at Giessen. He has been said to have been one of Liebig's most outstanding pupils. Vogt then turned to medicine, receiving his diploma in 1839. He next took up natural history and was the author of a treatise on freshwater fish (1842). Moving on to Paris, he wrote a series of *Lettres philosophiques et physiologiques* for his German colleagues, published in 1845–1847 as *Physiologische Briefe*. Concerning this work, Virchow wrote in his weekly, *Die medicinische Reform* (2 March 1849, p. 218), that "for the first time in modern times, one of our most important scholars is one of our most prominent political figures, Herr Carl Vogt, whom we have seen" do "through his *Letters in Physiology* what Herr Liebig has done for chemistry, and what is still to be done for medicine."

It was this revolutionary political activity in 1848, to which Virchow referred, that had caused Vogt to lose his post as Reichregent at the University of Giessen. Moving to Geneva, where he became a naturalized citizen, he gained a professorship of geology and wrote a number of works on this subject and in 1872 became professor of zoology and director of the Geneva Institute of Zoology. In Geneva he continued to be active in politics, gaining the post of *conseiller aux états* and later *conseiller national.* No longer a radical, he espoused 'liberal' causes, though he did not lose touch with other émigré radicals and ex-radicals. In his later years he pioneered in the new subject of physical anthropology and was active in support of such movements as scientific materialism and Darwinian evolution.

Like Virchow, Vogt had been an active revolutionary in 1848, when he lost his academic post. But unlike Virchow, Vogt was not a revolutionary scientist, nor a gifted or radical innovator. Whereas Virchow combined his scientific and political ideals and actions, Vogt seems to have kept them apart. His career as a revolutionary was not related to his scientific goals and ideas and affected his scientific career only to the extent of causing him to be transplanted from Giessen to Geneva, and to become—at least for a while—a geologist rather than a zoologist.

In his volume of *Zoologische Briefe* (1851, *1*: 9–19), Vogt expanded his ideas about the development of science. In this context, he did not refer to 'revolutions' in science, but rather 'rebirth' and 'new directions'. But he did mention both the French Revolution and the revolutions of 1848, expressing the opinion that political revolutions can provoke new directions in science as a secondary effect of a general upheaval. Although he did link political and scientific events, he did not do so here in a carefully reasoned way that would suggest a fully worked-out or deep philosophical view of scientific change. An example is Vogt's discussion of the way in which the followers of Linnaeus lost sight of the animal as a whole and in their classification schemes treated only a few external characteristics. The result was a "dry formularization" that drove out the lively quality of the days of Linnaeus: "The antiquated customs, which had ruled in political life before the French Revolution, extended also into science, and a time of desolation threatened, but was luckily overcome by the refreshing breath of the French Revolution." Another example: "The French Revolution, which so mightily shook all minds [Geister] and everywhere opened new paths, also brought new directions [neue Richtungen] into zoological science" (pp. 13–14). Cuvier, according to Vogt, found gaps in the record of the natural history of animals and began an intensive study of neglected classes which, in the end, provided new foundations for the classification of the whole of "animal creation." A result was that comparative anatomy became the basis of zoology in all lands.

Vogt (1882, 13) also linked scientific and political activity in a defense of vivisection, in which he discussed how experiments with animals have greatly benefited science, notably physiology, ever since the days of Harvey. Vogt then coupled the Harveyan revolution with the English Revolution. Harvey's proof of the circulation of the blood, he wrote, "brought about a revolution which was as significant as the one which his contemporary Cromwell brought about . . . in the political organization of his country." But Vogt hastened to add that Cromwell's revolution (the English Revolution) was "more far-reaching." Although Vogt presented these two revolutions together, as events that had occurred at the same time, he did not necessarily imply any causal link between them, nor did he even suggest that the scientific and the political revolution might be related to some more fundamental revolutionary spirit of the times.

Vogt achieved a kind of notoriety through a polemic with Karl Marx in the late 1850s. Marx accused Vogt of sympathizing with Bonaparte and in the ensuing debates (conducted in part in the pages of the press), Vogt published a "Warning" against "a clique of fugitives" known as "the Vagabonds," which now was centered in London under Marx's leadership (see Mehring 1962, 281). When Marx's colleagues publicly attacked Vogt, accusing him of charges that could not be substantiated, Vogt sued for libel. In 1860 Vogt published a book containing a report of the court proceedings plus other documents. In it Marx was described as "the leader of a band of blackmailers whose members lived by 'so compromising people in the Fatherland' that they were compelled to purchase the silence of the band" (ibid.). Marx tried legal action against Vogt, but failed. His only recourse was to write a book against Vogt which apparently has the single merit of being "the only one of Marx's independent works which has never been reprinted" (ibid., 294), except in the current edition of Marx's collected works. According to Marx's biographer, Franz Mehring, the text is livened by a constant comparison of Vogt and Falstaff, beginning on the first page: "The original of Karl Vogt is the immortal Sir John Falstaff and in his zoological resurrection he has lost nothing of his character."

On Vogt, see the article by Pilet in the *Dictionary of Scientific Biography* (1976, *14*: 57–58). Vogt published an autobiography in 1896.

23.1. Some Statements by Engels on Revolution in Science

Engels to Marx (14 July 1858)

"One has no conception of the progress which the natural sciences have made in the last thirty years. Decisive for physiology have been (1) the gigantic development of organic chemistry, (2) the microscope, which only in the past twenty years has been used correctly. The latter has led to more important results than has chemistry. The main point, by which all of physiology was revolutionized [revolutioniert] and a comparative physiology first became possible, is the discovery of cells, in plants by Schleiden, in animals by Schwann (ca. 1836). Everything is cell."

Engels to Marx (16 June 1867)

"[I have] read Hoffmann. Recent chemical theory with all its errors [represents] a great progress over the earlier atomistic theory. The molecule as the *smallest part of matter capable of independent existence* is a completely rational category, a "node," as Hegel says, in the infinite series of divisions, which does not conclude the series, but provides a qualitative difference. The atom — earlier represented as the limit of divisibility — is now merely a *relation*, although Mr. Hoffmann himself now and then falls back into the old notion that real, indivisible atoms exist. Moreover, the progress of chemistry which is established in the book [Hoffmann's] is truly enormous, and Schorlemmer says that this revolution [Revolution] still continues daily, so that one can expect new revolutions [Umwälzungen] every day."

Engels to Eduard Bernstein (27 February– 1 March 1883)

"For V[iereck], who understands absolutely nothing of the matter, the noise about the electrochemical revolution is purely an advertisement for his brochure. But in fact the matter is enormously revolutionary. The steam engine taught us to convert heat into mechanical motion; in the utilization of electricity the path is opened for all forms of energy — heat, mechanical motion, electricity, magnetism, light — to be converted one into another and back again and to be exploited industrially. The circle is closed. And Deprez's most recent discovery, that electrical currents of very high voltage propagate with relatively small loss of energy through a simple telegraph wire over distances undreamed of until now, and at the end can be converted — the matter is still embryonic — definitely frees industry from nearly all local restraints, makes possible the use of the most inaccessible water power, and if in the beginning it will be advantageous for cities, it must finally become the most powerful lever to overcome the opposition of city and country. It is also clear that the productive forces are thereby expanded, whereby they outgrow the leadership of the bourgeoisie with increasing speed. The narrow-minded Viereck only sees in this a new argument for his beloved nationalization. Bismarck should do what the bourgeoisie cannot do."

Engels on Revolutions in Terminology (preface to the English edition of vol. 1 of Das Kapital, *trans. Samuel Moore and Edward Aveling)*

"Every new aspect of science involves a revolution in the technical terms of that science. . . . [For example] the whole terminology of chemistry is radically changed every twenty years . . . [There is] hardly . . . a single organic compound that has not gone through a whole series of different names."

24.1. The 'Electra Complex'

It may seem anomalous that Freud should link his discovery of sexual fantasies of young girls (often involving their fathers) with the discovery of the Oedipus complex — since the latter term clearly refers to libidinal feelings of a male child for his female parent and a concomitant hostility toward his male parent. Freud was aware that the signification of the Oedipus complex might be extended beyond its onomastic origins to include females as well as males. For instance, in a manuscript of 1921, he wrote of "the female Oedipus complex" (1953, *18*: 214); in his *New Introductory Lectures on Psycho-Analysis* of 1933, Freud (1953, 22: 120) wrote of the "phantasy of being seduced by the father" as "the expression of the typical Oedipus complex in women" and (p. 133) referred to a girl's "attachment to her father" as an instance of "the Oedipus complex."

The name 'Oedipus complex', however, does suggest at once the libidinal feelings of a boy for his mother, accompanied by hostility toward his father. And so we can understand why an attempt was made to have a female cognate expression, 'Electra complex.' This name was first introduced by Jung in 1913 (Freud 1953, *23*: 194n), but it has never caught on. On one occasion Freud said rather specifically that he did "not see any advantage or gain in the introduction of the term 'Electra complex'" and so did "not advocate its use," preferring to write of "the feminine Oedipus complex" (1953, *18*: 155n). In an essay on "Female Sexuality" (1953, *21*: 229), Freud said he had been "right in rejecting the term 'Electra complex', which he said "seeks to emphasize the analogy between the attitude of the two sexes." He insisted that it is "only in the male child that we find the fateful combination of love for the one parent and simultaneous hatred for the other as a rival." The reason for such strong statements by Freud is that in psychoanalysis the Oedipus complex in males had become linked specifically with a young male's fear of castration; accordingly there did not seem to be any possibility of a complete analogy between males and females. Later on, however, Freud did recognize the role of the castration complex in both females and males and accepted the reality and significance of a kind of Oepidus complex among females, even if it was not the exact counterpart of the male form. In one of Freud's last writings, he compared and contrasted the Oedipus complex in boys (1953, *23*: 189–192) and in girls (192–194), including a discussion of a castration complex in both sexes. In this work, however, Freud refers both to the "Oedipus complex" of females and to a woman's "feminine Oedipus attitude," once again specifically rejecting "the term 'Electra complex'" (p. 194).

25.1. Some Additional Opinions of Twentieth-Century Scientists on Revolution in Science

Among the twentieth-century scientists who have accepted the existence of revolutions in science, and have even carefully stated which events in the history of science are to be considered in this category, is Jerrold Zacharias, physicist and reformer of science education. Zacharias (1963, 74–75) holds that there have been six "revolutions in physics." These are "Newtonian mechanics"

(Copernicus-Tycho-Galileo-Kepler-Newton), "atomicity" (Dalton-Avogadro-Maxwell and Boltzmann), "electricity" (Faraday-Maxwell-Hertz), "relativity" (Michelson-Lorentz-Einstein), "quanta" (Planck, Rutherford, Bohr, Schrödinger-Heisenberg-Born plus J. J. Thomson), and "a revolution still in the making" which "will probably have much to do with understanding the structure of the nuclei of atoms and of particles moving with gigantic energies" (Zacharias 1963, 74–75).

Another physicist who has written about revolutions in science is Charles Galton Darwin, a grandson of the famous naturalist, who wrote a projection of *The Next Million Years* (1952). Chapter three was devoted to "The Four Revolutions." These were steps forward, which were "never lost again," or "the *irreversible* stages." Darwin would call each one "a revolution, though the word is not meant to imply any extreme suddenness": "In each, the germs may be detected long before, and it may have been a long time before they spread over the world; in some cases the revolution has been made independently in different regions. The central feature of each revolution has been to make it possible for mankind largely to multiply in numbers" (p. 46). The "first revolution" was "the discovery of fire," the second "the invention of agriculture," the third "the urban revolution, the invention of living in cities." The "fourth revolution in human history" ("so recent that it has hardly been recognized" — "we are still in the middle of it") may be "called the Scientific Revolution." It is "based on the discovery that it is possible consciously to make discoveries about the fundamental nature of the world" so that "man can intentionally and deliberately alter his way of life." According to Darwin, "Our histories are so detailed, and run so uniformly through this period, that it has hardly been noticed as constituting a revolution." The "germinating idea" of this revolution is to be found "in the experiments of Galileo and in the writings of Bacon," but the actual birth of the revolution was "at the time of the English Industrial Revolution," and "in particular through the invention of railways." Finally, the "central fact of this revolution" was "the discovery that nature can be controlled and conditions modified intentionally" (p. 51).

Darwin makes conjectures about "other revolutions in store for humanity." One is "nearly a certainty"; it will be "the fifth revolution" and will come when we will have exhausted our supply of fossil fuels. Whereas the other four revolutions all led to an increase of the human population of the world, this one — it is prophesied — is apt to produce a decrease. Two other "possible revolutions" are proposed by Darwin: the discovery of "some new large source of human food," and the possibility of foreseeing the future "with substantially greater accuracy than we now can." Darwin also envisages an "internal revolution," which could center on the discovery of a means of "deliberately altering human nature itself." The last three revolutions occurred within a span of less than twenty thousand years. Will there be a revolution "at least every ten thousand years"? While Darwin doubted that there would be "so many revolutions in store for our descendants," he was aware that there could be little profit from such idle speculation.

J. B. S. Haldane, the biologist who later in life espoused political and social revolution, once wrote that "Arrhenius' ionic theory has transformed chemistry," but J. J. Thomson's "electron theory has revolutionised physics" (1940, x). Unfortunately, Haldane did not spell out carefully the distinction between the ionic and the electron theory that enabled him to consider one in terms of transformation and the other in terms of revolution. In an early work, *Daedalus, or Science and the Future* (1924, 80), Haldane stressed that biologists "do not see themselves as sinister and revolutionary figures." "They have no time to dream," he added, "But I suspect that more of them dream than would care to confess it." In this book, Haldane presented what we would call great 'revolutionizing' or 'revolutionary' discoveries without using the word 'revolution' or any of its derivatives.

We have earlier mentioned Rutherford's opinion that science increases incrementally. Yet we have seen that he boldly asserted that Bohr's quantum theory of atomic spectra was "revolutionizing physics." His complete statement about the development of science reads as follows (Needham and Pagel 1940, 73–74):

> [It] is not the nature of things for any one man to make a sudden violent discovery; science goes step by step, and every man depends on the work of his predecessors. When you hear of a sudden unexpected discovery — a bolt from the blue as it were — you can always be sure that it has grown up by the influence of one man on another, and it is this mutual influence which makes the enormous possibility of scientific advance. Scientists are not dependent on the ideas of a single man, but on the combined wisdom of thousands of men, all thinking of the same problem, and each doing his little bit to add to the great structure of knowledge which is gradually being erected.

Among scientists of Rutherford's era who have expressed somewhat different views is Werner Heisenberg, particularly noted for his many thoughtful and philosophical essays on modern physics. In his *Philosophical Problems of Nuclear Science* (1952, 13), he compared the origin of modern theories of physics with the classical revolutions:

> Modern theories did not arise from revolutionary ideas which have been, so to speak, introduced into the exact sciences from without. On the contrary they have forced their way into research which was attempting consistently to carry out the programme of classical physics — they arise out of its very nature. It is for this reason that the beginnings of modern physics cannot be compared with the great upheavals of previous periods like the achievements of Copernicus. Copernicus's idea was much more an import from outside into the concepts of the science of his time, and therefore caused far more telling changes in science than the ideas of modern physics are creating to-day.

Heisenberg made a striking comparison of the world of geography and modern physics. He was primarily concerned with a general question, "the finality of the changes wrought by modern physics on the foundations of exact science" (p. 16). In his discussion, he introduced the ancient notion of the world as a flat disc — a belief destroyed forever by the circumnavigation of the earth. The concept of an end to the 'world' had "a definite and clear meaning" until the "voyages of discovery of Columbus and Magellan." But, according to Heisenberg, "mankind did not renounce the idea of 'the end of the world' as a result of having explored the whole surface of the world" (because "even to-day there are some unexplored parts"), but these voyages of discovery and exploration "gave clear proofs of the necessity to make use of new lines of approach." In short, the acceptance of a spherical shape for the earth did not cause "the loss of the old concept" to be felt as "a loss." It is, says Heisenberg, the same with quantum and classical physics. The "new lines of thought introduced by quantum theory" have the result "that the loss of concepts of classical physics no longer appears a loss."

Heisenberg then carried his analogy a step further. The discoveries made by Columbus had no material effect on Mediterranean geography. Hence it would be an error to claim that Columbus's voyages "had made obsolete the positive geographical knowledge of the day." And this leads him to the following conclusion about the revolution in physics. "It is equally wrong," he writes, "to speak to-day of a revolution in physics" (p. 18):

> Modern physics has changed nothing in the great classical disciplines of, for instance, mechanics, optics, and heat. Only the conception of hitherto unexplored regions, formed prematurely from a knowledge of only certain parts of the world, has undergone a decisive transformation. This conception, however, is always decisive for the future course of research.

The analogy of geographical discovery was used somewhat differently by the Physics Survey Committee of the U.S. National Academy of Sciences – National Research Council in its report on *Physics in Perspective* (1973). This report began with a discussion of "The Nature of Physics." Here it was stated unequivocally (pp. 61–62):

> What has been learned in physics stays learned. People talk about scientific revolutions. The social and political connotations of revolution evoke a picture of a body of doctrine being rejected, to be replaced by another equally vulnerable to refutation. It is not like that at all. The history of physics has seen profound changes indeed in the way that physicists have thought about fundamental questions. But each change was a widening of vision, an accession of insight and understanding. The introduction, one might say the recognition, by man (led by Einstein) of relativity in the first decade of this century and the formulation of quantum mechanics in the third decade are such landmarks. The only intellectual casualty attending the discovery of quantum mechanics was the unmourned demise of the

patchwork quantum theory with which certain experimental facts had been stubbornly refusing to agree. As a scientist, or as any thinking person with curiosity about the basic workings of nature, the reaction to quantum mechanics would have to be: "Ah! So that's the way it really is!" There is no good analogy to the advent of quantum mechanics, but if a political-social analogy is to be made, it is not a revolution but the discovery of the New World.

What is most remarkable about this declaration is that no examples are provided and no basis of support is given to the bald statement, "It is not like that at all." Nor is it in any way obvious that the "picture of a body of doctrine being rejected, to be replaced by another equally vulnerable to refutation"— whether this is a valid image or not—is a necessary consequence of the alleged "social and political connotations of revolution." This "picture," be it noted, is a rather accurate description of the replacement of Aristotelian physics by Cartesian physics and the consequent replacement of the latter by Newtonian physics. And although it is dangerous to assume the role of seer, a historian of science—aware of the dramatic and basic changes that have occurred and keep on occurring in the interpretation of historical events—cannot help reaching for the salt cellar whenever encountering the sentence, "that's the way it really is."

25.2. A Bergsonian Revolution?

Writing in the ninth decade of the twentieth century, I have difficulty imagining that Bergsonian philosophy could ever have been considered of real significance as a revolutionary force in science. Bergson's most scientific book was *L'évolution créatrice* (1907), published when he was forty-eight years old. In this work he took a strong stand against the Darwinian theory of evolution by natural selection, particularly arguing against the notion of random variations. In presenting Bergson's position in 1970, the article on Bergson in the *Dictionary of Scientific Biography* argues as if Bergson's strictures are valid—that the key to the fundamental problems of organic evolution "is to be found not in biology, but in metaphysics." Forsooth! In any event, Bergson particularly opposed such concepts as random variations and selection pressure, and argued for a 'vital impulse' or 'élan vital' which "pervades the whole evolutionary process and accounts for its dominant features."

We are to believe (according to the *D.S.B.*) that, "in place of the theories of Darwin, Lamarck, and Spencer," Bergson "advanced a doctrine which owes much to the tradition of European vitalism and also draws inspiration from Plotinus." So be it! But what is of more concern is a topic omitted from the *D.S.B.* altogether, namely, Bergson's actual influence—if any—on the sciences. In this context, it should be noted that Bergson not only took a strong stand against Darwinian evolution but also against Einsteinian relativity.

It is somewhat paradoxical that Louis de Broglie, famous for his contributions to quantum mechanics (eponymously honored in the term 'de Broglie waves'), was both an ardent admirer of Bergson and a firm believer in Einstein's relativistic physics, the foundation on which his own work in physics was firmly based. Since Bergson was "a philosopher of anti-intellectualism, a protagonist of instinct against reason," it is certainly remarkable that Louis de Broglie (according to Feuer 1974, 219) "avowed that he found the most significant analogies between Bergson's ideas and those of quantum and wave mechanics." Louis de Broglie wrote an essay on "The Concepts of Contemporary Physics and Bergson's Ideas on Time and Motion," in which he argued specifically for "an analogy between Bergson's critique of the idea of motion and the conceptions of contemporary quantum theories" (Gunter, 1969, 52). Bergson himself claimed that he had anticipated such ideas and that he had in his last writings shown "the direct consequences of the theory of Louis de Broglie" (quoted in Feuer 1974, 221). A case can be made that some quantum physicists may have been impressed by Bergsonian ideas (see Feuer 1974, 206–214, 220–222; Gunter 1969; Čapek 1971), but there are hardly grounds here for asserting that there was a Bergsonian revolution.

In biology the Bergsonian impact was more direct. The *D.S.B.* article on Bergson repeats Bergson's claim that Darwinian evolution "fails to give a satisfactory explanation of the evolution of complex organs and functions, such as the eye [sic] of vertebrates" and even of "the evolution of complex organisms from relatively simple ones." But, of the Bergsonian critique, Julian Huxley (1974, 458) has written:

> To read *L'Evolution Créatrice* is to realize that Bergson was a writer of great vision but with little biological understanding, a good poet but a bad scientist. To say that an adaptive trend towards a particular specialization or towards all-round biological efficiency is explained by an *élan vital* is like saying that the movement of a railway train is "explained" by an *élan locomotif* of the engine. Molière poured ridicule on the similar pseudo-explanations in vogue in the official medical thought of his day.

Ernst Mayr tells me that when he was a young biologist, everyone was reading and talking about Bergson on 'creative evolution' and Hans Driesch on vitalism. So, when he went off on an expedition to New Guinea in 1928, Ernst Mayr took their books as reading matter. He chuckles as he tells the story today, because — he explains — in this way he had no choice but to read them through, since he had nothing else to read. If he had started to read them at home, he would all too quickly have laid them aside. As he wryly remarks, "I found there was nothing of any consequence in either of them."

Yet, in France at least, Bergsonianism became a really potent force in biology. This was no doubt associated with the continued (until all but now) rejection by French biologists of Darwinian evolution and natural selection. This aspect of science in France has been graphically portrayed for us in an essay by Ernest Boesiger (1980). The record of anti-Darwinism in France is long and dismal, punctuated by such statements as those made in 1937 that "no biological fact favors the theory of evolution"; "natural selection plays no role"; "evolution is a kind of dogma in which the priests no longer believe." Boesiger explains that "all French biologists were very heavily and directly influenced by Bergson" (p. 314). Among them was P. P. Grassé, president of the French Academy of Sciences, who still argued in 1973, in his *L'évolution du vivant*, against natural selection and population genetics. The French Nobelist Jacques Monod, in his *Chance and Necessity* (1970), stated unequivocally, "When I was young, there was no hope of getting the bachelor's degree if one had not read Bergson's *Creative Evolution.*" But even so strong an influence is not equivalent to a revolution in science and actually partakes more of a counterrevolution.

Yet the claim has been made for a Bergsonian revolution, notably by Charles Péguy, in his *Cahiers de la quinzaine*, a major publication for young Bergsonians (see Halévy 1946). As Romain Rolland explained (1944, 25–40), for the young idealists who were opposed to crass materialism, positivism, and dogmatic sociology, Bergson was "the magician of thought around whom all the revolts had grouped" (see Feuer 1974, 209).

In his joint study of Bergsonian and Cartesian philosophy, Péguy (1935, 13) wrote:

Just as in the natural sciences the revolutions of anatomy and physiology have never consisted of opposing the animal *kingdom* to the vegetable *kingdom*, but of pursuing *in parallel* in both kingdoms a certain resituation of thought across two parallel realities, so the revolution of the Bergsonian philosophy has never consisted of opposing nor of displacing the realm of thought or of being. It has consisted of pursuing *in parallel* in all realms, in all orders, and in all disciplines, a certain resituation of thought across these parallel realities.

In the preface to a collection of essays on *Bergson and the Evolution of Physics* (1969), Pete Gunter began with a pair of observations about revolutions. "The twentieth century," he wrote, "has witnessed a striking and as yet unfinished series of revolutions in every branch of physics." Henri Bergson, Gunter said next, "considered such conceptual revolutions in physics inevitable" and "was able to schematically foresee certain of their most important theoretical consequences." We may agree that this would be "a remarkable achievement," especially in the light of "Bergson's reputation as an 'anti-scientific' or 'literary intellect.'" The essays collected by Gunter explore this topic from both an affirmative and a negative point of view. Bergson's main work on a topic in physics was his *Durée et simultanéité: à propos de la théorie d'Einstein* (Paris, 1922). It must be noted, at the outset, that when Bergson's collected works (*Oeuvres*, ed. A. Robinet, Paris, 1970) were published some thirty years after his death, the editor tactfully omitted this anti-Einsteinian tract.

In one of the essays in Gunter's volume, Olivier Costa de Beauregard, physicist and philosopher—but, first and foremost, physicist—reluctantly concludes a historico-philosophical study of Bergson's anti-Einsteinian position by remarking (Gunter 1969, 250), "The technical arguments produced by Bergson in *Duration and Simultaneity* are absolutely erroneous and do not apply." Einstein's biographer, Abraham Pais (1982, 510), reports that Einstein "came to know, like, and respect Bergson," with whom he had been associated in 1922 as a member of the League of Nation's Committee for Intellectual Cooperation; but on Bergson's philosophy, Einstein would remark, "Gott verzeih ihm" (God forgive him). In a letter to André Metz, who had become embroiled in a public debate on relativity with Bergson in 1924, Einstein found it "regrettable that Bergson should be so thoroughly mistaken," noting that "his error is really of a purely physical nature," a misunderstanding concerning the "simultaneity . . . of two events" (Gunter 1969, 190).

25.3. Revolutions in the Social Sciences

In the twentieth century, revolutions are said to have occurred in the social sciences as well as in the natural and exact sciences. The subject of "'Revolutions' in economics" was discussed in 1976 by Sir John Hicks. He had been impressed by the way in which the "study of scientific 'revolution'" had become a "powerful tool in the methodology of natural science," and thought it would be fruitful to study the 'revolutions' in economics "in much the same manner." But he insisted that the significance of such revolutions for economics was very different from what he had found it to be for natural science. The difference between the two areas arose, as he saw it, from the fact that the history of science "is not important to the working scientist in the way that the history of economics is important to the working economist" (1976, 207). In economics he found that "big revolutions are (fortunately) rare"; an "obvious example" was the "Keynesian revolution," to which not more than two or three others can be compared. Hicks put forth the interesting hypothesis that big revolutions are most likely to originate "outside the ranks of academic economists in the narrow sense," whereas "small revolutions, that are revolutions . . . nevertheless, can more easily be made by academics."

According to Hicks, the first economic revolution was the "establishment of 'classical' economics" by Adam Smith (in part based on the ideas of the physiocrats). A second ("a minor 'revolution'") was the "transition from Smith to Ricardo." Then came two "revolutions, at about the same time; one made by Marx, the other by Jevons, Walras, and Menger." Hicks devotes much of his discussion to "the 'marginal revolution' (or 'catallactist revolution')," the development of an economics based on "exchange" rather than on "production and distribution." He proposed an older name of 'catallactist' in place of 'marginalist' to describe the new economists who based their activities primarily on the concept of marginal utility.

More recently, Martin Feldstein (economist and former Chairman of the Board of Economic Advisers to President Reagan) has written of a "major revolution in economics [that] is under way"—a "retreat from the Keynesian ideas" which have "dominated economic policy for the past 55 years." He predicted that "this revolution in economic thinking" will have "profound effects" on the national economy and on "our individual daily lives" (1981, 64).

26.1. Some Twentieth-Century Alternatives to Revolution in Science

This book is limited in its scope to the history and significance of the concept of revolution in science. But there are other expressions of scientific change that have been developed by twentieth-century historically-minded philosophers. Imre Lakatos (1978) has developed the twin concepts of 'methodology of scientific research programmes' and the 'rational reconstruction' of the history of science. Lakatos (and E. Zahar) have applied these concepts in an attempt to discover (1978, 168) "Why did Copernicus's research programme supersede Ptolemy's?" Lakatos (1978, 90) particularly objected to "Kuhn's view that there can be no logic, but only psychology of discovery." Larry Laudan (1977) has developed a theory of scientific change which centers on the concept of "principles of scientific rationality" that are not "permanently fixed, but have altered significantly through the course of science" (Hacking 1981, 144). There have also been major presentations of a philosophical approach to scientific change by Karl Popper (see the bibliography and Magee 1973), and by Paul Feyerabend, N. R. Hanson, Hilary Putnam, and others. We must take cognizance of these developments only to the extent of recognizing that current discussions do not all agree with T. S. Kuhn on the centrality of revolutions in considerations of the development of science. (See the 1981 volume of selections edited by Ian Hacking.)

The French school has not as yet had a significant influence in the English-speaking world on the mode of considering science's history. On the opening page of Alexandre Koyré's *Galilean Studies* (1939; Engl. trans. 1978) the great advances in scientific thinking are presented as if each required "superhuman effort," and each great step is said to be "a veritable 'mutation' in human thought." Koyré declared that the "scientific revolution in the seventeenth century was without doubt such a mutation," actually "one of the most important, perhaps even the most important, since the invention of the Cosmos by Greek thought." I quote this passage for a second time to stress the word 'mutation', which Koyré pointed out was introduced in this context of historical analysis by G. Bachelard (1934; 1938).

In *Les mots et les choses* (1966), translated as *The Order of Things* (1970), Michel Foucault has stressed the concept of 'episteme'. His aim has been summarized as an examination of "the continuities and transformations in European thought from the Renaissance to the present" (Sheridan 1980, 210). Foucault chose to concentrate on "the three essential areas concerned respectively with living beings, language, and wealth, [and] their relations with the philosophy of each period and their extension into the 'human sciences' that emerged in the nineteenth century." In Ian Hacking's analysis (1982, 3), Foucault's "'interdisciplinary' mutation in thought" is exemplified by "the story of the 'same' transition occurring in three domains at about the same time." That is, "Natural history becomes biology, general grammer becomes philology, and the theory of wealth becomes economics."

Quite a different approach is attempted by Stephen Toulmin in his *Human Understanding* (1972). One of the most interesting aspects of this work, in the present context, is the exploration of an evolutionary basis for "the historical process of conceptual change in intellectual disciplines," specifically "in terms of a populational model" (ch. 3). Toulmin thus seizes upon one of Charles Darwin's most original concepts. In an article entitled "Natural Selection and Other Models in the Historiography of Science" (1981), Robert J. Richards has explored some aspects of "Evolutionary Models of Scientific Development," in particular evaluating the 'models' of Popper and Toulmin and developing some analogies between historical and Darwinian thought (p. 58). In §18.2, above, there is a discussion of the use of evolution as an organizational concept for understanding the growth of science; here it may be pointed out that a number of scientists have opposed an evolutionary model of scientific change to a revolutionary one.

In an introductory address at the Congress of Arts and Science that was convened in conjunction with the world's fair (Universal Exposition) at St. Louis in 1904, the astronomer Simon Newcomb presented his views on "The Evolution of the Scientific Investigator." His main theme was that evolution and not revolution has characterized the advance of science over the centuries. This concept of history was in accordance with the current thought about "the process of development in nature" being "continuously forward" as a result of combining "the opposite processes of evolution and dissolution." The tendency of the times, according to Newcomb (1905, 137), was in "the direction of banishing cataclysms to the theological limbo" and seeing nature "as a sleepless plodder, endowed with infinite patience, waiting through long ages for results." But Newcomb was aware that history (both of nature and of man) is punctuated by major events (these are 'revolutions'), and he proposed an analogy in order to explain how a process of long slow evolution can produce an apparent momentous change: "The building of a ship from the time that her keel is laid until she is making her way across the ocean is a slow and gradual process; yet there is a cataclysmic epoch opening up a new era in her history. It is the moment when, after lying for months or years a dead, inert, immovable mass, she is suddenly endowed with the power of motion, and, as if imbued with life, glides into the stream, eager to begin the career for which she was designed." The same process, Newcomb believed, has been occurring "in the development of humanity." There may be long periods in which there seems to be "no real progress"; additions are made to knowledge, but there is nothing in the "sphere of thought" that can be "called essentially new." But "nature" may have been "slowly working" in "a way which evades our scrutiny," according to Newcomb, until "the result of her operations suddenly appears in a new and revolutionary movement, carrying the race to a higher plane of civilization." For Newcomb, the history of science was therefore only an example of "studying a special phase of evolution" (p. 143).

A little over a half-century later, Arthur Koestler developed the theme of "Evolution and Revolution in the History of Science" at the Cambridge meeting (1965) of the British Association for the Advancement of Science. Here he introduced the idea that "the history of any branch of science" shows "a rhythmic alternation between long periods of relatively peaceful evolution and shorter bursts of revolutionary change" (p. 37). It follows as a corollary that "the progress of science [is] continuous and cumulative in the strict sense"only during "the peaceful periods which follow after a major breakthrough."

26.2. Duhem on Evolution and Revolution in the Sciences

Pierre Duhem's views on evolution and revolution attract our attention because of his importance as a historian of science and his role as philosopher. His two major philosophical works are his ΣΩZEIN TA ΦAINOMENA: *essai sur la notion de théorie physique de Platon à Galilée* (1908) and his *La théorie physique: son objet, sa structure* (1906; 1914; 1954). Duhem's major historical works dealt with medieval physics, among them *Etudes sur Léonard de Vinci, ceux qu'il a lus et ceux qui l'ont lu* (Paris, 1906–1913), *Le système du monde* (1913–1959), and *Les origines de la statique* (Paris, 1905–1906). He also wrote many technical works, chiefly on chemical and physical thermodynamical and electromagnetic theory, and produced major monographs on J. Willard Gibbs, on the rise of the 'new mechanics' (1905; 1980), and on J. C. Maxwell's contribution to electromagnetic theory (1902). An ardent Catholic, and a conservative in political and social realms, he was author of a famous essay on "Physics of a Believer" (1954, 273–335). It was Duhem who called attention to the scientific importance of the writings of the scholastic thinkers, stressing their anticipation of many concepts of the science of motion and of rational mechanics that had been thought to be revolutionary innovations of the Scientific Revolution of the seventeenth century.

It is hardly surprising that such a man should adopt a nonrevolutionary stance with respect to the development of science. In the conclusion to the Preface to his *Origin of Statics* (1905, *1*: p. iv), dated 21 March 1905, Duhem stated that his historical studies had led him to a firm conclusion "that the sciences of mechanics and of physics, of which our modern times are rightly so proud, have arisen by an uninterrupted sequence of scarcely perceptible improvements of the doctrines taught in the medieval schools." Also, "the alleged intellectual revolutions have most often been only evolutions which have been slow and prepared over long periods, the so-called renaissances having been only reactions that were frequently unjust and sterile." Finally, "respect for tradition is an essential condition of scientific progress."

Despite his strong position against revolutions, Duhem used the name (if not the fully developed concept) of a Copernican revolution ("révolution copernicienne") to demarcate the periods of what he called (2: pp. vi–vii) "the evolution of ideas put forth by the scholastic masters," an evolution that "resulted in the celebrated principle of Torricelli." The text divided this evolution into a "first period: from Albert of Saxony to the Copernican Revolution" and a "second period: from the Copernican Revolution to Torricelli" (2: 1, 91). The latter began with a discussion of "The tradition of Albert of Saxony and the Copernican Revolution," in which Duhem said that Albert in 1508 was "a forerunner of the Copernican Revolution." In *The Aim and Structure of Physical Theory*, Duhem also introduced the name Copernican Revolution, but he did not — in either work — discuss any revolutionary qualities of Copernican astronomy. In fact, he was rather explicit in stating that the "Copernican revolution" was only a modification (1905, 2: p. vii).

In the conclusion to *The Origins of Statics* (2: 278–279), Duhem restated his theme of evolution of science, of the extreme rarity of any "sudden birth." Science moves ahead "progressively"; "it advances, but only one step at a time," never by "abrupt changes." It is, he said, an error of historians to "celebrate fulgurant discoveries which have replaced the dark night of ignorance and error by the bright day of truth." Rather, he maintained, whoever studies the history of science carefully will see almost always a "result of a host of imperceptible efforts and an infinitude of obscure tendencies." Thus the "two characteristics" that appear in each of the "phases of the evolution that leads science to its accomplishment" are "continuity and complexity." It is to be noted that in his scientific work, as in his writings on the history and philosophy of science, Duhem dealt exclusively with the physical sciences. There is no hint whatever that his use of the concept and name 'evolution' had any Darwinian overtones. (On this topic, see Paul 1979.)

Duhem emphasized the theme of evolution by entitling another of his works *The Evolution of Mechanics* (1905; Engl. tr. 1980). In the final paragraph of the conclusion, he stated categorically that "The development of Mechanics is therefore, properly speaking, an *evolution;* each of the stages of this evolution is the natural corollary of the stages that have preceded it; it is the chief part of the stages which will follow it" (1980, 188). It would be presumptuous, therefore, for any theoretical physicist to suppose "that the system for the achievement of which he works will escape the fate common to the systems that have preceded it and will merit lasting longer than they." But, since the development of scientific thought is an evolutionary process, every theoretician "has the right to believe that his efforts will not be sterile," that for centuries to come "the ideas that he has sown and germinated will continue to increase and to bear their fruit." Yet only two paragraphs earlier, Duhem wrote not only about "the Cartesian revolution," but even called "the New Mechanics" a "counter-revolution opposed to the Cartesian revolution." He did conclude, however, that "this [present] counter-revolution abandons nothing of the Cartesian conquests."

The theme of evolution in science appeared in many of Duhem's writings, for example, an article on "L'évolution des théories physiques du XVIIIe siècle jusqu'à nos jours" in the *Revue des Questions Scientifiques* (Oct. 1896) and the conclusion of a series of articles on "Les théories de la chaleur" in the *Revue des Deux Mondes* (1895, *129*: 869–901; 1895, *130*: 380–415, 851–868), which concluded with a discussion of "the evolution to which all physical theories have been subject from the 17th century to our days."

The contradiction in Duhem's thinking about evolution and revolution is most apparent in his essay on Maxwell and the electromagnetic theory. For here he saw no process of evolution but unambiguously hailed Maxwell's great achievement as a revolution. It is perhaps a sign of the extraordinary quality of Maxwellian physics that a man like Duhem, so actively committed to evolution and so opposed to revolution, should nevertheless have seen this advance as a revolution.

29.1. The Influence of T.S. Kuhn on the Historiography of Continental Drift

At the beginning of chapter 29 I mentioned the fact that a number of earth scientists have presented the history of ideas concerning continental drift as a revolution, using the terms set forth in Thomas S. Kuhn's *The Structure of Scientific Revolutions* (1962). In this context we must remember that the Tasmanian symposium of 1956 was a major event heralding the change in point of view with respect to a possible motion of continents. A major paper was contributed by the Australian structural geologist S. W. Carey, who advanced a new version of continent motion and expansion (see Hallam 1973, 53). The proceedings of the symposium were not published until 1959, the year in which J. Tuzo Wilson (together with Jacobs and Russell) published a textbook entitled *Physics and Geology*. Wilson informs me (priv. comm. 3 Aug. 1982) that before 1959 he "did not believe in continental drift. But by 1959 I did so and I was annoyed that I had dismissed the subject in such a brief fashion in the textbook." Wilson recalls that "the turning point was the receipt of a copy from Sam Carey of his mimeographed paper from the Tasmanian symposium" in 1956 which dealt with continental drift. Carey, however, while accepting the lateral motion of continents, continued to believe in his own "model of a rapidly expanding Earth in the face of adverse evidence" and, as late as 1973, rejected plate tectonics (Hallam 1973, 106). Wilson (in his letter to me, 3 Aug. 1982) says that he believes "it was the paleomagnetic evidence that convinced me." Wilson had spent a long time in the Canadian Shield and he "did not like Carey's ideas about any mobility of continents internally." But it was during Wilson's reflections on Carey's paper that he conceived "the idea of rigid plates which I put to good use about 1965 in the papers I wrote on transform faults and plate tectonics."

Recall that in 1960, one year after Wilson's conversion, Harry Hess circulated a preprint of his paper on sea-floor spreading; Dietz's article in *Nature* (proposing the actual name of 'sea-floor spreading') was published in 1961; and Hess's final version appeared in 1962. The Vine-Matthews hypothesis appeared in *Nature* in 1963; and as confirmation was made in the next years, the idea of continental displacement became "generally accepted as correct" in "just half a decade"—to quote from an unsigned article in *Nature* (2 Sept. 1967, 215: 1061–1062) entitled "Continental Drift Comes True." At this time, 1965, J. Tuzo Wilson published his seminal paper on transform faults (entitled "A New Class of Faults and Their Bearing on Continental Drift"), containing what has been called the "germinal idea behind the theory of plate tectonics" and introducing (in this context) the very word 'plates' (Hallam 1973, 68). The theory was elaborated in the next years in a series of major publications by Jason W. Morgan (in 1968), Xavier Le Pichun (in 1968), and Dan P. McKenzie (in 1969).

In 1963, the year of publication of the Vine-Matthews article, J. Tuzo Wilson—speaking out for continental drift in an address on "The Movement of Continents" at the Berkeley Symposium—boldly argued that a revolution in earth science was imminent. Wilson's expression of revolution is all the more prescient in that it just pre-dated the experimental verification of the Vine-Matthews hypothesis. But he did not put the mimeographed version of his address into print, nor did he refer to the concept of revolution in an article in *Scientific American* in April of that same year (208: 88–100; repr. in Wilson 1973). In 1968, however, Wilson called for the name 'Wegenerian revolution', arguing from the parallel to the Copernican revolution (1968, 317). In this presentation he called attention to some of the main points in Kuhn's "brilliant analysis." In a supplement (1968) Wilson wrote an elegant reply to a Russian geophysicist who refused to accept the idea of continental drift, who did not as yet believe that "what is happening in earth science is similar to what happened in chemistry about A.D. 1800, in biology when evolution was introduced, in physics when classical views were replaced by modern." Wilson said: "It's not new data, but a change in outlook that marks a scientific revolution, as T. S. Kuhn (1962) has so elegantly pointed out." In a presentation at the American Philosophical Society in April 1968, Wilson, discussing continental drift, which he said was "a major scientific revolution in our own time" (1968, 309), concluded by stressing Kuhn's insight (pp. 317–318). He explained how, for Kuhn, the essence of such a revolution as the Copernican "was not any improvement in techniques, not more or better data, not an advance in mathematics; it was a change in idēas." He stressed Kuhn's "point that a change in belief has been the essence of all the great scientific revolutions like those from phlogistics to modern chemistry, from caloric to modern thermodynamics, or from special creation to evolution." Like Cox, Wilson adopted Kuhn's basic idea that science advances by revolutions and that these revolutions are primarily a change in ideas, a new way of looking at the data, a mutation in the framework of general belief.

In the volume he edited, *Continents Adrift* (1973; rev. ed. 1976), Wilson does not use Kuhn's language or ideas in any of his editor's introductions to sections of his book, nor do these appear in his own contribution (the *Scientific American* article of April 1963). In his preface, however, he does refer to "the latest scientific revolution," and stresses the fact that Kuhn (like some other historians of science) has pointed out that there comes a stage in the development of science when "theoreticians reinterpret the lore of practical men by new and subtler formulations," many of which at first seem "to be contrary to reason" (1976, pp. v–vi). In time, however, they have been accepted. As a "classic example of a scientific revolution," Wilson then cites Kuhn's example of "the abandonment of the Ptolemaic belief that the earth is the center of the solar system" in favor of Copernican astronomy, and the change from "phlogistic alchemy to modern chemistry," the rise of Darwinian evolution, quantum mechanics, and relativity. It is this kind of scientific revolution that "many earth scientists believe . . . has occurred in their own subject."

J. Tuzo Wilson is a widely read and thoughtful man and so it is not at all surprising that in 1963 he should have been thinking of scientific revolutions, since that year was marked by many reviews and discussions of Kuhn's *Structure of Scientific Revolutions*, which had been published in 1962. Wilson has written to me (3 Aug. 1982) that he remembers "reading Kuhn's book and being impressed by it." Furthermore, he has the distinct impression "that I got my ideas about revolutions of the earth sciences from comparison with his revolutions" and no doubt "got the use of the word 'revolution' from him."

Considering the important place of Wilson in the community of earth scientists, it is hardly surprising that others should have followed his lead in seeing a revolution in their subject. For instance, in 1969 in an essay review of an English and an American reprint of the translation of the fourth edition of Wegener's book, and of two others, W. B. Harland (1969, 100) reviewed the "early background to the history of ideas concerning continental drift." He not only invoked the concept and term 'revolution', but discussed the revolution in earth science in Kuhnian language, 'paradigm' and 'normal science'. He also referred to John Ziman's discussion of continental drift: "Ziman (1968) usefully claimed that the complexities and ramifications of the continental drift problem provide a far better working model of science for its historians and philosophers than do the usual stock examples from physical science (from phlogiston to photons and neutrons) which are so stylized that they require relatively simple logic to handle" (Harland 1969, 103). Concerning Kuhn, he wrote:

> Indeed, this revolution of continental drift is unlike the textbook scientific revolution (*e.g.* Kuhn 1962) in which the new paradigm has only to be properly conceived and clearly enunciated to win out in a short sharp battle. The importance of the issue had long been seen and perhaps the

necessary evidence had long been available. Mutually contradictory hypotheses could co-exist for decades because there were many alternative ways of removing the contradictions. The difficulties have lain and still lie in the formulation of a complex of interdependent concepts, some of which may shift with the modification of assumptions or the application of new evidence. Perhaps only the exceptional revolution is over in a day. Normal science consists, more than Kuhn admits, in continuing revolution.

In a chapter of his entitled "Reflections on the Revolution," Hallam (1973, 106) discusses "the influential work of T. S. Kuhn," who "challenges the traditional view of scientific progress by the gradual accumulation of discoveries and inventions." Hallam notes, with evident approval, Kuhn's thesis that "revolutions occur by the replacement of one *paradigm* or world view by another," and he devotes two pages to examples from the history of physics that illustrate Kuhn's central thesis. But Hallam also takes note of John Ziman's analysis of science, in which it is shown that "progress in a given field is marked by a change in consensus of the scientific community" (p. 105), and he says "the immediate stimulus to writing this book" came from John Ziman (p. v). Hallam (p. 105) thus would see the revolution in the earth sciences as an illustration of how "a majority of what Ziman calls the 'invisible college' is converted from an old to a new set of beliefs about a particular phenomenon" (see Ziman 1968). Hallam's discussion of Kuhn's ideas and their applicability to continental drift occupies three of the ten pages of the final chapter of his book. He concludes (p. 108):

> Kuhn has been criticized for not defining the concept of the paradigm more clearly and for overdramatizing the contrast between 'revolutions' and 'normal science.' Certainly some might validly object to revolutions that persist unresolved for half a century *pari passu* with other more modest rates of change in thought and technique. It seems to me, however, that Kuhn has highlighted major features of science with a most illuminating conceptual model and has been perceptive in challenging the conventional view of cumulative progress. The Earth sciences do indeed appear to have undergone a revolution in the Kuhnian sense and we should not be misled by the fact that, viewed in detail, the picture may appear somewhat blurred at the edges.

Allan Cox, in 1973, introduced his book of readings on *Plate Tectonics and Geomagnetic Reversals* (pp. 4–5) with a discussion of Kuhn's theory of scientific revolutions, and he called section 1 of the book "Paradigm of Plate Tectonics." Cox was chiefly concerned with Kuhn's viewpoint as a contrast with "the traditional one that science grows in a continuous manner by a process of steady accretion" (1973, 4) and he expressly said that, in the introductory chapters of his book, he had "followed Kuhn in using the terms 'scientific revolution' and 'paradigm' to describe plate tectonics and sea-floor spreading, rather than the more traditional 'hypothesis' or 'theory.'" In this way, he said, "a sterile argument was avoided whether plate tectonics should be described as a hypothesis or a theory." He did note that the development of plate tectonics, while "describable in terms of several theories of the history of science," was found to fit "the pattern of Kuhn's scientific revolutions surprisingly well." It was for this reason that he regarded "developments in the earth sciences during the past decade as the emergence of a major new scientific paradigm." Following the short introductory chapter, however, Cox makes no direct mention of Kuhn or his theories.

Although Ursula Marvin entitled her book *Continental Drift: The Evolution of a Concept,* she concluded her account of the 'evolution' of the concepts and theories of continental drift with a discussion of 'revolution', with reference to Kuhn's "structure of scientific revolutions" and its applicability to the subject at hand. Here she merely points out that "the story of continental drift as a geologic concept . . . bears out in dramatic fashion a thesis developed by Thomas S. Kuhn" (1973, 189). She was primarily interested in the fact that plate tectonics introduced "a new way of viewing the earth, a new array of assumptions, intuitions, and problems, and new criteria for evaluating observational data and making predictions." She noted that "acceptance has required the rejection of an older corpus of belief centered upon the concept of an essentially stable configuration of the earth's surface features" (p. 190). And she called attention to Kuhn's insight "that to change one matrix of beliefs for another is tantamount to undergoing a conversion experience."

In a very thoughtful study of "Continental Drift and Scientific Revolution," David B. Kitts — geologist and historian of science — has called into question the use of Kuhnian theory in "uncritically labeling recent events in earth science as a revolution, and by implication a 'Kuhnian revolution'" (1974, 2490). Kitts fears that in this way "we may miss something significant about the history of geology and, more important, something fundamental about the very nature of geologic knowledge." The main thrust of Kitts's argument is that Kuhnian revolutions — in the strict sense — cannot occur in geology. The reason is that "the laws of physics are not questioned within the context of geologic inference. They are simply presupposed" (p. 2491). Kuhnian paradigms, according to Kitts, are general laws or theories of physics and "geologists never doubt, attack, or repudiate physical theory" (p. 2491). We may not accept this interpretation, nor the conclusion to which it leads him (p. 2492), that "geologists have had no role in the revolutions which have led to the overthrow of comprehensive theoretical paradigms." But he does admit, of course, that once "a theoretical paradigm is abandoned in favor of another it may have revolutionary consequences for geology." An example is the revolution of quantum mechanics, which "affected geology in an important way" but "was not accomplished with the participation of geologists or at their instigation."

In a carefully reasoned critique of Kitts's article, Rachel Laudan (1978) made the point that Kitts was "interpreting Kuhn too rigidly." For example, she said, "Kuhn himself is willing to allow a Darwinian revolution in biology, or a Lyellian revolution in geology, neither of which involved an overthrow of fundamental physical and chemical principles." And she found herself "uneasy" (as I do) "about Kitts' analysis of the nature of geology." Rachel Laudan, I may add, has produced an admirable concise history of the main issues in the revolution, which includes a very important criticism of the use of Kuhnian categories of revolutionary change. In particular she makes the important point that during the half-century between Wegener's revolution on paper and the revolution in science (Rachel Laudan does not use these expressions), there "was no dominant paradigm in which all the geological community was working." Rather, "there were conflicting theories, none of which had a hold on the majority of scientists." This would be a pre-paradigmatic state. Thus, there was during these some fifty years "nothing corresponding to Kuhnian 'normal' science." She also is careful to distinguish the "new lines of evidence" ("from paleomagnetism and from oceanography") that made a dramatic change in the situation during and after the 1950s. She concludes:

> the more flexible interpretation of Kuhn espoused by geologists and historians . . . seems to be too coarse-grained to do justice to the historical details. Furthermore, certain definitive features of Kuhn's account are lacking; there is, for example, no incommensurability between the pre- and post-tectonic geological theories; neither was plate tectonics proposed and advocated by a younger generation of geologists. Its proponents came from all stages of the career spectrum, including those who had earlier decisively rejected drift. If all Kuhn had meant by a revolution was a period of rapid theory change, then it would be appropriate to invoke his work. However, Kuhn surely had a great deal more in mind when he described scientific revolutions, almost none of which is exemplified in the construction and acceptance of plate tectonic theory.

Her final word is that she finds "no reason for terming the change in geology in the 1970's a Kuhnian revolution."

Attention should also be called to the valuable analyses by Henry Frankel (1979; 1979a), in which he applies to the history of this subject first Imre Lakatos's analysis of scientific growth (in terms of "research programmes") and then Larry Laudan's account of scientific growth and change (in terms of his concept and theory of "rationality"). This pair of studies, like Rachel Laudan's, is very helpful to anyone seeking to understand the revolution in the earth sciences. But a critical and comprehensive evaluation of Frankel's ideas and insights would take us far from the subject of continental drift and plate tectonics and lead us deep into the debates of philosophy of science. It is more to the point to note, with Rachel Laudan, that this particular revolution (and I shall indicate in a moment why I consider it to be a revolution) may not exactly follow the simplest Kuhnian patterns, since—as we have seen Harland saying — there was long a co-existence of "mutually contradictory hypotheses," without precipitating a "crisis" that led to a new paradigm. This has also been a feature of the recent history of evolution (see Mayr and Provine 1980; Greene 1971).

There can be no doubt that the recognition that there had been a revolution in earth science was doubly influenced by T. S. Kuhn. First, he had forcibly called the attention of scientists to the process of revolutionary change in science, causing them to question and even to abandon the view held by many twentieth-century scientists that science advances through an orderly process of accumulation, accretion, and linear progress. Second, he had given them a particular set of ideas about the 'structure' of scientific revolutions, which — in varying degrees — came to be adopted by earth scientists, following the lead of J. Tuzo Wilson.

A Note on Citations and References

———◆———

In the main text, and in the notes and supplements, all citations and references are given (usually within parentheses) by author, date, volume, and page. These may be readily identified from the reference list, which follows the notes. If the author's name is already given in a sentence, it is not repeated: "John Heilbron (1979: 169) said of William Gilbert . . . " Similarly, if the date of publication occurs in the sentence, it too is not repeated: "In 1926, in *Science and the Modern World,* Alfred North Whitehead (p. 212) expressed . . . " In the text, notes, and supplements, almost all titles of books and articles are given in English translation (often accompanied by the title in the original language); similarly, quoted phrases, sentences, and extracts are given in English versions, with some key words or phrases printed also in the original language. In text and notes, a title may be followed by a date or a place of publication as well as a date (or by a volume, date, and page numbers of a journal); this usually indicates that the work in question will not be found in the list of references.

For almost every chapter there is a final note that is intended to guide the uninitiated reader to some additional readings and to indicate some of my main sources, primary and secondary — especially for those chapters in which I have profited from the labors of others, who have provided factual matter, relevant texts and information, and insights and interpretations.

In the text, I have tried to keep the number of sources cited to a minimum, for ease of bibliographical reference, but that should in no case be interpreted to imply that the number of works consulted was either so limited or so small as the citations and references might indicate. Similarly, although I have consulted works in several languages — Latin, English, French, German, Italian, and Spanish — I have, wherever possible, cited an English version (for the convenience of English readers), even where I have used the work in the original language.

Notes

———◆———

1. Introduction

1. On a larger scale, one might ask whether the concept of revolution is generally useful in understanding any change — whether political, social, economic, artistic, intellectual, religious, or scientific. Stephen Toulmin (1972) has argued that revolution is a concept of limited worth in any historical inquiry and of doubtful value for understanding scientific change. But in political history there are great classic events (the Glorious Revolution, the French Revolution, the Russian Revolution of 1917) that have been universally referred to as revolutions. They have had both an existential and symbolic influence on later thought and action; they embody special qualities which are connoted by the commonly accepted name of revolution.

Peter Calvert (1970, 141) advocates that we "retain the term 'revolution' itself as a political term covering all forms of violent change of government or regime originating internally," as "a simple recognition of the fact that it is the meaning most widely used in the modern world."

2. In his *Revolution and the Transformation of Societies,* S. N. Eisenstadt indicates five "dimensions" of "the revolutionary process" ("developed in part by revolutionists and in part by modern intellectuals and sociologists in particu-

lar"). These encompass "violence, novelty, and totality of change" (1978, 2–3). The dimensions are: (1) a "violent change of the existing political regime, its bases of legitimation, and its symbols"; (2) the "displacement of the incumbent political elite or ruling class by another"; (3) "far-reaching changes in all major institutional spheres—primarily in economic and class relations—leading to the modernization of most aspects of social life, to economic development and industrialization, and to growing centralization and participation in the political sphere"; (4) "a radical break with the past" (Alexis de Tocqueville "pointed out the relativity of such discontinuity in his *The Old Regime and the French Revolution*); (5) a bringing about of "not only institutional and organizational transformations but also moral and educational changes" (thus under "the strong ideological and millennarian orientations of revolutionary imagery," it is assumed that revolutions "will create or generate a new man"). A somewhat different version of these "dimensions of revolution" may be found in A. S. Cohan 1975, 14–31.

3. The fact that revolutions in science preserve much of the pre-revolutionary system of belief has been stressed by many scientists. In the concluding sentence of *The Nature of the Physical World* (1928, 353), Sir Arthur Eddington wrote that "in each revolution of scientific thought new words are set to the old music, and that which has gone before is not destroyed but refocussed. Amid all our faulty attempts at expression the kernel of scientific truth steadily grows; and of this truth it may be said—The more it changes, the more it remains the same thing."

A somewhat different view is given in Werner Heisenberg's discussion of "revolutions in science" in *Physics and Beyond* (1971, 147–148). He writes of a conversation with a student who says to him that "relativity and quantum theory represent radical breaks with everything that has gone before." Heisenberg replies that Planck initially "had no desire to change classical physics in any serious way." His only ambition was "to solve a particular problem, namely the distribution of energy and the spectrum of a black body." Only when it proved to be impossible to find a solution that would fit into the framework of classical physics was Planck "led to a radical reconstruction of all physics." But, Heisenberg observes, even at that point "those realms of physics which can be described with concepts of classical physics remained quite unchanged." In other words, "only those revolutions in science will prove fruitful and beneficial whose investigators try to change as little as possible and limit themselves to the solution of a particular and clearly defined program." That is, "utter confusion" will result from an attempt to "make a clean sweep of everything or to change things quite arbitrarily." In short, in science, "only a crazed fanatic" would "try to overthrow everything, and needless to say, all such attempts are completely abortive."

4. Newton's activity in establishing a network of control over British scientific thought is a feature of Frank Manuel's *Portrait* (1968); the actual activity of the Newtonian Royal Society has been analyzed by John Heilbron (1983).

In a discussion of the corpuscular and wave theories of light as an instance of "the science and politics of a revolution in physics," Eugene Frankel (1976, 142) has gathered evidence to show that "the fight to establish the wave theory was carried out on a political as well as an intellectual level, and that both struggles were necessary for the final rapid, total victory of the undulationists. We shall see that the consistent strategy of the corpuscular school in this era was to ignore developments in the wave theory, and that it required the publicizing efforts of Arago as much as the research of Fresnel to bring the issue to a head in the Paris scientific community. Ultimately the controversy was resolved not through the conversion of the corpuscularians, but rather through the ascendancy of an anti-Laplace faction to dominant position in French science."

5. Here I am using the expressions 'scientific revolutionary' or 'scientific radical' to mean a scientist engaged in revolutionary or radical *scientific* activity, that is, one who is a radical in his or her science. I do not mean a political revolutionary or radical who believes there is a 'scientific' basis to his ideas, nor a scientist who happens to be a radical or a revolutionary in his political views or actions.

6. Some twentieth-century psychologists (notably Wolfgang Köhler and Kurt Lewin) claimed that their work derived from physical field theory (more directly from Einstein than from Faraday and Maxwell). Whether or not this is an example of an ideological (that is, extrascientific) component depends on the judgment of whether 'field theories' of psychology are or are not a part of 'science'.

7. That is, the *applications* of molecular biology (such as gene splitting) have aroused great social, ecological, and even economic problems and discussions. There has also been an ideological component to some questions raised by these applications: Should we 'tamper' with life? Have we the 'right' to alter a form of 'unborn life', as in a human ovum or fetus? But these are implications of the research and practice, not of the theory or intellectual content, of molecular biology.

8. Only a philosopher and certainly not a practicing scientist could have written (Meade 1936, 117): "We are pleased to have these revolutions take place in our theories, gratified to have our universe fall down so that it is replaced by a new one. We even erect institutions called universities and invite research professors at high salaries who wreck our universes and substitute others in their place. And it appears to us to be perfectly right and natural."

9. The Nobel Prizes in science have tremendous prestige, but it must be noted that they distort the character of the scientific enterprise as a whole since they are limited to physics, chemistry, and medicine. This excludes mathematics, astronomy (unless an award may be made under physics), earth science, and most of biology—especially natural history, plant science, and evolutionary biology. Although the Balzan Prize has been established to make up for the defect, at least in biology, the prize system downgrades much of science and sets up a skewed system of values. Generally speaking, the Nobel Prizes in the sciences have been awarded to deserving scientists, but sometimes the award is made many years after the event. In at least one case (the spin of the electron,

discovered by S. A. Goudsmit and G. E. Uhlenbeck), the physics committee passed by a significant contribution, and there is an egregious example of a Danish physician, Johannes Fibiger, who received the prize for an alleged but soon discredited discovery about the propagation of malignant tumors (see Zuckerman 1977).

Although almost all of the recipients of the Nobel Prize for physics, chemistry, and medicine have merited an award, the same can hardly be said for literature. Many awards have been made to relatively unknown and uninfluential writers who have soon thereafter been forgotten. Those who have failed make up a list of enormous distinction (like the 'salon des refusés'), including Guillaume Apollinaire, W. H. Auden, Berthold Brecht, Herman Broch, L.-F. Céline, Joseph Conrad, Hendrik Ibsen, Henry James, James Joyce, D. H. Lawrence, Federico García Lorca, Stéphane Mallarmé, Vladimir Mayakovsky, Robert Musil, Vladimir Nabokov, Ezra Pound, Marcel Proust, Rainer Maria Rilke, August Strindberg, Leo Tolstoy, Paul Valéry, and Virginia Woolf. These 'losers' constitute a veritable 'Who's Who' of modern literature.

10. Paul Ceruzzi has suggested that this oft-quoted statement could be a perversion of a remark by Howard H. Aiken, one of America's first great pioneers in the computer area.

11. In his *Ideas and Men* (1950, 12), Crane Brinton stressed the difference between "*cumulative* knowledge . . . best exemplified by the knowledge we call commonly natural science, or just science" and "noncumulative knowledge . . . illustrated best from the field of literature." But he warned that this distinction should not be taken to mean that art or music or literature "is somehow inferior to science." He argued that "from the beginnings of the study of astronomy and physics several thousand years ago in the eastern Mediterranean, our astronomical and physical ideas have accumulated, have gradually built up into the astronomy and physics we study in school and college. The process of building up has not been regular, but on the whole it has been steady. Some of the theories of the very beginning are still held true, such as the ideas of the ancient Greek Archimedes on specific gravity, but many, many others have been added to the original stock. Many have been discarded as false. The result is a discipline, a science, with a solid and universally accepted core of accumulated knowledge and a growing outer edge of new knowledge. Dispute—and scientists dispute quite as much as do philosophers and private persons—centers on this growing outer edge, not in the core. This core all scientists accept as true." Furthermore, "new knowledge can, of course, be reflected back through the whole core, and cause what may not unfairly be called a 'revolution' in the science. Thus quantum mechanics and relativity theories have been reflected back into the core of Newtonian physics."

12. Some of these that either mention the revolution in their titles or that deal explicitly with the Scientific Revolution as a historical event (thus excluding works that may deal with science during the time of the Scientific Revolution) are in chronological order: W. E. Knowles Middleton, *The Scientific Revolution* (1963), A. M. Duncan, *The Scientific Revolution of the Seventeenth Century* (1964),

Hugh F. Kearney, *Origins of the Scientific Revolution* (1964), J. F. West, *The Great Intellecutal Revolution* (1965), George Basalla, *The Rise of Modern Science, External or Internal Factors?* (1968), Robin Briggs, *The Scientific Revolution of the Seventeenth Century* (1969), Vern L. Bullough, *The Scientific Revolution* (1970), Paolo Rossi, *La rivoluzione scientifica: da Copernico a Newton* (1973), M. L. Righini-Bonelli and William R. Shea, *Reason, Experiment, and Mysticism in the Scientific Revolution* (1975), Robert Westman and J. G. McGuire, *Hermeticism and the Scientific Revolution* (1977), Carolyn Merchant, *The Death of Nature: Women, Ecology, and the Scientific Revolution* (1980), P. M. Harman, *The Scientific Revolution* (1983), A. R. Hall, *The Scientific Revolution 1500–1800* (2d ed., 1983).

13. Alistair Crombie's interesting article on "Historians and the Scientific Revolution" (1969) discusses the manner in which the science of the seventeenth century entered the record of history and not the treatment by historians of the concept or name 'scientific revolution'. In his "Discovery in Medieval Science and Its Contribution to the Scientific Revolution," Crombie sees the origins of modern science in a two-stage revolution, "the Aristotelian Revolution of the twelfth and thirteenth centuries and the so-called Scientific Revolution of the sixteenth and seventeenth" (p. 60).

14. There are three excellent general introductions to the topic of revolution: Hannah Arendt's *On Revolution* (1965), Peter Calvert's *Revolution* (1970), and A. S. Cohan's *Theories of Revolution* (1975). The last-named is especially valuable for its analysis (in ch. 6) of the 'functionalist' approach and of psychological approaches to revolution. A classic example of the latter is Chalmers Johnson's *Revolutionary Change* (1966; rev. ed., 1982), which stresses the idea of revolution as "social dysfunction." Two older classics are Pitirim A. Sorokin's *The Sociology of Revolution* (1925), which began "research, as opposed to speculation, in this field" (Calvert 1970, 127), and Lyford P. Edwards' *The Natural History of Revolution* (1927). Especially valuable are George S. Pettee's *The Process of Revolution* (1938) and Crane Brinton's *The Anatomy of Revolution* (1952). The latter compares and contrasts four 'great' revolutions: the English, American, French, and Russian revolutions. A convenient summary of many "theories of revolution" is given in Lawrence Stone's *The Causes of the English Revolution* (1972), first published in volume 18 of *World Politics* (1966).

On the psychological side, attention may be called to E. V. Wolfenstein's *The Revolutionary Personality* (New York, 1967) and Bruce Mazlich's *The Revolutionary Ascetic* (New York, 1976). Robert K. Merton's celebrated essay on "Social Structure and Anomie" (1957, ch. 4) proposes two sets of psychological responses (or escape routes) for individuals reacting to the pressure of social structure: ritualism and retreatism; innovation and rebellion. While Freud himself was not especially concerned with the question of revolutions and revolutionaries, he did discuss the nature of 'crowds' in relationship to their leaders in his *Group Psychology and the Analysis of the Ego* (1921; Freud 1953, *18*: 65–153), an attack on Gustave Le Bon's *Psychology of Crowds* (Paris, 1895; Engl. trans. London, 1920). See also Le Bon's *The Psychology of Revolution* (London, 1913), a translation of a work of 1912 entitled *La révolution française et la psychologie des révolutions*. The

student of revolutions should also examine the writings of Alexis de Toque-
ville and of Marx and Engels, as well as their followers Lenin and Marcuse.

Dirk Käsler's *Revolution und Veralltäglichung: Eine Theorie postrevolutionärer
Prozesse* (1977) offers a summary of the chief theories of revolution in diagram-
matic form and in particular explores the forces and groups that come into
power after a revolution. In many ways the most interesting discussion of revolution (and its changing historical significance) is Reinhart Koselleck's "Der
neuzeitliche Revolutionsbegriff als geschichtliche Kategorie" (1969). A very
thorough review of the literature is given in Hartmut Tetsch's *Die permanente
Revolution* (1973), where the subject of revolution is treated under four categories: psychology, history, law, political science. Tetsch's own model for revolutions is sociological.

The relations between scientific revolutions and political revolutions are
discussed in a very stimulating way by Wolf Lepenies in his *Das Ende der Naturgeschichte* (Munich, 1976), esp. pp. 106–114 on "Wissenschaftliche und politische
Revolution." F. Hayek (1955) and David Thomson (1955) explore the effects of
science on political and social movements.

2. *The Stages of Revolutions in Science*

1. In a reformulation of his concepts, Kuhn (1970, 182–187; 1974, 463) would
have a paradigm be composed of a "disciplinary matrix" and "exemplars."
Frank Sulloway, who has used these Kuhnian concepts in a particularly effective
way (1979, 358, 500), has given a neat summary of this later Kuhnian formulation. The 'disciplinary matrix' consists of a set of shared symbolic generalizations: Newton's second law, $f = ma$; models, such as comprise atomism; and
values, such as the goal of verifiable prediction. The 'exemplars' are particular
instances of solved problems, embodied in textbooks or handbooks.

Kuhn also writes of a pre-paradigm stage, in which the various phenomena
are explained from greatly differing points of view. The emergence of a paradigm is usually associated with the work of a few great scientists and is followed
by a period of normal science. Such an emergence of a paradigm is not considered by Kuhn to be a revolution.

I endorse the stand of Stephen Brush (1982, 1), who follows "some but not all
of the conclusions of Thomas S. Kuhn," that a revolution in science "involves
not only a radical change of specific theories and techniques but also a change in
the kinds of questions that theories are expected to answer, and the criteria for
judging those answers, as well as a basic transformation in the worldview of
those people who look to science for intellectual guidance. In short, a Scientific
Revolution involves a change from one *paradigm* to another, to use a term which
Kuhn himself has now abandoned but which many of us find indispensable."

2. I have discussed the ways in which new scientific ideas are disseminated in
twentieth-century fashion, but there are many points of identity and similarity
between these modes of communication in the period of the Scientific Revolution and in our own days. For instance, some of Newton's innovations in mathe-

matics and in physics became known and were put into circulation as a result of personal contacts. Edmond Halley had conferences with Newton in 1684, and his subsequent reports to the Royal Society gave public notice of Newton's great leap forward in celestial dynamics (Cohen 1971; Westfall 1980); in the 1690s, the Scots mathematician David Gregory had a number of consultations with Newton about mathematics and physics which led to extensive memoranda and notes that found their way into Gregory's commentary on the *Principia* (Cohen 1980).

Public dissemination by preprint also occurred in those days. Newton distributed manuscript copies of some of his mathematical tracts and even treatises (for which see Whiteside's commentaries in Newton 1967). An account of Newton's famous discovery of the composition of white light and the nature of colors was read at a meeting of the Royal Society and discussed there, before publication, as were many papers by others that were published in the *Philosophical Transactions*. At meetings of the Royal Society and of the Paris Academy of Sciences, communications were presented that were never printed or never printed in full. Many such papers from the Paris Academy were not published until years later. Newton, Huygens, and others of those days also spread their ideas through an international network of correspondence. This developed to such a degree that letters of this sort came to be known as 'epistolary discourses'. These examples show that the verb 'publish' should be widely construed as 'making public' and not limited to 'print'.

3. There are at least three historical uses of 'paper' which differ greatly from that used by me. The first occurs in Galileo's *Dialogue on the Two Great World Systems* (1632; 1964, 7: 139; trans. S. Drake 1953, 113), in which he says: "I discorsi nostri hanno a essere intorno al mondo sensibile, e non sopra un mondo di carta" ("Our discourses must be related to the sensible world and not based on a world of paper"). This text provides the title, *A World on Paper* (1976; 1980), of Enrico Bellone's "Studies on the Second Scientific Revolution." In a review of Bellone's book, Stephen Brush (*Isis* 1981, 72: 284) has called attention to a third example, P. G. Tait's discussion of "the modern manikins of *paper science*, who are always thrusting their crude notions on the world; the anatomists who have never dissected, the astronomers who have never used a telescope, or the geologists who have never carried a hammer!" Brush notes that Tait had in mind "cranks like antivivisectionists who pester legitimate scientists with irrelevant criticism and unwanted suggestions."

4. The revolution need not leave a permanent impress upon science. The Cartesian revolution of the *Principles of Philosophy*, though dramatic, was soon displaced by the Newtonian science of the *Principia*.

3. Evidence for the Occurrence of Revolutions in Science

1. Almost all scientists and historians of science agree that Einstein's relativity theory (or theories) constitutes one of the great revolutions of science in the twentieth century, a judgment expressed by many contemporaneous observers.

Yet Albert Einstein himself (see chapter 28) insisted that relativity was to be considered from the point of evolution rather than revolution.

2. I here discount the preliminary announcement in the *Journal of the Linnean Society* in 1858.

3. All but two of these statements of a self-proclaimed revolution occur elsewhere in the book. Liebig's remarks about his own revolution appear in a letter to Friedrich Mohr (24 May 1842), in reference to his recently published *Animal Chemistry*. "If you would have been a physician," Liebig wrote, "you would have led the revolution which is now being launched. You would have led it to its complete development" (Kahlbaum 1904, 71–72).

In 1931 the American biologist Ernest Everett Just referred to his own theory about the cortical cytoplasm as "aimed at nothing short of revolution," and five years later reiterated his claim that his "scientific work" was "revolutionary" (Manning 1983, 239, 289).

4. Transformations in the Concept of Revolution

1. In the *Principia*, Newton used the term 'revolutio' to denote both revolution (in an orbit) and rotation (about an axis). But in 1713, when Roger Cotes made an index for the second edition of the *Principia*, he introduced the term 'conversio' in its traditional meaning of 'revolutio' (as revolution [rotation] about an axis): Jupiter— "Its revolution [conversio] about its axis: in what time it is completed"; the planets— "Their daily revolutions [conversiones] are uniform." In the text which Cotes indexed, Newton had used 'revolutio' and 'revolutiones'.

2. Heloise's first letter to Abelard (1875, letter 2) speaks of a letter of Abelard's which told "the pitiful history of our conversion [nostrae conversionis miserabilem historiam]." An eighteenth-century English version (1729), first published in 1713 and translated from a French version of 1693, freely expands Heloise's phrases, representing the words quoted above as (p. 110) "a particular and melancholy Account of our Misfortunes" and "the Representation of our Sufferings and Revolutions."

A French version of 1697, presumably a reprint of the 1693 edition, which is generally closer to the English than to the Latin, does not use the word 'revolutions'. Here the text mentions "un long détail de nos malheurs" and "nos traverses & nôtre infortune." Consequently, although Hughes, the author of the English version of 1713, did look at the Latin edition of 1616, "our Sufferings and Revolutions" may represent "nos traverses & nôtre infortune" rather than "nostrae conversionis miserabilem historiam."

3. Hatto (1949, 504) observes that Rosenstock (1931, 100) not only was the first to take notice of "the Italian provenance of the term," but also had made the shrewd observation that "it was not adopted in other lands until the age of the great politicians, and then because of its impersonality and amorality, which so characteristically distinguish it from 'revolt', 'civil war', 'disorders' and other words tinged with ethical notions."

4. A chief mathematical use of this word has been to designate a solid produced by the turning of a plane figure through 360° about an axis in that plane so as to produce a 'solid of revolution'. Thus a sphere is the 'solid of revolution' produced when the plane figure is a circle and the axis a diameter of that circle; if the axis does not intersect or touch the circle, the 'solid of revolution' is a torus (that is, doughnut-shaped).

5. Another time-measurer, the hourglass, required the act of overturning or turning over, akin to the general concept of revolution.

6. The succession of 'revolutions of time' or 'of the ages' and hence 'changes through time' and then 'a sudden, surprizing change' can be followed in the *New English Dictionary (O.E.D.)*, although as Hatto (1949, 510) has noted "it neglects the political implications of the meaning of 'turning forward *or backward* to a starting point' (cf. Milton's 'Fear comes thundering back with a dreadful revolution in my defenceless head', *Paradise Lost*, x, 13 . . .) and the decisive part played by astrological conceptions in its semantic development."

7. Snow (1962, 169) suggests that Howell had reference to "various types of political and social changes which took place within the monarchical form of government."

8. The dates 1625–1660 begin with the reign of Charles I (1625–1649) and end with the restoration of the monarchy with Charles II.

9. The French historian and statesman François Guizot observed (1867, 1: 252) that the "great revolution" in England "which dethroned Charles I" is "sometimes improperly called the great rebellion," which was Clarendon's name. Although there are several authors of the late eighteenth and early nineteenth centuries who wrote of these mid-century events as a revolution, Guizot was the first major historian to do so and to do so prominently as the title of his work.

10. On 11 December 1688, James fled for France, but was captured and returned to London; he finally left England on 22 December. It was on 12 December 1688, between these two events, that John Evelyn wrote to Samuel Pepys about "this prodigious revolution."

In July 1685 John Evelyn (1906, 3: 166) witnessed the punishment by public whipping of Titus Oates. Evelyn's comment. "A strange revolution." Perhaps Evelyn was referring to Oates's strange career, with its rise and fall, or possibly he had in mind the alleged plot of the Jesuits (or 'popish plot') which Oates had fabricated and which had resulted in the 'judicial murder' of about thirty-five men. A few weeks after William of Orange had landed at Torbray, Evelyn wrote in his diary about the way in which various major communities in England returned to the earlier condition and "declare for the Protestant religion and laws," while the "papists in offices lay down their commissions, and fly." There is "universal consternation amongst them," he noted; "it looks like a revolution" (p. 247).

11. See chapter 5 below, on the use of 'revolution' to describe the calculus and to indicate the profound change in physical science associated with Newton's *Principia*.

12. For example, Goulemot has studied forty-two tragedies produced be-

tween 1689 and 1715, and the "ideological and political significance" of the thirty-eight of them that were published, showing the effect of the events in England. Of these, the best known is Racine's *Athalie*.

13. In a conversation recorded by John Conduitt (King's College, Cambridge, Keynes M.S. 130, no. 11) Newton used the term 'revolution' in the sense of a cycle. This cycle, "a sort of revolution in the heavenly bodies," could take different forms. In the revolution discussed with Conduitt, Newton envisaged that "the vapours and light emitted by the sun" could coalesce and attract further matter "from the planets" so as to form a satellite or "secondary planett (viz. one of those that go round another planet)"; then by "attracting more matter," the satellite could become "a primary planet"; and "then by increasing" still more become "a comet"; finally, the comet, revolving and "coming nearer and nearer the sun," would have "all its volatile parts condensed" so as to become "matter set to recruit and replenish the Sun." Such, he said, "would probably be the effect of the comet in 1680, sooner or later." Thus there was for Newton a "cyclical cosmos" (Kubrin 1967, 341), in which the satellites of Jupiter or Saturn could "take the places of the Earth, Venus, Mars if they are destroyed, and be held in reserve for a new Creation," as Newton explained his cyclical conception to David Gregory (5–7 May 1694; Newton 1961, *3*: 336). The force, Gregory explained, would come from a comet's gravitational attraction exerted on a satellite so "as to make it leave its Primary Planet and itself become a Primary Planet about the Sun."

Leibniz also used the general term 'révolution', but it is difficult to tell whether in fact he intended the sense of cycle. The passage occurs in his *Nouveaux essais* (iv, ch. 16, §4 = 1896, 536): "But these persons will possibly experience themselves the evils they think reserved for others. If, however, this disease of an epidemic mind whose bad effects begin to be visible is corrected, these evils will perhaps be prevented; but if it goes on increasing, Providence will correct men by the revolution itself which must spring therefrom: for whatever may happen, everything will always turn out for the better in general at the end of the account, although that ought not and cannot happen without the punishment of those who have contributed even to the good by their bad acts." Evidently, this 'revolution' is a correction of a "disease of an epidemic mind," with the result that "everything will turn out for the better." The text gives no basis for deciding whether this improvement corresponds to a return to an earlier and healthier state or whether the correction entails the establishment of something absolutely new.

14. The major study of the history of the concept of revolution is Karl Griewank's *The Modern Concept of Revolution* (1973), a work that is imperfect, as all such pioneering studies must be. It is a pity that the author did not live to produce a revised and corrected version. A valuable collection of materials relating to the use of the word 'revolution' is given in Seidler's unpublished doctoral dissertation (1955). A number of articles on the history of the concept and term 'revolution' contain important information and insights: their authors are, in historical order, Eugen Rosenstock (1931), Arthur Hatto (1949), Vernon

F. Snow (1962), and Felix Gilbert (1973). An encyclopaedic study of this topic is being completed by Reinhart Koselleck, who has given us an earnest of his endeavors in his article of 1969.

5. *The Scientific Revolution*

1. Galileo proved by the correlation of the phases of Venus with the apparent size of Venus (under constant magnification) that the orbit of Venus encircles the sun and not the earth. Hence, Ptolemy was wrong. But Venus could encircle the sun in the Tychonian system (in which the sun moves around a stationary earth while the planets move around the sun) as well as in the Copernican system.

2. A late work in the Quarrel between the Ancients and the Moderns, Louis Dutens's *Recherches sur l'origine des découvertes attribuées aux modernes* . . . (1766; later editions in 1776, 1796, 1812) contains a postil to the second paragraph reading "Révolution dans les sciences," a phrase that also appears in the index in the second and later editions. In context, however, it is evident that Dutens was referring to a return, a finding again of the scientific truths known — at least in principle — in Antiquity.

Although Sprat does not refer to revolutions in the sciences, he does use the term (1667 [1958], 383).

3. The most complete, general work on the Scientific Revolution is A. R. Hall's *The Scientific Revolution 1500–1800* (London, 1957; rev. ed. 1983), which may be supplemented by Allen G. Debus's *Man and Nature in the Renaissance* (Cambridge, 1978) and R. S. Westfall's *The Construction of Modern Science: Mechanism and Mechanics* (New York, 1971; repr. Cambridge 1977). Carolyn Merchant's *The Death of Nature* (San Francisco, 1980), among other topics, stresses the shift from an organic to a mechanical world view. E. J. Dijksterhuis's *The Mechanization of the World Picture* (Oxford, 1961) is still a classic. A recent review is provided by P. M. Harmon's pamphlet *The Scientific Revolution* (London, 1983).

The pioneering study on the ancients versus the moderns is Hippolyte Rigault's *Histoire de la querelle des anciens et des modernes* (Paris, 1856), which may be supplemented by Ferdinand Brunetière's essay on "La formation de l'idée de progrès au XVIIIᵉ siecle" in his *Etudes critiques sur l'histoire de la littérature française*, cinquième série (Paris, 1893) and by H. Gillot's *La querelle des anciens et des modernes* (Paris, 1914). John B. Bury's *The Idea of Progress* (New York, 1932; repr., New York, 1955) set the problem in a larger historical perspective. The major study that relates the battle of the books to the rise of the new science and the Baconian philosophy is Richard Foster Jones's *Ancients and Moderns: A Study of the Background of the Battle of the Books* (St. Louis, 1936), which was revised and amplified and reprinted under the title *Ancients and Moderns: A Study of the Rise of the Scientific Movement in Seventeenth-Century England* (St. Louis, 1961; repr., New York, 1982). Two important additional studies were published in the *Journal of the History of Ideas:* Edgar Zilsel's "The Genesis of the Concept of Scientific

Progress" (1945, *6*: 325–349) and Hans Baron's "The *Querelle* of the Ancients and the Moderns as a Problem for Renaissance Scholarship" (1959, *20*: 3–22).

6. A Second Scientific Revolution and Others?

1. My periodization into four Scientific Revolutions is presented only as a first attempt and eventually must need revision or refinement. But I did want to suggest that there are revolutions of a type very different from the 'revolutions in science' that are the subject of this book.

2. My ideas concerning the simultaneity of intellectual and institutional changes were forged during discussions with Ian Hacking and the reading of his challenging essay (1983).

3. I warmly endorse Kuhn's tentative conclusions about the nature of these distinguishing characteristics (1977, 220); but this is a topic for research and discussion that would take us far beyond our present purpose.

4. Kuhn informs me that the first version of this paper was read at a Berkeley Social Science Colloquium a few years earlier, but apparently did not contain any reference to a second Scientific Revolution.

5. Brush (1979, 141–142) published an earlier statement on the "different time-scales of the first and second Scientific Revolutions."

6. In *Isis* 72:(1981) 284.

7. The Copernican Revolution

1. We do not have firm and unambiguous evidence as to the exact title chosen by Copernicus for his book. In his own copy, Rheticus (who was Copernicus's only pupil, and who brought the manuscript from Frauenberg [Frombork] to Nuremberg for publication) deleted the last two words of the title (see Rosen 1943; Gingerich 1973, 514).

2. Copernicus not only endowed the earth with two motions (a daily rotation on its axis and an annual revolution in its orbit about the sun), but also introduced a third motion in order to account for the fact that the positions of the earth's axis, as the earth moves in orbit, are parallel to one another. We would attribute this parallelism to a nonmotion of the earth's axis, but Copernicus said that this constant parallelism required motion, the third motion of the earth. Dreyer (1906, 328) compares this point of view to the ancient opinion regarding the moon, which always shows the same face toward the earth. We would say that this is caused by the moon having two motions: its motion of rotation having a period equal to its period of revolution. But the ancients said that the reason for this constancy in the appearance of the moon was that it does *not* rotate. Copernicus, according to Dreyer,

> would have expected the axis of the earth to have continued during the year to be directed to a point a long way above the sun, as if the earth were

the bob of a gigantic conical pendulum. This would have made the celestial pole in the course of a year describe a circle parallel to the ecliptic, and as it does nothing of the kind but remains fixed, Copernicus had to postulate a third motion of the earth, a "motion of the inclination," as he calls it [Lib. I. cap. 11, p. 31], whereby the axis of the earth describes the surface of a cone in a year, moving in the opposite direction to that of the earth's centre, i.e. from east to west. Hereby the axis continues to point in the same direction in space. But the period is not exactly a year, it is slightly less, and this slight difference produces a slow backward motion of the points of intersection of the ecliptic and the equator—the precession of the equinoxes. This was now at last correctly explained as a slow motion of the earth's axis and not as hitherto as a motion of the whole celestial sphere, and this almost reconciles us to the needless third motion of the earth, which certainly had its share in the unpopularity against which the Copernican system had to do battle for a long time, as it seemed bad enough to give the earth one motion—but three!

3. Although Copernicus made quite a production of the evils of the equant and of his own claims to have eliminated it from his astronomy, it has been shown by Neugebauer (1968, 92ff.) that in a very real sense "it was the goal of Copernicus's cinematic arrangements [for the planets] to *maintain* the equant, by no means to eliminate it." According to Neugebauer, who has made a detailed mathematical analysis of Copernican planetary theory in relation to the Ptolemaic, it is not correct to say that Copernicus truly eliminated the equant from his astronomy (p. 95):

> Since we shall find that Copernicus preserved the equant also for Mercury we can now say that his aim was by no means to abolish the concept of the equant, but, exactly as his Islamic predecessors, to demonstrate that a secondary epicycle is capable of producing practically the same results (thanks to the smallness of the eccentricities) as Ptolemy's equant. Though the resultant deferent is unfortunately not a circle, each component motion is uniform and circular. Both ash-Shāṭir and Copernicus considered this as their main achievement, even if the model had become more complicated than Ptolemy's.
>
> Kepler was less philosophically prejudiced, and he not only reintroduced the Ptolemaic equants in the planetary theory but took the heliocentric approach seriously and hence provided also the (circular) orbit of the Earth with an equant, an improvement which increased the accuracy in his determinations of the positions of Mars. Ptolemy's discovery of the 'equant' was not only never abandoned but proved of the greatest importance for the construction of the 'oval' orbit of Mars and hence of the Kepler ellipse.

4. "Copernicus's only real objection to Ptolemy's model, therefore, is that the motion of the center of the epicycle is not uniform with respect to the center from which it maintains a constant distance. The reason for this objection is that

Copernicus considers the motion of a planet to be directed by the revolution of a material sphere or spheres in which the planet is fixed. The only motion permitted to the sphere is a simple uniform rotation about its diameter; it cannot move uniformly with respect to any other straight line passing through it" (quoted from Swerdlow 1973, 435).

5. Much has been made of the distortion of Copernicus's point of view by Osiander, who actually wrote the preface to the first edition of *De Revolutionibus*. (On this point, see Rosen 1971, 22ff.; Duhem 1969, 66ff.) Despite the claims that *De Revolutionibus* expressed Copernicus's beliefs about the 'real' external world, there are examples in that work that can be only models or computing schemes. According to Neugebauer (1968, 100):

> Neither Ptolemy's nor Copernicus's machinery for the motion of Mercury could be considered plausible representations of physical facts; Copernicus himself had speculated in a loose fasion about an alternative mechanism [*Revol.* v, 32]. It is difficult to see how devices of this type ever could have been taken for more than mathematical similes of no other significance than to guide the computations. I realize that one is supposed to be disgusted with Osiander's preface which he added to the *De Revolutionibus* (in keen anticipation of the struggle of the next generations) in which he, in the traditional fashion of the ancients, speaks about mere 'hypotheses' represented by the cinematic models adopted in this work. It is hard for me to imagine how a careful reader could reach a different conclusion.

Following Kepler's revelation of the true authorship of the preface concerning hypotheses in *De Revolutionibus*, the information spread rapidly. In Sherburne's history of astronomy (1675, 50), published together with his translation of Manilius, he wrote (under the date 1536):

> Andreas Osiander took not only care in publishing the first Edition of *Copernicus* his Book *De Revolutionibus*, but condescended to be Overseer of the Press, while it was Printing, to which he added a brief Preface of his own, therein chiefly endeavouring, because of the seeming Novelty of the Opinion, to perswade the Reader, to look upon it as an assumed Hypothesis, rather than an asserted Tenet. To which purpose, about that time, was published this Distich,
> *Quid tum si mihi Terra movetur, Solque quiescit*
> *Et Coelum? Constat Calculus inde Mihi.*
> Of which *Gassendus, in Vita Copernicii.*

> This elegaic couplet may be translated as follows:

> What then if for me the Earth moves, and the Sun is at rest
> And the Heavens? I can make a computation on that basis.

We don't have any study of the actual effect on readers of Osiander's "Ad lectorem." Since it refers to Copernicus in the third person, a careful reader could readily have concluded that it was not by Copernicus. Kepler did not say that readers had been led astray by believing the "Ad lectorem" to have been written by Copernicus. In those days many books had a preliminary note by someone other than the author, although such notes did not usually present a point of view at odds with the author's.

6. I do not want to get into a discussion of Copernicus's actual nationality; he is said to have been (Rosen 1971, 315) "a German-speaking Pole." In the concluding section of Rheticus's *Narratio Prima* (ib., 188), "In Praise of Prussia," Copernicus is presented as a Prussian.

7. In 1535 Bernard Wapowski, the secretary to the Polish king, sent to Sigismund Herberstein in Vienna a copy (in manuscript) of an almanac prepared by Copernicus, containing "the most authentic and best explained movements of the planets, figured out on the basis of new tables" (Biskup 1973, 155). Wapowski explained that these tables were superior to any then in circulation and could be used "even by a poorly educated person." It was Wapowski's hope that Copernicus's tables be printed and become "widespread, especially among specialists in celestial matters [i.e., astrologers] writing almanacs" so that they would be able to "write more correct ones and acknowledge their own errors." There is no information concerning this almanac and its tables, other than Wapowski's letter of transmittal, and there seems no way of telling on what basis Wapowski made his judgment. This example does show, however, that what improvements Copernicus made to the method of working astronomers (and astrologers) lay in his tables and not his planetary cosmology.

8. In the *Commentariolus* (trans. Rosen, 1971, 58), Copernicus referred to his own system as requiring "fewer and much simpler constructions than were formerly used." In *De Revolutionibus*, book 1, ch. 10, he wrote that he would find it easier to grant certain aspects of the distances between the earth and the planetary orbs or spheres than "to perplex the intellect by an almost infinite multitude of spheres — as those are forced to do, who keep the earth at the center of the world." In the original Latin: "quod facilius concedendum puto quam in infinitam pene orbium multitudinem distrahi intellectum quod coacti sunt facere, qui terram in medio mundi detinuerunt." Thus Copernicus himself began the myth that his own system was characterized by a much greater simplicity than Ptolemy's.

Robert Palter (1970, 127, n.7) cites two extreme examples of suppositions concerning the simplicity of the Copernican system. The philosopher A. Kaplan (1952) discusses the alleged "triumph of the modern world view inaugurated by Copernicus" as follows: "Scientists like Kepler accepted the Copernican hypothesis because it made the cosmos more mathematically elegant than Ptolemy's cumbersome epicycles had been able to do" (p. 275). The same opinion, that Copernicus used (or needed) no epicycles, is found in David Bohm (1965, 5).

Owen Gingerich (1975*b*, 93) calls attention to an extravagance of the reverse kind, making the Ptolemaic system much more complex than it was. In the *Encyclopaedia Britannica* (1969), vol. 2, p. 645, in relation to King Alfonso X of Castile, it is said that by the end of the thirteenth century "each planet had been provided with from 40 to 60 epicycles to represent after a fashion its complex movement among the stars."

9. O. Neugebauer, the outstanding scholar in our times in the field of early astronomy, has summed up his analysis as follows: "The popular belief that Copernicus's heliocentric system constitutes a significant simplification of the Ptolemaic system is obviously wrong. The choice of the reference system has no effect whatever on the structure of the model, and the Copernican models themselves require about twice as many circles as the Ptolemaic models and are far less elegant and adaptable" (1957, 204).

10. A notorious example of the errors and unnecessary complications that arose from Copernicus's uncritical dependence on Ptolemy may be seen in the matter of the precession of the equinoxes. This could have been easily explained by the difference "between the periods of the orbital motion and of the axial motion of the earth" (Dreyer 1906, 329). But Copernicus had "boundless confidence in the accuracy and integrity of his predecessors in astronomy" (Dijksterhuis 1961, 295), and so he believed that the precession is a variable quantity and accordingly introduced a complex mechanism to account for this inequality and an irregular change in the inclination of the ecliptic.

11. The *De Revolutionibus* of Copernicus is available in a number of editions and facsimile reprints. The most recent one, sponsored by the Polish Academy of Sciences, is accompanied by an English version, edited by Jerzy Dobrzycki, translation and commentary by Edward Rosen (1978). There are two English versions of the *Commentariolus:* by Noel Swerdlow (1973), with an important and extensive technical commentary, and an earlier version by Edward Rosen (1971 [1939]).

Edward Rosen has provided a survey of Copernicus's life and works in his *D.S.B.* article (1971), which many be supplemented by the volumes published at the time of the five hundredth anniversary of Copernicus's birth (19 Feb. 1473), edited by Arthur Beer and K. Aa. Strand (1975), Owen Gingerich (1975), Jerzy Neyman (1974), Nicholas Steneck (1975), and Robert Westman (1975).

A general introduction to *The Copernican Revolution* is available in Thomas S. Kuhn's volume with that title (1957). The comparative reception of Copernicus's ideas is presented in vol. 1 of *Colloquia Copernicana* (1972). Still valuable, although to be used with care, is Dorothy Stimson's *The Gradual Acceptance of the Copernican System of the Universe* (1917). It is to be wished that Hans Blumenberg's *Die Genesis der kopernikanischen Welt* (1975) were available in English. On Copernicus's astronomy, see Derek J. de S. Price's "Contra-Copernicus: A Critical Re-estimation of the Mathematical Planetary Theory of Ptolemy, Copernicus, and Kepler" (pp. 197–218 of Marshall Clagett, ed., *Critical Problems in the History of Science,* Madison [Wisconsin], 1959) and O. Neugebauer's "On the Planetary Theory of Copernicus" (*Vistas in Astronomy* 1968, 10: 89–103; reprinted in Neugebauer 1983).

8. Kepler, Gilbert, and Galileo

1. In the introduction (1937, 3: 18), Kepler wrote of the difficulties in reading scientific or mathematical books in Latin, "a language which has no articles and which lacks the felicity of Greek."

2. Galileo quite properly conceived that his struggle was to achieve a victory for Copernican versus Ptolemaic astronomy, that the Tychonic system was a compromise that should not be taken too seriously. Venus encircles the sun in both the Copernican and the Tychonic systems. But if the earth is like the planets, should it not move around the sun like the other planets? In other words, Galileo's discoveries did in their totality seem to favor a kind of Copernicanism.

3. None of Kepler's most important writings has ever been published in a full English version, but his *Mysterium Cosmographicum* has been translated by Duncan (1981) with a commentary by Eric Aiton, while Edward Rosen has given us English texts of Kepler's *The Dream* (Madison, 1967) and *Conversations with Galileo's Sidereal Messenger* (New York, 1965). A good introduction to Kepler's work is given in Alexandre Koyré's *Astronomical Revolution* (1973) and in Owen Gingerich's account of Kepler in the *D.S.B.* (1973). A massive volume edited by Arthur and Peter Beer (*Vistas in Astronomy*, 1975, *18*), based on a large number of Kepler symposia, gives the texts or abstracts of major articles on every aspect of Kepler's life and work.

Gilbert's great book *On the Magnet* is available in two English translations: by P. Fleury Mottelay (1893) and by Silvanus Thomson (1900; 1958). Two scholarly monographs on Gilbert are Suzanne Kelly's *The De Mundo of William Gilbert* (Amsterdam, 1965) and Duane H. D. Roller's *The De Magnete of William Gilbert* (Amsterdam, 1959). For an up-to-date critical account of Gilbert and his work, see Heilbron 1979.

The literature on Galileo is enormous—in all languages. A good survey of various currents of Galileo research, based on an international symposium in 1983, is being edited for publication by Paolo Galuzzi. This will supplement the earlier collection edited by Ernan McMullin (1967), which contains a bibliography of Galileo studies from 1940 to 1964. Galileo's *Two New Sciences* and his *Dialogue* concerning the two chief systems of the world are available in translations by Stillman Drake, who has also produced (1957) a volume of Galileo's shorter works. Drake has summarized a lifetime of Galileo research in his massive *Galileo at Work* (1978). Although partly out-dated by recent research, Alexandre Koyré's *Galilean Studies* (1939 [1978]) is still of extraordinary value. A simple introduction to the work of Galileo in its context is given in my *Birth of a New Physics* (Garden City, 1960 [1985]) and Ludovico Geymonat's *Galileo Galilei* (New York, 1965).

9. Bacon and Descartes

1. Among Descartes's significant contributions to philosophy are his formulation of the duality of 'mind' and 'body' (often known as 'Cartesian dualism'). The body is a machine, he held, and the mind (soul) is a pure thinking substance.

2. The writings of Bacon are available in several printings of the *Works* (7 vols., 1857–1859; 1887–1892; 1963), edited by J. Spedding, R. L. Ellis, and D. D. Heath, of which the English versions have been assembled in a convenient volume, *The Philsophical Works* (1905), edited by John M. Robertson. *The New*

Organon has been edited (1960) by Fulton M. Anderson, who has also produced a comprehensive summary of Bacon's philosophy (1948). An edition of the Latin text with extensive English notes was published by Thomas Fowler (1878; 1889); Marta Fattori has compiled an extraordinarily useful lexicon (1980). Hugh Dick (1955) has assembled a collection of *Selected Writings*. Anthony Quinton (1980) has produced an admirable short introduction to Bacon's thought and influence, which should be supplemented by Thomas Fowler's classic study of 1881. Challenging new ideas are presented in Paolo Rossi's *Francis Bacon: From Magic to Science* (1968); the practical side is stressed in Benjamin Farrington's introduction to Bacon (1949).

The writings of Descartes are available in a standard 12-volume edition (1897–1913; 1971–1976) by Charles Adam and Paul Tannery. English collections of texts are available in translations by Elizabeth Haldane and G. R. T. Ross (1911–12; 1931; 1958), by John Veitch (1912), by Norman Kemp Smith (1952), and by Elizabeth Anscombe and Peter Thomas Geach (1954). *The Discourse on Method* is available in translations by Laurence Lafleur (1956), F. E. Sutcliffe (1968), as well as in the Haldane-Ross, Veitch, Smith, and Anscombe-Geach volumes. It also appears in a volume containing Descartes's *Optics, Geometry, and Meteorology,* translated by Paul J. Olscamp. The *Geometry* is available in an English version by David Eugene Smith and Marcia Latham (1925; 1954); *Le monde* has been translated with an introduction by Michael Mahoney (1979). The latter two translations are published together with facsimiles of the original French printings, as is the *Treatise on Man,* translated (1972) with a commentary by Thomas S. Hall.

The best study of Cartesian physics is still Mouy 1934, supplemented by Geoffrey Sutton's important doctoral dissertation (1982). The collection of studies by Gaston Milhaud, *Descartes savant* (Paris, 1921), is still a major work more than six decades after it was published. A good up-to-date general study on Cartesian physics, its history, and its influence is greatly to be desired. Two very perceptive works on Descartes that discuss his science are Rée 1974 and Williams 1978. On Descartes's philosophy as a whole Keeling 1968 (1934) is still a good introduction; Smith 1952 is extremely important. Also recommended is Alexandre Koyré's essay on "Newton and Descartes" with 13 supplements in Koyré 1965.

10. The Newtonian Revolution

1. They were found independently also by Wren and Huygens (see Dugas 1955, ch. 5).

2. Newton, in fact, was not able to carry out this program fully, although he claimed to have done so. He was really successful only in accounting for what is called the variation and the nodal motion (see Cohen 1980, 76–77; Waff 1975; Chandler 1975). The review appeared in the Berlin *Acta Eruditorum.*

3. This presentation of Newton's development of the law of universal gravity is an abridgment of the fuller presentation given in my *Newtonian Revolution* (1980), §§5.4–5.6. Here there may be found also an exposition of the stages of transformation of the concept of inertia leading to Newton's first law of motion.

4. I assume here, in the absence of any contradictory evidence, that in the successive versions of his preliminary tract *De Motu* and in the *Principia*, Newton was presenting his ideas and results more or less in the logical-chronological order in which he had developed them. See Cohen 1980, 248ff., 258ff.

5. The literature on Newton and the Newtonian revolution is vast, encompasing the findings of a large Newton research industry. A good entry into this area is by way of R. S. Westfall's monumental biography, *Never at Rest* (1980). On the developments in astronomy and mathematical physics that led to Newtonian science and on Newton's work in these areas, see the two books by René Dugas, *History of Mechanics* (1905 [1955]), *Mechanics in the Seventeenth Century* (1958 [1954]), and Westfall's *Force in Newton's Physics: The Science of Dynamics in the Seventeenth Century* (1971). Some of the ideas in this chapter have been presented by me in greater detail in my *The Newtonian Revolution* (1980), and in two articles: "The *Principia*, universal gravitation, and the 'Newtonian style,'" in Zev Bechler, ed., *Contemporary Newtonian Research* (1982), and "Newton's Discovery of Gravity" in the March 1981 issue of *Scientific American*, 1244: 166–179.

Major documents on the development of Newton's ideas on dynamics and celestial mechanics are available, with valuable commentaries, in *Unpublished Scientific Papers of Isaac Newton* (1962), ed. A. Rupert Hall and Marie Boas Hall; John W. Herivel's *The Background to Newton's Principia* (1965); and vol. 6 of the great edition of *The Mathematical Papers of Isaac Newton* (1974), ed. D. T. Whiteside. On the history of the *Principia*, see my *Introduction to Newton's 'Principia'* (1971). Anne Miller Whitman and I have completed a new English translation of the *Principia*, scheduled for publication in the near future.

Newton's *Opticks* is available in a convenient paperback edition; a full edition, with commentary and variant readings, prepared by Henry Guerlac, is to be published in 1985. The *Lectiones Opticae* is currently being published with a translation and commentary by Alan Shapiro. *Isaac Newton's Papers and Letters on Natural Philosophy and Related Documents* (1978 [1958]), ed. I. B. Cohen and Robert E. Schofield, contains facsimile reprints (with commentaries) of Newton's articles and related letters.

On the influence of Newton in the Enlightenment and after, the best introduction is still *The Making of the Modern Mind* (1940) by John Herman Randall, Jr. Another good general source is Herbert Butterfield's *The Origins of Modern Science* (1957 [1949]). See also Henry Guerlac's *Newton on the Continent* (1981), Margaret C. Jacob's *The Newtonians and the English Revolution* (1976), and Alexandre Koyré's *Newtonian Studies* (1965). Especially useful for the Enlightenment are Crane Brinton's *Ideas and Men: The Story of Western Thought* (1950) and Peter Gay's two volumes on *The Enlightenment* (New York, 1966–1969).

11. Vesalius, Paracelsus, and Harvey

1. Vesalius was a pioneer and hence his own work is far from error-free. Scarborough (1968, 209) has argued that "Vesalius's anatomical errors indicate that he used Galen as much as possible."

2. Some historians and medical doctors have written in terms of a Vesalian revolution. Sir William Osler said that "the works of Vesalius, of Fallopius, and of Fabricius effected a revolution in anatomy" (1906, 18), but he at once limited the scope of the revolution by adding that "there was not at the close of the sixteenth century a new physiology." John Scarborough has written a study of "The Classical Background of the Vesalian Revolution" (1968).

3. At that time the Roman physician Celsus was respected as one of the greatest authorities on medicine. The name 'Paracelsus' literally means 'beside' Celsus or 'greater than' Celsus. It was much as if a scientist today would announce publicly that he was greater than Einstein.

4. There is no want of modern scholars who have hailed Harvey's work as a revolution. Thus John G. Curtis (1915, 111): Harvey's "discovery of the circulation" was "the most modern and revolutionary achievement of his time"; Charles Singer (1956, 42): Harvey's discovery "was destined to change all the current ideas of the working of the body, and to revolutionise the practice of medicine and of all the sciences connected with it"; Sten Lindroth (1957, 209): Harvey's "new revolutionary physiology"; Sir Thomas Lewis (1933, 4): "Revolutionary . . . was his [Harvey's] discovery in its time, far reaching in its subsequent development"; John F. Fulton (1931, 29): Harvey's "revolutionary doctrines"; Walter Pagel (1967, 698; see 1969, 6): "the revolutionising role [of] Harvey's discovery"; Herbert Butterfield (1957, 64): "The revolution that he [Harvey] brought about was like the one we have seen in the realm of mechanics, or the one that Lavoisier was to achieve in chemistry"; J. D. Bernal (1969, 2: 438): "What Harvey established by his close reasoning from experiment had the same revolutionary effect on . . . physiology as the discoveries of Galileo and Kepler had on . . . astronomy."

5. In the dedication of *De Motu Cordis,* Harvey claimed "to learn and teach anatomy not from books but from dissections, not from the tenets of Philosophers but from the fabric of Nature," but he made it clear that he did not "think it honourable to attack or fight those who excelled in anatomy" and who had been his "own teachers." Nor would he "willingly charge with falsehood any searcher after truth, or besmirch any man with a stigma of error." When he did find it necessary to correct errors, there was "no acerbity in his words" (Keynes 1966, 179).

Harvey was essentially conservative, and although his findings directly contradicted Galenic fundamentals, he never flatly attacked or denigrated Galen, whom he reverently called (ch. 7) a "vir divinus, pater medicorum," that is, a "divine man, father of physicians." Harvey's embarrassed translators have rendered this phrase as "that great man, that father of physicians" (Willis), "that great Prince of Physicians" (Leake), and "the reverend Sire of Physicians" (Franklin). In this instance, Harvey quoted at length from Galen and then said that Galen's argument was intended "to explain the passage of blood from the vena cava through the right ventricle into the lungs," but we have only "to change the terms" to apply this argument "more properly to the transfer of blood from the veins through the heart to the arteries."

In the beginning of his "Second Disquisition to John Riolan," Harvey described an experiment which, he says, "Vesalius only prescribed" and "Galen advises" to "those anxious to discover the truth." Harvey actually performed the experiment and concludes that "if the experiment which Galen recommends were properly performed by any one, its results would be found in opposition to the views which Vesalius believed they would support."

6. Although Galen was in fact a pioneer in experiment, dissection, and critical observation, he also created a "highly ingenious physiology," which (Singer and Rabin 1946, p. xxxix) is not "derived from an investigation of human anatomy." Often, insofar as Galen's physiology "is related to actual structures, they are those of animals and not of men." An example is the 'marvelous network' or 'rete mirabile' in the brain, where Galen said the 'vital spirit' is elaborated into a finer and higher 'spirit', namely, 'animal spirit'; but no such 'rete mirabile' is to be found in a human brain, although it may be found in the brain of a calf.

7. Harvey's work illustrates the way in which a revolutionary new discovery changes even the language of a science. Prior to the publication of his book in 1628, 'veins' were defined as those blood vessels carrying venous blood, while an 'artery' was similarly defined as a blood vessel carrying 'spiritous' blood. The veins, according to this classification, generally would have thin, flexible walls, whereas the walls of arteries would be thick and resisting. But in the case of the lungs, this situation had to be reversed. Thus the vessel thought to be carrying venous blood to the lungs has thick, resisting walls and was called 'vena arteriosa' (or 'arterial vein'); the vessel carrying 'spiritous' blood to the lungs has thin, flexible walls and was called the 'arteria venosa' (or 'venous artery'). But Harvey's doctrine showed that all arteries are the same in texture and always carry blood away from the heart to the organs of the body. Hence the 'vena arteriosa' was misnamed; it is the pulmonary artery, carrying blood from the heart to the lungs. The veins are also all alike in texture and all carry blood from the organs of the body to the heart. As a result, the 'arteria venosa' became simply the 'pulmonary vein' (see Dalton 1884, 190).

8. There seems to be no reason to suppose that Columbus's discovery was not independent. But the discovery had been announced at least twice earlier: by Ibn al-Nafīs in the twelfth century and by Michael Servetus in the sixteenth. It is doubtful, however, that this idea had become generally known to the scientists and physicians of Columbus's day.

9. In Kenneth Franklin's translation of Harvey's *De Motu Cordis*, he has Harvey exclaim: "But, damme, there are no pores and it is not possible to show such" (p. 19). This may sound stronger than was intended by Harvey. Franklin used the archaism 'damme' as an English equivalent for Harvey's 'mehercule' —a difficult expression to translate. In Latin, this was a very mild expletive, not so strong as 'damn it' and perhaps closer to the British 'by George' (which has a linkage through St. George to Hercules). This is certainly not in the class with the statement by Harvey a few paragraphs earlier, when he wrote: "Good God! How do the mitral valves hinder the return of air, and not of blood?" Here

'Good God!' is an exact equivalent of Harvey's 'Deus bone!' Franklin's 'damme' is to be preferred to Chauncey D. Leake's 'damn it' and is even better than Robert Willis's 'in faith'.

10. Sir William Osler (1906, 40–41) discussed the practicality of Harvey's discovery. He claimed that in "a less conspicuous manner" Harvey's "triumph was on the same high plane" as the "truths . . . proclaimed by Copernicus, by Kepler, by Darwin, and others." In defense of Harvey's lack of practicality, Osler pointed out that "Newton's great work influenced neither the morals nor the manners of his age, nor was there any immediately tangible benefit that could be explained to the edification or appreciation of the 'ordinary man' of his day." Harvey, Copernicus, Kepler, Darwin, and Newton, he said, all "set forward at a bound the human mind." There "was nothing" in Harvey's work, he concluded, "which could be converted immediately into practical benefit, nothing that even the Sydenhams of his day could take hold of and use." For Osler, therefore, "the true merit of Harvey's work" lay "not so much really in the demonstration of the fact of the circulation as in the demonstration of the method."

11. The standard biography of Vesalius by C. D. O'Malley (1964) traces Vesalius's influence but does not altogether replace Moritz Roth's *Andreas Vesalius Bruxellensis* (1892). See also Singer and Rabin (1946) and Charles Singer's *A Short History of Anatomy* (London, 1929). O'Malley has written "A Review of Vesalian Literature," *History of Science* (1965, 4: 1–14). Especially valuable is J. J. Bylebyl's *Cardiovascular Physiology in the Sixteenth and Early Seventeenth Centuries* (1969).

Fundamental to understanding the enigmatic scientific career of Paracelsus is Walter Pagel's monograph (1958). The influence of Paracelsus has been studied in two works by Allen Debus (1965; 1977).

Two works of fundamental importance in understanding Harvey's life and work are Geoffrey Keynes's biography (1966) and Walter Pagel's study of William Harvey's biological ideas (1967), supplemented by his articles of 1969–1970. J. J. Bylebyl's articles are very illuminating as is his doctoral dissertation (1969). See also Gwenyth Whitteridge's *William Harvey and the Circulation of the Blood* (London, 1971). A good brief introduction is provided by Kenneth Keele's *William Harvey* (1965). The older books of John G. Curtis (1915) and John Call Dalton (1884) are still valuable. Especially recommended is Audrey Davis's "Some implications of the circulation theory for disease theory and treatment in the seventeenth century," *Journal of the History of Medicine and Allied Sciences*, 26 (1971): 28–39. See also *William Harvey and his age*, ed. J. J. Bylebyl (Baltimore, 1979); Robert J. Frank's *Harvey and The Oxford physiologists* (Berkeley, 1980). A collection of *Studies on William Harvey* (New York, 1981), ed. I. B. Cohen in an Arno Press Collection, makes available in a single volume the studies on Harvey and quantification by F. R. Jevons and F. G. Kilgour, and also Walter Pagel on "William Harvey and the Purpose of Circulation," Charles Webster on "William Harvey's Conception of the Heart as a Pump," William Hale-White's Bacon, Gilbert and Harvey," and H. D. Bayon on "William Harvey, Physician and Biologist."

12. *Transformations during the Enlightenment*

1. The events in question (Swift 1939, *1*: 75) occur after Peter's two younger brothers obtain a copy of their father's will; they discover that their rights as equal heirs have been disregarded by Peter, who has established himself as their father's sole heir. The brothers then break into the cellar to "get a little good *Drink*"; they "discard their Concubines and send for their Wives"; and they give the name of "Coxcomb" to "a Sollicitor from *Newgate,* desiring *Lord Peter* would please to procure a *Pardon* for a *Thief*" about to be hanged. Swift has a set of footnotes to explain that these refer to acts of the Reformation. Getting the will, which has been kept secret, signifies translating "the Scriptures into the vulgar Tongues"; getting a drink "to spirit and comfort their Hearts" means administering "the Cup to the Laity at the Communion"; and replacing concubines by wives refers to allowing "the Marriages of Priests." Each such step could be considered to be 'revolutionary', but to denote each of these as a 'revolution' indicates that Swift did not have in mind a single shattering event that effected a large-scale radical transformation.

2. During the next decades, Swift often used the term 'revolution' to denote radical political change and particularly the Glorious Revolution of 1688 (e.g., 1939, *1*: 227; *2*: 149, *3*: 46–47, 146–147, 163, *7*: 68–69, *8*: 92, *9*: 244, 230, *11*: 229), which he was apt to call "the late Revolution." He contrasted this revolution with the "rebellion" of "the Puritans" (*11*: 230). This theme was elaborated in an article in *The Examiner* of 28 December 1740 (*3*: 47). During the reign of Charles I, he wrote, there was a call "for a *thorow Reformation,* which ended at last in the Ruin of the Kingdom"; after James II ascended to the throne, "there was a restless Cry from Men of *the same Principles,* for a *thorow Revolution.*" He wrote also (p. 189) of those who defended "the Rebellion; and the Murder of King Charles I, which they asserted to altogether as justifiable as the late Revolution." In a "Letter to Mr Pope," Swift defined a revolution as "a violent change of Government" (*9*: 31), with direct reference to "the Prince of Orange."

3. Buffon was not the only author in whose writings Herder encountered the idea of revolutions occurring on the face of the earth. Another was the French civil engineer and geologist Nicholas Antoine Boulanger, author of *L'antiquité dévoilée* (1766). Herder (1887, *32*: 153) wrote a critique of Boulanger's central thesis, in which he held that Boulanger had relied more on history than on physics to prove "the old revolutions." Another writer on 'revolutions' of the earth between Buffon and Herder was August Ludwig von Schlözer, a professor at Göttingen, whose *Vorstellung seiner Universal-Historie* (1772) opened with a discussion of "die Revolutionen des Erdbodens" (the revolutions of the surface of the earth), in the style of Buffon (1772, 1, 4, 8, 10, 23, 350, etc.).

4. A clever witness to these years, Georg Christoph Lichtenberg, physicist and wit, discussed this question as follows: "I would like to know the ratio of the numbers expressing how often the word *Revolution* is uttered and printed in Europe in the 8 years from 1781 to 89 and in the 8 years from 1789 to 97." His reply (1801, 2: 253): "The ratio would hardly be less than 1:1,000,000."

5. A common expression in the late seventeenth and in the eighteenth century was 'revolution of empires'—as if the history of empires was a scroll unwound in the course of time. But this phrase seems to have been used in this context more for the succession of history than for particular dramatic events. An example is: "it is impossible to trace the progress of chemistry amid the revolutions of empire" (Fourcroy 1790, ch. 2, §2, p. 23).

In a dialogue between Molière and Paracelsus, Fontenelle refers to "certain revolutions" (1971, 365) in a way that seems to indicate that he probably had in mind the great revolutions of the wheel of time. In the passage in question ('spoken' by Molière), it is said that in "the order of the universe such as it is," there are "neither virtues of numbers, nor properties of planets, nor fatalities attached to certain times or to certain revolutions." Molière (ibid., 370) has the last word: "I know perfectly well what revolutions there can be in the Empire of Letters" ("quelles peuvent être les révolutions de l'Empire des Lettres"), but I nevertheless "guarantee the lasting quality of my plays." Here Fontenelle is paraphrasing one of the maxims of La Rochefoucauld; "Il y a une révolution générale qui change le goût des esprits, aussi-bien que les fortunes du monde" (There is a general revolution which changes the taste of minds, as well as the fortunes of the world).

13. Eighteenth-Century Conceptions of Scientific Revolution

1. We shall see a number of examples, from the eighteenth and nineteenth centuries, of a 'complete' revolution (for example, Playfair, Huxley), and we shall find one author (J. S. Mill) recognizing that a 'complete' revolution is no revolution at all.

2. R. N. Schwab points out (d'Alembert, 1963, 80, n.26) that in the revised edition of 1764, d'Alembert replaced "more difficult to perform" by "more essential than all those contributed."

3. Volumes 1–17, A-Z, were published from 1751 to 1765, followed by eleven volumes of plates (1762–1772). A four-volume *Supplément* and a volume of additional plates and a two-volume *Table générale* were published from 1776 to 1780. The fact that I may have found more mid-century references to the new meaning in French sources than in English ones may have no significance. I have not made an equally systematic search of the German or Italian writings.

4. For example, revolutions in science and in mathematics are mentioned in Jean Etienne Montucla, *History of Mathematics*, 2 vols. (Paris, 1758), and in the revised edition, 4 vols. (Paris, 1799); in MM. d'Alembert, l'Abbé Bossut, de la Lande, le Marquis de Condorcet, et al., *Dictionnaire encyclopédique des mathématiques*, vol. 1 (Paris, 1789); and in Charles Bossut, *Histoire générale des mathématiques, depuis leur origine jusqu'à l'année 1808*, 2 vols. (Paris, 1810), "Récapitulation succincte," p. 497.

5. At least once, in his *Lectures on History and General Policy* (new ed., 1826, 407), Priestley used the word 'revolution' in what may possibly be the cyclical sense: "Few observations remain to be made on the subject of science, as an

object of attention to an historian, after the account which has already been given of the progress and revolutions of it." Perhaps Priestley here means by 'revolutions' no more than great events.

6. As would be expected, Miller uses the word 'revolution' not only in his general summary but in his analysis or description of specific scientific advances, for example, "the signal revolution in chemical theory" (*1*: 92).

7. The fifth edition, 2 vols. (Paris, 1811), differs only slightly from the previous edition (Paris and Lyon, 1793). In the earlier edition, the opening paragraph is all but identical (only "Révolution périodique" being missing). The paragraph just quoted is likewise all but identical in the two editions, save for a few details and the fact that the final example, including "Révolution dans . . . les sciences" is not present in the early edition. Furthermore, where the 1811 edition lists the Roman, the Swedish, and the English revolutions as instances of "changemens mémorables et violens qui ont agité ces Pays," the 1793 edition referred rather to "un changement subit et violent dans le gouvernement d'un peuple," with examples being "La révolution Françoise de 1789," "Les révolutions d'Angleterre," and "Les révolutions Romaines."

14   *Lavoisier and the Chemical Revolution*

1. This extract is printed in full by Berthelot (1890, 48) as of 20 February 1772, translated by Meldrum (1930, 9); but Grimaux (1896, 104) assigns the date 20 February 1773, as does Guerlac.

2. As late as 1774, Lavoisier was still confused concerning which component of the air is responsible for the gain in weight when a substance such as sulphur or phosphorous undergoes combustion or a metal is calcined; but he had no doubt whatever that in combustion or calcination there is a gain in weight because there is a chemical combination with some component of the air (he tended to think this component was 'fixed air' or carbon dioxide, studied by Joseph Black). A good discussion of how Lavoisier hit upon oxygen as the active component of the air in these two processes, and the role of Priestley who reported his work on oxygen (he called it "dephlogisticated air") to Lavoisier in the autumn of 1774, is given by Henry Guerlac (1975, 83ff.).

3. An excellent introduction to the problems of the Chemical Revolution may be had in the writings of Henry Guerlac (1961; 1975; 1977). See also W. A. Smeaton's book on Fourcroy (1962) and the valuable papers by Andrew Meldrum (conveniently collected into a single volume, 1981). A very stimulating work is Archibald and Nan Clow's *The Chemical Revolution* (London, 1952).

15. *Kant's Alleged Copernican Revolution*

1. The discussions of this topic tend to leave out a fourth reference to Copernicus in Kant's *Critique of Pure Reason,* very likely because it is lacking in the index to Norman Kemp Smith's edition, which erroneously lists only the men-

tion of Copernicus in the preface to the second edition. Kant (1929, 273 = A 257, B 312–313) contrasts the expressions 'mundus sensibilis' and 'mundus intelligibilis'—making clear that the latter should not be rendered as "'an *intellectual* world', as is commonly done in German exposition."'"Observational astronomy," he writes, "which teaches merely the observation of the starry heavens, would give an account of the former." But "theoretical astronomy . . . as taught according to the Copernican system, or according to Newton's laws of gravitation, would give an account of the second, namely, of an intelligible world."

2. My own awareness of this problem goes back to 1959, when I was discussing Kant's *Critique* with my colleague, the late Philippe Le Corbeiller. He had been wondering whether the reference to Kant's 'Copernican revolution' occurred in the first or the second edition. We found that the relevant mention of Copernicus occurred in the second edition, but that even there Kant did not speak of a Copernican revolution.

3. A reply to Cross's article in *Mind* (1937, 214–217) is most instructive. Paton (1937) evidently considered himself to be a primary target of Cross's, since he had entitled a section of his two-volume commentary on the *Critique of Pure Reason* (1936), "The Copernican Revolution." Paton takes issue with a number of statements made by Cross about Kant, and particularly Kant's knowledge of Copernicus, to such a degree that the question of a Kantian 'Copernican revolution' becomes almost secondary. For example, he takes Cross to task for having said, in a footnote (p. 214, n.1), that "Kant does not appear to use the expression 'Copernican hypothesis.'" In strict fact, this expression was *not* used by Kant as such. Paton says that "Kant describes . . . what he regards as the Copernican hypothesis and, I will now make bold to add, the Copernican revolution." Hence he asks whether we are "justified in speaking of Kant's Copernican revolution," even though "it is true that Kant does not use the phrase" and that "the hypothesis of Copernicus . . . is not itself an example of the revolutions ascribed by Kant to mathematics and physics" (pp. 370–371).

Finally, Paton asks whether we are to "demand that Kant should actually have said in so many words 'my Copernican revolution.'" But the point is, surely, that Kant does not say there was a 'Copernican revolution', much less that he was making one. Surely we need not take seriously Paton's suggestion that 'Revolution der Denkart' and 'Umänderung der Denkart' are "equivalent . . . phrases"; any dictionary would indicate that 'Umänderung' means change, alteration, modification, and rearrangement—it is not as extreme an expression as 'Revolution'.

16. The Changing Language of Revolution in Germany

1. In the Grimms' dictionary, examples are given to show that the 'Umwälzung' (in the form 'Umbwalzung') was used in an astronomical/astrological sense for 'revolution'. Thurneysser, in his *Alchymia* of 1583, wrote of the contin-

ual 'Umbwalzung' of the eight spheres that carry the planets and stars around the earth, a usage that continues up to modern times. Instances are Gottsched in 1751 on the rotation of Venus and F.Th.v. Schubert in 1823 on the daily rotation of the earth. Like 'revolution', this word was also used for occurrences set in the succession of time: Pegius in 1570 on the 'Umbwalzungen' of the years and Thurneysser in 1583 on the 'Umbwalzung' of the months.

2. The case of Abraham Gotthelf Kästner is particularly interesting, since he eschewed the word 'revolution' even in its astronomical sense. A teacher of mathematics and physics at the University of Leipzig, later at Göttingen, during the latter part of the eighteenth century, Kästner produced a four-volume bibliographical record and "history of mathematics" from the "renewal of knowledge" up to "the end of the eighteenth century" (1796–1800). Kästner was so opposed to the use of the term 'revolution' that instead of discussing the 'revolutions' of the planets, he used the words 'Umwälzungen' and 'Umlaufszeiten' (4: 363–364, 379).

3. The Grimm brothers' dictionary gives examples of the scientific use of 'Umwälzung' to indicate a transformation of the earth (Oken), a disturbance or transformation during the diluvial period that suddenly destroyed vertebrates (von Humboldt), and the varied revolutions or complete transformations that concepts in theoretical physics have undergone (Boltzmann). There is also a reference to the nineteenth-century usage of 'Umwälzung' by one of the Grimm brothers in a statement that the way of handling Greek and Latin grammar must undergo a revolution.

17. The Industrial Revolution

1. E. J. Hobsbawm (1962, 48) has wisely remarked that "few intellectual refinements were necessary to make the Industrial Revolution," since "its technical inventions were exceedingly modest, and in no way beyond the scope of intelligent artisans experimenting in their workshops, or of the constructive capacities of carpenters, millwrights, and locksmiths." See, further, Schofield (1957; 1963), Gillispie (1957; 1980).

2. On this topic, see Fleming (1964), Schofield (1957; 1963), Gillispie (1957; 1980), Musson and Robinson (1969), and the collection of articles edited by Musson (1972).

3. Quoted by Anna Bezanson from C. L. and A. B. Berthollet, *Eléments de l'art de la teinture* (2nd ed., 1804), introd. Another example cited by Anna Bezanson (p. 348) is the discovery of sugar crystals in beet fibers, referred to in 1837 as "a revolution in the commercial relations between the two worlds"; to which the comment was made that it was a "genuine revolution."

4. In his *Principles of Political Economy* (1848), John Stuart Mill wrote of the "industrial revolution" in countries "whose resources were previously undeveloped."

5. Toynbee did not refer to any specific event of 1760 to justify his selection

of this date as the starting point of the revolution, although he said (ch. 4) that "in 1760 Carron iron was first manufactured in Scotland" and that "in 1760 the manufacturers of Lancashire began to use the fly-shuttle."

In an attempt to find out what significance the year 1760 might have for the development of technology, I turned to three world chronologies. S. H. Steinberg's *Historical Tables* (3d ed., 1949, 173) informed me that in 1760 Josiah Wedgwood established his famous "pottery works in Etruria, Staffs." Neville Williams's *Chronology of the Expanding World 1492–1762* (1960, 543) has no mention of Wedgwood's pottery, but informed me that in 1760 the "silk hat" was "invented in Florence," William Oliver first began to make "'Bath Oliver' biscuits," and John Smeaton devised "a cylindrical cast-iron bellows for smelting iron." Finally, Bernard Grun's *The Timetables of History* (1975, 351) repeated the information about Wedgwood. Surely the silk hat no more epitomized a revolution of industry than the biscuits, and I doubt whether Wedgwood's pottery (the only item to receive two votes) could possibly qualify as the major industry involved in the revolution. It is a curious fact that none of these chronologies even mentions innovations in the textile industry, which are stressed in most histories. This latter topic is presented in an admirable fashion in chapter 8 ("The Chronology of Invention") of Phyllis Deane's *The First Industrial Revolution* (1969) and in Samuel Lilley's chapter on "Technological Progress and the Industrial Revolution 1700–1914," in C. M. Cipolla, ed., *The Industrial Revolution* (1973).

6. Good summaries of these attempts to redate the Industrial Revolution are given in Deane (1962, ch. 1) and Lane (1978). T. S. Ashton in 1955 found that "after 1782 almost every available statistical series of industrial output reveals a sharp upward turn." W. W. Rostow concluded in 1960 that 1783–1802 was "the great watershed in the life of modern societies," the period of the "take-off into sustained growth" for the British economy. (The quotations from Ashton and Rostow are given in Deane 1969, 3.)

7. Among them, T. S. Ashton, *The Eighteenth Century* (1955); Carlo M. Cipolla, ed., *The Industrial Revolution* (1973; "The Fontana Economic History of Europe," 3); J. M. Clapham, *An Economic History of Modern Britain* (1939, 1); Phyllis Deane, *The First Industrial Revolution* (1969); R. M. Hartwell, ed., *The Causes of the Industrial Revolution in England* (1967); E. J. Hobsbawm, *Industry and Empire* (1968; 1969; "The Pelican Economic History of Britain"); David Landes, *The Unbound Prometheus* (1969); A. E. Musson, ed., *Science, Technology and Economic Growth in the Eighteenth Century* (1972); A. E. Musson and Eric Robinson, *Science and Technology in the Industrial Revolution* (1969); Eric Pawson, *The Early Industrial Revolution* (1979); L. S. Presnell, *Studies in the Industrial Revolution* (1960); Charles Singer et al., eds., *The Industrial Revolution* (1959; "A History of Technology," 4); Allan Thompson, *The Dynamics of the Industrial Revolution* (1973).

8. There are a number of excellent introductions to the Industrial Revolution, most of which have discussions of changing views on when the revolution occurred and on what it may have been. They include, among others, Ashton

(1948), Deane (1969), Landes (1969), Mathias (1969), and Pawson (1978), and the volume edited by Cipolla (1973). The classic work is Arnold Toynbee's posthumous book of 1884 based on a course of historical lectures delivered in 1881–1882 and (Clark 1953, 15) "printed partly from the author's manuscripts and partly from the imperfect notes of his undergraduate hearers," as part of a collection under the general title: *Lectures on the Industrial Revolution of the Eighteenth Century in England, Popular Addresses, Notes and Other Fragments*. A good summary, as of 1932, is given in Herbert Heaton's article, "Industrial Revolution," in vol. 8 of the *Encyclopaedia of the Social Sciences*. Hobsbawm (1968) provides a useful perspective (ch. 1–4) on the Industrial Revolution in the context of other revolutions and of European history; this may be supplemented by Hobsbawm (1962), ch. 2.

A key to the debates concerning science and technology and their relation to economic growth in the eighteenth century is available in the volume edited by A. E. Musson (1972). On the technological innovations, see vol. 4 (1959) of *A History of Technology*, edited by Charles Singer et al. The early use of the name 'Industrial Revolution' in France is documented in Bezanson (1921–2), and the later usage is analyzed in Clark (1953).

18. By Revolution or Evolution?

1. There were, of course, exceptions. In John Stuart Mill's *Principles of Political Economy* (1848), it is said that in a country whose resources are "underdeveloped for want of energy and ambition in the people," the "opening of foreign trade . . . sometimes works a complete industrial revolution" (1848, 2: 119). G. N. Clark (1953, 12) has commented shrewdly that "Mill was accustomed to weigh his words." When, therefore, he saw the phrase "a complete industrial revolution" on the printed page, this combination would "have appeared to him infelicitous," since the word 'complete' is strictly applicable only in the original or literal sense of the word, like the revolution of a planet or the turning of a wheel, a return to the starting point. Because in this instance Mill was "concerned with breaches of continuity, not with recurrent cycles," Clark concludes, "Mill crossed out the word 'complete' and in the later editions we read 'a sort of industrial revolution.'" Mill was, in this instance, sensitive to the Latin origins of the word 'revolution'; but as we saw in chapters 12 and 13, above, Fontenelle, Playfair, Huxley, and others wrote of a 'complete' or a 'total' revolution without paying any heed to the etymological implication that to 'come full circle' is to end up at the starting point.

2. In a very well researched and scholarly study of permanent revolution, Hartmut Tetsch traces the history of this concept, beginning with the French Revolution and the thoughts of Hegel and finding the first systematic elaboration in Trotsky. He notes that Alexis de Tocqueville wrote of "la révolution toujours la même" and used the words "révolution permanente" (1973, 220, n.354) for a trend in societies; he also cites Proudhon. He finds that the "real

theory" of permanent revolution begins with Marx and Engels. In Tetsch's analysis, Marx and Engels and their colleagues used this concept to stress that revolutionary action by the workers in any single country is only part of a world-wide movement and that "the revolution will be held in permanence until communism is realized as the last form of organization of the human family" (p.73). Tetsch summarizes Trotsky's views (developed in collaboration with Alexander Helphand) that even after the "socialist phase of revolution," society would remain somewhat divided into different groups that would clash and thus prevent the society from reaching a "final equilibrium." Revolutions, Trotsky held, would continue "in the economic system, in technology, in customs and family life, and in science." Tetsch, unfortunately, does not elaborate the nature of these revolutions in science, and I have not found any indication that this was a central theme for Trotsky, who believed that these "continual upheavals" would eventually result in "progress toward a more perfect Communist order." This conclusion indicates that at least one brand of Marxist thought both saw revolutions continuing in the ideal society and recognized that there would be further (and maybe even permanent) ongoing revolutions in technology and in science.

3. Volumes 1 and 2 of Merz's history are devoted to science, 3 and 4 to philosophy (and the social sciences). In volume 3 (on "Philosophical Thought"), Merz referred to science, the principles of science itself, and not to a "purely philosophical point of view" that was preparing what Merz unabashedly called "the great revolution regarding . . . the problem of knowledge" (p. 404), the neo-Kantian rise of the subject of epistemology.

4. The literature on science in the nineteenth century grows by leaps and bounds. Theodore Merz's classic volumes (1896–1903) are still unsurpassed, and may be supplemented by the writings of Walter F. (= Susan F.) Cannon, William Coleman, Frederic Gregory, Peter Harmon, and others.

19. The Darwinian Revolution

1. This sentence appears in all editions of the *Origin*, altered only (in the second and later editions) by the addition of a phrase, so as to have the sentence begin: "When the ideas advanced by me in this volume, and by Mr. Wallace, . . . "

2. Of course, despite extensive research over many years, and many discussions of this question with colleagues, I cannot claim to have examined *every* major published first presentation of a new scientific idea or method or theory.

3. Bentham later wrote (Darwin 1887, 2: 294) that as soon as he heard Darwin's paper, he "felt bound" to "defer" his own and to "entertain doubts of the subject" of "a fixity in species." (He later gave up, "however reluctantly," his "long cherished convictions, the results of much labour and study" and eventually came to a "full adoption of Mr. Darwin's views.")

4. In his autobiography Darwin (1887, *1*: 88) said that one element in the

success of his book was "its moderate size." This he owed to Wallace's essay. For, had he "published on the scale in which [he] began to write in 1856," the book would have become "four or five times as large as the 'Origin,' and very few would have had the patience to read it."

5. Wallace gave the following evaluation (29 May 1864) of Darwin's share in natural selection (Darwin 1903, 2: 36): "As to the theory of Natural Selection itself, I shall always maintain it to be actually yours and yours only. You had worked it out in details I had never thought of, years before I had a ray of light on the subject, and my paper would never have convinced anybody or been noticed as more than an ingenious speculation, whereas your book has revolutionised the study of Natural History." He added, "All the merit I claim is the having been the means of inducing you to write and publish at once."

After describing the events leading up to the 1858 joint publication, Wallace (1898, 141) said that "the theory of Natural Selection . . . received little attention till Darwin's great and epoch-making book appeared at the end of the following year." Wallace (1891) gave his own "estimates of Darwin's work" in an essay entitled "Debt of Science to Darwin."

6. Julian Huxley, in his classic *Evolution: The Modern Synthesis* (1974, 13–14) used the term 'Darwinism' for "that blend of induction and deduction which Darwin was the first to apply to the study of evolution"—a procedure which Huxley codified into "three observable facts of nature and two deductions from them." Mayr (1982, 479ff.) presents a different analysis, in which he sees Darwin's theory consisting of "three inferences based on five facts derived in part from population ecology and in part from phenomena of inheritance."

7. Huxley wrote, "There cannot be a doubt that the method of inquiry which Mr. Darwin has adopted is not only rigorously in accordance with the canons of scientific logic, but that it is the only adequate method."

8. Today there is a mammoth Darwin research industry, comparable only to the Newton industry. An edition of Darwin's correspondence is in progress, but there is as of now no program for publishing the collected works in a uniform scholarly edition. The first edition of *The Origin of Species* (1859) has been published in facsimile (1964), with an introduction by Ernst Mayr and an upgraded index. The early *Sketch* of 1842, the *Essay* of 1844, and the Darwin–Wallace papers of 1858 have been reprinted (1958). A mine of scholarly information is available in the five volumes of Darwin's letters (1887; 1903).

On Darwin's ideas and their development see Ernst Mayr's *The Growth of Biological Thought* (1982) and his essays, collected in *Evolution and the Diversity of Life* (1976). Highly recommended are the volumes by Michael Ghiselin (1969) and Sir Gavin de Beer (1965). Julian Huxley's *Evolution: The Modern Synthesis* (1963; 1974) is an updated classic; still valuable is Edward Poulton's volume of 1896. On the precursors, see *Forerunners of Darwin* (1959), ed. Bentley Glass, Owsei Temkin, and William L. Strauss, Jr. On the background of Darwin's ideas, see Howard Gruber's *Darwin on Man* (1974) and the studies by S. S. Schweber. For the controversies over Darwinian ideas and for the Darwinian revolution, see the books by Michael Ruse (1979; 1982), D. R. Oldroyd (1980), and

David Hull (1973). The later "synthesis" and the history of Darwinian ideas in various countries are presented in the 1980 volume edited by Ernst Mayr and William Provine; the latter's *Origins of Population Genetics* (1971) is exceptionally valuable. Highly recommended is M. J. S. Rudwick's *The Meaning of Fossils* (1972). Michael Scriven's paper (1959) is a major contribution to the topic of Darwinian evolution and scientific prediction. On Darwin's conversion to evolution, see Frank Sulloway's "Darwin's Conversion: The *Beagle* Voyage and Its Aftermath" (1982).

20. *Faraday, Maxwell, and Hertz*

1. These are: "On Faraday's Lines of Force" (1855–1856), "On Physical Lines of Force" (1861–1862), "Elementary Relations of Electrical Quantities" (with H. C. F. Jenkin, 1863), "Dynamical Theory of the Electromagnetic Field" (submitted 1864, published 1865), "Note on the Electromagnetic Theory of Light" (1868). C. W. F. Everitt points out that the third paper (with H. C. F. Jenkin, 1863) "is nearly always overlooked because it is not in the *Scientific Papers*, but is where Maxwell develops the dimensional analysis of the problem and also introduces the term 'field' in something approaching its modern usage."

2. With respect to the influence of Thomson on Maxwell, C. W. F. Everitt informs me that he has "recently come to a slightly clearer understanding of Thomson's role in this transition, namely that he arrived at his treatment of magneto-optical rotation by considering the precession of the Foucault pendulum (invented 1851). See Thomson's paper, *Proc. Roy. Soc.* 1856, vol. 8, pp. 150–158."

3. While Hertz's presentation of Maxwell's equations was of fundamental importance in making the theory more accessible on the Continent, he cannot be given priority for the simplification of the equations, which, C. W. F. Everitt points out, originates with Oliver Heaviside, as may be seen in the Hertz–Heaviside correspondence of 1888–1889.

4. In 1893 J. J. Thomson published what he "intended as a sequel to Professor Clerk-Maxwell's Treatise on Electricity and Magnetism." On page 1 he said that "one of the chief reasons why his [Maxwell's] views did not sooner meet with the general acceptance they have since received" was the "descriptive hypothesis, that of displacement in a dielectric, used by Maxwell to illustrate his mathematical theory." (On this topic, see Duhem 1902, 8.)

5. Helmholtz's lecture was not published until 1907 (see 1907, 2). As we have seen, Hertz's experiments were expected not only to demonstrate the existence of electromagnetic waves having the speed of light, as Maxwell had predicted, but to decide between Maxwell's own formulation and Helmholtz's revision of it. Hertz's first 'experimentum crucis' seemed to disprove *both* Maxwell's theory and Helmholtz's, although later experiments reversed this decision and favored Maxwell. Perhaps this was the crisis that either precipitated or was precipitated by the revolution.

6. A concise description and evaluation of Faraday's ideas and Thomson's, and their relation to Maxwell's, is given in C. W. F. Everitt's article on Maxwell in the *Dictionary of Scientific Biography* (1974, 9: 205–207), reprinted in book form (1975). On Faraday, see L. Pearce William's biography (1965) and also the classic account of Faraday's science by Silvanus P. Thompson (1901). A general view of the main features and the significance of the Maxwellian revolution is given by Peter Harman (1982), which may be supplemented by John T. Merz's *History* (1896–1903). Maxwell's *Scientific Papers,* an oft-reprinted two-volume set, omits some twenty articles. There are three editions of his *Treatise on Electricity and Magnetism* (1873, 1881, 1891); Maxwell himself only partly completed the revision of the second edition. On Maxwell's ideas, R. T. Glazebrook's *James Clerk Maxwell and Modern Physics* (1896) may still be recommended, along with Henri Poincaré's nonmathematical essay (1904) and Edmond Bauer's survey of *L'électromagnétisme hier et aujourd'hui* (Paris, 1949). On a simpler level, there are D. K. C. MacDonald's *Faraday, Maxwell, and Kelvin* (Garden City, 1964) and R. A. R. Tricker's *The Contributions of Faraday & Maxwell to Electrical Science* (Oxford 1966).

On field theory and the ether, see G. N. Cantor and M. J. S. Hodge, eds., *Conceptions of Ether: Studies in the History of Ether Theories 1740–1900* (New York, 1981), and E. T. Whittaker, *A History of the Theories of Aether and Electricity: The Classical Theories* (Edinburgh, 1951). There are important studies of the development of Maxwell's ideas by J. Z. Buchwald, P. M. Heimann, M. Norton Wise, and others; these are listed on pp. 166–171 of Peter Harmon's *Energy, Force and Matter* (1982). An incisive study is Peter Galison's "Re-reading the Past from the End of Physics: Maxwell's Equations in Retrospect" (1983).

21. Some Other Scientific Developments

1. In this chapter I have discussed only some selected examples of revolutionary nineteenth-century science. Although all other revolutions in biology pale on comparison with the Darwinian revolution, there were others which should be examined to see whether they pass the tests given in chapter 3. In addition to those mentioned here, consideration should be given, among others, to the cell theory, the developments in mammalian embryology, chemical and physical physiology, and paleontology. Subjects of this kind in chemistry would include the atomic molecular theory, the concept of valence and isomers, the rise of organic chemistry, and chemical dynamics and thermodynamics. In physics, there are similar topics other than the Faraday-Maxwell-Hertz concepts of electromagnetism, including the laws of electric currents and of electromagnetic forces, the concept of energy and the laws of thermodynamics, the wave theory of light, spectroscopy (and its applications to chemistry and astrophysics), molecular theory and the general theory of gases, statistical mechanics, and applied physics (notably the electrical industry). And there are other areas in astronomy and in earth sciences. William Coleman's introductory book (1977)

on biology in the nineteenth century finds a parallel in Peter Harmon's work (1982) on physics. We need studies in these areas that will explore the degrees to which they pass the tests for revolutions in science. Michael Heidelberger has discussed one such topic in his analysis of the revolutionary aspects of Ohm's law (1980, 104).

2. In an extensive reply to his critics, Wilson (1980, 202) again asserted that "the profound change in geology between 1822 and 1841" was "brought about almost entirely by Lyell," and that this change "involved such a complete and fundamental reinterpretation of the meaning of geological phenomena that it constituted a revolution in the science."

3. I have omitted from this discussion any consideration of the internal development of mathematical subjects of statistics and probability, and of whether they might possibly constitute a revolution of that subject, on the grounds that this book is devoted almost exclusively to the natural and physical sciences. But the internal history of probability does show major revolutions by the end of the nineteenth century and early years of the twentieth and again in the mid-twentieth century (as Ian Hacking has noted, 1983).

22. *Three French Views*

1. Trained in mathematics, physical science, and engineering, rather than in the classics, Comte concocted the new term 'sociology' by a barbarous combination of Latin and Greek roots.

2. Comte's further comments on this subject indicate the degree to which his views on development and classification of science and social science were linked with his positivism. In the study of "moral phenomena" (Fletcher 1974, 191), he wrote that:

> the positive method is admitted on all hands to be that alone admissible. Everyone recognises that the only legitimate aim is the combination of the anatomical with the physiological point of view. Theology and metaphysics are, by common consent, eliminated from the question; at least they never play any important part, and, whatever may be the final result of the discussion, it can only diminish their influence. In a word, these discussions being confined within the domain of science, philosophy has no further concern with them.
>
> I have particularly insisted on this last philosophical fact for two reasons. First, because it has hitherto been hardly observed and, not unfrequently, even disputed. Secondly because it furnishes to everyone who has rightly comprehended my classification of the sciences, at once a new proof, indirect indeed but unanswerable, and a clear *résumé* of the entire intellectual transformation.

3. It must be borne in mind, in this context, that Comte is endlessly repetitive and prolix, repeating himself almost word for word (but with later embel-

lishments) in work after work, so that extensive reading of Comte produces a continual sense of *déjà vu*. (On Comte's style, see John Stuart Mill's essay.) Furthermore, there are no adequate indexes to help locate Comte's views on particular topics such as revolution in science. No doubt, the prolixity and difficult style must rank high among the reasons why Comte is not more universally appreciated.

4. On Saint-Simon see Frank Manuel's two books: *The New World of Henri Saint-Simon* (1956) and *The Prophets of Paris* (1962). A volume of selections in translation has been edited by Felix Markham: Henri de Saint-Simon, *Social Organization* . . . (1964 [1952]).

Most studies of Comte lean heavily on his contributions to sociology. Of these the chapters on Comte in vol. 1 of Ronald Fletcher's *The Making of Sociology* (1971) and in vol. 1 of Raymond Aron's *Main Currents in Sociological Thought* (1965) are excellent. Major selections, plus commentaries, have been made by Stanislav Andreski (1974), Ronald Fletcher (1974), Gertrud Lenzer (1975), and Kenneth Thompson (1975). A brilliant study of positivist thought has been written by Leszek Kolakowski (1968). On Comte's philosophy, there is still nothing better than L. Lévy-Bruhl's *The Philosophy of Auguste Comte* (New York, 1903), E. Littré's *Auguste Comte et la philosophie positive* (Paris, 1863), and J. S. Mill's *Auguste Comte and Positivism* (London, 1865). Also, George Sarton on "Auguste Comte, Historian of Science," *Osiris*, 10 (1952): 328–357.

Cournot's major work, *On the Foundations of Our Knowledge*, translated with an introduction by Merritt H. Moore (1956), contains a valuable bibliography.

23. The Influence of Marx and Engels

1. It is possible that there are ideas about the development of mathematics, and even revolutions in mathematics, in Marx's largely unpublished mathematical papers. (This was suggested to me, in a personal communication, by Tom Bottomore.) In his speech at the graveside of Marx, Engels alleged that Marx "made independent discoveries" in "mathematics" (Marx and Engels 1962, 2: 168).

2. I may add that I have searched in vain in the available writings not only of Marx but also of scholars who have written on Marx and science (including J. D. Bernal, N. Bukharin, T. Carver, J. Diner-Dienes, D. Lecourt, H. and S. Rose, N. Rosenberg, A. Schmidt, J. Zeleny) and I have corresponded on this topic with S. Avineri, T. Bottomore, and D. McLellan.

3. But Marx did say that Darwin "developed" the subject "in a crude English way."

4. This law was "the simple fact, hitherto concealed by an overgrowth of ideology, that mankind must first of all eat, drink, have shelter and clothing, before it can pursue politics, science, art, religion, etc; that therefore the production of the immediate material means of subsistence and consequently the degree of economic development attained by a given people or during a given

epoch form the foundation upon which the state institutions, the legal conceptions, art, and even the ideas on religion, of the people concerned have been evolved, and in the light of which they must, therefore, be explained, instead of *vice versa*, as had hitherto been the case" (Marx and Engels 1962, 2: 167).

In this speech at Marx's graveside, Engels (ibid., 167–168) described Marx's "two . . . discoveries": "the law and development of human history" and the theory "of surplus value." He then referred to the "very many fields" in which Marx "made independent discoveries," even "in . . . mathematics.""Such," he said, "was the man of science." The implication is clearly that Marx's science was mainly (if not entirely) what we would call social science.

5. Aveling not only compared Marx and Darwin but drew three contrasts (pp. ix–x)—which put "the advantage on the side of the economic philosopher." Marx, unlike Darwin, "was not only a philosopher, but a man of action." Marx, again unlike Darwin, had "a huge sense of humour, and a singularly brilliant style, even in dealing with abstruse problems." Darwin, finally, was "a man given up to biological, or, at most, to scientific work, in the restricted sense of the term." But Marx was "master, in the fullest sense, not only of his special subject, but of all branches of science, of seven or eight different languages, of the literature of Europe."

6. Engels was antagonized by Dühring's brand of "elaborate socialist theory" and "complete practical plan for the reorganisation of society" and Dühring's direct attacks on Marx (1935, 7).

7. These are *Kursus der Philosophie* (1875), *Kursus der National-und Sozial-ökonomie* (1876), *Kritische Geschichte der National-ökonomie und des Sozialismus* (1875).

Engels later rewrote a portion of *Anti-Dühring* and published it as a tract entitled *Socialism, Utopian and Scientific*, first published in French (1880). For the English edition in 1892, Engels wrote an introduction which he translated into German, in which he described Dühring's "attempt at a complete revolution," rendered into German as a "Versuch einer kompleten 'Umwälzung der Wissenschaft.'"

8. For Engels' theories of scientific development, the factors he considered to be of importance in analyzing the history of science, see the article on Engels in the *Dictionary of Scientific Biography* (R. S. Cohen 1978, 135ff.).

9. Among many studies on Marx, the ones most related to this chapter include Shlomo Avineri's *The Social and Political Thought of Karl Marx* (1968), Isaiah Berlin's classic *Karl Marx* (4th ed., Oxford, 1978), vol. 1 of Leszek Kolakowski's *Main Currents of Marxism* (Oxford, 1978), and especially Alfred Schmidt's *The Concept of Nature in Marx* (1971). Much of the literature on Marx (as on Engels) does not make a distinction between natural science (the physical and biological sciences) and social science (economics). Engels's ideas on revolution and science are to be found chiefly in his *Anti-Dühring* and *Dialectics of Nature*. Robert S. Cohen has written articles on Marx and on Engels for the supplement to the *Dictionary of Scientific Biography;* the article on Engels stresses his theories of scientific development and the factors he considered of importance in analyzing the history of science.

24. The Freudian Revolution

1. There are two very different modes of criticizing or evaluating psychoanalysis. One is to gauge its relative success and failure as a treatment of mental disorders, to determine its clinical value as a therapeutic procedure. The other is to examine the methodology and the logical structure of the theory, to put psychoanalysis to a philosophical test. I would interpret the large and ever-growing literature of this second kind as a continuing proof of the revolutionary quality of psychoanalysis.

2. The primary critic of Freud's seduction theory is Jeffrey M. Masson, quondam Project Director of the Freud Archives and editor-designate of Freud's unpublished papers. Masson bases his revisionist position to some degree on a careful reading of hitherto unpublished manuscripts. He holds that Freud made a serious error, one that endangers the very foundation of psychoanalysis, when he changed his mind about the seduction of young girls by older members of their families.

Masson's attempt to build a new foundation for psychoanalysis has not met with favorable response, even though he has established that Freud had a deeper knowledge of child abuse than his writings would indicate. What seems significant is not so much whether some of Freud's patients (and others) may have actually experienced attempted seduction as children, but rather that their lives have been affected by conscious or unconscious fantasies about real or imagined seduction. Hence Masson's campaign is not likely to make much of a dent on psychoanalytic theory or treatment. Masson has described his research and findings in a book entitled *The Assault on Truth: Freud's Suppression of the Seduction Theory* (New York, 1984). An acerbic account of Masson's ideas and activities has been published by Janet Malcolm in two articles in *The New Yorker* ("Annals of Scholarship: Trouble in the Archives": 5, 12 December 1983): these have been incorporated into her book *In the Freud Archives* (New York, 1984).

3. On the history and prehistory of *das Es* and *das Ich* see Freud 1953, *19*: 7–8, 23.

4. The influence of Freudian ideas on academic psychology, notably personality theory, can be seen by examining the literature, for example, Henry Murray's *Explorations in Personality* (1938); Gardner Lindzey and Calvin Hall's *Theories of Personality* (1957); the volume on *Personality in Nature, Society and Culture* (1948, rev. 1967), ed. Clyde Kluckholm and Henry A. Murray; *The Study of Lives* (1963), ed. Robert. W. White; and Robert M. Liebert and Michael D. Spiegler's *Personality: An Introduction to Theory and Research* (1970); see also the works of such authors as Erik Erikson, Kenneth Keniston, David McClelland, Robert White, and Frederick Wyatt.

5. The significance of Freud's self-analysis was apparently first put forth during Freud's lifetime by Fritz Wittels and has been mentioned by almost all of his biographers (including Jones 1953, *1*: 320, 325). Reuben Fine (1963, 31) wrote of this episode as "revolutionary," precipitating "the decisive change in his

interest from neurology to psychology" and thereby creating "a whole new science, psychoanalysis."

6. Freud's conclusion that "men in the mass behaved to psycho-analysis in precisely the same way as individual neurotics under treatment for their disorders" led to a startling result. Men and women under treatment could, "by patient work," be convinced "that everything happened as we maintained it did: we had not invented it ourselves but had arrived at it from a study of other neurotics covering a period of twenty or thirty years. The position was at once alarming and consoling: alarming because it was no small thing to have the whole human race as one's patient, and consoling because after all everything was taking place as the hypotheses of psycho-analysis declared that it was bound to" (1955, *19*: 221).

7. The purpose of this essay was to analyze the factors that made it difficult for people to understand and accept the main tenets of psychoanalysis. Freud was thinking less of "an intellectual difficulty" than "an affective one," but he shrewdly observed that the two "amount to the same thing in the end": "Where sympathy is lacking, understanding will not come very easily." In particular, Freud explained and defended the "theory [that] has at last shaped itself in psycho-analysis, and . . . is known as the 'libido theory.'" The libido, he wrote, is the "force by which the sexual instinct is represented in the mind"; this "force . . . we call 'libido'—sexual desire—and we regard it as something analogous to hunger, the will to power, and so on, where the ego-instincts are concerned." Freud also discussed the criticism leveled against psychoanalysis of a "one-sidedness in our estimate of the sexual instincts." The "unintelligent opposition" had argued that "human beings have other interests besides sexual ones." Freud explained that "our one-sidedness" was to be likened to "that of the chemist, who traces all compounds back to the force of chemical attraction." But such a chemist, Freud argued, "is not on that account denying the force of gravity," but "leaves that to the physicist to deal with."

8. Walker's statement that Freud "once referred to himself as the Copernicus of the mind, and on another occasion compared himself to Darwin" is replete with errors. Freud never compared himself to Copernicus and Darwin, but compared psychoanalysis to their theories—and not once but several times. Furthermore Freud never used the phrase "Copernicus of the mind." Perhaps Walker misremembered Ernest Jones's evaluation of Freud (1913, p. xii) as the "Darwin of the mind."

9. For example, in a rejoinder to Walker, Jones said that "nothing would ["could" in the original version in *The Listener*] have been more unlike Freud "than to compare himself overtly with Copernicus or Darwin"; see Nelson 1957 (Jones's riposte appears in Nelson's book.)

10. I am puzzled by this statement. Jones, after all, invented and continued to use the phrase "Darwin of the mind."

11. The easiest entry into the essentials of Freudian science is a general article on "Psycho-analysis" written by Freud for the 13th edition of the *Encyclopaedia Britannica* (1926), reprinted in vol. 20 of *The Standard Edition of the Complete Psychological Works* (24 vols., 1953–1974). Freud's *An Autobiographical Study*

(1952; 1953, 22: 7–74) should be supplemented by Ernest Jones's three-volume biography (1953–1957). Especially recommended is Robert Holt's article on Freud (1968) in the *International Encyclopaedia of the Social Sciences,* a rewrite of a chapter in Robert White's collection, *The Study of Lives* (1966). A valuable study of the stages of the revolution, more general than the title indicates, is *The Influence of Freud on American Psychology* (1964) by David Shakow and David Rapaport. An extremely challenging work, especially for understanding how the Freudian revolution was related to the Darwinian revolution, is Frank Sulloway's *Freud: Biologist of the Mind* (1979). Two important collections of essays are Benjamin Nelson's *Freud and the 20th Century* (1957) and Md. Mujeeb-ur-Rahman's *The Freudian Paradigm* (1977); both volumes contain Alfred Kazin's essay on "The Freudian Revolution Analyzed"; the second of these volumes also contains a portion of the Shakow and Rapaport book dealing with Darwin and Freud. On the prehistory of the subject, see José M. Lopez Piñero's *Orígines históricos del concepto de neurosis* (1963) and Henri F. Ellengerger's *The Discovery of the Unconscious* (1970), which also carries the subject into post-Freudian times. Especially helpful is Paul F. Cranefield's article on "Some Problems in Writing the History of Psychoanalysis," pp. 41–55 of *Psychiatry and Its History: Methodological Problems in Research,* ed. George Moria and Jeanne L. Brand (Springfield, Ill., 1970); the whole volume is of great interest. A very complete bibliography, encompassing the major historical primary and secondary sources is given in Sulloway's *Freud: Biologist of the Mind.*

Freud himself wrote several short introductions to his subject, always referring to his predecessors. The article mentioned above, written for the *Encyclopaedia Britannica,* filled a gap because there was no article on psychoanalysis in either the 11th edition (1910–1911) or the 12th edition (1922, a reprint of the 11th edition with three "New Volumes"). The 13th edition (1926) was yet another reprint with three "New Supplementary Volumes"; the latter contains Freud's article, which was printed without alteration in the 14th edition (1929). Freud wrote another general article for a two-volume compilation (1924) published by the Encyclopaedia Britannica Publishing Co. under the title *These Eventful Years: The Twentieth Century in the Making, as Told by Many of Its Makers.* Freud's article (vol. 2, ch. 73, pp. 511–523) was originally entitled (in the German version) "A Short Account of Psychoanalysis," but when published in English it was headed "Psychoanalysis: Exploring the Hidden Recesses of the Mind." This article is sometimes confused with the one in the *Encyclopaedia.* The *Encyclopaedia* article was entitled by Freud "Psychoanalysis," but it was published under the title "Psychoanalysis: Freudian School" (and — according to the editorial note in Freud 1953, *20*: 262 — "an uncomplimentary reference" to Jung and Adler was "expunged"). Freud wrote yet another general article on psychoanalysis for an encyclopedia in 1922 (Freud 1953, *18*: 235–254) paired with one on "The Libido Theory" (pp. 255–259). All of these are interesting to read and are first-class introductions to the subject. Of the greatest value is Freud's own history, written in 1914, and entitled "On the History of the Psycho-Analytic Movement" (1953, *14*: 7–66).

Three other significant works are *Studies on Hysteria* by Josef Breuer and

Sigmund Freud (1957), *The Origins of Psycho-Analysis: Letters to Wilhelm Fliess, Drafts and Notes, 1887–1902* by Sigmund Freud (1954), and *The Freud/Jung Letters* (1974).

25. *The Scientists Speak*

1. In *Materialism and Empirio-Criticism,* Lenin was concerned not only with writers about dialectical materialism but with some recent trends in "idealist" philosophy, typified by the ideas of Ernst Haeckel and Ernst Mach (ch. 5, 6). Mach was the "most popular representative" of the new philosophy and hence Lenin's primary target (intro., 13–14). The title of Lenin's chapter on "The Recent Revolution in Natural Science and Philosophical Idealism" was adapted from a German article in *Die Neue Zeit* (1906–07) by Joseph Diner-Dénes, entitled "Marxism and the Recent Revolution in the Natural Sciences." In his discussion, Lenin refers to "the 'new' physics," primarily "the epistemological conclusions" to be drawn from "the recent discoveries in science, and especially in physics (x-rays, Becquerel rays [radioactivity], radium, etc.)." Lenin, however, did not develop a theory of revolution in science, nor was the theme of revolution in science a conspicuous feature of his other writings on revolution and its theory and practice, for instance, his *The State and Revolution.*

2. This issue was raised again in the famous debate on behaviorism of the late 1920s, in which John B. Watson and William McDougall used intemperate language that would be inconceivable in science today. For example, McDougall (1928, 44–45) began by "confessing" that he had an "unfair" advantage, since "all persons of common-sense will of necessity be on my side." But Watson, he claimed, had his own advantages: there was "a considerable number of persons," according to McDougall, who "will inevitably be on Dr. Watson's side." The reason, he said, was that they are "so constituted that they are attracted by whatever is bizarre, paradoxical, preposterous, and outrageous, whatever is 'agin the government,' whatever is unorthodox and opposed to accepted principles." Not only did McDougall claim that there was a correspondence between support of the new 'behavioristic' psychology and being opposed to the government; he carried the association of political radicalism with scientific radicalism a step further by asserting that "Dr. Watson's views are attractive to those . . . who are born Bolshevists."

3. Eddington had developed the theme of revolution more fully in his Romanes Lecture of 1922 on *The Theory of Relativity and Its Influence on Scientific Thought.* This work opens with a presentation of the Copernican "revolution [which] consisted in changing the view-point from which the phenomena were regarded" (1922, 3). This was, he said, "one of the great revolutions of scientific thought." "It has been left to Einstein," he then declared, "to carry forward the revolution begun by Copernicus" (pp. 1–2, 11, 26–27, 30–31). Invoking the idea that science advances by a series of revolutions, he referred to Einstein's two theories of relativity (the special and the general) as a "present revolution of

scientific thought," which "follows in natural sequence on the great revolutions at earlier epochs in the history of science."

4. Early in the century, during a discussion of the problem of 'whole-time' clinical professors, William Osler (1911, 5) said that advances in medical practice could "be done by practitioners," if they were "thoroughly trained" in research "methods." He saw no reason why "the most revolutionary researches of modern times" had to come from "private laboratories." We are here less interested in the argument concerning possible advances being made in medical schools and hospitals than in Osler's simple statement about "revolutionary researches."

5. This example was brought to my attention by Victor Hilts at the History of Science Society meeting in October 1983.

6. This symposium was triggered by Garland Allen's book, *Life Science in the Twentieth Century* (1978). A discussion was held in New York in 1979 at a meeting of the History of Science Society. The papers (by Jane Maienschein, Keith R. Benson, Ronald Rainger, Garland Allen, and Frederick B. Churchill) were published in 1978 in the *Journal of the History of Biology*.

7. This aspect of the problem of revolutions was suggested to me during a discussion of the subject with Peter Galison.

26. The Historians Speak

1. That is, if the 'pre-conditions' are realized, then a revolution may occur; and it may succeed (as did the Russian Revolution of 1917) or it may fail (as did the 1905 Revolution in Russia). But a 'postponed' revolution implies the concept of a revolution that simply did not occur when it should have, and so is a purely fictive or supposititious event. Possibly Butterfield's expression was intended to carry the implication that there should have been a revolution in chemistry at the time of the Scientific Revolution, the era of Galileo, Kepler, Descartes, Harvey, and Newton. (For a critique of Butterfield's concept, see Thackray 1970, viii.)

2. Guerlac (1977, 63) has recorded that Koyré once told him "that his reading of Burtt's remarkable book . . . played an essential role" in leading him "back to what he later called his 'first love,' the history of science."

3. Koyré pointed out (1939, 1; 1978, 39) that "given the scientific revolution of the last ten years it seems preferable to reserve for this revolution the word 'modern' and to label prequantum physics as 'classical.'"

To keep the chronological record straight, it should be observed that this first part of the *Galilean Studies* entitled "A l'aube de la science classique (La jeunesse de Galilée)" — "The Dawn of Classical Science (The Youth of Galileo)" — had been partially published in the *Annales de l'Université de Paris* in 1935.

4. In 1954 Rupert Hall published the first book that I know of in which the Scientific Revolution is so prominent a theme as to be part of the title.

5. In Guerlac's volume of essays (1977), comprising works written and pub-

lished in the post-Butterfield era, the index contains eleven seperate entries for Scientific Revolution.

6. Koyré often referred to this work as his "ill-fated" book. Not only was its influence postponed for a decade as a result of the war, but three attempted translations into English failed to meet the standards of the author and his colleagues. Only in 1978, fourteen years after Koyré's death, did an English version finally appear.

7. Kuhn (1957, p. xiii) graciously acknowledges that his "own understanding of the transformation of early modern science" was "greatly influenced by the writings of Alexandre Koyré," as was the case for Herbert Butterfield.

27. Relativity and Quantum Theory

1. Because there are two theories of relativity, there are essentially two revolutions. Thus, we have seen that Leopold Infeld (1950) devoted separate chapters of his book to "The First Einstein Revolution" and "The Second Einstein Revolution." More recently, Karl Popper (in a lecture given in Washington in 1979) advanced the thesis that in addition to Einstein's "revolutions in the fields of physics and of cosmology," there were "two Einsteinian revolutions in the field of epistemology: an early one, dating from 1905, which is very famous, and which had a tremendous impact; and a late one, which he formulated about twenty years later, and which has been widely ignored."

2. I am not concerned here with the origins of the theory or with Einstein's predecessors and contemporaries for whom priority has been claimed (for example, Lorentz and Poincaré) for special relativity. On this topic, see the encyclopedic account of the theory in Miller (1981), supplemented by Pais (1982), and the convenient summary by Bergia in French (1979, 77–82).

3. Stark, the first person to request an article on relativity from Einstein, later became a leading figure in the movement to purge 'German physics' of an 'un-German' predilection for theory over experiment. In the Nazi period he explained that the reason "so many German physicists accepted the relativity theory even though it was repugnant to the German spirit" was owing to the "circumstance that so many physicists had Jewish wives" (Frank 1947, 238).

4. Today Einstein's fundamental revisions of physical thought have become so universally accepted that it is difficult to imagine how radical they once were. P. W. Bridgman (1946, 1, 2, 4) stressed that Einstein's "greatest contribution" was that he "modified our view of what the concepts useful in physics are and should be." Wholly apart from our opinion about the "details of Einstein's restricted and general theories of relativity," he wrote, "there can be no doubt that through these theories physics is permanently changed." Bridgman hoped that a correct philosophical posture would have the result that "another change in our attitude, such as that due to Einstein, shall be forever impossible."

5. In 1938 it was shown that Bucherer's results were faulty, but that Einstein's predictions had been correct. Two years after Hupka's experiments,

when Einstein was proposed for a Nobel Prize in 1912, the letter of recommendation said that such "experiments on cathode and β rays" did not have "decisive power of proof" (Pais 1982, 159).

6. The judgment of physicists is that the "physical theory" proposed by Lorentz to account for the equations of the 'Lorentz transformation' was "very ingenious," but "not quite satisfactory" (Kronig 1954, 331). It was "too much in the nature of a hypothesis *ad hoc.*" What Einstein did was "to clarify the problem entirely, by a thorough criticism of the traditional conceptions of space and time."

It may be noted that around 1910, when von Laue made this statement, there did exist tests that would have enabled a distinction to be made between Einstein's and Lorentz's theories (see Miller 1981, ch. 7).

7. In his Washington lecture in 1979 Karl Popper stressed the fact that in 1905 Einstein had made an epistemological revolution that had an enormous influence on physics, the epistemology of physics, and "on almost all the other sciences." He saw that this revolution "was responsible for the intrusion of the observer into physics, and years later, more especially, for the intrusion of the observer into quantum mechanics." This thesis is also stated in Popper 1976, 96–87.

8. It is not relevant to my purpose here to choose between two conflicting viewpoints concerning the events of 1900/01. One, expressed by Martin Klein (1962, 459) asserts: "On December 14, 1900, Max Planck presented his derivation of the distribution law for black-body radiation to the German Physical Society, and the concept of energy quanta made its first appearance in physics." The other is stated by Thomas Kuhn (1978, 126) as follows: "My point is not that Planck doubted the reality of quantization or that he regarded it as a formality to be eliminated during the further development of his theory. Rather, I am claiming that the concept of restricted resonator energy played no role in his thought until after the *Lectures* (Planck [1906]) were written." For an insightful appraisal, see Galison (1981).

9. It is a remarkable fact of history that Einstein published three great papers in the single year 1905 in the same journal, *Annalen der Physik.* The third was on molecular physics (the so-called Brownian motion).

10. The best discussion of Millikan's research on this topic and his expressed predilection for what he called the "least revolutionary form of quantum theory," is given in Stuewer 1975, 71–75. Stuewer (p. 88, n. 25) finds it "rather shocking" that in his *Autobiography* Millikan wrote that his "complete verification of the validity of Einstein's equation . . . scarcely permits of any other interpretation than that which Einstein had originally suggested, namely, that of the . . . photon theory of light itself."

11. In 1909 Einstein expressed his opinion that "the next phase in the development of theoretical physics" would introduce "a new theory of light that can be interpreted as a kind of fusion of the wave and the emission theory" (Pais 1982, 404–405). But two years later he admitted to his friend Besso, "I do not ask any more whether these quanta really exist. Nor do I attempt any longer to

construct them, since I now know that my brain is incapable of fathoming the problem this way." But in 1916 he could report that a "splendid light has dawned on me about the absorption and emission of radiation." In 1919 he wrote that "I do not doubt any more the *reality* of radiation quanta, although I still stand quite alone in this conviction" (p. 411).

12. On this topic see, among others, Bohr (1949); Born (1949; 1971); Infeld (1950); Jammer (1966); Pais (1982); and Slater (1955).

13. A central problem in the Bohr theory has been described by von Laue (1950, 138) as follows: "Despite its great and lasting successes, the Bohr theory contained, nevertheless, a systematic defect. It employed classical mechanics for determining electron orbits, but thereupon, without inner connection with this calculation, discarded, with the aid of the quanta conditions, the overwhelming majority of these orbits as not being realized. Wave or quantum mechanics, founded in 1924–1926, is more uniform and somewhat more successful in accounting for spectra. It has now completely replaced its predecessor."

14. Heisenberg (1958, 40) reminds us that although the conclusion of Bohr, Kramers, and Slater (that the laws of conservation of energy and of momentum are not necessarily true for single events and are "true only in the statistical average") was proved to be incorrect, they did introduce the line of thought that led to "the concept of the probability wave," which was to become a cornerstone of quantum mechanics. It is also to be noted that the 'old' quantum theory was a poor fusion of quantum and classical physics that had developed flaws and contradictions which required a new kind of physics—quantum mechanics.

15. In his autobiography, Slater (1975, 11) wrote of his initial "consternation" on learning that Bohr and Kramers "completely refused to admit the real existence of the photons. It had never occurred to me that they would object to what seemed like so obvious a deduction from many types of experiments. The result was that they insisted on our writing a joint paper in which the electromagnetic field was described as having a continuously distributed energy density whose intensity determined the probability of transition of an atom from one stationary state to another. One had then to assume only a statistical conservation of energy between the continuously distributed energy density in the electromagnetic field and the quantized energy of the atoms. They grudingly allowed me to send a note to *Nature* indicating that my original idea had included the real existence of the photons, but that I had given that up at their instigation."

16. There is no adequate full-scale history of the special and general theory of relativity. There are, however, a number of collections and essays relating to Einstein's theories and their influence, edited by P. Schilp (1949), G. J. Whitrow (1967), P. C. Aichelburg and R. U. Sexl (1979), A. P. French (1979), H. Woolf (1980), and G. Holton and Y. Elkana (1982). Among many biographical studies of Einstein, the most useful have been written by P. Frank (1947), C. Seelig (1954) and B. Hoffmann (1972), B. Hoffmann and H. Dukas (1979), and especially A. Pais (1982). Particularly valuable are articles by G. Holton, M. J. Klein, and

A. Miller. The latter's documented history of special relativity (1981) calls for a sequel on general relativity. For quantum theory, there are the important articles by Martin J. Klein, Max Jammer's two histories (1966; 1974), and a host of articles and books, of which the major ones are cited in the text. Also an unpublished doctoral dissertation on "Werner Heisenberg and the Crisis in Quantum Theory 1920–1924" by David Charles Cassidy (Purdue University, 1976). I have profited greatly from an unpublished doctoral dissertation, "A History of the Problem of Atomic Structure from the Discovery of the Electron to the Beginning of Quantum Mechanics," by John L. Heilbron (University of California, Berkeley, 1965), and his many articles, some of which have been collected in his book of 1981. Especially noteworthy is Roger Stuewer's study of the Compton effect (1975).

28. Einstein on Revolution in Science

1. Einstein's statement on revolution is dated 6–7 February, 1947, exactly one week after the publication of the article on Schrödinger in the *New York Times*. The article, however, neither uses the word 'revolution' nor makes extravagant claims for Schrödinger's achievements. It is therefore puzzling that Einstein's reply contains so strong a statement about revolution.

29. Continental Drift and Plate Tectonics

1. Wegener did admit that in some aspects Taylor's "viewpoint . . . differs only quantitatively from my own, but not in crucial or novel ways." He observed that "Americans have called the drift theory the Taylor-Wegener theory," but he insisted that "in Taylor's train of thought continental drift in our sense played only a subsidiary role and was given only a very cursory explanation" (1966, 4).

Wegener's earliest paper on continental drift was published in 1912. In it he referred (pp. 185, 194–195) to Taylor's earlier publication of 1910. Years later, in the fourth edition of his book (1929), Wegener wrote up a history of the development of his ideas, claiming that he had never heard of any other work on this subject until after he had himself conceived of the main ideas of the hypothesis of continental drift. In particular, he alleged (p. 3) not to have heard of Taylor's ideas. Wegener said that he formulated his ideas during the months from autumn 1911 to early January 1912. Taylor's paper of 1910 was published in June, but had been read at a meeting of the Geological Society of America in December 1908. A review of Taylor's published paper, by Jesse Hyde, appeared in the *Geologische Zentralblatt* for 15 April 1911, which should have attracted Wegener's attention.

In a recently discovered letter by Taylor to the editor of *Popular Science Monthly*, written on 4 December 1931, it is claimed (see Totten 1981, 214) that

Wegener published "a very brief note" in "the spring of 1911," reviewing Taylor's paper, "partly in terms of approval, but putting forth some suggestions of his own." Taylor had either "mislaid or lost" his copy of the note, which he had "not been able to find . . . for three or four years." He had tried unsuccessfully, he wrote, to find this article in "the German scientific journals in the library of the University of Michigan at Ann Arbor." Harold Totten, who edited and published Taylor's letter, reported that "attempts by me and others to uncover the note have been unsuccessful."

Taylor described this note, which he says Wegener published in 1911, as "about twenty or twenty-five lines long," in "fine type." It is possible that Taylor's memory was at fault, and that the published report he had in mind was not by Wegener at all, but was the review by Jesse Hyde in that same year 1911, and which Taylor did not cite in his letter.

In an analysis of the texts of Taylor and Wegener, Totten (1981) finds strong evidence of Taylor's likely influence on Wegener. Not only are there "many similarities," but "in the case of Greenland, the reconstructions are nearly identical." Totten finds Taylor "justified in his claim" that "Wegener got at least some of his ideas" from him. What is perhaps of equal interest is the fact that Taylor had an early concept of plate tectonics, moving crustal sheets which, in collision, formed Tertiary mountain belts.

2. Wegener's thesis (1905, 1905a) consisted of a modernization of the "Alfonsine Tables" of planetary motion, which were named after King Alfonso el Sabio (1221–1284) of Castille, who sponsored a redaction and new edition in Spanish of the "Toledan Tables" of the Cordoban astronomer al-Zarqālī or Arzachel. Wegener converted the old sexagesimal numbers in the original tables into current decimal numbers, so that today's astronomers and chronologists could have more ready access to them in making their computations. The thesis was entitled *The Alfonsine Tables for the Use of Modern Computers*, which in 1905 meant 'for the use of men and women doing computations'. Wegener was deeply interested in the history of astronomy, but at that time no doctorates were being awarded in the history of science and there were no academic posts for specialists in the history of astronomy (Marvin 1973, 66). Wegener's edition of the Alfonsine Tables for modern users has proved to be a valuable tool for twentieth-century astronomers needing numerical planetary and lunar data of an earlier period, and there has been a recent reprint. But today's historians of astronomy prefer to use the originals, in their sexagesimal notation, and consider Wegener's efforts to have been misguided.

3. Suess had hypothesized that the earth consists of a central very heavy core (the Nife), a middle envelope of basic rock (the Sima), and an outer envelope primarily made up of lighter and more acid rocks (the Sal). Wegener transformed this schema (writing Sial for Sal), supposing that the Sial does not everywhere cover the globe, but rather occurs in large "independent patches" which float in the heavier or denser envelope of Sima. These patches are the continents or land masses, and ocean floors are Sima. The continents are like huge barges pushing through a lake-surface of Sima.

4. Hallam (1973, 9) has observed that "the subsequently coined term *conti-*

nental drift caught on universally in the English-speaking world presumably because people would rather not utter seven syllables when five will suffice."

5.   Thus, R. T. Chamberlin of the University of Chicago: can geology be considered a science if it is "possible for such a theory as this to run wild?" (p. 83). In a review of the published proceedings in *Geological Magazine* for September 1928, "P. L." (presumably Philip Lake) stressed the unfavorable comments, quoting with approval R. T. Chamberlin's remarks that the "appeal" of the Wegener hypothesis "seems to lie in the fact that it plays a game in which there are few restrictive rules, and no sharply drawn code of conduct. So a lot of things go very easily." Bailey Willis of Stanford University said simply that Wegener's book was "written by an advocate rather than by an impartial investigator." According to Edward W. Berry, Wegener's method "is not scientific"; Wegener ends up "in a state of auto-intoxication in which the subjective idea comes to be considered as an objective fact."

6.   The analogy between Wegener's theory and the Copernican theory (and Galileo's condemnation for advocating it) was a familiar theme of the 1920s (used by Daly [1926] and Chamberlin [1928]) and revived by many writers in the 1960s and 1970s.

7.   In the 1928 conference, Schuchert—after taking 41 pages to show that there are serious misfits in putting together the continental "jig saw puzzle" and to question the paleontological evidence used by Wegener, challenging him on methodological grounds—nevertheless concluded fairly by contrasting what he called his own "iconoclastic" position toward "the Wegener hypothesis" with his "open-minded" stance on "the idea that the continents may have moved slowly, very slowly indeed, laterally, and differently at different times" (p. 141). He thus was willing to go along with Daly and Emile Argand in accepting the general idea of mobility and, after reminding his readers that land masses must indeed have moved on the earth, he said that one "begins to remember the statement of Galileo in regard to the earth: 'And yet it does move.'" In thus going along with the general idea of mobilism, but nevertheless rejecting the specific form of mobilism proposed by Wegener, Schuchert admitted that he had been influenced by Daly, whose just published book "sounds the keynote for the attempt to save the germ of truth in the displacement theory and reconcile it with the facts that geology already has at hand" (pp. 142, 144).

8.   This letter was found by Kristie Macrakis among Wegener's Nachlass, in the Library of the Deutsches Museum. It is published inaccurately in Else Wegener's biography (1960, 75). A careful transcription of the letter, together with a commentary on its context and significance, can now be found in "Alfred Wegener: Self-Proclaimed Scientific Revolutionary" (Macrakis 1984).

9.   According to Hallam (1973, 68), the "germinal idea behind the theory of plate tectonics" is "clearly present" in a paper of 1965 by the Canadian geologist J. Tuzo Wilson on "transform faults," as is "the first use in this context of the term 'plates.'" But, according to Marvin (1973, 165), the "first paper to use the term 'plate' in this connection was published in 1967 by D. P. McKenzie and R. L. Parker, then at the University of California at San Diego."

10.   It seems to be generally agreed that the concept of sea-floor spreading

was invented by Harry Hess of Princeton in 1960 (Marvin 1973, 154–156; Hallam 1973, 54–67; Hess 1962; Cox 1973, 14–16), but not formally published by Hess until 1962, though privately circulated in 1960 in a "preprint." The first printed publication concerning sea-floor spreading was in 1961 (Dietz 1961). It has been suggested that Arthur Holmes was the "originator of [the] spreading ocean floor hypothesis" (Meyerhoff 1968). But Robert S. Dietz, in reply to the suggestion that Holmes was the "originator," said that Hess "deserves full credit for the concept . . . by reason of priority and for fully and elegantly laying down the basic premises. I have done little more than introduce the term *sea-floor spreading* . . . and apply it to such things as geosynclinal theory and the tectonic passiveness of the continental plates" (Dietz 1968; see Dietz 1963; 1966). Hess replied to the suggestion by pointing out that the "idea of sea-floor spreading was derived largely from new geophysical and topographic data on mid-ocean ridges, none of which were available to Holmes" (Hess 1968). He concluded, "The cogent term 'sea-floor spreading' which so nicely summed up my concept, was coined by Dietz (1961) after he and I had discussed the proposition at length in 1960." Dietz also substituted the eclogite-basalt phase change for the ocean floors for Hess's penidotite-serpentine composition, as I am informed by Ursula Marvin; Dietz was right, basaltic rather than serpentine.

11. The major lines of research that, among others, have not been mentioned include the pinpointing of the foci of earthquakes to a small number of belts girding the earth. These belts are now recognized as the fracture lines of the earth's surface, marking the boundaries of plates coming together. Another is the study of the phenomena along the sides of plates, where one plate may move along another, which are called 'transform faults'; this topic was explored by J. Tuzo Wilson and was of great significance (see Hallam 1973, 56–59; Uyeda 1978, 65–67, 74–79; Glen 1982, 304–307, 372–375).

12. The recent revolution in earth science is unusual in that many first-rate histories have been written by professional geologists and geophysicists soon after the event. A remarkable history of the actual post-World War II revolution is W. Glen's *The Road to Jaramillo* (1982), based on extensive reading in primary sources (both manuscript and printed) plus taped interviews with most of the major figures in the revolution. Glen comes to the subject with a background in geology and experience as author of textbook, *Continental Drift and Plate Tectonics* (1975). In 1973 Allan Cox produced an anthology of papers on *Plate Tectonics and Geomagnetic Reversals*, illuminated by the historical insights of a pioneer in the field; that same year saw the publication of Ursula B. Marvin's comprehensive *Continental Drift: The Evolution of a Concept*, especially rich for the Wegenerian and pre-Wegenerian periods, and Arthur Hallam's briefer but incisive *Revolution in the Earth Sciences*. Walter Sullivan's *Continents in Motion* (1974) is a readable presentation, based on wide reading in primary sources. Seiya Uyeda's *New View of the Earth* (1978) provides a clear historically oriented overview. Another valuable historical presentation is D. P. McKenzie's extensive article on "Plate Tectonics and Its Relationship to the Evolution of Ideas in the Geological Sciences" (1977). Extremely useful are the two editions of read-

ings from *Scientific American* edited by J. Tuzo Wilson (1976; 1973), who has also written many articles with a historical component. I have greatly profited from many discussions of this subject with the later Sir Edward (Teddy) Bullard, who wrote at least three major historical articles relating to the revolution and his own involvement in it (1965, 1975, 1975a). Finally, it should be recorded that both Wegener and du Toit included historical discussions in their respective treatises, and that the records of various symposia (e.g., the ones held in Tasmania in 1958, at the Royal Society in 1964, and at the American Philosophical Society in 1976), and the anthology (1962) edited by S. K. Runcorn on *Continental Drift*, are of enormous value for anyone studying the history of this subject.

In addition to the above-mentioned works by earth scientists, at least three historians of science (or historically minded philosophers of science — see §29.1, above) have written important studies on this revolution: David Kitts, Henry Frankel, and Rachel Laudan.

References

Abbud, Fuad. 1962. "The planetary theory of Ibn al-Shāṭir: reduction of the geometric models to numerical tables." *Isis* 53:492–499.

Abelard, *see* Heloise.

Ackerknecht, Erwin H. 1953. *Rudolf Virchow: doctor, statesman, anthropologist.* Madison: The University of Wisconsin Press. Photo-reprint, 1981, New York: Arno Press.

Acton, John Emerich Edward Dalberg-Acton, First Baron. 1906. *Lectures on modern history.* Ed. John Neville Figgis and Reginald Vere Laurence. London: Macmillan.

Adams, Henry. 1918. *The education of Henry Adams.* Published by Massachusetts Historical Society "as it was printed in 1907, with only such marginal corrections as the author made." Boston, New York: Houghton Mifflin Company.

Adelung, Johann Christoph. 1774–1786. *Versuch eines vollständigen grammatisch-kritischen Wörterbuches der hochdeutschen Mundart.* 4 vols. Leipzig: B. C. Breitkopf und Sohn.

———. 1798. *Grammatisch-kritisches Wörterbuch der hochdeutschen Mundart.* 6 vols. Leipzig: Breitkopf und Härtel.

Aichelburg, Peter C., and Roman U. Sexl, eds. 1979. *Albert Einstein: his influence on physics, philosophy and politics.* Braunschweig: Friedr. Vieweg & Sohn.

Aiken, Henry D., ed. 1957. *The age of ideology: the nineteenth-century philosophers.* Boston: Houghton Mifflin Company.

Aiton, Eric J. 1954. "Galileo's theory of the tides." *Annals of Science* 10:44–57.

———. 1969. "Kepler's second law of planetary motion." *Isis* 60:75–90.

———. 1972. *The vortex theory of planetary motions.* London: Macdonald; New York: American Elsevier.

———. 1975. "The elliptical orbit and the area law." *Vistas in Astronomy* 18:573–583.

———. 1976. "Johannes Kepler in the light of recent research." *History of Science* 14:77–100.

Aldrich, Michele L. 1976. "Taylor, Frederick Winslow." *Dictionary of Scientific Biography* 13:269–271.

Alembert, Jean le Rond d'. 1751. *Preliminary discourse to the Encyclopedia of Diderot.* Trans. Richard N. Schwab. Indianapolis: The Bobbs-Merrill Company [The Library of Liberal Arts], 1963.

———. 1853. *Oeuvres de d'Alembert: sa vie, ses oeuvres, sa philosophie.* Paris: Eugène Didier.

Alexander, S. 1909. "Ptolemaic and Copernican views of the place of mind in the universe." *Hibbert Journal* 8:47–66.

Allen, Garland E. 1978. *Life science in the twentieth century.* Cambridge: Cambridge University Press.

———. 1978a. *Thomas Hunt Morgan: the man and his science.* Princeton: Princeton University Press.

———. 1981. "Morphology and twentieth-century biology: a response." *Journal of the History of Biology* 14:159–176.

Anderson, Fulton H. 1948. *The philosophy of Francis Bacon.* Chicago: University of Chicago Press.

Arago, François. 1855. *Oeuvres complètes.* Vol. 3. Paris: Gide et J. Baudry; Leipzig: T. O. Weigel.

———. 1859. *Biographies of distinguished scientific men.* 1st series. Trans. W. H. Smyth, Baden Powell, and Robert Grant. Boston: Ticknor and Fields.

Arendt, Hannah. 1965. *On revolution.* New York: The Viking Press.

Argand, Emile. 1977. *Tectonics of Asia.* Trans., ed. Albert V. Carozzi. New York: Hafner Press; London: Collier Macmillan Publishers.

Armitage, Angus. 1957. *Copernicus: the founder of modern astronomy.* New York, London: Thomas Yoseloff.

Aron, Raymond. 1965, 1977. *Main currents in sociological thought.* 2 vols. New York: Basic Books.

Ascham, Anthony. 1975. *Of the confusions and revolutions of governments* (1649). Delmar, N.Y.: Scholars' Facsimiles & Reprints.

Aveling, Edward. 1892. *The students' Marx: an introduction to the study of Karl Marx' Capital.* London: Swan Sonnenschein & Co.

Aylmer, G. E., ed. 1975. *The Levellers in the English Revolution*. Ithaca, N.Y.: Cornell University Press.

Babbage, Charles. 1838. *The ninth Bridgewater treatise: a fragment*. 2nd ed. London: John Murray.

Bachelard, Gaston. 1934. *Le nouvel esprit scientifique*. Paris: F. Alcan.

———. 1938. *La formation de l'esprit scientifique*. Paris: Vrin.

Bacon, Francis. 1838. *The works of Lord Bacon*. 2 vols. London: William Ball.

———. 1857–1874. *Works*. 14 vols. Ed. J. Spedding, R. L. Ellis, and D. D. Heath. London: Longman and Co. Facsimile reprint, 1963, Stuttgart-Bad Cannstatt: Friedrich Frommann Verlag (Günther Holzboog).

———. 1878. *Bacon's Novum organum*. Ed., intro., notes by Thomas Fowler. Oxford: at the Clarendon Press.

———. 1889. *Bacon's Novum organum*. Ed. Thomas Fowler. 2nd ed. Oxford: at the Clarendon Press.

———. 1905. *The philosophical works of Francis Bacon*. Ed. John M. Robertson. London: George Routledge and Sons; New York: E. P. Dutton & Co. Reprint from the texts and translations, with the notes and prefaces, of Ellis and Spedding.

———. 1911. *The physical and metaphysical works of Lord Bacon including The advancement of learning and Novum organum*. Ed. Joseph Devey. London: G. Bell and Sons.

———. 1960. *The New organon and related writings*. Ed., intro. by Fulton H. Anderson. Indianapolis, New York: The Bobbs-Merrill Company [The Library of Liberal Arts].

Badash, Lawrence. 1972. "The completeness of nineteenth-century science." *Isis* 63:48–58.

Bailly, Jean-Sylvain. 1779–1782. *Histoire de l'astronomie moderne depuis la fondation de l'école d'Alexandrie, jusqu'à l'époque de M.D.CC.XXX*. 3 vols. Paris: chez de Bure. New ed., 1785.

———. 1781. *Histoire de l'astronomie ancienne, depuis son origine jusqu'à l'établissement de l'école d'Alexandrie*. 2nd ed. Paris: chez de Bure fils aîné.

Barber, Bernard. 1952. *Science and the social order*. Foreword by Robert K. Merton. Glencoe, Ill.: The Free Press.

———. 1961. "Resistance by scientists to scientific discovery." *Science* 134:596–602.

Basalla, George, ed. 1968. *The rise of modern science: external or internal factors?* Lexington, Mass.: D. C. Heath.

Bauer, Edmond. 1949. *L'électromagnétisme hier et aujourd'hui*. Paris: Editions Albin Michel.

Beck, Lewis White. 1963. *Studies in the philosophy of Kant*. New York: The Bobbs-Merrill Company.

———, ed. 1972. *Proceedings of the Third International Kant Congress*. Dordrecht: D. Reidel Publishing Company.

Beer, Arthur, and K. Aa. Strand, eds. 1975. *Copernicus: yesterday and today*. Oxford, New York: Pergamon Press [Vistas in Astronomy 17].

Beer, Arthur, and Peter Beer, eds. 1975. *Kepler: four hundred years.* Oxford, New York: Pergamon Press [Vistas in Astronomy 18].

Bell, E. T. 1937. *Men of mathematics.* New York: Simon & Schuster.

———. 1945. *The development of mathematics.* 2nd ed. New York, London: McGraw-Hill Book Company.

Bellone, Enrico. 1980. *A world on paper: studies on the second scientific revolution.* Trans. Mirella and Riccardo Giacconi. Cambridge, Mass., London: The MIT Press.

Ben-David, Joseph. 1971. *The scientist's role in society: a comparative study.* Englewood Cliffs, N.J.: Prentice-Hall.

Bennett, John Hughes. 1845. "Introductory address to a course of lectures on histology, and the use of the microscope." *Lancet* 1:517–522.

Bergia, Silvio. 1979. "Einstein and the birth of special relativity." Pp. 65–89 of French 1979.

Berlin, Isaiah. 1980. *Personal impressions.* Ed. Henry Hardy. London: The Hogarth Press.

Bernal, J. D. 1939. *The social function of science.* London: George Routledge & Sons.

———. 1954. *Science in history.* London: C. A. Watts & Co. 3rd ed., 1965; illustrated ed., 4 vols., 1969, Harmondsworth, Middlesex, England: Penguin Books.

Bernard, Claude. 1867. *Rapport sur les progrès de la physiologie.* Paris: Librairie de L. Hachette et Cie.

———. 1872. *Leçons de pathologie expérimentale.* Paris: J. B. Baillière.

———. 1878. *La science expérimentale.* 2nd ed. Paris: Librairie J. B. Ballière.

———. 1927. *An introduction to the study of experimental medicine.* Trans. Henry Copley Greene. New York: The Macmillan Company. Reprint, 1957, intro. by I. Bernard Cohen, New York: Dover Publications.

———. 1963. *Principes de médecine expérimentale.* Geneva: Alliance Culturelle du Livre [Les Classiques de la Médecine 5].

———. 1965. *Cahier de notes, 1850–1860.* Paris: Editions Gallimard.

———. 1966. *Leçons sur les phénomènes de la vie communs aux animaux et aux végétaux.* Paris: Librairie Philosophique J. Vrin.

———. 1967. *The cahier rouge.* Trans. Hebbel Hoff, Lucienne Guillemin, and Roger Guillemin. Cambridge, Mass.: Schenkman Publishing Co.

Berry, Arthur. 1898. *A short history of astronomy.* London: John Murray.

Berthelot, Marcellin P. E. 1890. *La révolution chimique: Lavoisier.* Paris: Félix Alcan.

Beyerchen, Alan D. 1977. *Scientists under Hitler: politics and the physics community in the Third Reich.* New Haven: Yale University Press.

Bezanson, Anna. 1921–1922. "The early use of the term industrial revolution." *The Quarterly Journal of Economics* 36:343–349.

Bilaniuk, Olexa Myron. 1973. "Lichtenberg, Georg Christoph." *Dictionary of Scientific Biography* 8:320–323.

Bird, Graham. 1973. *Kant's theory of knowledge*. New York: Humanities Press.

Biskup, Marian. 1973. *Regesta Copernicana* (calendar of Copernicus' papers). Warsaw: The Polish Academy of Sciences Press [Studia Copernicana 7].

Blainville, Henri de. 1845. *Histoire des sciences de l'organisation et de leurs progrès, comme base de la philosophie*. 3 vols. Paris, Lyons: Librairie Classique de Perisse Frères.

Blake, Ralph M., Curt J. Ducasse, and Edward J. Madden. 1960. *Theories of scientific method: the Renaissance through the nineteenth century*. Seattle: University of Washington Press.

Blondlot, R. 1905. *"N" rays: a collection of papers communicated to the Academy of Sciences*. Trans. J. Garcin. London, New York: Longmans, Green, & Co.

Boesiger, Ernest. 1980. "Evolutionary biology in France at the time of the evolutionary synthesis." Pp. 309–328 of Mayr and Provine 1980.

Bohm, David. 1965. *The special theory of relativity*. New York: W. A. Benjamin.

Bohr, Niels. 1913. "On the constitution of atoms and molecules." *Philosophical Magazine* 26:1–25, 476–502, 857–875.

———. 1924. *The theory of spectra and atomic constitution*. Cambridge: at the University Press.

———. 1949. "Discussion with Einstein on epistemological problems in atomic physics." Pp. 199–241 of Schilpp 1949.

———. 1963. *On the constitution of atoms*. New York: W. A. Benjamin. A reprint of Bohr's papers of 1913, intro. by Leon Rosenfeld.

———. 1965. "The structure of the atom." Nobel Lecture, Dec. 11, 1922. Pp. 7–43 of *Nobel Lectures, Physics, 1922–1941*. Amsterdam, London, New York: Elsevier Publishing Company.

Boltzmann, Ludwig. 1905. *Populäre Schriften*. Leipzig: Johann Ambrosius Barth.

———. 1905. "The relations of applied mathematics." Pp. 591–603 of Rogers 1905, vol. 1.

———. 1974. *Theoretical physics and philosophical problems: selected writings*. Trans. Paul Foulkes, ed. Brian McGuinness. Dordrecht, Boston: D. Reidel.

Borel, Emile. 1914. *Le hasard*. Paris: F. Alcan.

———. 1960. *Space and time*. Foreword by Banesh Hoffmann. New York: Dover Publications.

Born, Max. 1946. *Atomic physics*. 4th ed. New York: Hafner Publishing Company.

———. 1948. "Max Karl Ernst Ludwig Planck." *Obituary Notices of Fellows of the Royal Society* 6:161–188.

———. 1949. "Einstein's statistical theories." Pp. 161–177 of Schilpp 1949.

———. 1962. *Einstein's theory of relativity*. Rev. ed. New York: Dover Publications.

———. 1965. *My life and my views*. Intro. by I. Bernard Cohen. New York: Charles Scribner's Sons.

———. 1968. *My life and my views*. New York: Charles Scribner's Sons.

———. 1969. *Physics in my generation*. 2nd rev. ed. New York: Springer-Verlag.

————, ed. 1971. *The Born-Einstein letters: correspondence between Albert Einstein and Max and Hedwig Born from 1916 to 1955.* Trans. Irene Born. New York: Walker and Company.

Bossut, John. 1803. *A general history of mathematics: from the earliest times, to the middle of the eighteenth century.* London: J. Johnson.

Boulanger, Nicholas Antoine. 1766. *L'antiquité dévoilé par ses usages.* Amsterdam: chez Marc Michel Rey.

Bowler, Peter J. 1983. *The eclipse of Darwinism: anti-Darwinian evolution theories in the decades around 1900.* Baltimore, London: The Johns Hopkins University Press.

Boyer, Carl B. 1968. *A history of mathematics.* New York, London: John Wiley & Sons.

Boyle, Robert. 1744. *The works of the Honourable Robert Boyle.* 5 vols. London: printed for A. Millar. A new edition in 6 vols., 1772, London: printed for J. and F. Rivington, L. Davis, W. Johnston.

Bredvold, Louis I. 1934. *The intellectual milieu of John Dryden.* Ann Arbor: The University of Michigan Press.

Breuer, Josef, and Sigmund Freud. 1957. *Studies on hysteria.* Trans., ed. James Strachey. New York: Basic Books. Based on vol. 2 of Freud 1953; the original German version was published in 1895.

Bridgman, P. W. 1946. *The logic of modern physics.* New York: The Macmillan Company.

Brinton, Crane. 1950. *Ideas and men: the story of western thought.* New York: Prentice-Hall.

————. 1952. *The anatomy of revolution.* New York: Prentice-Hall.

————. 1963. *The shaping of modern thought.* Englewood Cliffs, N.J.: Prentice-Hall.

Broad, C. D. 1978. *Kant: an introduction.* Ed. C. Lewy. Cambridge, New York: Cambridge University Press.

Broglie, Louis de. 1951. "The concept of time in modern physics and Bergson's pure duration." Pp. 46–62 of Gunter 1969.

Brunck, H. 1901. *The history of the development of the manufacture of indigo.* New York: Kuttroff, Pickhardt & Co.

Brush, Stephen G. 1979. "Scientific revolutionaries of 1905: Einstein, Rutherford, Chamberlin, Wilson, Stevens, Binet, Freud." Pp. 140–171 of Mario Bunge and William R. Shea, eds., *Rutherford and physics at the turn of the century.* New York: Science History Publications.

————. 1981. [Review of Enrico Bellone 1980.] *Isis* 72:284–286.

————. 1982. "The second scientific revolution 1800–1950." Unpublished typescript. Development of ideas expressed in Brush 1979, 140–142.

Buchdahl, Gerd. 1961. *The image of Newton and Locke in the age of reason.* London, New York: Sheed and Ward.

————. 1963. "The relevance of Descartes's philosophy for modern philosophy of science." *The British Journal for the History of Science* 1:227–249.

Buchler, Justus, ed. 1940. *The philosophy of Peirce: selected writings.* New York: Harcourt, Brace; London: K. Paul, Trench, Trubner.

Buffon, Georges-Louis Leclerc, Comte de. 1954. *Oeuvres philosophiques.* Ed. Jean Piveteau. Paris: Presses Universitaires de France.

Bullard, Edward C. 1965. "Historical introduction to terrestrial heat flow." Chap. 1 of *Terrestrial heat flow.* Washington, D.C.: American Geophysical Union [Geophysical Monograph no. 8].

————. 1975. "The effect of World War II on the development of knowledge in the physical sciences." *Royal Society of London Proceedings,* ser. A, 342:519–536.

————. 1975*a.* "The emergence of plate tectonics: a personal view." *Annual Review of Earth and Planetary Sciences.* 3:1–30.

Bullen, K. E. 1976. "Wegener, Alfred." *Dictionary of Scientific Biography* 14:214–217.

Bullock, Alan, and Oliver Stallybrass, eds. 1977. *The Harper dictionary of scientific thought.* New York: Harper & Row.

Bunge, Mario, and William R. Shea, eds. 1979. *Rutherford and physics at the turn of the century.* New York: Dawson & Science History Publications.

Buonafede, Appiano (*pseudonym:* Agatapisto [Agatopisto] Cromaziano). 1791. *Kritische Geschichte der Revolutionen der Philosophie in den drey letzten Jahrhunderten.* 3 pts. Trans. Karl Heinrich Heydenreich. Leipzig: in der Weygandschen Buchhandlung. Facsimile reprint, 1968, Brussels: Culture et Civilisation.

Burckhardt, Jacob. 1942. *Historische Fragmente.* Basel: Benno Schwabe & Co.

Burdach, Konrad. 1963. *Reformation, Renaissance, Humanismus.* Darmstadt: Wissenschaftliche Buchgesellschaft.

Burke, Edmund. 1959. *Reflections on the revolution in France, and on the proceedings in certain societies in London.* Ed. William B. Todd. New York, Chicago, San Francisco: Holt, Rinehart & Winston.

Burkhardt, Richard W. 1983. [Review of Peter J. Bowler's *The Eclipse of Darwinism.*] *Science* 222:156–157.

Burns, Edward McNeill, Philip Lee Ralph, Robert E. Lerner, Standish Meacham. 1982. *World civilizations, their history and their culture.* 6th ed. New York: W. W. Norton & Company.

Burtt, E. A. 1925. *The metaphysical foundations of physical science.* New York: Harcourt, Brace & Company; London: Kegan Paul, Trench, Trubner & Co. 2nd rev. ed., 1932.

Bury, John B. 1920. *The idea of progress; an inquiry into its origins and growth.* London: Macmillan & Co. Reprint, 1932, New York: Macmillan. Reprint, 1955, New York: Dover Publications.

Butterfield, Herbert. 1949. *The origins of modern science, 1300–1800.* London: G. Bell and Sons. 2nd ed., rev., 1957, New York: The Macmillan Company.

————. 1965. *The present state of historical scholarship: an inaugural lecture.* Cambridge: at the University Press.

Bylebyl, Jerome Joseph. 1969. "Cardiovascular physiology in the sixteenth and early seventeenth centuries." Doctoral dissertation, Yale University.

———. 1972. "Harvey, William." *Dictionary of Scientific Biography* 6:150–162.

———. 1978. "William Harvey, a conventional medical revolutionary." *The Journal of the American Medical Association* 239:1295–1298.

Bynum, W. F., E. J. Browne, and Roy Porter, eds. 1981. *Dictionary of the history of science.* Princeton: Princeton University Press.

Calvert, Peter. 1970. *Revolution.* London: Macmillan and Company; New York: Praeger Publishers.

Camden, William. 1605. *Remaines of a greater worke, concerning Britaine . . .* London: printed by G. E. for Simon Waterson.

———. 1614. *Remaines concerning Britaine . . .* Rev. ed. London: printed by John Legatt for Simon Waterson.

Campe, Joachim Heinrich. 1801. *Wörterbuch zur Erklärung und Verdeutschung der unserer Sprache aufgedrungenen fremden Ausdrücke.* Vol. 1. Braunschweig: Schulbuchhandlung.

Cannizzaro, Stanislao. 1891. *Abriss eines Lehrganges der theoretischen Chemie.* Trans. Arthur Miolati, ed. Lothar Meyer. Leipzig: Verlag von Wilhelm Engelmann [Die Klassiker der Exakten Wissenschaften].

Cannon, Walter F. 1961. "The bases of Darwin's achievement: a revaluation." *Victorian Studies* 1:109–134.

Čapek, Milič. 1971. *Bergson and modern physics: a re-interpretation and re-evaluation.* Dordrecht: Reidel [Boston Studies in the Philosophy of Science 7].

Carey, John. 1981. *John Donne: life, mind, and art.* New York: Oxford University Press.

Carozzi, Albert V. 1970. "A propos de l'origine de la théorie des dérivés continentales: Francis Bacon (1620), François Placet (1668), A. von Humboldt (1801) et A. Snider (1858)." *Compte Rendu des Séances de l'Académie des Sciences* (Paris) 4:171–179.

Caspar, Max. 1959. *Kepler.* Trans., ed. C. Doris Hellman. London, New York: Abelard-Schuman.

Cassidy, David Charles. 1976. "Werner Heisenberg and the crisis in quantum theory, 1920–1926." Doctoral dissertation, Purdue University.

Cassirer, Ernst. 1950. *The problem of knowledge: philosophy, science, and history since Hegel.* Trans. William H. Woglom and Charles W. Hendel. New Haven: Yale University Press; London: Oxford University Press.

———. 1963. *The individual and the cosmos in Renaissance philosophy.* Trans. Mario Domandi. New York: Barnes and Noble.

———. 1981. *Kant's life and thought.* Trans. James Haden; intro. Stephan Körner. New Haven and London: Yale University Press.

Cedarbaum, Daniel Goldman. 1983. "Paradigms." *Studies in History and Philosophy of Science* 14:173–213.

Chamberlain, Houston Stewart. 1914. *Immanuel Kant.* Trans. from German by Lord Redesdale. 2 vols. London: John Lane, the Bodley Head; New York: John Lane Company.

Chambers, Robert. 1844. *Vestiges of the natural history of creation.* London: John Churchill. Originally published anonymously. 1st ed., reprinted, 1969, intro. by Gavin de Beer, Leicester University Press; dist. in North America by Humanities Press, New York.

———. 1845. *Explanations: a sequel to "Vestiges of the natural history of creation."* London: John Churchill.

Chandler, Philip P., II. 1975. "Newton and Clairaut on the motion of the lunar apse." Doctoral dissertation, University of California, San Diego.

Charleton, Walter. 1654. *Physiologia Epicuro-Gassendo-Charltoniana.* London: printed by Tho. Newcomb for Thomas Heath. Facsimile reprint, 1966, indexes, intro. by Robert Hugh Kargon, New York and London: Johnson Reprint Corporation [The Sources of Science 31].

Chevalier, Jacques. 1961. *Histoire de la pensée.* Vol. 3: *La pensée moderne.* Paris: Flammarion, Editeur.

Churchill, Frederick B. 1981. "In search of the new biology: an epilogue." *Journal for the History of Biology* 14:177–191.

Cipolla, Carlo M., ed. 1973. *The industrial revolution.* London, Glasgow: Collins/Fontana Books [The Fontana Economic History of Europe 3].

Clairaut, Alexis-Claude. 1749. "Du système du monde dans les principes de la gravitation universelle." Pp. 329–364 of *Histoire de l'Académie Royale des Sciences, année MDCCXLV, avec les mémoires de mathématique et de physique pour la même année.* Paris: de l'Imprimerie Royale. Clairaut's paper is said to have been read "à l'Assemblée publique du 15 Nov. 1747."

Clarendon, Edward, Earl of. 1888. *The history of the rebellion and civil wars in England begun in the year 1641.* 6 vols. Oxford: at the Clarendon Press.

Clark, George N. 1937. *Science and social welfare in the age of Newton.* Oxford: at the Clarendon Press.

———. 1953. *The idea of the industrial revolution.* Glasgow: Jackson, Son & Company [Glasgow University Publications 95].

Clark, Joseph T. 1959. "The philosophy of science and the history of science." Pp. 103–140 of Marshall Clagett, ed., *Critical problems in the history of science.* Madison: University of Wisconsin Press.

Clarke, Frank Wigglesworth. 1905. "The progress and development of chemistry during the nineteenth century." Pp. 221–240 of Rogers, 1905, vol. 4.

Cohan, A. S. 1975. *Theories of revolution: an introduction.* London: Nelson.

Cohen, Carl. 1975. "Revolutions and Copernican revolutions." Pp. 86–103 of Steneck 1975.

Cohen, I. Bernard. 1956. *Franklin and Newton . . .* Philadelphia: American Philosophical Society. Reissue, 1966, Cambridge, Mass.: Harvard University Press.

———. 1971. *Introduction to Newton's Principia.* Cambridge, Mass.: Harvard University Press; Cambridge: at the University Press.

———. 1972. *Benjamin Franklin: scientist and statesman.* New York: Charles Scribner's Sons.

———. 1980. *The Newtonian revolution: with illustrations of the transformation of scientific ideas.* Cambridge, London: Cambridge University Press.

———. 1981. "Newton's discovery of gravity." *Scientific American* 244:166–179.

————. 1982. "The *Principia*, universal gravitation, and the 'Newtonian style.'" Pp. 21–108 of Zev Bechler, ed., *Contemporary Newtonian research*, Dordrecht: D. Reidel Publishing Company.

————, ed. 1981. *Studies on William Harvey*. New York: Arno Press.

Cohen, I. Bernard, and Robert E. Schofield, eds. 1978. *Isaac Newton's papers and letters on natural philosophy and related documents*. 2nd ed. Cambridge, Mass., London: Harvard University Press.

Cohen, Robert S. 1978. "Engels, Friedrich." *Dictionary of Scientific Biography* 15:131–147.

Coleman, William. 1977. *Biology in the nineteenth century: problems of form, function, and transformation*. Cambridge, London: Cambridge University Press.

Comte, Auguste. 1855. *The positive philosophy of Auguste Comte*. Trans. Harriet Martineau. New York: Calvin Blanchard.

————. 1970. *Introduction to positive philosophy*. Ed. Frederick Ferre. Indianapolis, New York: The Bobbs-Merrill Company [The Library of Liberal Arts].

————. 1975. *Physique sociale: cours de philosophie positive, leçons 46 à 60*. Ed. Jean-Paul Enthoven. Paris: Hermann.

Conant, James B. 1947. *On understanding science: an historical approach*. New Haven: Yale University Press; London: Oxford University Press.

————. 1951. *Science and common sense*. New Haven: Yale University Press; London: Oxford University Press.

Condillac, Etienne Bonnet de. 1798. *Oeuvres*. Vol. 6. Paris: de l'Imprimerie de Ch. Houel.

————. 1947. *Oeuvres philosophiques*. 2 vols. Paris: Presses Universitaires de France.

Condorcet, Antoine-Nicolas, Marquis de. 1792. *Reflections on the English revolution of 1688, and that of the French, August 10, 1792*. Trans. from the French. London: printed for James Ridgeway, St. James Square.

————. 1804. *Oeuvres complètes de Condorcet*. Vol. 18. Paris: A. Brunswick.

————. 1955. *Sketch for a historical picture of the progress of the human mind*. Trans. June Barraclough. New York: The Noonday Press. Hyperion reprint, 1979, Westport, Conn.: Hyperion Press.

————. 1968. *Oeuvres*. Stuttgart-Bad Cannstatt: Friedrich Frommann Verlag (Günther Holzboog). Facsimile reprint of 1847–1849 edition.

Conn, H. W. 1888. "The germ theory as a subject of education." *Science* 11:5–6.

Copernicus, Nicolaus. 1978. *On the revolutions*. Ed. Jerzy Dobrzycki. Trans., commentary by Edward Rosen. Baltimore: The Johns Hopkins University Press.

Copleston, Frederick. 1960. *A history of philosophy*. Vol. 6, *Wolff to Kant*. New York, London: The Newman Press.

Cotgrave, Randle. 1611. *A dictionarie of the French and English tongues*. London: printed by Adam Islip.

Cournot, Antoine-Augustin. 1851. *Essai sur les fondements de nos connaissances et sur les caractères de la critique philosophique*. Vol. 1. Paris: Librairie de L.

Hachette et Cie. Published, 1975, as vol. 2 of A. A. Cournot, *Oeuvres complètes*, ed. Jean-Claude Pariente, Paris: Librairie Philosophique J. Vrin.

————. 1861. *Traité de l'enchaînement des idées fondamentales dans les sciences et dans l'histoire*. Vol. 1. Paris: Librairie de L. Hachette et Cie. Published, 1982, as vol. 3 of A. A. Cournot, *Oeuvres complètes*, ed. Nelly Bruyère, Paris: Librairie Philosophique J. Vrin.

————. 1872. *Considérations sur la marche des idées et des événements dans les temps modernes*. Vol. 1. Paris: Librairie Hachette et Cie. Published, 1973, as vol. 4 of A. A. Cournot, *Oeuvres complètes*, ed. André Robinet, Paris: Librairie Philosophique J. Vrin.

————. 1956. *An essay on the foundations of our knowledge*. Trans. Merritt H. Moore. New York: The Liberal Arts Press.

Cousin, Victor. 1841. *Cours d'histoire de la philosophie moderne pendant les années 1816 et 1817*. Paris: Librairie de Ladrange.

————. 1842. *Cours d'histoire de la philosophie morale au dix-huitième siècle pendant l'année 1820*. 3rd pt., *Philosophie de Kant*. Vol. 1. Paris: Librairie Philosophique de Ladrange.

————. 1846. *Cours de l'histoire de la philosophie moderne*. New ed., 1st series. Paris: Ladrange, Didier.

————. 1854. *The philosophy of Kant: lectures*. Trans. A. G. Henderson. London: Trubner & Co.

————. 1857. *Philosophie de Kant*. 3rd ed. Paris: Librairie Nouvelle.

Cox, Allan. 1973. *Plate tectonics and geomagnetic reversals*. San Francisco: W. H. Freeman and Company.

Crombie, Alastair C., ed. 1963. *Scientific change: historical studies in the intellectual, social and technical conditions for scientific discovery and technical invention, from antiquity to the present*. New York: Basic Books.

————. 1969. *Augustine to Galileo*. 2 vols. Harmondsworth, Middlesex, England: Penguin Books. Reprint of rev. ed. (1959) of a work first published in 1952.

————. 1969a. "Historians and the scientific revolution." *Physis* 11:162–180.

Cromwell, Oliver. 1937–1947. *The writings and speeches of Oliver Cromwell*. 4 vols. Ed. Wilbur Cortez Abbott. Cambridge, Mass.: Harvard University Press.

Cropsey, Joseph. 1975. "Commentary" on C. Cohen. Pp. 103–107 of N. Steneck, ed., *Science and society: past, present, and future*. Ann Arbor: University of Michigan Press.

Cross, F. L. 1937. "Kant's so-called Copernican revolution." *Mind* 46:214–217.

————. 1937a. "Professor Paton and 'Kant's so-called Copernican revolution.'" *Mind* 46:475–477. A reply to Paton 1937.

Crowe, Michael. 1975. "Ten 'laws' concerning patterns of change in the history of mathematics." *Historia Mathematica* 2:161–166.

Cunningham, E. 1919. "Einstein's relativity theory of gravitation." *Nature* (4 Dec.):354–356.

Curie, Marie. 1923. *Pierre Curie*. Trans. Charlotte and Vernon Kellogg. New York: The Macmillan Company.

Curry, Charles Emerson. 1897. *Theory of electricity and magnetism.* With a preface by Ludwig Boltzmann. London: Macmillan and Co.; New York: The Macmillan Company.

Curtis, John G. 1915. *Harvey's views on the use of the circulation of the blood.* New York: Columbia University Press.

Cushing, Harvey. 1943. *A bio-bibliography of Andreas Vesalius.* New York: Schuman's [Yale Medical Library, Historical Library, Publication 6].

Cuvier, Georges. 1812. *Discours sur les révolutions du globe.* Paris: Berché et Tralin.

————. 1812. *Recherches sur les ossemens fossiles des quadrupèdes* . . . Vol. 1. Paris: chez Deterville, Libraire.

Dalton, John Call. 1884. *Doctrines of the circulation: a history of physiological opinion and discovery in regard to the circulation of the blood.* Philadelphia: Henry C. Lea's Son & Co.

Daly, Reginald A. 1926. *Our mobile earth.* New York: Scribner.

Dampier, William C. 1929. *A history of science and its relations with philosophy and religion.* Cambridge: at the University Press.

Darnton, Robert. 1968. *Mesmerism and the end of the Enlightenment in France.* Cambridge, Mass.: Harvard University Press.

————. 1974. "Mesmer, Franz Anton." *Dictionary of Scientific Biography* 9:325–328.

Darrow, Karl Kelchner. 1937. *The renaissance of physics.* New York: The Macmillan Company.

Darwin, Charles. 1859. *On the origin of species by means of natural selection, or the preservation of favoured races in the struggle for life.* London: John Murray. Facsimile reprint, 1964, intro. by Ernst Mayr. Cambridge, Mass.: Harvard University Press.

————. 1887. *The life and letters of Charles Darwin.* 3 vols. Ed. Francis Darwin. London: J. Murray.

————. 1903. *More letters of Charles Darwin.* 2 vols. Eds. Francis Darwin and A. C. Seward. New York: D. Appleton and Company.

————. 1958. *The autobiography of Charles Darwin (1809–1882).* Ed. Nora Barlow. London: Collins.

————. 1971. *The origin of species.* London: J. M. Dent & Sons Ltd.; New York: E. P. Dutton & Co. Everyman's University Library ed. of the 6th ed. (1882).

Darwin, Charles, and Alfred Russell Wallace. 1958. *Evolution by natural selection.* Ed. Gavin de Beer. Cambridge: Cambridge University Press.

Darwin, Charles Galton. 1952. *The next million years.* London: Rupert Hart-Davis.

Dauben, Joseph Warren. 1969. "Marat: his science and the French Revolution." *Archives Internationales d'Histoire des Sciences* 22:235–261.

————. 1979. *Georg Cantor: his mathematics and philosophy of the infinite.* Cambridge, Mass., London: Harvard University Press.

Davies, P. A., and S. K. Runcorn, eds. 1980. *Mechanisms of continental drift and plate tectonics.* London, New York: Academic Press.

Davies, Paul. 1980. *Other worlds: a portrait of nature in rebellion: space, superspace and the quantum universe.* New York: Simon and Schuster.

Davis, William Morris. 1906. "The relations of the earth-sciences in view of their progress in the nineteenth century." Pp. 488–503 of Rogers 1905, vol. 4.

de Beer, Sir Gavin. 1965. *Charles Darwin, a scientific biography.* Garden City, N.Y.: Doubleday & Company [The Natural History Library; Anchor Books]. Original ed., 1965, Thomas Nelson and Sons.

de Milt, Clara. 1948. "Carl Weltzein and the congress at Karlsruhe." *Chymia* 1:153–169.

De Quincey, Thomas. 1859. *The logic of political economy and other papers.* Boston: Ticknor and Fields.

Deane, Phyllis, and W. A. Cole. 1962. *British economic growth, 1688–1959: trends and structure.* Cambridge: at the University Press.

Debus, Allen G. 1965. *The English Paracelsians.* London: Oldbourne.

———. 1976. "The pharmaceutical revolution of the Renaissance." *Clio Medica* 11:307–317.

———. 1977. *The chemical philosophy: Paracelsian science and medicine in the sixteenth and seventeenth centuries.* 2 vols. New York: Science History Publications.

Deleuze, Gilles. 1971. *La philosophie critique de Kant.* Paris: Presses Universitaires de France.

Descartes, René. 1911–1912. *The philosophical works.* 2 vols. Ed., trans. Elizabeth S. Haldane and G. R. T. Ross. Cambridge: at the University Press. Rev. reprint, 1931; photo-reprint, 1958, New York: Dover Publications.

———. 1952. *Descartes' philosophical writings.* Selected and trans. by Norman Kemp Smith. London: Macmillan & Co.

———. 1954. *Philosophical writings.* Trans., ed. Elizabeth Anscombe and Peter Thomas Geach, intro. by Alexandre Koyré. Edinburgh, London: Thomas Nelson and Sons Ltd.

———. 1954a. *The geometry of René Descartes.* Trans. David Eugene Smith and Marcia L. Latham. New York: Dover Publications.

———. 1956. *Discourse on method.* Trans., intro. by Laurence J. Lafleur. Indianapolis, New York, Kansas City: The Bobbs-Merrill Company [The Library of Liberal Arts].

———. 1965. *Discourse on method, Optics, Geometry, and Meteorology.* Trans., introd. by Paul J. Olscamp. Indianapolis, New York, Kansas City: The Bobbs-Merrill Company [The Library of Liberal Arts].

———. 1970. *Philosophical letters.* Trans., and ed. Anthony Kenny. Oxford: Clarendon Press.

———. 1971–1976. *Oeuvres.* 11 vols. Ed. Charles Adam and Paul Tannery. Paris: Librairie Philosophique J. Vrin. Rev. photo-reprint of the 1897–1910 ed., Paris: Léopold Cerf.

———. 1972. *Treatise of man.* French text with trans. and commentary by Thomas Steele Hall. Cambridge, Mass.: Harvard University Press.

————. 1979. *Le monde (The world)*. Trans. Michael Sean Mahoney. New York: Abaris Books.

Devaux, Philippe. 1955. *De Thalès à Bergson: introduction historique à la philosophie européenne*. Liège: Sciences et Lettres.

Dewey, John. 1929. *The quest for certainty: a study of the relation of knowledge and action*. Gifford Lectures, 1929. New York: Minton, Balch & Co.

Diderot, Denis. 1818. *Oeuvres*. Vol. 1. Paris: A. Belin.

Diderot, Denis, and Jean le Rond d'Alembert, eds. 1751–1780. *Encyclopédie, ou dictionnaire raisonné des sciences, des arts et des métiers*. 21 vols. Paris: chez Briasson, David l'aîné, Le Breton, Durand; Neuchâtel: chez Samuel Faulèbe & Compagnie. Facsimile reprint, 1967, Stuttgart-Bad Cannstatt: Friedrich Frommann Verlag (Günther Holzboog).

Dietz, Robert S. 1961. "Continent and ocean basin evolution by spreading of the sea floor." *Nature* 190:854–857.

————. 1963. "Collapsing continental rises: an actualistic concept of geosynclines and mountain building." *Journal of Geology* 71:314–333.

————. 1966. "Passive continents, spreading sea floors, and collapsing continental rises." *American Journal of Science* 264:177–193.

————. 1968. "Reply." *Journal of Geophysical Research* 73:6567.

Dijksterhuis, Eduard Jan. 1961. *The mechanization of the world picture*. Trans. C. Dikshoorn. Oxford: at the Clarendon Press.

Donne, John. 1969. *Ignatius his conclave*. An ed. of the Latin and English texts with intro. by T. S. Healy. Oxford: at the Clarendon Press.

Drake, Stillman. 1957. *Discoveries and opinions of Galileo*. Garden City, N.Y.: Doubleday & Co. [Doubleday Anchor Books].

————. 1978. *Galileo at work: his scientific biography*. Chicago, London: The University of Chicago Press.

Dreyer, J. L. E. 1906. *History of the planetary systems from Thales to Kepler*. Cambridge: at the University Press. Reprint, 1953, foreword by W. M. Stahl, *A history of astronomy from Thales to Kepler*, New York: Dover Publications.

DuBois-Reymond, Emil. 1912. *Reden*. 2 vols. 2nd ed. Leipzig: Von Veit & Co.

Dugas, René. 1955. *A history of mechanics*. Trans. J. R. Maddox. New York: Central Book Company.

Duhem, Pierre-Maurice-Marie. 1902. *Les théories électriques de J. Clerk Maxwell: étude historique et critique*. Paris: Librairie Scientifique A. Hermann.

————. 1905. *Les origines de la statique*. Paris: Librairie Scientifique A. Hermann.

————. 1906. *Etudes sur Léonard de Vinci: ceux qu'il a lus et ceux qui l'ont lu*. 1st series. Paris: Librairie Scientifique A. Hermann.

————. 1908. Σώζειν τὰ φαινόμενα: *Essai sur la notion de théorie physique de Platon à Galilée*. Paris: A. Hermann et Fils. Reprinted from *Annales de Philosophie Chrétienne* 6:113–138, 277–302, 352–377, 482–514, 576–592.

————. 1914. *La théorie physique: son objet, sa structure*. Paris: Marcel Rivière.

————. 1954. *The aim and structure of physical theory*. Trans. Philip P. Wiener. Princeton: Princeton University Press.

———. 1969. *To save the phenomena: an essay on the idea of physical theory from Plato to Galileo*. Trans. Edmund Doland and Chaninah Maschler. Chicago: The University of Chicago Press.

———. 1980. *The evolution of mechanics*. Trans. Michael Cole. Alphen aan den Rijn, Netherlands, Germantown, Md.: Sijthoff & Noordhoff.

Dukas, Helen, and Banesh Hoffman. 1979. *Albert Einstein: the human side*. Princeton: Princeton University Press.

Dupree, A. Hunter. 1959. *Asa Gray, 1810–1888*. Cambridge, Mass.: The Belknap Press of Harvard University Press.

Dutens, Louis. 1766. *Recherches sur l'origine des découvertes attribuées aux modernes*. Paris: veuve Duchesne.

Du Toit, Alex L. 1937. *Our wandering continents: an hypothesis of continental drifting*. Edinburgh, London: Oliver and Boyd.

Duveen, Denis I., and Herbert S. Klickstein. 1955. "Benjamin Franklin (1706–1790) and Antoine Laurent Lavoisier (1743–1794)." *Annals of Science* 11:126–128.

Eddington, Arthur Stanley. 1920. *Report on the relativity theory of gravitation*. 2nd ed. The Physical Society of London. London: Fleetway Press.

———. 1920. *Space, time and gravitation: an outline of the general relativity theory*. Cambridge: at the University Press.

———. 1922. *The theory of relativity and its influence on scientific thought*. The Romanes Lecture, 1922. Oxford: at the Clarendon Press.

———. 1923. *Mathematical theory of relativity*. Cambridge: at the University Press.

———. 1928. *The nature of the physical world*. The Gifford Lectures, 1927. New York: The Macmillan Company; Cambridge: at the University Press.

Edelstein, Ludwig. 1943. "Andreas Vesalius, the humanist." *Bulletin of the History of Medicine* 14:547–561.

Edward, Lyford P. 1970. *The natural history of revolution*. Chicago, London: The University of Chicago Press. First pub. 1927.

Edwards, Paul, ed. 1967. *The encyclopedia of philosophy*. 8 vols. New York: The Macmillan Company and The Free Press; London: Collier Macmillan.

Einstein, Albert. 1934. *Mein Weltbild*. Amsterdam: Querido Verlag.

———. 1934a. *The world as I see it*. New York: Covici Friede Publishers.

———. 1949. "Autobiographical notes." Pp. 1–95 of Schilpp 1949.

———. 1950. "Paul Langevin in memoriam." Pp. 231–232 of *Out of my later years*. Rev. reprint ed., Westport, Conn.: Greenwood Press.

———. 1953. *Mein Weltbild*. Ed. Carl Seelig. Zurich: Europa Verlag.

———. 1954. *Ideas and opinions*. New York: Crown Publishers.

———. 1956. *Lettres à Maurice Solovine*. Paris: Gauthier-Villars.

———. 1961. *Relativity: the special and the general theory*. Trans. Robert W. Lawson. New York: Crown Publishers.

Einstein, Albert, and Leopold Infeld. 1938. *The evolution of physics: the growth of ideas from early concepts to relativity and quanta*. New York: Simon and Schuster.

Einstein, Albert, H. A. Lorentz, H. Minkowski, and H. Weyl. 1952. *The principle of relativity: a collection of original memoirs on the special and general theory of relativity.* Trans. W. Perrett and G. B. Jeffery. New York: Dover Publications.

Eisenstadt, S. N. 1978. *Revolution and the transformation of societies: a comparative study of civilizations.* New York: The Free Press; London: Collier Macmillan Publishers.

Eisenstein, Elizabeth L. 1979. *The printing press as an agent of change: communications and cultural transformations in early-modern Europe.* 2 vols. Cambridge, London, New York: Cambridge University Press.

Eldredge, Niles, and Steven J. Gould. 1972. "Punctuated equilibria: an alternative to phyletic gradualism." Pp. 82–115 of T. J. M. Schopf and J. M. Thomas, eds., *Models in paleobiology,* San Francisco: Freeman, Cooper.

Eliade, Mircea. 1954. *Cosmos and history: the myth of the eternal return.* New York: Harper & Row.

Ellenberger, Henri F. 1970. *The discovery of the unconscious: the history and evolution of dynamic psychology.* New York: Basic Books; London: Allen Lane.

Engle, S. Morris. 1963. "Kant's Copernican analogy: a re-examination." *Kant-Studien* 54:243–251.

Engels, Friedrich. 1932. *M. E. Dühring bouleverse la science.* Trans. Bracke (A. M. Desrousseaux). Paris: Alfred Costes, Editeur. An earlier translation, 1911, by Edmond Laskine was entitled *Philosophie, économie politique, socialisme (contre Eugène Dühring),* Paris: V. Giard & E. Brière.

———. 1935. *Socialism: utopian and scientific.* Trans. Edward Aveling. New York: International Publishers.

———. 1939. *Anti-Dühring: Herr Eugen Dühring's revolution in science.* Trans. Emile Burns. New York: International Publishers. 2nd ed., 1959, Moscow: Foreign Languages Publishing House.

———. 1940. *Dialectics of nature.* Trans. and ed. Clemens Dutt. New York: International Publishers.

———. 1948. *Herrn Eugen Dührings Umwälzung der Wissenschaft (Anti-Dühring).* Berlin: Dietz Verlag.

———. 1968. *The condition of the working class in England.* Trans. W. O. Henderson and W. H. Chaloner. Stanford: Stanford University Press.

———. 1975. *Dialektik der Natur.* Berlin: Dietz Verlag.

———. 1980. *Herrn Eugen Dührings Umwälzung der Wissenschaft.* Berlin: Dietz Verlag.

Erxleben, Johann Christian Polycarp. 1794. *Anfangsgründe der Naturlehre.* 6th ed., rev. by G. C. Lichtenberg. Göttingen: Johann Christian Dieterich.

Evelyn, John. 1906. *The diary of John Evelyn.* 3 vols. Ed. Austin Dobson. London: Macmillan and Co.

Everitt, C. W. F. 1974. "Maxwell, James Clerk." *Dictionary of Scientific Biography* 9:198–230.

Ewing, A. C. 1938. *A short commentary on Kant's Critique of pure reason.* Chicago: The University of Chicago Press.

Farrington, Benjamin. 1949. *Francis Bacon: philosopher of industrial science.* New York: Henry Schuman.

Fayet, Joseph. 1960. *La révolution française et la science: 1789–1795.* Paris: Librairie Marcel Rivière.

Feldstein, Martin. 1981. "The retreat from Keynesian economics." *The Public Interest* 64:92–105.

Ferrand, Antoine-François-Claude. 1817. *Théorie des révolutions, rapprochée des principaux événemens qui en ont été l'origine, le développement ou la suite; avec une table générale et analytique.* 4 vols. Paris: chez L. G. Michaud (de l'Imprimerie Royale).

Feuer, Lewis S. 1971. "The social roots of Einstein's theory of relativity." *Annals of Science* 27:227–298, 313–344.

———. 1974. *Einstein and the generations of science.* New York: Basic Books.

Feuerbach, Ludwig. 1969–1972. *Gesammelte Werke: kleinere Schriften.* 4 vols. Berlin. "Die Naturwissenschaft und die Revolution" (1850) appears in vol. 3, pp. 347–368, a review of Jacob Moleschott's *Die Physiologie der Nahrungsmittel* (1850).

Feynmann, Richard. 1965. *The character of physical law.* A series of lectures recorded by the BBC at Cornell University and televised on BBC-2. London: British Broadcasting Corporation.

Figuier, Louis. 1876. *Vies des savants illustres.* Vol. 4, *Savants du XVIIᵉ siècle.* 2nd ed. Paris: Librairie Hachette.

———. 1879. *Vies des savants illustres.* Vol. 5, *Savants du XVIIIᵉ siècle.* 3rd ed. Paris: Librairie Hachette.

———. 1881. *Vies des savants illustres.* Vol. 3, *Savants de la Renaissance.* 3rd ed. Paris: Librairie Hachette.

Findlay, Alexander. 1916. *Chemistry in the service of man.* London, New York: Longmans, Green and Co.

Fine, Reuben. 1963. *Freud: a critical re-evaluation of his theories.* London: George Allen & Unwin.

Fisher, Arthur. 1979. "Grand unification: an elusive grail." *Mosaic* 10:2–21.

Fletcher, Ronald. 1971. *The making of sociology.* Vol. 1, *Beginnings and foundations.* New York: Charles Scribner's Sons.

———. 1974. *The crisis of industrial civilization: the early essays of Auguste Comte.* London: Heinemann Educational Books.

Fleming, Donald. 1964. See Loeb 1964.

Flew, Antony. 1971. *An introduction to western philosophy: ideas and argument from Plato to Sartre.* Indianapolis, New York: The Bobbs-Merrill Company.

Fontenelle, Bernard le Bovier de. 1683. *Nouveaux dialogues des morts.* Paris: C. Blageart. Reprint, 1971, ed. Jean Dagen, Paris: Librairie Marcel Didier.

———. 1688. *Digression sur les anciens et les modernes.* In *Poésies pastorales de M. D. F., avec un traité sur la nature de l'églogue, et une digression sur les anciens et les modernes.* Paris: Michel Guerout.

———. 1708. "The usefulness of mathematical learning." In *Miscellanea curiosa,* vol. 1, 2nd ed., London: by J. M. for R. Smith.

Fontenelle, Bernard le Bovier de. 1708a. *Dialogues of the dead*. Trans. John Hughes. London: printed for Jacob Tonson.

———. 1730. *Dialogues of the dead*. Trans. John Hughes. 2nd ed. London: printed for J. Tonson.

———. 1734. "Préface sur l'utilité des mathématiques et de la physique, et sur les travaux de l'Académie des Sciences." In *Histoire de l'Académie Royale des Sciences. Année M.DC.XCIX. Avec les mémoires de mathématique et de physique, pour la même année. Tirés des registres de cette Académie.* 2nd ed. Amsterdam: chez Pierre Mortier.

———. 1790–1792. *Oeuvres*. New ed. Vols. 6, 7. Paris: chez Jean-François Baslieu et Jean Servière.

———. 1955. *Entretiens sur la pluralité des mondes; digression sur les anciens et les modernes*. Ed. Robert Shackleton. Oxford: at the Clarendon Press.

Föppl, A. 1894. *Einführung in die Maxwell'sche Theorie der Elektricität*. Leipzig: Druck und Verlag von B. G. Teubner.

Foucault, Michel. 1970. *The order of things: an archaeology of the human sciences.* Trans. from *Les mots et les choses*, 1966. Ed. R. D. Laing. New York: Pantheon Books.

Fougeyrolles, Pierre. 1972. *Marx, Freud et la révolution totale*. Paris: Editions Anthropos.

Fourcroy, Antoine François de. 1790. *Philosophie chimique, ou vérités fondamentales de la chimie moderne, disposées dans un nouvel ordre*. Paris: [de l'imprimerie de Cl. Simon].

Frank, Philipp. 1947. *Einstein, his life and times*. New York: Alfred A. Knopf.

Frankel, Eugene. 1976. "Corpuscular optics and the wave theory of light: the science and politics of a revolution in physics." *Social Studies of Science* 6:141–184.

Frankel, Henry. 1978. "The non-Kuhnian nature of the recent revolution in the earth sciences." Pp. 197–214 of *PSA 1978, Proceedings of the 1978 Biennial Meeting of the Philosophy of Science Association*, vol. 2, ed. Peter D. Asquith and Ian Hacking, East Lansing, Mich.: Philosophy of Science Association.

———. 1979. "The career of continental drift theory: an application of Imre Lakatos' analysis of scientific growth to the rise of drift theory." *Studies in History and Philosophy of Science* 10:21–66.

———. 1979a. "The reception and acceptance of continental drift theory as a rational episode in the history of science." Pp. 51–89 of Mauskopf 1979.

———. 1981. "The paleobiogeographical debate over the problem of disjunctively distributed life forms." *Studies in History and Philosophy of Science* 12:211–259.

Franks, Felix. 1981. *Polywater*. Cambridge, Mass., London: MIT Press.

Freind, John. 1750. *The history of physick from the time of Galen to the beginning of the sixteenth century*. 2 vols. 4th ed. London: M. Cooper.

French, A. P., ed. 1979. *Einstein: a centenary volume*. London: Heinemann Educational Books; Cambridge, Mass.: Harvard University Press.

Freud, Sigmund. 1952. *An autobiographical study.* Authorized trans. by James Strachey. New York: W. W. Norton.

———. 1953–1974. *The standard edition of the complete psychological works of Sigmund Freud.* 24 vols. Trans. under the general editorship of James Strachey, in collaboration with Anna Freud, assisted by Alix Strachey and Alan Tyson. London: The Hogarth Press and the Institute of Psycho-analysis.

———. 1954. *The origins of psycho-analysis: letters to Wilhelm Fliess, drafts and notes, 1887–1902.* Ed. Marie Bonaparte, Anna Freud, Ernst Kris. New York: Basic Books.

Friedrich, Carl, ed. 1949. *The philosophy of Kant: Immanuel Kant's moral and political writings.* New York: The Modern Library.

Fulton, John F. 1931. *Physiology.* New York: Paul B. Hoeber [Clio Medica 5].

Gage, A. T. 1938. *A history of the Linnean Society of London.* London: Taylor & Francis.

Galilei, Galileo. 1953. *Dialogue concerning the two chief world systems—Ptolemaic and Copernican.* Trans. Stillman Drake. Berkeley, Los Angeles: The University of California Press.

———. 1957. *Discoveries and opinions of Galileo.* Trans. Stillman Drake. New York: Doubleday & Company [Doubleday Anchor Books].

———. 1964–1966. *Le opere di Galileo Galilei.* 20 vols. Ed. Antonio Favaro. Florence: G. Barbèra-Editore. Reprint of original ed., 1890–1909.

———. 1974. *Two new sciences, including centers of gravity and force of percussion.* Trans. Stillman Drake. Madison: The University of Wisconsin Press.

Galison, Peter J. 1979. "Minkowski's space-time: from visual thought to the absolute world." *Historical Studies in the Physical Sciences* 10:85–121.

———. 1981. "Kuhn and the quantum controversy." *British Journal for the Philosophy of Science* 32:71–84.

Gamow, George. 1966. *Thirty years that shook physics: the story of quantum theory.* Garden City, N.Y.: Doubleday & Co.

Gardiner, Samuel Rawson. 1886. *The first two Stuarts and the Puritan Revolution, 1603–1660.* New York: Charles Scribner's Sons. 1st ed., London, 1876.

———. 1906. *The constitutional documents of the Puritan Revolution, 1625–1660.* 3rd ed. Oxford: at the Clarendon Press. 1st ed., 1889.

Ghiselin, Michael T. 1969. *The triumph of the Darwinian method.* Berkeley, Los Angeles: University of California Press.

Gide, Charles, and Charles Rist. 1914 (?). *A history of economic doctrines.* Trans. R. Richards. Boston, New York: D. C. Heath.

———. 1947. *Histoire des doctrines économiques.* 7th ed. Paris: Librairie du Recueil Sirey.

Gilbert, Felix. 1973. "Revolution." Pp. 152–167 of *Dictionary of the history of ideas,* vol. 4, ed. Philip P. Wiener, New York: Charles Scribner's Sons.

Gilbert, William. 1900. *On the magnet, magnetick bodies also, and on the great magnet the earth: a new physiology, demonstrated by many arguments and experi-*

ments. London: Chiswick Press. Facsimile reprint, 1958, New York: Basic Books.

Gillispie, Charles C. 1957. "The Natural History of Industry." *Isis* 48:398–407.

———. 1980. *Science and polity in France at the end of the Old Regime.* Princeton: Princeton University Press.

Gillot, H. 1914. *La querelle des anciens et des modernes.* Nancy: A. Crépin-Leblond.

Gingerich, Owen. 1972. "Johannes Kepler and the new astronomy." *Quarterly Journal of The Royal Astronomical Society* 13:346–373.

———. 1973. "The role of Erasmus Reinhold and the Prutenic Tables in the dissemination of Copernican theory." Pp. 43–63, 123–125 of *Studia Copernicana* (containing "Colloquia Copernicana II. Etudes sur l'audience de la théorie héliocentrique [Torún, 1973]"), Wroclaw, Warsaw, Krakow, Gdansk: Ossolineum, The Polish Academy of Sciences Press.

———. 1973a. "From Copernicus to Kepler: heliocentrism as model and as reality." *Proceedings of the American Philosophical Society* 117:513–522.

———. 1975. " 'Crisis' versus aesthetic in the Copernican revolution." *Vistas in Astronomy* 17:85–93.

———. 1975a. "Reinhold, Erasmus." *Dictionary of Scientific Biography* 11:365–367.

———, ed. 1975b. *The nature of scientific discovery: a symposium commemorating the 500th anniversary of the birth of Nicolaus Copernicus.* Washington: Smithsonian Institution Press.

Glanvill, Joseph. 1668. *Plus ultra: or, the progress and advancement of knowledge since the days of Aristotle.* London: for James Collins. Facsimile reprint, 1958, Gainesville, Fla.: Scholars' Facsimiles and Reprints.

———. 1676. "Modern improvements of useful knowledge." Essay III of *Essays on several important subjects in philosophy and religion.* London: printed by J. D. for John Baker and Henry Mortlock.

Glass, Bentley, Owsei Temkin, and William L. Straus, Jr., eds. 1959. *Forerunners of Darwin, 1745–1859.* Baltimore: The Johns Hopkins Press.

Glazebrook, R. T. 1896. *James Clerk Maxwell and modern physics.* London: Cassell and Company.

Glen, William. 1982. *The road to Jaramillo: critical years of the revolution in earth science.* Stanford: Stanford University Press.

Goethe, Johann Wolfgang von. 1902–1907. *Sämtliche Werke, Jubiläums-Ausgabe.* 40 vols. Ed. Eduard von der Hellen. Stuttgart, J. G. Cotta.

———. 1947– . *Die Schriften zur Naturwissenschaft.* Ed. G. Schmid et al. Weimar: Böhlau.

Goldsmith, Donald, ed. 1977. *Scientists confront Velikovsky.* New York, London: W. W. Norton & Company. Essays by Norman W. Storer, J. Derral Mulholland, Carl Sagan, Peter J. Huber, and David Morrison.

Gough, Jerry B. 1970. "Blondlot, René-Prosper." *Dictionary of Scientific Biography* 2:202–203.

———. 1982. "Some early references to revolutions in chemistry." *Ambix* 29:106–109.

Gould, Stephen Jay, and Niles Eldredge. 1977. "Punctuated equilibria: the tempo and mode of evolution reconsidered." *Paleobiology* 3:115–151.

Goulemot, Jean Marie. 1975. *Discours, révolutions et histoire*. Paris: Union Générale d'Edition.

Grande, Francisco, and Maurice B. Visscher, eds. 1967. *Claude Bernard and experimental medicine*. Cambridge, Mass.: Schenkman Publishing Company.

Granger, G. 1971. "Cournot, Antoine-Augustin." *Dictionary of Scientific Biography* 3:450–454.

Grant, Edward. 1962. "Hypotheses in early science." *Daedalus*, 599–612.

———. 1962a. "Late medieval thought, Copernicus, and the scientific revolution." *Journal of the History of Ideas* 23:197–220.

Grattan-Guiness, I. 1970. "An unpublished paper by Georg Cantor." *Acta Mathematica* 70:65–107.

Gray, Asa. 1846. [Review of *Explanations—a sequel to the Vestiges of the natural history of creation*. New York: Wiley & Putnam, 1846.] *North American Review* 62:465–506.

Green, Alice (Mrs. J. R.). 1894. *Town life in the fifteenth century*. 2 vols. New York, London: Macmillan.

Greenberg, Jay R., and Stephen A. Mitchell. 1983. *Object relations in psychoanalytic theory*. Cambridge, Mass., London: Harvard University Press.

Greene, John C. 1971. "The Kuhnian paradigm and the Darwinian revolution in natural history." Pp. 3–25 of *Perspectives in the History of Science and Technology*, ed. D. Roller, Norman, Okla.: University of Oklahoma Press. Reprinted in Gutting 1980 and in Greene 1981.

———. 1981. *Science, ideology, and world view: essays in the history of evolutionary ideas*. Berkeley, London: University of California Press.

Greene, Mott T. 1982. *Geology in the nineteenth century: changing views of a changing world*. Ithaca: Cornell University Press.

Griewank, Karl. 1973. *Der neuzeitliche Revolutionsbegriff: Entstehung und Geschichte*. Ed. Ingeborg Horn-Staiger. Frankfurt: Suhrkamp.

Grimaux, Edouard. 1896. *Lavoisier, 1743–1794*. 2nd ed., Paris: Félix Alcan.

Groth, Angelika. 1972. *Goethe als Wissenschaftshistoriker*. Munich: W. Fink [Münchener Germanistische Beiträge].

Gruber, Howard E. 1974. *Darwin on man: a psychological study of scientific creativity*. Together with *Darwin's early and unpublished notebooks*, transcribed and annotated by Paul H. Barrett. New York: E. P. Dutton & Co.

Guerlac, Henry. 1952. *Science in western civilization, a syllabus*. New York: The Ronald Press Company.

———. 1975. *Antoine-Laurent Lavoisier: chemist and revolutionary*. New York: Charles Scribner's Sons.

———. 1976. "The chemical revolution: a word from Monsieur Fourcroy." *Ambix* 23:1–4.

———. 1977. *Essays and papers in the history of modern science*. Baltimore, London: The Johns Hopkins University Press.

———. 1981. *Newton on the continent*. Ithaca: Cornell University Press.

Guerlac, Henry, and Margaret C. Jacob. 1969. "Bentley, Newton, and Providence (the Boyle Lectures once more)." *Journal of the History of Ideas* 30:307–318.

Guicciardini, Francesco. 1970. *Opere.* Vol. 1. Ed. Emanuella Lugnani Scarano. Turin: Unione Tipografico-Editrice.

Guizot, François. 1826–1856. *Histoire de la révolution d'Angleterre.* Paris: A. Leroux & C. Chantpie.

———. 1854–1856. *Histoire de la révolution d'Angleterre.* 6 vols. Paris: Didier.

———. 1867. *The history of civilization from the fall of the Roman Empire to the French Revolution.* 2 vols. Trans. William Hazlitt. New York: D. Appleton & Company.

Gunter, Pete A. Y., ed. 1969. *Bergson and the evolution of physics.* Knoxville: The University of Tennessee Press.

Gutting, Gary, ed. 1980. *Paradigms and revolutions: applications and appraisals of Thomas Kuhn's philosophy of science.* Notre Dame, Ind.: University of Notre Dame Press.

Haber, L. F. 1958. *The chemical industry during the nineteenth century: a study of the economic aspect of applied chemistry in Europe and North America.* Oxford: at the Clarendon Press.

Hacking, Ian. 1983. "Was there a probabilistic revolution, 1800–1930?" Pp. 487–506 of Michael Heidelberger, Lorenz Krüger, and Rosemarie Rheinwald, eds., *Probability since 1800: interdisciplinary studies of scientific development,* Bielefeld: B. Kleine Verlag.

———, ed. 1981. *Scientific revolutions.* Oxford, London, New York: Oxford University Press.

Hahn, Paul. 1927. *Georg Christoph Lichtenberg und die exakten Wissenschaften.* Göttingen: Dandenhoeck & Ruprecht.

Hahn, Roger. 1971. *The anatomy of a scientific institution: the Paris Academy of Sciences, 1666–1803.* Berkeley, Los Angeles: University of California Press.

Haldane, J. B. S. 1924. *Daedalus or science and the future.* London: Kegan Paul, Trench, Trübner & Co.

———. 1940. *Keeping cool, and other essays.* London: Chatto & Windus.

Haldane, Viscount. 1921. *The reign of relativity.* New Haven: Yale University Press.

Hall, A. Rupert. 1952. *Ballistics in the seventeenth century.* Cambridge: at the University Press.

———. 1954. *The scientific revolution, 1500–1800: the formation of the modern scientific attitude.* London: Longmans, Green and Co.

———. 1970. "Merton revisited, or science and society in the seventeenth century." *History of Science* 2:1–16.

———. 1970a. "On the historical singularity of the scientific revolution of the seventeenth century." Pp. 199–221 of J. H. Elliott and H. G. Koenigsberger, eds., *The diversity of history: essays in honour of Sir Herbert Butterfield,* London: Routledge & Kegan Paul.

————. 1981. "Scientific revolution." Pp. 378–380 of W. F. Bynum, E. J. Browne, and Roy Porter, eds., *Dictionary of the history of science*, Princeton: Princeton University Press.

Hallam, Anthony. 1973. *A revolution in the earth sciences: from continental drift to plate tectonics*. Oxford: at the Clarendon Press.

Halévy, Daniel. 1946. *Péguy and "Les cahiers de la quinzaine."* Trans. Ruth Bethell. London: D. Dobson.

Hanfling, Oswald. 1972. "Kant's Copernican revolution: moral philosophy." The Age of Revolutions, Units 17–18, Arts: a second level course. Prepared for the course team. Bletchley, Bucks., England: The Open University Press.

Hankel, Hermann. 1884. *Die Entwickelung der Mathematik in den letzten Jahrhunderten*. 2nd ed. Tübingen: F. Fues. 1st ed., 1869.

Hankins, Thomas L. 1972. "Hamilton, William Rowan." *Dictionary of Scientific Biography* 6:85–93.

————. 1980. *Sir William Rowan Hamilton*. Baltimore, London: The Johns Hopkins University Press.

Hanson, Norwood Russell. 1959. "Copernicus' role in Kant's revolution." *Journal of the History of Ideas* 20:274–281.

————. 1961. "The Copernican disturbance and the Keplerian revolution." *Journal of the History of Ideas* 22:169–184.

Harland, W. B. 1969. "Essay review: the origins of continents and oceans." *Geological Magazine* 106:100–104.

Harman, P. M. 1982. *Energy, force, and matter: the conceptual development of nineteenth-century physics*. Cambridge, London, New York: Cambridge University Press.

————. 1983. *The scientific revolution*. London: Methuen.

Harré, Rom, ed. 1975. *Problems of scientific revolution: progress and obstacles to progress in the sciences*. The Herbert Spencer Lectures 1973. Oxford: at the Clarendon Press.

Hartner, Willy. 1974. "Ptolemy, Azarquiel, Ibn al-Shāṭir, and Copernicus on Mercury: a study of parameters." *Archives Internationales d'Histoire des Sciences* 24:5–25.

Harvey, James. 1720. *Praesagium medicum*. 2nd ed. Preface by William Cockburn. London: G. Strahan.

Harvey, William. 1957. *Movement of the heart and blood in animals*. Trans. Kenneth J. Franklin. Oxford: Blackwell Scientific Publications.

————. 1958. *The circulation of the blood: two anatomical essays by William Harvey, together with nine letters written by him*. Trans. Kenneth J. Franklin. Oxford: Blackwell Scientific Publications. A translation of *Exercitationes duae anatomicae de circulatione sanguinis ad Joannem Riolanum, filium, Parisiensem*.

————. 1963. *The circulation of the blood and other writings*. New York: Dutton [Everyman's Library]. Reprint of Harvey 1957 and Harvey 1958.

Hatto, Arthur. 1949. "'Revolution': an inquiry into the usefulness of an historical term." *Mind* 58:495–517.

Hawkins, Thomas W. 1982. [Unpublished lecture notes on history of mathematics.] Boston: Boston University.

Hayek, F. A. 1955. *The counter-revolution of science: studies on the abuse of reason.* London: Collier-Macmillan.

Hegel, Georg Wilhelm Friedrich. 1927–1940. *Sämtliche Werke, Jubiläumsausgabe.* 26 vols. Ed. Hermann Glockner. Stuttgart: F. Frommann.

———. 1969. *Hegel's science of logic.* Trans. A. V. Miller. New York: Humanities Press; London: Allen & Unwin.

———. 1970. *Hegel's philosophy of nature.* 3 vols. Trans. M. J. Petry. New York: Humanities Press; London, Allen & Unwin.

Heidelberger, Michael. 1976. "Some intertheoretic relations between Ptolemean and Copernican astronomy." *Erkenntnis* 10:323–336. Reprint, pp. 271–283 of Gutting 1980.

———. 1980. "Towards a logical reconstruction of revolutionary change: the case of Ohm as an example." *Studies in History and Philosophy of Science* 11:103–121.

———. 1981. "Some patterns of change in the Baconian sciences of early nineteenth century Germany." Pp. 3–18 of H. N. Jahnke and M. Otte, eds., *Epistemological and social problems of the sciences in the early nineteenth century,* Dordrecht: D. Reidel Publishing Company.

Heidelberger, Michael, Lorenz Krüger, and Rosemarie Rheinwald, eds. 1983. *Probability since 1980: interdisciplinary studies of scientific development.* Bielefeld: Universität Bielefeld. (Workshop at the Centre for Interdisciplinary Research of the University of Bielefeld, 16–20 Sept. 1982.)

Heilbron, John Lewis. 1965. "A history of the problem of atomic structure from the discovery of the electron to the beginning of quantum mechanics." Ph.D. dissertation, University of California, Berkeley.

———. 1976. "Symmer, Robert." *Dictionary of Scientific Biography* 13:224–225.

———. 1976a. "Robert Symmer and the two electricities." *Isis* 67:7–20.

———. 1979. *Electricity in the seventeenth and eighteenth centuries: a study of early modern physics.* Berkeley, Los Angeles: University of California Press.

———. 1981. *Historical studies in the theory of atomic structure.* New York: Arno Press.

———. 1983. *Physics at the Royal Society during Newton's presidency.* Los Angeles: William Andrews Clark Memorial Library, University of California.

Heilbron, John Lewis, and Thomas S. Kuhn. 1969. "The genesis of the Bohr atom." *Historical Studies in the Physical Sciences* 1:211–290.

Heisenberg, Werner. 1952. *Philosophic problems of nuclear science.* Eight lectures. Trans. F. C. Hayes. London: Faber and Faber; New York: Pantheon.

———. 1958. *Physics and philosophy: the revolution in modern science.* New York: Harper & Brothers.

———. 1971. *Physics and beyond: encounters and conversations.* Trans. Arnold J. Pomerans. New York: Harper & Row.

Helmholtz, H. von. 1907. *Vorlesungen über Elektrodynamik und Theorie des Magnetismus.* Leipzig: Verlag von Johann Ambrosius Barth.

Heloise. 1697. *Lettre d'Heloïse à Abailard.* Amsterdam: chez Pierre Chayer.

————. 1729. *Letters of Abelard and Heloise.* 5th ed. Trans. John Hughes. London: J. Watts. First pub. 1713.

————. 1875. *Lettres d' Abélard et d' Héloïse.* 2nd ed. Trans. Octave Gréard. Paris: Garnier Frères.

Herbert, Sandra. 1971. "Darwin, Malthus, and selection." *Journal of the History of Biology* 4:209–217.

Herder, Johann Gottfried. 1800. *Outlines of a philosophy of the history of man.* Trans. from *Ideen zur Philosophie der Geschichte der Menschheit* by T. Churchill. London: Johnson. Facsimile reprint, 1966[?], New York: Bergman Publishers.

————. 1887–1913. *Sämtliche Werke.* Ed. Bernhard Suphan, Carl Redlich, Reinhold Steig, et al. 33 vols. Vols. 13–14, *Ideen zur Philosophie der Geschichte der Menschheit.* Berlin: Weidmannsche Buchhandlung. Facsimile reprint, 1967, Hildesheim: Georg Olms Verlags-Buchhandlung.

————. 1968. *Reflections on the philosophy of the history of mankind.* Abridged, with intro. by Frank E. Manuel. Chicago, London: The University of Chicago Press.

Hermann, Armin. 1973. "Lenard, Philipp." *Dictionary of Scientific Biography* 8:180–183.

———— . 1975. "Stark, Johannes." *Dictionary of Scientific Biography* 12:613–616.

Herschel, John F. W. 1857. *Essays from the Edinburgh and Quarterly Reviews, with addresses and other pieces.* London: Longman, Brown, Green, Longmans & Roberts. Photo-reprint, 1981, New York: The Arno Press.

Hertz, Heinrich. 1893. *Electric waves.* Trans. D. E. Jones. London: Macmillan.

Hess, H. H. 1962. "History of the ocean basins." Pp. 599–620 of A. E. J. Engel, H. L. James, and B. F. Leonards, eds., *Petrologic studies: a volume in honor of A. F. Buddington,* New York: Geological Society of America.

————. 1968. "Reply." *Journal of Geophysical Research* 73:6569.

Hessen, Boris. 1931. "The social and economic roots of Newton's 'Principia.'" Pp. 149–209 of N. I. Bukharin et al., *Science at the cross roads: Papers presented to the International Congress of the History of Science and Technology, London, 29 June–3 July 1931, by the delegates of the U.S.S.R.* London: Kniga. Reprint, 1971, foreword by Joseph Needham, intro. by P. G. Werskey, London: Frank Cass & Co.

————. 1971. *The social and economic roots of Newton's 'Principia.'* Intro. by Robert S. Cohen. New York: Howard Fertig. Facsimile reprint of Hessen 1931.

Hicks, Sir John. 1976. "'Revolutions' in economics." Pp. 207–218 of Spiro Latsis, ed., *Method and appraisal in economics,* Cambridge: Cambridge University Press.

Hill, Christopher. 1965. *Intellectual origins of the English Revolution.* Oxford: Oxford University Press.

————. 1966. *The century of revolution, 1603–1714.* New York: W. W. Norton & Company. First pub., 1961, London, Edinburgh: Thomas Nelson and Sons.

————. 1972. *The century of revolution, 1603–1714.* 2nd ed. London: Sphere Books.

————. 1972a. *The world turned upside down: radical ideas during the English Revolution.* New York: The Viking Press.

————. 1974. *Change and continuity in seventeenth-century England.* London: Weidenfeld & Nicolson.

Himmelfarb, Gertrude. 1959. *Darwin and the Darwinian revolution.* Garden City, N.Y.: Doubleday & Company.

Hirschfield, John Milton. 1981. *The Académie Royale des Sciences, 1666–1683.* New York: Arno Press.

Hobbes, Thomas. 1962. *Behemoth: the history of the causes of the civil wars of England.* Ed. William Molesworth. New York: Burt Franklin.

————. 1969. *Behemoth or the Long Parliament.* Ed. Ferdinand Tönnies. 2nd ed. London: Frank Cass and Co.

Hobsbawm, Eric J. 1954. "The general crisis of the seventeenth century." *Past and Present* 5:33–53; 6:44–65.

————. 1959. *Primitive rebels: studies in the archaic forms of social movement in the nineteenth and twentieth centuries.* New York: W. W. Norton & Company.

————. 1962. *The age of revolution, 1789–1848.* New York, Toronto: Mentor.

————. 1975. *The age of capital, 1848–1875.* New York: Charles Scribner's Sons.

Hoffmann, Banesh, and Helen Dukas. 1972. *Albert Einstein: creator and rebel.* New York: The Viking Press.

Holland, Sir Henry. 1848. [Review of Edward Turner's *Elements of Chemistry* (8th ed., ed. Baron Liebig and Professor Gregory, London, 1847) and Thomas Graham's *Elements of Chemistry* (2nd ed., London, 1847).] *Quarterly Review* 83:37–70.

Holmes, A. 1928. "Continental drift." *Nature* 122:431–433.

Holmes, Frederick Lawrence. 1974. *Claude Bernard and animal chemistry: the emergence of a scientist.* Cambridge, Mass.: Harvard University Press.

Holt, Robert H. 1968. "Freud, Sigmund." *International Encyclopedia of the Social Sciences* 6:1–12. An earlier version (1966) in Robert W. W. White, ed., *The study of lives,* New York: Atherton Press.

Holton, Gerald. 1973. *Thematic origins of scientific thought: Kepler to Einstein.* Cambridge, Mass.: Harvard University Press.

————. 1978. *The scientific imagination: case studies.* Cambridge, London, New York: Cambridge University Press.

————. 1981. "Einstein's search for the *Weltbild.*" *Proceedings of the American Philosophical Society* 125:1–15.

Holton, Gerald, and Yehuda Elkana, eds. 1982. *Albert Einstein, historical and cultural perspectives: the centennial symposium in Jerusalem.* Princeton: Princeton University Press.

Hooke, Robert. 1665. *Micrographia: or some physiological descriptions of minute bodies made by magnifying glasses, with observations and inquiries thereupon.* London: printed by Jo. Martyn and Ja. Allestry.

l'Hospital [l'Hôpital], Guillaume-François-Antoine de. 1696. *Analyse des infiniment petits pour l'intelligence des lignes courbes.* Paris: de l'Imprimerie Royale. 2nd ed, 1715, Paris: chez François Montalant.

Howell, James. 1890–1892. *Epistolae Ho-Elianae: the familiar letters.* 2 vols. Ed. Joseph Jacobs. London: David Nutt.

Hull, David L. 1973. *Darwin and his critics: the reception of Darwin's theory of evolution by the scientific community.* Cambridge, Mass.: Harvard University Press.

Humboldt, Alexander von. 1845–1862. *Kosmos.* 5 vols. Stuttgart: J. G. Cotta.

———. 1848–1865. *Cosmos.* 5 vols. Trans. E. C. Otté. London: H. G. Bohn.

———. 1849–1851. *Cosmos: a sketch of a physical description of the universe.* 3 vols. Trans. Mrs. Edward Sabine. London: Longman, Brown, Green, and Longmans; and John Murray.

Hume, David. 1888. *A treatise of human nature.* Ed. L. A. Selby-Bigge. Oxford: at the Clarendon Press. Reprint, 1967.

Hurley, Patrick M. 1959. *How old is the earth?* Garden City, N.Y.: Doubleday.

Hutchings, Donald, ed. 1969. *Late seventeenth century scientists.* Oxford, London, New York: Pergamon Press.

Huxley, Julian. 1942. *Evolution: the modern synthesis.* London: George Allen & Unwin. Rev. ed., 1974, ed. John R. Baker, London: George Allen & Unwin.

Huxley, Thomas H. 1881. "On the hypothesis that animals are automata, and its history." Pp. 199–245 of *Science and culture and other essays,* London: Macmillan and Co.

———. 1887. "On the reception of the 'Origin of Species.'" Pp. 179 ff. of Darwin 1887, vol. 2.

———. 1894. *Discourses biological and geological.* New York: D. Appleton and Company. [Collected Essays, vol. 8.]

Hyman, Anthony. 1982. *Charles Babbage: pioneer of the computer.* Oxford, New York: Oxford University Press.

Ihde, Aaron J. 1964. *The development of modern chemistry.* New York, Evanston, London: Harper and Row.

Infeld, Leopold. 1950. *Albert Einstein: his work and influence on our world.* New York: Charles Scribner's Sons.

Ives, E. W., ed. 1968. *The English revolution: 1600–1660.* New York, London: Harper & Row; New York: Barnes & Noble.

Jacob, James R. 1978. *Robert Boyle and the English Revolution.* New York: B. Franklin.

Jacob, Margaret C. 1976. *The Newtonians and the English Revolution, 1689–1720.* Ithaca: Cornell University Press.

———. 1981. *The radical enlightenment: pantheists, Freemasons and republicans.* London, Boston: George Allen & Unwin.

Jacobs, J. A., R. D. Russell, and J. Tuzo Wilson. 1959. *Physics and geology.* New York: McGraw-Hill Book Company.

Jammer, Max. 1966. *The conceptual development of quantum mechanics.* New York: McGraw-Hill Book Company.

———. 1974. *The philosophy of quantum mechanics: the interpretations of quantum mechanics in historical perspective.* New York, London: John Wiley & Sons.

Jastrow, Robert. 1979. "Velikovsky, a star-crossed theoretician of the cosmos." *New York Times,* 2 Dec. 1979.

Jeans, Sir James. 1943. *Physics and philosophy.* Cambridge: at the University Press; New York: The Macmillan Company.

———. 1948. *The growth of physical science.* New York: The Macmillan Company.

Jeffreys, Harold. 1952. *The earth: its origin, history and physical constitution.* Cambridge: at the University Press.

Johnson, Chalmers. 1964. *Revolution and the social system.* Stanford: The Hoover Institution on War, Revolution, and Peace, Stanford University.

———. 1966. *Revolutionary change.* Boston: Little, Brown and Company.

Johnson, Samuel. 1755. *A dictionary of the English language.* . . . 2 vols. London: printed by W. Strahan for J. and P. Knapton, T. and T. Longman, C. Hitch and L. Hawes, A. Millar, and R. and J. Dodsley. Photo-reprint, 1979, New York: Arno Press.

———. 1969, *The rambler.* Ed. W. J. Bate and Albrecht B. Strauss. Vol. 2. New Haven: Yale University Press.

Jones, Bessie Zaban, ed. 1966. *The golden age of science: thirty portraits of the giants of nineteenth-century science by their scientific contemporaries.* Intro. by Everett Mendelsohn. New York: Simon and Schuster [in cooperation with the Smithsonian Institution].

Jones, Ernest. 1913. *Papers on psycho-analysis.* London: Ballière, Tindall and Cox; New York: William Wood and Co.

———. 1940. "Sigmund Freud." *International Journal of Psychoanalysis* 21:2–26.

———. 1953–1957. *The life and work of Sigmund Freud.* 3 vols. New York: Basic Books; London: The Hogarth Press.

Jones, J. R. 1972. *The revolution of 1688 in England.* New York: W. W. Norton & Company.

Jones, Richard Foster. 1936. *Ancients and moderns: a study of the background of the battle of the books.* St. Louis: Washington University.

———. 1961. *Ancients and moderns: a study of the rise of the scientific movement in seventeenth-century England.* St. Louis: Washington University. Reprint, 1982, New York: Dover Publications.

Jonson, Ben. 1941. "Newes from the new world discover'd in the moone." Ed. C. H. Herford, Percy and Evelyn Simpson. Pp. 511–525 of *Ben Jonson,* vol. 7. Oxford: at the Clarendon Press.

Kahan, Théo. 1962. "Un document historique de l'académie des sciences de Berlin sur l'activité scientifique d'Albert Einstein (1913)." *Archives Internationales d'Histoire des Sciences* 15:337–342.

Kahlbaum, Georg W. A. 1904. *Justus von Liebig und Friedrich Mohr in ihren Briefen von 1834–1870.* Leipzig: Johann Ambrosius Barth. [Monographien aus der Geschichte der Chemie 8.]

Kant, Immanuel. 1787. *Kritik der reinen Vernunft.* Zweite hin und wieder verbesserte Auflage. Riga: Johann Friedrich Hartnoch.

———. 1855. *Critique of pure reason.* Trans. J. M. D. Meiklejohn. London: Henry G. Bohn.

———. 1902–1955. *Gesammelte Schriften.* 23 vols. Berlin: German Academy of Sciences.

———. 1926. *Kritik der reinen Vernunft.* Ed. Raymund Schmidt. Leipzig: Verlag von Felix Meiner. [Der Philosophischen Bibliothek, Band 37a.]

———. 1929. *Immanuel Kant's Critique of pure reason.* Trans. Norman Kemp Smith. London: Macmillan and Co. A corrected second impression appeared in 1933. Reprints 1950, 1973.

———. 1960. *Religion within the limits of reason alone.* Trans. T. M. Greene and H. H. Hudson. 2nd ed. La Salle, Ill.: The Open Court Publishing Co.

———. 1967. See Martin 1967.

Kaplan, A. 1952. "Sociology learns the language of mathematics." *Commentary* 14:274–284. Reprint, 1953, pp. 394–412 of Philip Wiener, ed., *Readings in philosophy of science,* New York: Charles Scribner's Sons.

Käsler, Dirk. 1977. *Revolution und Veralltäglichung: eine Theorie postrevolutionärer Prozesse.* Munich: Nymphenburger Verlagshandlung.

Kastner, Abraham Gotthelf. 1796–1800. *Geschichte der Mathematik seit der Wiederherstellung der Wissenschaften bis an das Ende des achtzehnten Jahrhunderts.* 4 vols. Göttingen: J. G. Rosenbusch.

Kaufmann, Walter. 1980. *Discovering the mind: Goethe, Kant, and Hegel.* New York: McGraw-Hill Book Company.

Kazin, Alfred. 1957. "The Freudian revolution analyzed." Pp. 65–74 of Mujeeb-Ur-Rahman 1977; also in Nelson 1957.

Kearney, Hugh. 1964. *Origins of the scientific revolution.* London: Longmans, Green and Co.

Keele, Kenneth D. 1965. *William Harvey: the man, the physician, and the scientist.* London, Edinburgh: Nelson.

Keeling, S. V. 1968. *Descartes.* 2nd ed. London, Oxford, New York: Oxford University Press.

Kemp, John. 1968. *The philosophy of Kant.* Oxford, New York: Oxford University Press.

Kennedy, Edward S., and Imad Ghanem, eds. 1976. *The life and work of Ibn al-Shāṭir, an Arab astronomer of the fourteenth century.* A memorial volume published on the occasion of the opening of the Institute for the History of Arabic Science of the University of Aleppo. Aleppo: Institute for the History of Arabic Science, University of Aleppo.

Kenyon, J. P. 1978. *Stuart England.* Harmondsworth, Middlesex, England: Penguin Books [The Pelican History of England 6].

Kepler, Johannes. 1929. *Neue Astronomie.* Trans., ed. Max Caspar. Berlin: Verlag R. Oldenbourg.

———. 1937–. *Gesammelte Werke.* Herausgegeben im Auftrag der Deutschen

Forschungsgemeinschaft und der Bayerischen Akademie der Wissenschaften. Munich: C. H. Beck'sche Verlagsbuchhandlung.

———. 1981. *Mysterium cosmographicum: the secret of the universe*. Trans. A. M. Duncan, intro. and commentary by E. J. Aiton. New York: Abaris Books.

Keynes, Sir Geoffrey. 1966. *The life of William Harvey*. Oxford: at the Clarendon Press.

Kidd, Benjamin. 1894. *The evolution of society*. New York, London.

Kilgour, Frederick G. 1954. "William Harvey's use of the quantitative method." *The Yale Journal of Biology and Medicine* 26:410–421.

Kitts, David. 1974. "Continental drift and scientific revolution." *The American Association of Petroleum Geologists Bulletin* 58:2490–2496.

Klein, Martin J. 1962. "Max Planck and the beginnings of the quantum theory." *Archive for History of Exact Sciences* 1:459–472.

———. 1964. "Einstein and the wave-particle duality." *The Natural Philosopher* 3:1–49.

———. 1965. "Einstein, specific heats, and the early quantum theory." *Science* 148:173–180.

———. 1970. "The first stage of the Bohr-Einstein dialogue." *Historical Studies in the Physical Sciences* 2:1–39.

———. 1975. "Einstein on scientific revolutions." *Vistas in Astronomy* 17:113–133.

———. 1977. "The beginnings of the quantum theory." Pp. 1–39 of C. Weiner, ed., *History of twentieth century physics*, New York: Academic Press.

Kline, Morris. 1972. *Mathematical thought from ancient to modern times*. New York: Oxford University Press.

Koestler, Arthur. 1959. *The sleepwalkers: a history of man's changing vision of the universe*. London: Hutchinson.

———. 1965. "Evolution and revolution in the history of science." *Encounter* 25:32–38.

———. 1971. *The case of the midwife toad*. New York: Random House.

Koselleck, Reinhart. 1969. "Der neuzeitliche Revolutionsbegriff als geschichtliche Kategorie." *Studium Generale* 22:825–838.

Kottler, M. J. 1974. "Alfred Russel Wallace, the origin of man, and spiritualism." *Isis* 65:145–192.

Koyré, Alexandre. 1939. *Etudes galiléennes*. Paris: Hermann & Cie. Reprint, 1966.

———. 1961. *La révolution astronomique: Copernic, Kepler, Borelli*. Paris: Hermann [Histoire de la Pensée 3].

———. 1965. *Newtonian studies*. Cambridge, Mass.: Harvard University Press.

———. 1973. *The astronomical revolution: Copernicus, Kepler, Borelli*. Trans. R. E. W. Maddison. Paris: Hermann; Ithaca, N.Y.: Cornell University Press.

———. 1978. *Galileo studies*. Trans. John Mepham. Atlantic Highlands, N.J.: Humanities Press.

Kramnick, Isaac. 1972. "Reflections on revolution: definition and explanation in recent scholarship." *History and Theory* 11:26–63.

Kremer, Richard L. 1981. "The use of Bernard Walther's astronomical observations: theory and observations in early modern astronomy." *Journal for the History of Astronomy* 12:124–132.

Kristeller, Paul Oskar. 1943. "The place of classical humanism in Renaissance thought." *Journal of the History of Ideas* 4:59–63.

Kronig, R., ed. 1954. *Textbook of physics*. London: Pergamon Press.

Kubrin, David. 1967. "Newton and the cyclical cosmos." *Journal of the History of Ideas* 28:325–346.

Kuhn, Thomas S. 1957. *The Copernican revolution: planetary astronomy in the development of western thought*. Cambridge, Mass.: Harvard University Press.

———. 1961. "The function of measurement in modern physical science." *Isis* 52:161ff.

———. 1962. *The structure of scientific revolutions*. Chicago. London: The University of Chicago Press. 2nd ed., rev., 1970.

———. 1970. "Reflections on my critics." Pp. 231–278 of Lakatos and Musgrave 1970.

———. 1974. "Second thoughts on paradigms." Pp. 459–482 of Suppe 1974.

———. 1977. *The essential tension: selected studies in scientific tradition and change*. Chicago, London: The University of Chicago Press.

———. 1978. *Black-body theory and the quantum discontinuity, 1894–1912*. Oxford: Clarendon Press.

Lakatos, Imre. 1978. *The methodology of scientific research programmes*. Ed. John Worrall and Gregory Currie. Cambridge: Cambridge University Press.

Lakatos, Imre, and Alan Musgrave, eds. 1970. *Criticism and the growth of knowledge: Proceedings of the International Colloquium in the Philosophy of Science, London, 1965*. Cambridge: at the University Press.

Lake, Philip. 1922. "Wegener's displacement theory." *The Geological Magazine* 59:338–346.

———. 1928. [Review of van der Gracht 1928.] *The Geological Magazine* 65:422–424.

Lalande, Joseph Jérôme le Français de. 1764. *Astronomie*. 2 vols. Paris: chez Desaint & Saillant.

Landes, David S. 1969. *The unbound Prometheus*. Cambridge: at the University Press

———. 1983. *Revolution in time: clocks and the making of the modern world*. Cambridge, London: The Belknap Press of Harvard University Press.

Lane, Peter. 1978. *The industrial revolution: the birth of the modern age*. London: Weidenfeld and Nicolson.

Langer, William L. 1969. *Political and social upheaval, 1832–1852*. New York, Evanston, London: Harper & Row.

———. 1969a. *The revolutions of 1848*. New York, Evanston: Harper & Row.

———, ed. 1968. *An encyclopedia of world history*. 4th ed. Boston: Houghton Mifflin.

Langmuir, I. 1968. "Pathological science." Ed. R. N. Hall from a talk given at the Knolls Research Laboratory, Dec. 18, 1953. [General Electric Report 68–C–035.]

Laplace, Pierre Simon, Marquis de. 1966. *Celestial mechanics.* 4 vols. Trans. Nathaniel Bowditch. New York: Chelsea Publishing Company. Corrected facsimile reprint of the volumes published in Boston, 1829, 1832, 1834, 1839.

Lasky, Melvin J. 1976. *Utopia and revolution.* Chicago. London: The University of Chicago Press.

Laslett, P. 1956. "The English Revolution and Locke's 'Two treatises of government.'" *The Cambridge Historical Journal* 12:40–55.

Latsis, Spiro J., ed. 1976. *Method and appraisal in economics.* Cambridge, New York: Cambridge University Press.

Laudan, Larry. 1977. *Progress and its problems: toward a theory of scientific growth.* Berkeley, Los Angeles: University of California Press.

———. 1981. "A problem-solving approach to scientific progress." Pp. 144–155 of Hacking 1981.

Laudan, Rachel. 1978. "The recent revolution in geology and Kuhn's theory of scientific change." Pp. 227–239 of *Proceedings of the 1978 biennial meeting of the Philosophy of Science Association,* ed. Peter D. Asquith and Ian Hacking, East Lansing, Mich.: Philosophy of Science Association. Reprinted, pp. 284–296 of Gutting 1980.

Laue, Max von. 1950. *History of physics.* Trans. Ralph Oesper. New York: Academic Press.

Lavoisier, Antoine-Laurent, et al. 1789. *Nomenclature chimique ou synonymie ancienne et moderne.* Paris: Cuchet.

Le Clerc, Daniel. 1723. *Histoire de la médecine.* New ed. Amsterdam: aux dépens de la Compagnie.

Lederer, Emil. 1936. "On revolutions." *Social Research* 3:1–18.

Leibnitz, Gottfried W. 1896. *New essays concerning human understanding.* Trans. A. G. Langley. La Salle, Ill.: The Open Court Publishing Co.

Lenin, V. I. 1908. *Materialism and empirio-criticism.* Moscow: Foreign Languages Publishing House.

Lenzer, Gertrud, ed. 1975. *Auguste Comte and positivism: the essential writings.* New York, Evanston: Harper & Row [Harper Torchbooks].

Lepenies, Wolf. 1976. *Das Ende der Naturgeschichte: Wandel kultureller Selbstverständlichkeiten in den Wissenschaften des 18. und 19. Jahrhunderts.* Munich: Carl Hauser Verlag. Reprint, 1978, Frankfurt am Main: Suhrkamp.

Leslie, Sir John. 1835. "Dissertation fourth: exhibiting a general view of the progress of mathematical and physical science, chiefly during the eighteenth century." Pp. 575–677 of *Dissertations on the history of metaphysical and ethical and of mathematical and physical science,* by Dugald Stewart, Sir James Mackintosh, John Playfair, and Sir John Leslie. Edinburgh: Adam and Charles Black.

Leuliette, Jean-Jacques. 1804. *Discours qui a eu la mention honorable, sur cette question proposée par l'Institut National: quelle a été l'influence de la réformation* de Luther, sur les lumières et la situation politique des différens états de l'Europe? Paris: chez Gide, chez J.-P. Jacob.

Lewis, Bernard. 1972. "Islamic concepts of revolution." Pp. 30–40 of P. S. Vatikiotis, ed., *Revolution in the Middle East,* London: George Allen & Unwin, 1972.

Lewis, Sir Thomas. 1933. *The Harveian oration on "clinical science."* London: M. K. Lewis & Co.

Lichtenberg, Georg Christoph. 1800–1806. *Vermischte Schriften.* 9 vols. Ed. Ludwig Christian Lichtenberg and Friedrich Kries. Göttingen: Dieterichsche Buchhandlung. Vols. 1–5 make up the *Vermischte Schriften* (1800–1804); vols. 6–9 are entitled *Physikalische und mathematische Schriften;* later volumes bear the imprint Heinrich Dieterich.

———. 1967–1972. *Schriften und Briefe.* Ed. Wolfgang Promies. 4 vols. Munich: Carl Hanser Verlag.

Liebig, Justus von. 1863. "Lord Bacon as natural philosopher." *MacMillan's Magazine* 8:237–267.

———. 1874. *Reden und Abhandlungen.* Ed. M. Carriere. Leipzig, Heidelberg: C. F. Winter'sche Verlagshandlung. Reprint, Wiesbaden, Dr. Martin Sändig.

———. 1891. "An autobiographical sketch." Trans. J. Campbell Brown. *Chemical News* 63:265–267, 276–277.

———. 1891a. "Eigenhändige biographische Aufzeichnungen." *Deutsche Rundschau* 66:30–39.

Limoges, Camille. 1970. *La sélection naturelle: étude sur la première constitution d'un concept (1837–1859).* Paris: Presses Universitaires de France.

Lindroth, Sten. 1957. "Harvey, Descartes, and young Olaus Rudbeck." *Journal of the History of Medicine and Allied Sciences* 12:209–219.

Lindsay, A. D. 1934. *Kant.* London: Ernest Benn.

Littré, E. 1881–1883. *Dictionnaire de la langue française.* 4 vols. Paris: Librairie Hachette.

Loeb, Jacques. 1964. *The mechanistic conception of life.* Ed. Donald Fleming. Cambridge, Mass.: The Belknap Press of Harvard University Press.

Longwell, Chester R. 1944. "Some thoughts on the evidence for continental drift." *American Journal of Science* 242:218–231.

López Piñero, José M. 1963. *Orígenes históricos del concepto de neurosis.* Valencia: Cátedra e Instituto de História de la Medicina.

Lorentz, H. A., A. Einstein, H. Minkowski, and H. Weyl. 1923. *The principle of relativity: a collection of original memoirs on the special and general theory of relativity.* Trans. W. Perrett and G. B. Jeffery. London: Methuen and Company. Reprint, 1952, New York: Dover Publications.

Losee, John. 1972. *A historical introduction to the philosophy of science.* London, Oxford: Oxford University Press.

Lukács, Georg. 1923. *History and class consciousness: studies in Marxist dialectics.* Trans. Rodney Livingstone. Cambridge, Mass.: The MIT Press.

Lyell, Charles. 1914. *The geological evidence of the antiquity of man*. London: J. M. Dent & Sons; New York: E. P. Dutton & Co.

MacDonald, D. K. C. 1964. *Faraday, Maxwell, and Kelvin*. Garden City, N.Y.: Doubleday & Company [Anchor Books].

Macey, Samuel L. 1980. *Clocks and the cosmos: time in western life and thought*. Hamden, Conn.: Archon Books.

Mach, Ernst. 1896. *Populär-Wissenschaftliche Vorlesungen*. Leipzig: Johann Ambrosius Barth.

———. 1911. *History and root of the principle of the conservation of energy*. Trans. Philip E. B. Jourdain. Chicago: The Open Court Publishing Co.; London: Kegan Paul, Trench, Trubner & Co.

———. 1914. *The analysis of sensations and the relation of the physical to the psychical*. Trans. C. M. Williams. Rev. by Sydney Waterlow. Chicago, London: The Open Court Publishing Company.

———. 1943. *Popular scientific lectures*. Trans. Thomas J. McCormack. 5th ed. La Salle, Ill.: The Open Court Publishing Company.

———. 1960. *The science of mechanics: a critical and historical account of its development*. Trans. Thomas J. McCormack. 6th ed. La Salle, Ill.: The Open Court Publishing Company.

Maclaurin, Colin. 1748. *An account of Sir Isaac Newton's philosophical discoveries, in four books*. London: printed for the author's children and sold by A. Millar and J. Nourse. Facsimile reprint, 1968, New York, London: Johnson Reprint Corporation.

Macrakis, Kristie. 1984. "Alfred Wegener: self-proclaimed scientific revolutionary." *Archives Internationales d'Histoire des Sciences* 112:182–195.

Magee, Bryan. 1973. *Popper*. London: Fontana/Collins.

Magie, William Francis. 1912. "The primary concepts of physics." *Science* 35:281–293.

Maienschein, Jane. 1981. "Shifting assumptions in American biology: embryology, 1890–1910." *Journal of the History of Biology* 14:89–113.

Maleville [fils]. 1804. *Discours sur l'influence de la réformation de Luther*. Paris: chez Le Normant.

Mandelbrot, Benoît B. 1982. "Des monstres de Cantor et de Peano à la géométrie fractale de la nature." Pp. 226–251 of Roger Apéry et al., eds., *Penser les mathématiques*, Paris: Editions du Seuil.

———. 1983. *The fractal geometry of nature*. Updated and augmented. San Francisco: W. H. Freeman and Company. Earlier eds., 1977, 1982.

Manning, Kenneth R. 1983. *Black Apollo of science: the life of Ernest Everett Just*. New York, Oxford: Oxford University Press.

Manuel, Frank E. 1956. *The new world of Henri Saint-Simon*. Cambridge, Mass.: Harvard University Press.

———. 1962. *The prophets of Paris*. Cambridge, Mass.: Harvard University Press.

———. 1968. *A portrait of Isaac Newton*. Cambridge, Mass.: The Belknap Press of Harvard University Press.

————. 1974. *The religion of Isaac Newton: the Freemantle Lectures 1973*. Oxford: at the Clarendon Press.

Marichal, Juan. 1970. "From Pistoia to Cadiz: a generation's itinerary." Pp. 97–110 of Alfred O. Aldridge, ed., *The Ibero-American Enlightenment*, Urbana: The University of Illinois Press.

Marmontel, Jean François. 1787. *Oeuvres complettes*. Vol. 9. Paris: chez Née de la Rochelle.

Martin, Gottfried, ed. 1967–1969. *Allgemeiner Kantindex zu Kants Gesammelten Schriften*. 3 vols. Berlin: de Gruyter.

Marvin, F. S., ed. 1923. *Science and civilization*. London: Oxford University Press.

Marvin, Ursula B. 1973. *Continental drift: the evolution of a concept*. Washington, D.C.: Smithsonian Institution Press.

————. 1980. "Continental drift." Pp. 108–115 of *The New Encyclopaedia Britannica, Macropaedia*, vol. 5, 15th ed.

Marx, Karl. 1954–1962. *Capital*. 3 vols. Trans. Samuel Moore and Edward Aveling, ed. Friedrich Engels. Moscow: Foreign Languages Publishing House.

————. 1963–1971. *Theories of surplus-value*. 3 vols. Moscow: Progress Publishers.

————. 1971. *On revolution*. Trans., ed., intro. by Saul K. Padover. New York: McGraw-Hill Book Company.

————. 1977. *Selected writings*. Ed. David McLellan. Oxford: Oxford University Press, 1977.

Marx, Karl, and Frederick Engels. 1962. *Selected works in two volumes*. Moscow: Foreign Languages Publishing House.

————. 1962a. *On Britain*. 2nd ed. Moscow: Foreign Languages Publishing House.

Mason, S. F. 1953. *Main currents of scientific thought: a history of the sciences*. New York: Henry Schuman.

Masterman, Margaret. 1970. "The nature of a paradigm." Pp. 59–89 of Lakatos & Musgrave 1970.

Maupertuis, Pierre Louis Moreau de. 1736. "Sur les loix de l'attraction." Pp. 473–504 of *Suite des mémoires de mathématique et de physique, tirés des registres de l'Académie Royale des Sciences de l'année M.DCCXXXII*. Amsterdam: chez Pierre Mortier.

Mauskopf, Seymour H., ed. 1979. *The reception of unconventional sciences*. Washington: American Association for the Advancement of Science [AAAS Selected Symposium 25].

Mautner, Franz H., and Franklin H. Miller, Jr. 1952. "Remarks on G. C. Lichtenberg." *Isis* 43:223–231.

Maxwell, James Clerk. 1873. *Treatise on electricity and magnetism*. Oxford: Oxford University Press. 2nd ed. 1881.

————. 1890. *The scientific papers of James Clerk Maxwell*. 2 vols. Ed. W. D. Niven. Cambridge: at the University Press.

Mayr, Ernst. 1961. "Cause and effect in biology." *Science* 134:1501–1506.

―――. 1972. "The nature of the Darwinian revolution." *Science* 176:981–989. Reprint, pp. 276–296 of Mayr 1976.

―――. 1976. *Evolution and the diversity of life: selected essays.* Cambridge, Mass., London: The Belknap Press of Harvard University Press.

―――. 1977. "Darwin and natural selection." *American Scientist* 65:321–327.

―――. 1982. *The growth of biological thought: diversity, evolution, and inheritance.* Cambridge, Mass., London: The Belknap Press of Harvard University Press.

Mayr, Ernst, and William B. Provine, eds. 1980. *The evolutionary synthesis or the unification of biology.* Cambridge, Mass., London: Harvard University Press.

Mazlish, Bruce, 1976. *The revolutionary ascetic: evolution of a political type.* New York: Basic Books.

Mazzini, Joseph. 1907. *The duties of man and other essays.* London: J. M. Dent & Sons; New York: E. P. Dutton & Co.

McCormmach, Russell. 1972. "Hertz, Heinrich Rudolph." *Dictionary of Scientific Biography* 6:340–350.

McKenzie, D. P. 1977. "Plate tectonics and its relationship to the evolution of ideas in the geological sciences." *Daedalus* 106:97–124.

McKie, Douglas. 1935. *Antoine Lavoisier: the father of modern chemistry.* Philadelphia: J. B. Lippincott Co.

McLellan, David, ed. 1977. *Karl Marx: selected writings.* Oxford, London: Oxford University Press.

McMullin, Ernan, ed. 1967. *Galileo, man of science.* New York, London: Basic Books.

Mead, George Herbert. 1936. *Movements of thought in the nineteenth century.* Chicago: University of Chicago Press.

―――. 1949. *Movements of thought in the nineteenth century.* Chicago: University of Chicago Press.

Medawar, P. B. 1977. "Fear and DNA." *The New York Review of Books,* 27 Oct. 1977, pp. 15–20.

―――. 1979. *Advice to a young scientist.* New York: Harper & Row.

Medawar, P. B., and J. S. Medawar. 1983. *Aristotle to zoos: a philosophical dictionary of biology.* Cambridge, Mass.: Harvard University Press.

Mehra, Jagdish, and Helmut Rechenberg. 1982. *The historical development of the quantum theory.* Vol. I, pt. 1. New York, Heidelberg, Berlin: Springer–Verlag.

Mehring, Franz. 1962. *Karl Marx: the story of his life.* Ann Arbor: The University of Michigan Press.

Mehrtens, Herbert. 1976. "T. S. Kuhn's theories and mathematics: a discussion paper on the 'new historiography' of mathematics." *Historia Mathematica* 3:297–320.

Meissner, Walter. 1951. "Max Planck, the man and his work." *Science* 113:75–81.

Meldrum, Andrew N. 1930. *The eighteenth century revolution in science—the first phase.* London, New York, Toronto: Longmans, Green and Co. Reprint,

1981, in Andrew Meldrum, *Essays in the history of chemistry*, ed. I. Bernard Cohen, New York: Arno Press.

Melnick, Arthur. 1973. *Kant's analogies of experience*. Chicago, London: The University of Chicago Press.

Mendelsohn, Everett. 1966. "The context of nineteenth-century science." Pp. xiiiff. of Bessie Zaban Jones, ed., *The golden age of science: thirty portraits of the giants of nineteenth-century science by their scientific contemporaries*, New York: Simon and Schuster.

Merchant, Carolyn. 1980. *The death of nature: woman, ecology, and the scientific revolution*. San Francisco: Harper & Row.

Merriman, Roger B. 1938. *Six contemporaneous revolutions*. Oxford: at the Clarendon Press.

Merton, Robert K. 1938. "Science, technology and society in seventeenth century England." *Osiris* 4:360–632. Reprint, 1970, with a new introduction by the author, New York: Howard Fertig; New York: Harper & Row [Harper Torchbooks].

———. 1957. *Social theory and social structure*. Glencoe, Ill.: The Free Press.

Merz, John Theodore. 1896–1914. *A history of European thought in the nineteenth century*. 2 vols. Edinburgh, London: William Blackwood and Sons.

Meschkowski, H. 1971. "Cantor, Georg." *Dictionary of Scientific Biography* 3:52–58.

Meyerhoff, A. A. 1968. "Arthur Holmes: originator of spreading ocean floor hypothesis." *Journal of Geophysical Research* 73:6563–6565.

Meyerson, Emile. 1931. *Du cheminement de la pensée*. 3 vols. Paris: Librairie Félix Alcan.

Michelson, Albert A. 1903. *Light waves and their uses*. Chicago: The University of Chicago Press.

Milhaud, Gaston. 1921. *Descartes savant*. Paris: Librairie Félix Alcan.

Mill, John Stuart. 1843. *A system of logic, ratiocinative and inductive, being a connected view of the principles of evidence, and the methods of scientific investigation*. 2 vols. London: J. W. Parker.

———. 1845. [Review of *The logic of political economy*, by Thomas De Quincey.] *The Westminster Review* 43:319–331.

———. 1848. *Principles of political economy with some of their applications to social philosophy*. 2 vols. London: John W. Parker.

———. 1889. *An examination of Sir William Hamilton's philosophy*. 6th ed. London: Longmans, Green, and Co.

———. 1973–1974. *A system of logic ratiocinative and inductive, being a connected view of the principles of evidence and the methods of scientific investigation*. Ed. J. M. Robson. Toronto, Buffalo: University of Toronto Press; London: Routledge & Kegan Paul. *Collected Works of John Stuart Mill*, vol. 7 (Books 1–3, 1973) and vol. 8 (Books 4–6 and appendices, 1974).

Miller, Arthur I. 1981. *Albert Einstein's special theory of relativity: emergence (1905) and early interpretation (1905–1911)*. Reading, Mass.: Addison-Wesley Publishing Company.

———. 1982. "The special relativity theory: Einstein's response to the physics of 1905." Pp. 3–26 of Holton and Elkana 1982.

———. 1984. *Creating twentieth-century physics: imagery in scientific thought.* Cambridge, Mass.: Birkhauser.

Miller, Samuel. 1803. *A brief retrospect of the eighteenth century.* Part first; in 2 vols.; containing a sketch of the revolutions and improvements in science, arts, and literature, during that period. New York: printed by T. and J. Swords.

Millikan, Robert A. 1912. "New proofs of the kinetic theory of matter and the atomic theory of electricity." *The Popular Science Monthly* 80:417–440.

———. 1917. *The electron: its isolation and measurement and the determination of some of its properties.* Chicago: University of Chicago Press.

———. 1918. "Twentieth century physics." Pp. 169–184 of *Annual Report of the Board of Regents of the Smithsonian Institution.*

———. 1947. *Electrons (+ and —) protons, photons, neutrons, mesotrons, and cosmic rays.* Chicago: The University of Chicago Press.

———. 1949. "Albert Einstein on his seventieth birthday." *Reviews of Modern Physics* 21:343–345.

———. 1950. *The autobiography of Robert A. Millikan.* New York: Prentice-Hall.

Monardes, Nicholas. 1925. *Joyfull newes out of the newe founde worlde.* Trans. John Frampton. London: Constable; New York: A. A. Knopf.

Monod, Jacques. 1970. *Le hasard et la nécessité.* Paris: Editions du Seuil.

Montaigne, Michel de. 1595. *Les essais.* New ed. Paris: chez Abel l'Angelier.

———. 1603. *The essayes.* Trans. John Florio. London: Printed by Val. Sims for Edward Blount.

———. 1906. *Les essais.* Vol. 1. Ed. Fortunat Strowski. Bordeaux: F. Pech.

———. 1958. *The complete essays.* Trans. Donald M. Frame. Stanford: Stanford University Press.

Montesquieu, Charles de Secondat, Baron de la Brède et de. 1949. *The spirit of the laws.* Trans. Thomas Nugent, intro. by Franz Neumann. 2 vols. New York: Hafner Press.

Montucla, Jean-Etienne. 1758. *Histoire des mathématiques.* 2 vols. Paris: chez Ch. Ant. Jombert. New ed., 4 vols., 1799, Paris: chez Henri Agasse.

More, Louis T. 1912. "The theory of relativity." *The Nation* 94:370–371.

———. 1934. *Isaac Newton: a biography.* New York: Charles Scribner's Sons.

Morton, A. L., ed. 1975. *Freedom in arms: a selection of Leveller writings.* New York: International Publishers.

Moszkowski, Alexander. 1921. *Einstein the searcher: his works explained from dialogues with Einstein.* Trans. Henry L. Brose. London: Methuen & Co. Reprint, 1972, entitled *Conversations with Einstein,* London: Sidgwick & Jackson.

Mujeeb-ur-Rahman, Md., ed. 1977. *The Freudian paradigm: psychoanalysis and scientific thought.* Chicago: Nelson-Hall.

Mullinger, James Bass. 1884. *The University of Cambridge from the royal injunctions of 1535 to the accession of Charles the First.* 2 vols. Cambridge: at the University Press.

Murray, Robert H. 1925. *Science and scientists in the nineteenth century.* London: The Sheldon Press. New York, Toronto: The Macmillan Co.

Musson, A. E., and E. Robinson. 1969. *Science and technology in the Industrial Revolution.* Manchester: Manchester University Press.

National Research Council, Physics Survey Committee. 1973. *Physics in perspective: the nature of physics and the subfields of physics.* Washington, D.C.: National Academy of Sciences.

Needham, Joseph, and Walter Pagel, eds. 1938. *Background to modern science.* New York: The Macmillan Company; Cambridge: at the University Press. Reprint, 1975, of 1940 ed., New York: Arno Press.

Nelson, Benjamin, ed. 1957. *Freud and the twentieth century.* New York: Meridian Books.

Neugebauer, O. 1957. *The exact sciences in antiquity.* Providence: Brown University Press.

————. 1968. "On the planetary theory of Copernicus." *Vistas in Astronomy* 10:89–103.

————. 1975. *A history of ancient mathematical astronomy.* 3 vols. Berlin, Heidelberg, New York: Springer-Verlag.

————. 1983. *Astronomy and history: selected essays.* New York, Berlin: Springer-Verlag.

Newcomb, Simon. 1905. "The evolution of the scientific investigator." Pp. 135–147 of Rogers 1905, vol. 1.

Newton, Alfred. 1888. [Review of *The life and letters of Charles Darwin, including an autobiographical chapter.*] *The Quarterly Review* 166:1–30.

Newton, Isaac. 1961–1977. *The correspondence of Isaac Newton.* 8 vols. Ed. H. W. Turnbull, J. F. Scott, A. Rupert Hall, and Laura Tilling. Cambridge: at the University Press.

————. 1967–1981. *The mathematical papers of Isaac Newton.* 8 vols. Ed. D. T. Whiteside. Cambridge: at the University Press.

————. 1972. *Isaac Newton's Philosophiae naturalis principia mathematica.* The 3rd ed. (1726) with variant readings, assembled by Alexandre Koyré, I. Bernard Cohen, and Anne Whitman. 2 vols. Cambridge: at the University Press; Cambridge, Mass.: Harvard University Press.

Neyman, Jerzy, ed. 1974. *The heritage of Copernicus: theories "pleasing to the mind."* Cambridge, Mass., London: The MIT Press.

Nicolson, Marjorie. 1976. *Science and imagination.* Hamden, Conn.: Archon Books.

Niebuhr, Barthold Georg. 1828–1832. *Römische Geschichte.* 3rd ed. 3 vols. Berlin: Reimer.

Nordenskiöld, Erik. 1928. *The history of biology: a survey.* Trans. Leonard Bucknall Eyre. New York: Tudor Publishing Co.

Norlind, Wilhelm. 1953. "Copernicus and Luther: a critical study." *Isis* 44:273–276.

Nussbaum, Frederick L. 1953. *The triumph of science and reason, 1660–1685.* New York: Harper & Brothers.

Nye, Mary Jo. 1980. "N-rays: an episode in the history and psychology of science." *Historical Studies in the Physical Sciences* 11:125–156.

O'Malley, C. D. 1964. *Andreas Vesalius of Brussels, 1514–1564.* Berkeley, Los Angeles: University of California Press.

———. 1976. "Vesalius, Andreas." *Dictionary of Scientific Biography* 14:3–12.

Oiserman, T. I. 1972. "Kant und das Problem einer wissenschaftlichen Philosophie." Pp. 121–127 of Beck 1972.

Olby, Robert C. 1966. *Origins of Mendelism.* Intro. by C. D. Darlington. New York: Schocken Books.

Oldroyd, D. R. 1980. *Darwinian impacts: an introduction to the Darwinian revolution.* Milton Keynes, Eng.: The Open University Press.

Olmsted, J. M. D. 1938. *Claude Bernard, physiologist.* 2nd ed. New York: Harper & Brothers.

Olschki, Leonardo. 1922. *Bildung und Wissenschaft im Zeitalter der Renaissance in Italien.* Leipzig and Florence: Leo S. Olschki.

Opdyke, N. D., B. P. Glass, J. D. Hays, and J. H. Foster. 1966. "Paleomagnetic study of Antarctic deep-sea cores." *Science* 154:349–357.

Orléans, F. J. [=Pierre Joseph] d'. 1722. *The history of the revolutions in England under the family of the Stuarts.* 2nd ed. Intro. Laurence Echard. London: printed for E. Bell.

Ornstein, Martha. 1928. *The rôle of the scientific societies in the seventeenth century.* Chicago: The University of Chicago Press. Reprint ed., 1975, New York: Arno Press [author's name given as Martha Ornstein Bronfenbrenner].

Ortega y Gasset, José. 1923. *El tema de nuestro tiempo.* Madrid: Espasa Calpe.

———. 1961. *The modern theme.* Trans. James Cleugh. New York: Harper & Row [Harper Torchbooks].

Osler, William. 1906. *The growth of truth: as illustrated in the discovery of the circulation of the blood.* London: Henry Frowde, Oxford University Press Warehouse.

———. 1911. *Whole time clinical professors.* A letter to President Remson, Johns Hopkins University.

Ovington, J. 1929. *A voyage to Surat, in the year 1689.* Ed. H. G. Rawlinson. London: Humphrey Milford, Oxford University Press.

Padover, Saul K., ed. 1978. *The essential Marx: the non-economic writings — a selection.* New York, Scarborough, Ont.: New American Library; London: The New English Library.

Pagel, Walter. 1958. *Paracelsus.* Basel, New York: S. Karger.

———. 1967. *William Harvey's biological ideas.* Basel, New York: S. Karger.

———. 1969–1970. "William Harvey revisited." *History of Science* 8:1–29; 9:1–41.

———. 1974. "Paracelsus, Theophrastus Philippus Aureolus Bombastus von Hohenheim." *Dictionary of Scientific Biography* 10:304–313.

Pagel, Walter, and Pyarali Rattansi. 1964. "Vesalius and Paracelsus." *Medical History* 8:309–328.

Paine, Thomas. 1791. *Rights of man: being an answer to Mr. Burke's attack on the French Revolution.* London: J. S. Jordan.

Pais, Abraham. 1982. *"Subtle is the Lord . . . ": the science and the life of Albert Einstein.* New York: Oxford University Press.

Palter, Robert. 1970. "An approach to the history of early astronomy." *Studies in History and Philosophy of Science* 1:93–133.

Partington, J. R., and Douglas McKie. 1938. "Historical studies on the phlogiston theory." *Annals of Science* 2:361–404; 3:1–58, 337–371; 4:113–149.

Passmore, J. A. 1952. *Hume's intentions.* Cambridge: at the University Press.

Paton, H. J. 1936. *Kant's metaphysic of experience: a commentary on the first half of the 'Kritik der reinen Vernunft.'* 2 vols. London: George Allen & Unwin.

———. 1937. "Kant's so-called Copernican revolution." *Mind* 46:365–371. A rejoinder to Cross 1937.

———. 1946. *The categorical imperative: a study in Kant's moral philosophy.* London, New York: Hutchinson's University Library.

Paul, Charles B. 1980. *Science and immortality: the éloges of the Paris Academy of Sciences (1699–1791.)* Berkeley, Los Angeles, London: University of California Press.

Paul, Harry W. 1979. *The edge of contingency: French Catholic reaction to scientific change from Darwin to Duhem.* Gainesville: The University Presses of Florida.

Peirce, Charles Sanders. 1934. *Collected papers of Charles Sanders Peirce.* Vol. 5, *Pragmatism and pragmaticism;* vol. 6, *Scientific metaphysics.* Ed. Charles Hartshorne and Paul Weiss. Cambridge, Mass.: Harvard University Press. Reprint, 1978.

Pepys, Samuel. 1879. *Diary and correspondence of Samuel Pepys.* Ed. Richard Lord Braybrooke and Mynors Bright. London: Bickers and Son.

Péguy, Charles. 1935. *Note sur M. Bergson et la philosophie bergsonienne. Note conjointe sur M. Descartes et la philosophie cartésienne.* Paris: Gallimard.

Perrault, Charles. 1688. *Parallèle des anciens et des modernes en ce qui regarde les arts et les sciences.* Paris: Jean Baptiste Coignard. Facsimile reprint, 1964, Munich: Eidos Verlag.

Pettee, George Sawyer. 1938. *The process of revolution.* New York: Harper & Brothers.

Pilet, P. E. 1976. "Vogt, Carl." *Dictionary of Scientific Biography* 14:57–58.

Planck, Max. 1922. *The origin and development of the quantum theory.* Trans. H. T. Clarke and L. Silberstein. Oxford: at the Clarendon Press.

———. 1931. "Maxwell's influence in Germany." Pp. 45–65 of *James Clerk Maxwell: a commemoration volume, 1831–1931,* Cambridge: at the University Press.

———. 1949. *Scientific autobiography and other papers.* Trans. Frank Gaynor. New York: Philosophical Library.

Playfair, John. 1802. *Illustrations of the Huttonian theory of the earth.* Edinburgh: William Creech; London: Cadell and Davies.

———. 1819. *Outlines of natural philosophy. Being heads of lectures delivered in the University of Edinburgh.* In 2 vols. 3rd ed. Edinburgh: printed for Archibald Constable and Co.

———. 1820[?]. *Dissertation second: exhibiting a general view of the progress of mathe-*

matical and physical science, since the revival of letters in Europe. 2 pts. [separate pagination], n.p., n.d. Reprint 1835, Edinburgh: Adam and Charles Black (as part of *Dissertations on the history of metaphysical and ethical, and of mathematical and physical science,* by Dugald Stewart, James Mackintosh, John Playfair, and John Leslie).

————. 1822. *The works of John Playfair, Esq.* With a memoir of the author. 4 vols. Edinburgh: printed for Archibald Constable & Co.

Plot, Robert. 1677. *Natural history of Oxford-shire, being an essay toward the natural history of England.* Oxford: printed at the Theater in Oxford "and are to be had there"; London: at Mr. S. Miller's.

Poincaré, Henri. 1890. *Les théories de Maxwell et la théorie électromagnétique de la lumière.* Leçons professées pendant le second semestre 1888–89. Paris: G. Carré, Editeur.

————. 1907. *The value of science.* Trans. George Bruce Halsted. New York: The Science Press.

————. 1963. *Mathematics and science: last essays.* Trans. John W. Bolduc. New York: Dover Publications.

Poincaré, Henri, and Frederick K. Vreeland. 1904. *Maxwell's theory and wireless telegraphy.* Pt. 1, *Maxwell's theory and Hertzian oscillations,* by H. Poincaré, trans. F. K. Vreeland; pt. 2, *The principles of wireless telegraphy,* by F. K. Vreeland. New York: McGraw Publishing Company.

Poincaré, Lucien. 1907. *The new physics and its evolution.* London: Kegan Paul, Trench, Trubner, & Co. The authorized translation of *La physique moderne, son évolution.*

Pomey, François, S.J. 1691. *Le dictionaire royal.* Dernière éd. Lyon: Molin.

Poor, Charles Lane. 1922. *Gravitation versus relativity.* With a preliminary essay by Thomas Chrowder Chamberlin. New York, London: G. P. Putnam's Sons.

Popper, Karl R. 1962. *Conjectures and refutations: the growth of scientific knowledge.* New York, London: Basic Books.

————. 1975. "The rationality of scientific revolutions." Pp. 72–101 of Harré 1975.

————. 1979. "The revolution in our idea of knowledge." Unpublished manuscript.

————. 1983. *Realism and the aim of science.* From the *Postscript to the logic of scientific discovery.* Ed. W. W. Bartley III. London: Hutchinson.

Poulton, Edward B. 1896. *Charles Darwin and the theory of natural selection.* London, Paris: Cassell and Company.

Power, Henry. 1664. *Experimental philosophy.* London: printed by T. Roycroft for John Martin and James Allestry. Facsimile reprint, 1966, New York and London: Johnson Reprint Corporation [The Sources of Science 21].

Price, Derek J. de Solla. 1963. *Little science, big science.* New York: Columbia University Press.

Priestley, Joseph. 1790. *Experiments and observations on different kinds of air, and other branches of natural philosophy, connected with the subject.* Vol. 1. Birmingham: printed by Thomas Pearson.

————. 1796. *Considerations on the doctrine of phlogiston and the decomposition of water.* Philadelphia: Th. Dobson. Reprint, 1929, together with John Maclean, *Two lectures on combustion and an examination of Doctor Priestley's Considerations on the doctrine of phlogiston,* ed. William Foster, Princeton: Princeton University Press.

————. 1826. *Lectures on history, and general policy; to which is prefixed an essay on a course of liberal education for civil and active life.* A new ed. with numerous enlargements by J. T. Rutt. London: printed for Thomas Tegg.

————. 1966. *The history and present state of electricity.* Reprint of the 3rd (London 1755) ed., intro. by Robert E. Schofield. New York, London: Johnson Reprint Corporation.

Prigogine, Ilya. 1980. *From being to becoming: time and complexity in the physical sciences.* San Francisco: W. H. Freeman and Company.

Proudhon, P. -J. 1923. *Idée générale de la révolution au XIXe siècle.* Intro. by Aimé Berthod. Nouvelle éd. de C. Bougle et H. Moysset. Paris: Marcel Rivière [*Oeuvres complètes de P.-J. Proudhon,* vol. 3].

Provine, William B. 1971. *The origins of theoretical population genetics.* Chicago, London: The University of Chicago Press.

Prowe, Leopold. 1883. *Nicolaus Coppernicus.* Vol. 1, *Das Leben;* pt. 2, *1512–1543.* Berlin: Weidmannsche Buchhandlung.

Quinton, Anthony. 1980. *Francis Bacon.* Oxford, Toronto: Oxford University Press.

Rabinowitch, Eugene. 1963. "Scientific revolution." *Bulletin of the Atomic Scientists,* Sept., pp. 15–18; Oct., pp. 11–16; Nov., pp. 9–12; Dec., pp. 14–17.

Rainger, Ronald. 1981. "The continuation of the morphological tradition: American paleontology, 1880–1910." *Journal of the History of Biology* 14:129–158.

Raman, V. V., and Paul Forman. 1969. "Why was it Schrödinger who developed de Broglie's ideas?" *Historical Studies in the Physical Sciences* 1:294–314.

Randall, John Herman. 1926. *The making of the modern mind.* Boston: Houghton Mifflin Co. Rev. ed., 1940.

Reclus, Elisée. 1891. *Evolution et révolution.* 6th ed. Paris: Au bureau de la Révolte.

————. 1902. *L'évolution, la révolution et l'idéal anarchique.* 5th ed. Paris: P. V. Stock.

————. [n.d.] *Evolution and revolution.* 7th ed. London: W. Reeves.

Redwood, John. 1977. *European science in the seventeenth century.* New York: Barnes and Noble.

Rée, Jonathan. 1974. *Descartes.* New York: Pica Press.

Reingold, Nathan. 1980. "Through paradigm-land to a normal history of science." *Social Studies of Science* 10:475–496.

Reinhold, Karl Leonhard. 1784. "Gedanken über Aufklärung." *Der Teutsche Merkur* 3:3–22.

————. 1786. "Briefe über die Kantische Philosophie. Erster Brief. Bedürfniss einer Kritik der Vernunft." *Der Teutsche Merkur* 3:99–127.

————. 1794. *Beyträge zur Berichtigung bisheriger Missverständnisse der Philosophen.* Vol. 2. Jena: bey Johann Michael Mauke.

Rey, Abel. 1927. *Le retour éternel et la philosophie de la physique.* Paris: Ernest Flammarion, Editeur.

Rice, Eugene F., Jr. 1970. *The foundations of early modern Europe, 1460–1559.* New York, London: W. W. Norton & Company.

Richards, Robert J. 1981. "Natural selection and other models in the historiography of science." Pp. 37–76 of Marilyn B. Brewer and Barry E. Collins, eds., *Scientific inquiry and the social sciences,* San Francisco, Washington, London: Jossey-Bass Publishers.

Rideing, W. H. 1878. "Hospital life in New York." *Harper's New Monthly Magazine* 57:171–189.

Rigault, Hippolyte. 1856. *Histoire de la querelle des anciens et des modernes.* Paris: L. Hachette et Cie.

Robert, Marthe. 1964. *La révolution psychanalytique: la vie et l'oeuvre de Sigmund Freud.* 2 vols. Paris: Petite Bibliothèque Payot.

————. 1966. *The psychoanalytic revolution: Sigmund Freud's life and achievement.* Trans. Kenneth Morgan. New York: Harcourt, Brace & World.

Rolland, Romain. 1944. *Péguy.* Paris: Editions Albin Michel.

Roberts, V. 1957. "The solar and lunar theory of Ibn al-Shāṭir." *Isis* 48:428–432.

Robertson, Priscilla. 1952. *Revolutions of 1848: a social history.* Princeton: Princeton University Press.

Robinson, James Harvey. 1915. *An outline of the history of the intellectual class in Western Europe.* 3rd ed. New York: The Marion Press.

————. 1921. *The mind in the making: the relation of intelligence to social reform.* New York, London: Harper and Brothers.

Roger, Jacques. 1971. *Les sciences de la vie dans la pensée française du XVIIIᵉ siècle: la génération des animaux de Descartes à l'Encyclopédie.* 2nd ed. Paris: Armand Colin.

Rogers, Howard J., ed. 1905–1907. *Congress of arts and science: Universal Exposition, St. Louis, 1904.* 8 vols. Boston, New York: Houghton, Mifflin and Company.

Rohault, Jacques. 1672. *Traité de physique.* 2nd ed. 2 vols. Paris: chez la Veuve de Charles Savreux . . . et chez Guillaume Desprez.

————. 1723. *Rohault's system of natural philosophy, illustrated with Dr. Samuel Clarke's notes.* 2 vols. Trans. John Clarke. London: printed for James Knapton. Reprint, 1969, New York, London: Johnson Reprint Corporation.

Rosen, Edward. 1971. *Three Copernican treatises.* 3rd ed., with a biography of Copernicus and Copernicus bibliographies, 1939–1958. New York: Octagon Books.

————. 1971a. "Copernicus, Nicolaus." *Dictionary of Scientific Biography* 3:401–410.

————. 1975. "The impact of Copernicus on man's conception of his place in the world." Pp. 52–66 of Steneck 1975.

Rosenstock, Eugen. 1931. "Revolution als politischer Begriff in der Neuzeit."

Pp. 83–124 of *Festgabe Paul Heilbron. Abhandlungen der Schlesischen Gesell-schaft für vaterländische Cultur, Geisteswissenschaftliche Reihe.* Heft 5. Breslau: M. and H. Marcus.

Rosmorduc, Jean. 1972. "Une erreur scientifique au début de siècle: 'les rayons N.'" *Revue d'Histoire des Sciences* 25:13–25.

Rostand, Jean. 1960. *Error and deception in science: essays on biological aspects of life.* Trans. A. J. Pomerans. New York: Basic Books.

Roth, Leon. 1937. *Descartes' Discourse on method.* Oxford: at the Clarendon Press.

Roth, Mathias. 1892. *Andreas Vesalius Bruxellensis.* Berlin: Reimer.

Rousseau, G. S., and Roy Porter, eds. 1980. *The ferment of knowledge: studies in the historiography of eighteenth-century science.* Cambridge, London, New York: Cambridge University Press.

Rousseau, Jean-Jacques. 1896. *Du contrat social.* Ed. Edmond Dreyfus-Brisac. Paris: F. Alcan.

———. 1913. *The social contract and discourses.* Trans. G. D. H. Cole. London: J. M. Dent & Sons; New York: E. P. Dutton & Co. [Everyman's Library].

———. 1946. *Discours sur les sciences et les arts.* Ed. George R. Havens. New York: Modern Language Association.

———. 1959–1964. *Oeuvres complètes.* 3 vols. Paris: Gallimard [Bibliothèque de la Pléiade]. *Du contrat social,* ed. Robert Derathé, appears in vol. 3.

———. 1964. *The first and second discourses.* Trans. Roger D. and Judith R. Master. New York: St Martin's Press.

———. 1972. *Du contrat social.* Ed. Ronald Grimsley. Oxford: at the Clarendon Press.

Rudwick, Martin J. S. 1972. *The meaning of fossils: episodes in the history of palaeonto-logy.* London: Macdonald; New York: American Elsevier.

Runcorn, S. K., ed. 1962. *Continental drift.* New York, London: Academic Press.

Ruse, Michael. 1975. "Darwin's debt to philosophy: an examination of the influ-ence of the philosophical ideas of John R. W. Herschel and William Whe-well on the development of Charles Darwin's theory of evolution." *Studies in History and Philosophy of Science* 6:159–181.

———. 1979. *The Darwinian revolution: science red in tooth and claw.* Chicago, London: The University of Chicago Press.

———. 1982. *Darwinism defended: a guide to the evolution controversies.* Reading, Mass., London: Addison-Wesley Publishing Company.

Russell, Bertrand. 1945. *A history of western philosophy and its connection with politi-cal and social circumstances from the earliest times to the present day.* New York: Simon and Schuster.

———. 1948. *Human knowledge: its scope and limits.* London: George Allen & Unwin.

Rutherford, Lord. 1938. "Forty years of physics." Pp. 47–74 of Needham and Pagel 1938.

Sagan, Carl. 1979. "Immanuel Velikovsky's unlikely collisions." Letter to the editor, *New York Times,* 29 Dec. 1979.

Said, Edward W. 1978. *Orientalism.* New York: Pantheon Books.

Saint-Simon, Claude Henri de Rouvroy, Comte de. 1858. *Science de l'homme, physiologie religieuse.* Paris: Librairie Victor Masson.

———. 1865–1878. *Oeuvres de Saint-Simon et d'Enfantin, précédées de deux notices historiques et publiées par les membres du conseil institué par Enfantin pour l'exécution de ses dernières volontés.* 47 vols. Paris: Dentu (1865–1876); Paris: E. Leroux (1877–1878). Saint-Simon's writings appear in vols. 15, 18, 19–23, 37–40.

———. 1964. *Social organization, the science of man and other writings.* Ed. Felix Marlcham. New York, Evanston: Harper & Row, Publishers [Harper Torchbooks]. An earlier ed., 1952, under the title, *Henri de Saint-Simon: selected writings,* Oxford: Basil Blackwell.

Sarton, George. 1936. *The study of the history of science.* Cambridge, Mass.: Harvard University Press.

———. 1937. *The history of science and the new humanism.* Cambridge, Mass.: Harvard University Press.

Sauter, Eugen. 1910. *Herder und Buffon.* Inaugural-Dissertation zur Erlangung der Doktorwürde bei der hohen philosophischen Fakultät der Universität Basel. Rixheim: Buchdruckerei von F. Sutter & Cie.

Savioz, Raymond. 1948. *Mémoires autobiographiques de Charles Bonnet de Genève.* Paris: Librairie Philosophique J. Vrin.

Scarborough, John. 1968. "The classical background of the Vesalian revolution." *Episteme* 2:200–218.

Scheffler, Israel. 1967. *Science and subjectivity.* Indianapolis, New York, Kansas City: The Bobbs-Merrill Company.

Schieder, Theodor. 1950. "Das Problem der Revolution im 19. Jahrhundert." *Historische Zeitschrift* 170:233–271.

Schiller, Joseph. 1967. *Claude Bernard et les problèmes scientifiques de son temps.* Paris: Les Editions du Cèdre.

Schilpp, Paul Arthur, ed. 1949. *Albert Einstein: philosopher-scientist.* Evanston, Ill.:The Library of Living Philosophers.

Schlözer, August Ludwig von. 1772. *Vorstellung seiner Universal-Histoire.* Göttingen, Gotha: Johann Christian Dieterich.

Schmeck, Harold M. 1983. "DNA's code: 30 years of revolution." *New York Times, Science Times,* 12 Apr. 1983.

Schmidt, Alfred. 1971. *The concept of nature in Marx.* Trans. Ben Fowkes. London: NLB.

Schneer, Cecil. 1973. "Critical years in geology." [Review of Leonard G. Wilson's *Charles Lyell. The years to 1841: the revolution in geology.*] *Science* 179:57–58.

Schofield, Robert E. 1957. "The industrial orientation of science in the Lunar Society of Birmingham." *Isis* 48:408–415.

———. 1963. *The lunar society of Birmingham: a social history of provincial science in eighteenth-century England.* London: Oxford University Press.

———, ed. 1966. *A scientific autobiography of Joseph Priestley, 1773–1804: selected scientific correspondence, with commentary.* Cambridge, Mass., London: The M.I.T. Press.

Schrecker, Paul. 1967. "Revolution as a problem in the philosophy of history." *Nomos* 8:34–53.

Schuster, Arthur. 1911. *The progress of physics during 33 years (1875–1908).* Cambridge: at the University Press. Reprint, 1975, New York: Arno Press.

Schuster, John A. 1975. "Rohault, Jacques." *Dictionary of Scientific Biography* 11:506–509.

Schütt, Hans Werner. 1979. "Lichtenberg als 'Kuhnianer.'" *Sudhoffs Archiv: Zeitschrift für Wissenschaftsgeschichte* 63:87–90.

Schweber, Silvan S. 1977. "The origins of the *Origin* revisited." *Journal of the History of Biology* 10:229–316.

Sciama, D. W. 1959. *The unity of the universe.* London: Faber and Faber.

———. 1969. *The physical foundations of general relativity.* Garden City, N.Y.: Doubleday & Company [Anchor Books].

Scriven, Michael. 1959. "Explanation and prediction in evolutionary theory." *Science* 130:477–482.

Scruton, Roger. 1982. *Kant.* Oxford: Oxford University Press.

Sears, Paul. 1952. "The assimilation of science into general education." Pp. 31–45 of I. Bernard Cohen and Fletcher A. Watson, eds., *General education in science,* Cambridge, Mass.: Harvard University Press.

Seelig, Carl. 1960. *Albert Einstein.* 2nd ed. Zurich: Europa Verlag.

Segrè, Emilio. 1976. *From x-rays to quarks: modern physicists and their discoveries.* San Francisco: W. H. Freeman & Company.

Seidler, Franz Wilhelm. 1955. *Die Geschichte des Wortes Revolution: ein Beitrag zur Revolutionsforschung.* Doctoral dissertation, Ludwig-Maximilians-Universität, Munich; available in the Bayerische Staatsbibliothek, Munich, u.56.7037.

Shakow, David, and David Rapaport. 1964. *The influence of Freud on American psychology.* New York: International University Press [*Psychological Issues* 4, no. 1.; Monograph 13].

Shapere, Dudley. 1964. "The structure of scientific revolutions." *Philosophical Review* 7:383–394; reprint, pp. 27–38 of Gutting 1980.

Sherburne, Edward. 1675. [Astronomical appendix, pp. 1–221, plus index, to *The sphere of Marcus Manilius made an English poem.*] London. printed for Nathanael Brooke.

Sheridan, Alan. 1980. *Michel Foucault: the will to truth.* London, New York: Tavistock Publications.

Sherrington, Sir Charles. 1940. *Man on his nature.* Cambridge: at the University Press.

Shirley, John W., ed. 1981. *A source book for the study of Thomas Harriot.* New York: Arno Press.

Shryock, Richard H. 1947. *The development of modern medicine.* New York: Alfred A. Knopf.

————. 1980. *American medical research, past and present.* New York: Arno Press. Reprint of the 1947 ed. published by New York Academy of Medicine.

Simpson, George Gaylord. 1978. *Concession to the improbable: an unconventional autobiography.* New Haven, London: Yale University Press.

Singer, Charles. 1956. *The discovery of the circulation of the blood.* London: Wm. Dawson & Sons. Reprint of 1922 ed., London: A. Bell and Sons.

Singer, Charles, and C. Rabin. 1946. *A prelude to modern science: being a discussion of the history, source, and circumstance of the "Tabulae anatomicae sex" of Vesalius.* Cambridge: at the University Press.

Singer, Charles, and E. Ashworth Underwood. 1962. *A short history of medicine.* New York: Oxford University Press; Oxford: at the Clarendon Press.

Slater, John. 1955. *Modern physics.* New York, Toronto, London: McGraw-Hill Book Company.

————. 1975. *Solid-state and molecular theory: a scientific biography.* New York, London: John Wiley & Sons.

Smeaton, W. A. 1962. *Fourcroy, chemist and revolutionary, 1755–1809.* Cambridge, Eng.: printed for the autor by W. Heffer & Sons.

Smith, Anthony. 1973. *The concept of social change: a critique of the functionalist theory of social change.* London, Boston: Routledge & Kegan Paul.

Smith, Edgar F. 1927. *Old chemistries.* New York: McGraw-Hill Book Co.

Smith, Norman Kemp. 1923. *A commentary to Kant's 'Critique of pure reason.'* 2nd ed. London: Macmillan & Co. 1st ed., 1918.

————. 1952. *New studies in the philosophy of Descartes: Descartes as pioneer.* London: Macmillan & Co.

————, ed. 1952a. *Descartes' philosophical writings.* Selected and trans. by N. K. Smith. London: Macmillan & Co.

Smith, Preserved. 1930. *A history of modern culture.* Vol. 1, *The great renewal, 1543–1687.* New York: Henry Holt and Company.

Snelders, H. A. M. 1974. "The reception of J. H. van't Hoff's theory of the asymmetric carbon atom." *Journal of Chemical Education* 51:2–6.

Snider-Pellegrini, Antonio. 1859. *La création et ses mystères dévoilés.* Paris: A. Franck.

Snow, Vernon F. 1962. "The concept of revolution in seventeenth-century England." *The Historical Journal* 2:167–174.

Sommerfeld, Arnold. 1923. *Atomic structure and spectral lines.* Trans. H. L. Brose. New York: E. P. Dutton and Co.

Sorokin, Pitirim A. 1925. *The sociology of revolution.* Philadelphia, London: J. B. Lippincott Company.

Sprat, Thomas, 1667. *History of the Royal Society.* London: printed for J. Martyn, and J. Allestry. Facsimile reprint, 1958, ed. Jackson I. Cope and Harold Whitmore Jones, Saint Louis: Washington University Studies.

Spronsen, Johannes W. van. 1969. *The periodic system of chemical elements: a history of the first hundred years.* Amsterdam: Elsevier.

Staël Holstein, Mme La Baronne de. 1813. *De l'Allemagne.* 3 vols. London: John Murray.

Stapfer, Philipp Albert. 1818. "Kant." *Biographie Universelle* 22:229–257. Paris: chez L. G. Michaud.

Stark, Johannes. 1922. *Die gegenwärtige Krisis in der deutschen Physik.* Leipzig: Verlag J. A. Barth.

[State tracts.] 1692. *State tracts: being a farther collection of several choice treatises relating to the government. From the year 1660, to 1689. Now published in a body, to shew the necessity, and clear the legality of the late Revolution . . .* London: printed, and are to be sold by Richard Baldwin.

Stearns, Peter N. 1974. *1848: the revolutionary tide in Europe.* New York: W. W. Norton.

Steneck, Nicholas H., ed. 1975. *Science and society: past, present, and future.* Ann Arbor: The University of Michigan Press.

Stern, J. P. 1959. *Lichtenberg: a doctrine of scattered occasions reconstructed from his aphorisms and reflections.* Bloomington: Indiana University Press.

Stillman, John Maxson. 1920. *Theophrastus Bombastus von Hohenheim called Paracelsus: his personality and influence as physician, chemist and reformer.* Chicago: The Open Court Publishing Company.

Stone, Lawrence. 1972. *The causes of the English Revolution, 1529–1642.* New York, Evanston: Harper & Row.

Straka, Gerald M. 1971. "Sixteen eighty-eight as the year one: eighteenth-century attitudes towards the Glorious Revolution." Pp. 143–167 of Louis T. Milic, ed., *The modernity of the eighteenth century,* Cleveland, London: The Press of the Case Western Reserve University.

Struik, Dirk J. 1954. *A concise history of mathematics.* London: G. Bell and Sons.

Stuewer, Roger H. 1975. *The Compton effect: turning point in physics.* New York: Science History Publications.

Sullivan, Walter. 1974. *Continents in motion: the new earth debate.* New York: McGraw-Hill Book Company.

Sulloway, Frank J. 1979. *Freud, biologist of the mind: beyond the psychoanalytic legend.* New York: Basic Books.

———. 1982. "Darwin's conversion: the *Beagle* voyage and its aftermath." *Journal of the History of Biology* 15:325–396.

Suppe, Frederick, ed. 1974. *The structure of scientific theories.* Urbana: University of Illinois Press. 2nd ed., 1977.

Sutton, Geoffrey Vincent. 1982. "A science for a polite society: Cartesian natural philosophy in Paris during the reigns of Louis XIII and XIV." Doctoral dissertation, Princeton University.

Swerdlow, Noel. 1973. "The derivation and first draft of Copernicus's planetary theory: a translation of the Commentariolus with commentary." *Proceedings of the American Philosophical Society* 117:423–512.

———. 1976. "Pseudodoxica Copernicana: or, enquiries into very many received tenents and commonly presumed truths, mostly concerning spheres." *Archives Internationales d'Histoire des Sciences* 26:108–158.

Swift, Jonathan. 1886. *The battle of the books, and other short pieces.* London: Cassell & Company.

————. 1939–1974. *The prose works.* Ed. Herbert Davis. 16 vols. Oxford: Basil Blackwell.

Tait, Peter G. 1890. *Properties of matter.* 2nd ed. Edinburgh: Adam and Charles Black.

Takeuchi, H., S. Uyeda, and H. Kanamori. 1970. *Debate about the earth: approach to geophysics through the analysis of continental drift.* Rev. ed. San Francisco: Freeman, Cooper, and Co.

Tannery, Paul. 1934. *Mémoires scientifiques.* Vol. 13, *Correspondence.* Toulouse: Edouard Privat; Paris: Gauthier-Villars & Cie.

Temkin, Owsei. 1961. "A Galenic model for quantitative physiological reasoning?" *Bulletin of the History of Medicine* 35:470–475.

————. 1973. *Galenism: rise and decline of a medical philosophy.* Ithaca, London: Cornell University Press.

Temple, Sir William. 1821. *Essays.* 2 vols. London: John Sharpe.

————. 1963. *Five miscellaneous essays.* Ed. S. H. Monk. Ann Arbor: The University of Michigan Press.

Tetsch, Hartmut. 1973. *Die permanente Revolution: ein Beitrag zur Soziologie der Revolution und zur Ideologiekritik.* Opladen: Westdeutscher Verlag.

Thackray, Arnold. 1970. *Atoms and powers: an essay on Newtonian matter-theory and the development of chemistry.* Cambridge, Mass.: Harvard University Press.

Thackray, Arnold, and Robert K. Merton. 1972. "On discipline building: the paradoxes of George Sarton." *Isis* 63:673–695.

Thompson, Silvanus P. 1901. *Michael Faraday, his life and work.* London, New York, Paris: Cassell and Company.

Thomson, David. 1955. "Scientific thought and revolutionary movements." *Impact of Science on Society* 6:3–29.

Thomson, Sir J. J. 1893. *Notes on recent researches in electricity and magnetism.* Oxford: at the University Press.

————. 1936. *Recollections and reflections.* London: G. Bell and Sons.

Thomson, William [Lord Kelvin]. 1904. *Baltimore lectures.* Delivered in 1884 at Johns Hopkins University. Baltimore: Publishing Agency of the Johns Hopkins University.

Totten, Stanley M. 1981. "Frank B. Taylor, plate tectonics, and continental drift." *Journal of Geological Education* 29:212–220.

Toulmin, Stephen. 1963. *Foresight and understanding: an enquiry into the aims of science.* New York, Evanston: Harper & Row. First published, 1961, by Indiana University Press.

————. 1968. "Conceptual revolutions in science." *Boston Studies in the Philosophy of Science* 3:331–347.

————. 1972. *Human understanding: the collective use and evolution of concepts.* Princeton: Princeton University Press.

Toynbee, Arnold. 1884. *Lectures on the Industrial Revolution in England.* London: Rivingtons. Reprinted as *The Industrial Revolution,* Boston: Beacon Press, 1956.

Trevelyan, George M. 1939. *The English Revolution, 1688–1689.* London: T.

Butterworth. A later ed., 1976, London, Oxford, New York: Oxford University Press.

Trevor-Roper, H. R. 1959. "The general crisis of the seventeenth century." *Past and Present* 16:31–64.

Tricker, R. A. R. 1966. *The contributions of Faraday and Maxwell to electrical science.* Oxford, London, New York: Pergamon Press.

Truesdell, C. 1968. *Essays in the history of mechanics.* New York: Springer Verlag.

Turgot, Anne-Robert-Jacques, Baron de l'Aulne. 1913–1923. *Oeuvres de Turgot et documents le concernant.* 5 vols. Paris: F. Alcan.

―――. 1973. *Turgot on progress, sociology and economics: A philosophical review of the successive advances of the human mind; On universal history; Reflections on the formation and the distribution of wealth.* Trans. Ronald L. Meek. Cambridge: at the University Press.

Turner, R. Steven. 1972. "Helmholtz, Hermann von." *Dictionary of Scientific Biography* 6:241–253.

Ulam, Adam. 1981. *Russia's failed revolutions: from the Decembrists to the dissidents.* New York: Basic Books.

Unamuno, Miguel. 1951. *Ensayos.* 2 vols. Madrid: Aguilar.

Uyeda, Seiya. 1978. *The new view of the earth: moving continents and moving oceans.* San Francisco: W. H. Freeman and Company.

Von der Gracht, Willem A. J. M. van Waterschoot, et al. 1928. *Theory of continental drift: a symposium on the origin and movement of land masses.* Tulsa: American Association of Petroleum Geologists.

Van Helden, Albert. 1977. *The invention of the telescope.* Philadelphia: American Philosophical Society [Transactions of the American Philosophical Society 67, no. 4].

Vatikiotis, P. J., ed. 1972. *Revolution in the Middle East and other case studies.* London: George Allen and Unwin.

Vertot, René Aubert. 1695. *Histoire des révolutions de Suède: où l'on voit les changemens qui sont arrivés dans ce royaume au sujet de la religion & du gouvernement.* Vol. 1. Paris: chez Michel Brunet.

―――. 1761. *The history of the revolution in Sweden, occasioned by the changes of religion and alteration of the government in that kingdom.* Glasgow: printed for R. Urie.

―――. 1825. *Histoire des révolutions de Portugal.* Paris: chez Dabo-Butschert. Contains prefaces to the 1689 and 1711 eds.

―――. 1833. *Histoire des révolutions arrivées dans le gouvernement de la république romaine.* 4 vols. Paris: Librairie de Lecointe.

Vesalius, Andreas. 1932. "The preface of Andreas Vesalius to De Fabrica Corporis Humani 1543." Trans. Benjamin Farrington. *Proceedings of the Royal Society of Medicine* 25:1357–1368 (Section of the History of Medicine, pp. 25–38).

Vesey, Godfrey. 1972. "Kant's Copernican revolution: speculative philosophy." The Age of Revolutions, Units 15–16. Arts: A Second Level Course, Bletchley, Bucks.: The Open University Press.

Viëtor, Karl. 1950. *Goethe the thinker*. Trans. Bayard Q. Morgan. Cambridge, Mass.: Harvard University Press.

Villani, Matteo. 1848. "Cronica di Matteo Villani." *Croniche storiche di Giovanni, Matteo e Filippo Villani*. Vols. 5 and 6. Milan: Borroni e Scotti.

Villers, Charles de. 1799. "Critique de la raison pure." *Le Spectateur du Nord* 10:1–37.

———. 1801. *Philosophie de Kant ou principes fondamentaux de la philosophie transcendentale*. Metz: chez Collignon.

———. 1804. *Essai sur l'esprit et l'influence de la Réformation de Luther*. Paris: chez Henrichs, Libraire; Metz: chez Collignon, Imprimeur-Libraire.

Virchow, Rudolf. 1848. "Was die 'medicinische Reform' will." *Die medicinische Reform* 1:1–2.

———. 1849. "Die sociale Stellung des Arztes." *Die medicinische Reform* 35.

———. 1858. *Die Cellularpathologie in ihrer Begründung auf physiologische und pathologische Gewebelehre*. Berlin: August Hirschwald.

———. 1860. *Cellular pathology as based upon physiological and pathological histology*. Trans. Frank Chance. 2nd ed. London: John Churchill.

Vlachos, Georges. 1962. *La pensée politique de Kant*. Foreword by Marcel Prélot. Paris: Presses Universitaires de France.

Vleeschauwer, H. J. de. 1937. *La déduction transcendantale dans l'oeuvre de Kant*. Vol. 3. Paris: Librairie Ernest Leroux; Antwerp: De Sikkel.

Vogt, Carl. 1851. *Zoologische Briefe. Naturgeschichte der lebenden und untergegangenen Thiere, für Lehrer, höhere Schulen und Gebildete aller Stände*. 2 vols. Frankfurt: Literarische Anstalt.

———. 1882. *Ein frommer Angriff*. Breslau: S. Schottlaender.

Voltaire, François Marie Arouet de. 1733. *Letters concerning the English nation*. London: printed for C. Davis and A. Lyon.

———. 1792. *Oeuvres*. Nouvelle éd., avec des notes et des observations critiques, par M. Palissot. Vols. 16–20. Paris: chez Stoupe, Imprimeur; Servière, Libraire. [*Essai sur les moeurs et l'esprit des nations . . . depuis Charlemagne jusqu' à Louis XIII*, vols. 1–5.]

———. 1926. *The age of Louis XIV*. Trans. Martyn P. Pollack. London: J. M. Dent & Sons; New York: E. P. Dutton & Co.

———. [193–]. *Siècle de Louis XIV*. Ed. Emile Bourgeois. Paris: Librairie Hachette.

———. 1964. *Lettres philosophiques*. Ed. Gustave Lanson. Nouveau tirage revu et complété par André M. Rousseau. 2 vols. Paris: Librairie Marcel Didier.

Vuillemin, Jules. 1954. *L'héritage kantien et la révolution copernicienne*. Paris: Presses Universitaires de France.

———. 1955. *Physique et métaphysique kantiennes*. Paris: Presses Universitaires de France.

Waff, Craig. 1975. "*Universal gravitation and the motion of the moon's apogee: the establishment and reception of Newton's inverse-square law, 1687–1749*." Ph.D. dissertation, The Johns Hopkins University.

Walker, Nigel D. 1957. "A new Copernicus?" Pp. 75–79 of Nelson 1957. A reprint, with revisions, of "Freud and Copernicus," in *The Listener* 1956. Reprinted, pp. 35–42 of Mujeeb-Ur-Rahman 1977.

Wallace, Alfred R. 1891. *Natural selection and tropical nature: essays on descriptive and theoretical biology.* New ed., with corrections and additions. London, New York: Macmillan and Co.

———. 1898. *The wonderful century: its successes and its failures.* New York: Dodd, Mead and Company.

Wallis, John. 1669. "A summary account given by Dr. John Wallis, of the general laws of motion . . . " *Philosophical Transactions* 3:864–866.

———. 1670. *Mechanica: sive, de motu, tractatus geometricus.* London: typis Gulielmi Godbid, impensis Mosis Pitt.

Watson, James D. 1980. *The double helix: a personal account of the discovery of the structure of DNA.* Text, commentary, reviews, original papers, ed. Gunther S. Stent. New York, London: W. W. Norton & Company.

Watson, John B. 1924. *Behaviorism.* The People's Institute Publishing Company. Reprint, 1930, 1970, New York: W. W. Norton & Company.

Watson, John B., and William McDougall. 1928. *The battle of behaviorism: an exposition and an exposure.* London: Kegan Paul, Trench, Trubner & Co.

Webster, Charles. 1965. "William Harvey's conception of the heart as a pump." *Bulletin of the History of Medicine* 39:508–517.

———, ed. 1974. *The intellectual revolution of the seventeenth century.* London, Boston: Routledge & Kegan Paul.

———. 1975. *The great instauration: science, medicine and reform, 1626–1660.* London: Duckworth.

Wegener, Alfred. 1905. *Die Alphonsinischen Tafeln für den Gebrauch eines modernen Rechners.* Inaug.-diss., Berlin.

———. 1915. *Die Entstehung der Kontinente und Ozeane.* Braunschweig: Friedrich Vieweg & Sohn.

———. 1924. *The origin of continents and oceans.* Trans. from 3rd German ed. by J. G. A. Skerl., intro. by John W. Evans. London: Methuen.

———. 1966. *The origins of continents and oceans.* Trans. from 4th rev. ed. by John Biram. New York: Dover Publications.

Wegener, Else. 1960. *Alfred Wegener: Tagebücher, Briefe, Erinnerungen.* Wiesbaden: F. A. Brockhaus.

Weinberg, Stephen. 1977. "The search for unity: notes for a history of quantum field theory." *Daedalus* 106:17–35.

———. 1981. "Einstein and spacetime: then and now." *Proceedings of the American Philosophical Society* 125:20–24.

Weiss, F. E. 1922. "The displacement of continents: a new theory." *Manchester Guardian,* 16 March 1922.

Weisskopf, Victor F. "The impact of quantum theory on modern physics." *Die Naturwissenschaften* 60:441–446.

Weldon, T. D. 1945. *Introduction to Kant's Critique of pure reason.* Oxford: at the Clarendon Press.

Westfall, Richard S. 1971. *The construction of modern science: mechanism and mechanics.* New York, London, Sydney, Toronto: John Wiley & Sons.

————. 1971a. *Force in Newton's physics: the science of dynamics in the seventeenth century.* London: Macdonald; New York: American Elsevier.

————. 1980. *Never at rest: a biography of Isaac Newton.* Cambridge, London, New York: Cambridge University Press.

Westman, Robert S., ed. 1975. *The Copernican achievement.* Berkeley, Los Angeles, London: University of California Press.

Wheaton, Bruce R. 1983. *The tiger and the shark: empirical tests of wave-particle duality.* Cambridge: Cambridge University Press.

Whewell, William. 1837. *History of the inductive sciences, from the earliest to the present times.* 3 vols. London: John W. Parker. 3rd ed., 1865, with additions, 2 vols., New York: D. Appleton and Company.

————. 1840. *The philosophy of the inductive sciences, founded upon their history.* 2 vols. London: John W. Parker. 2nd ed., London, 1847. Reprint, 1967, New York, London: Johnson Reprint Corporation.

Whitehead, Alfred North. 1923. "The first physical synthesis." Pp. 161–178 of F. S. Marvin, ed., *Science and civilization,* London: Oxford University Press.

————. 1925. *Science and the modern world.* Lowell lectures, 1925. New York: The Macmillan Company.

Whiteside, Derek T. 1961. "Patterns of mathematical thought in the later seventeenth century." *Archive for History of Exact Sciences* 1:179–388.

Whitmore, P. J. S. 1967. *The order of Minims in seventeenth-century France.* The Hague: Martinus Nijhoff.

Whitrow, G. J., ed. 1967. *Einstein: the man and his achievement.* A series of broadcast talks. London: British Broadcasting Corporation.

Whyte, Lancelot Law. 1960. *The unconscious before Freud.* New York: Basic Books.

Williams, Bernard. 1967. "René Descartes." Pp. 344–354 of Edwards 1967, vol. 2.

————. 1978. *Descartes: the project of pure inquiry.* New York: Penguin Books.

Williams, L. Pearce. 1965. *Michael Faraday, a biography.* New York: Basic Books.

————, ed. 1968. *Relativity theory: its origins and impact on modern thought.* New York, London: John Wiley & Sons.

Williams, Trevor I., ed. 1969. *A biographical dictionary of scientists.* London: Adam & Charles Black.

Wilson, J. Tuzo. 1963. "Continental drift." *Scientific American* 208:86–100.

————. 1966. "Some rules for continental drift." Pp. 3–17 of G. D. Garland, ed., *Continental drift,* The Royal Society of Canada Special Publications, no. 9, Toronto: University of Toronto Press.

————. 1968. "Static or mobile earth: the current scientific revolution." *Proceedings of the American Philosophical Society* 112:309–320.

————. 1968a. "Reply to V. V. Beloussov." *Geotimes* 13(12):20–22.

————. 1968b. "A revolution in earth science." *Geotimes* 13(10):10–17.

————, ed. 1973. *Continents adrift.* Readings from *Scientific American.* San Francisco: W. H. Freeman and Company. A second edition, 1976, entitled *Continents adrift and continents aground*.

Wilson, Leonard G. 1972. *Charles Lyell: the years to 1841: the revolution in geology*. New Haven, London: Yale University Press.

————. 1980. "Geology on the eve of Charles Lyell's first visit to America, 1841." *Proceedings of the American Philosophical Society* 124:168–202.

Wilson, Woodrow. 1917. *Constitutional government in the United States*. New York: Columbia University Press [The Columbia University Lectures of 1907].

Wirgman, Thomas. 1825. "Philosophy." *Encyclopaedia Londinensis* 20:109–261.

Witt, O. N. 1913. "Wechselwirkungen zwischen der chemischen Forschung und der chemischen Technik." *Die Kultur der Gegenwart*, pt. 3, sec. 3, vol. 2.

Wittmer, Louis. 1908. *Charles de Villers, 1765–1815*. Geneva: George.

Wohlwill, Emil. 1904. "Melanchthon und Copernicus." *Mitteilungen zur Geschichte der Medizin und der Naturwissenschaften* 3:260–267.

Wolf, A. 1935. *A history of science, technology, and philosophy in the sixteenth and seventeenth centuries*. London: George Allen & Unwin.

————. 1938. *A history of science, technology, and philosophy in the XVIIIth century*. London: George Allen & Unwin.

Wolfe, Bertrand. 1965. *Marxism: one hundred years in the life of a doctrine*. New York: The Dial Press.

Wolff, Robert Paul. 1963. *Kant's theory of mental activity*. Cambridge, Mass.: Harvard University Press.

Woodbridge, Homer Edwards. 1940. *Sir William Temple: the man and his works*. New York: The Modern Language Association of America; London: Oxford University Press.

Woolf, Harry, ed. 1980. *Some strangeness in the proportion: a centennial symposium to celebrate the achievements of Albert Einstein*. Reading, Mass.: Addison-Wesley Publishing Company.

Wotton, William. 1694. *Reflections upon ancient and modern learning*. London: printed by J. Leake for Peter Buck.

Wren, Matthew. 1659. *Monarchy asserted, or the state of monarchicall and popular government in vindication of the considerations upon Mr. Harrington's Oceana*. Oxford: W. Hall.

————. 1781. "Of the origin and progress of the revolutions in England." Pp. 228–253 of John Gutch, ed., *Collectanea curiosa; or miscellaneous tracts relating to the history and antiquities of England and Ireland*, vol. 1, Oxford: at the Clarendon Press.

Wright, W. D. 1923. "The Wegenerian Hypothesis." *Nature* 111:30–31.

Wundt, Wilhelm. 1902. *Grundzüge der physiologischen Psychologie*. Vol. 1, 5th rev. ed. Leipzig: Verlag von Wilhelm Engelmann.

Youschkevitch, A. P. 1968. "Sur la révolution en mathématique des temps modernes." *Acta historiae rerum naturalium necnon technicarum*, Czechoslovak Studies in the History of Science, special issue 4, pp. 5–33.

Zacharias, Jerrold. 1963. "Teaching and machines." Pp. 73–86 of P. M. S. Blacket, A. J. Ayer, and Jerrold Zacharias, *The British Association / Granada*

Guildhall lectures 1963, Manchester: Granada TV Network [distr. by Mac-Gibbon & Kee].

Zagorin, Perez. 1954. *A history of political thought in the English Revolution*. London: Routledge & Kegan Paul.

————. 1973. "Theories of revolution in contemporary historiography." *Political Science Quarterly* 88:23–52.

————, ed. 1980. *Culture and politics from Puritanism to the Enlightenment*. Berkeley, Los Angeles, London: University of California Press.

Zedler, Johann Heinrich. 1742. *Grosses Universal-Lexicon*. 64 vols. Leipzig and Halle, 1732–1750.

Zilboorg, Gregory. 1935. *The medical man and the witch during the Renaissance*. Baltimore: The John Hopkins Press.

———— and George W. Henry. 1941. *A history of medical psychology*. New York: W. W. Norton.

Zilsel, Edgar. 1945. "The genesis of the concept of scientific progress." *Journal of the History of Ideas* 6:325–349.

Ziman, J. M. 1968. *Public knowledge: an essay concerning the social dimension of science*. Cambridge: at the University Press.

————. 1970. "Some pathologies of the scientific life." Extracts from the presidential address to section X of the British Association meeting. *Nature* 227:996–997.

Zinner, Ernst. 1943. *Entstehung und Ausbreitung der Coppernicanischen Lehre*. Erlangen: Mencke [Sitzungsberichte der Physikalisch-medizinischen Sozietaet zu Erlangen, 74].

Zuckerman, Harriet. 1977. *Scientific elite: Nobel Laureates in the United States*. New York: The Free Press; London: Collier Macmillan.

Index